BREEDING PLANTS FOR LESS FAVORABLE ENVIRONMENTS

BREEDING PLANTS FOR LESS FAVORABLE ENVIRONMENTS

Edited by

M. N. CHRISTIANSEN

U.S. Department of Agriculture, Science and Education Administration, Agricultural Research Plant Physiology Institute

CHARLES F. LEWIS

U.S. Department of Agriculture, Science and Education Administration, Agricultural Research, National Program Staff (Retired)

A WILEY-INTERSCIENCE PUBLICATION

JOHN WILEY & SONS

New York · Chichester · Brisbane · Toronto · Singapore

Library of Congress Cataloging in Publication Data:

Christiansen, M. N.
Lewis, Charles F.
    Breeding plants for less favorable environments.

    "A Wiley-Interscience publication."
    Includes index.
1. Plant-breeding. 2. Plants—Hardiness.
3. Plants, Effect of stress on. 4. Plants,
Cultivated—Ecology. I. Christiansen, M. N.
(Meryl N.), 1925–   . II. Lewis, Charles
Franklin, 1917–   .

SB123.B686   631.5′3   81-10346
ISBN 0-471-04483-0   AACR2

Printed in the United States of America

10 9 8 7 6 5 4 3 2 1

PREFACE

Efforts to improve the productivity and quality of crop plants have continued since the beginning of agricultural husbandry. Selection of seed from the most productive plants to produce succeeding crops was in reality an effort to improve plant adaptation to the environment.

Modern crop production, as developed during this century, has been primarily concerned with modification and control of environmental factors that determine productivity. Drought is overcome by irrigation; pests, mineral deficiency, and adverse pH are controlled with chemicals; and low temperature is ameliorated with thermal inputs. The resulting high-energy-utilizing systems, tenable with abundant low-cost fossile energy sources, are uneconomical under today's escalating energy costs. We, therefore, are placing more emphasis on suiting the plant to the environment with less energy input. Scientists are deliberately developing varieties of plants resistant to the stresses encountered in less favorable environments.

As the population of the world doubles, food production must increase proportionately. Efforts to produce enough food will push farming on to land less favorable for growing crops than most land now in cultivation. We will be forced to utilize these marginal environments while investing less energy in production. A combination of genetic and cultural management is essential for maintaining and improving world food levels.

Much information has developed concerning the physiological responses of plants to adverse environments. A number of good reviews of the physiology of stress effects on plants are available. However, a compilation of literature on the genetic variance of plant response to stress is currently not available. Our combined effort to present both physiological and genetic knowledge should be useful, particularly in fostering information exchange and cooperative effort between disciplines to develop crop plants with improved environmental resiliency.

M. N. CHRISTIANSEN
C. F. LEWIS

Beltsville, Maryland
Bryan, Texas
December 1981

CONTENTS

BREEDING PLANTS FOR LESS FAVORABLE ENVIRONMENTS

Chapter 1

WORLD ENVIRONMENTAL LIMITATIONS TO FOOD AND FIBER CULTURE

M. N. CHRISTIANSEN

U.S. Department of Agriculture, Science and Education Administration, Agricultural Research, Plant Physiology Institute, Beltsville, Maryland

INTRODUCTION

Frequently one hears Jean Meyer (12) quoted as saying, "We have to find, in the next 25 years, food for as many people again as we have been able to develop in the whole history 'til now." The subject of feeding a swelling world population has been discussed repeatedly in the 1970s. Agronomy and other research disciplines often are given the charge of increasing food production before the world faces disaster. Perhaps sometime soon the emphasis will shift from agriculture's ability to increase food supplies to a consideration of population control. However, in the time span required to develop serious efforts for zero population growth, extreme efforts must be made to buy time by increasing agricultural productivity. As stated at the 1974 World Food Conference (6):

> Although it was clear that the food needs entailed by inevitable population growth must be met, it was equally clear that population could not indefinitely continue to double every generation, and that at some point the earth's capacity to sustain life would inevitably be overstretched.

Crises such as the 1972 world imbalance of population and food supply are not unique. The 1972 situation demonstrated how closely the world borders on severe food shortages. The 1972 crisis began with poor world grain crops. A few fluctuations in weather contributed to the immediate problem

1

of cereal grain supplies upon which many developing and some developed countries rely to feed increasing populations. The Soviet Union then purchased about one-fourth of the U.S. grain production. Other countries listed orders with the United States, and in the same year the U.S. government had paid wheat farmers 860 million dollars to set aside 20 million acres of wheat land. The world grain supply only decreased 1% but compounded by an increase of 2% in world population, a crisis situation developed.

Since the 1972 crisis, the contributing factors of population increases, fertilizer shortages, energy cost increases, and large purchases of basic grain supplies by the USSR and The Peoples Republic of China have continued to cause world food reserves to fluctuate. The food problems of the 1970s have plainly demonstrated how precariously demand and food supply are balanced. However, in the United States the low grain balance of 1973–1974 was followed by three massive wheat crops. Prices have reflected the "famine-to-glut" situation. Wheat was selling at a record $5.50 per bushel in early 1974, three crop years later in mid-1977 the price was $2.00. The reserve of U.S. wheat was 340 million bushels in mid-1974—less than six months' supply for U.S domestic needs. By 1977 reserves were in excess of 1.2 billion bushels; the highest in 16 years. With lower demand and resultant lower prices, production controls were again considered to prevent massive surpluses. These data indicate that the U.S. power to produce, while altered by weather, is in part mediated by commodity prices influencing planting volume. United States agricultural output has great capacity to respond to favorable economic conditions in spite of adverse climatic fluctuation!

Various U.S. and international agencies continually assess the world food situation (22). The general conclusion of the reports of these organizations is that the short-term situation (1985) is not disastrous if no serious climatic disturbances occur to reduce agricultural productivity. A report by Buringh et al. (3) computed the absolute total food-producing capacity of the world in terms of a standard cereal crop at almost 40 times the present production provided maximum technology inputs were applied.

By contrast, at the 1972 Food Conference (6), Rodionov of the USSR delegation insisted that the world has sufficient land to support 40 billion people. Revelle (16) of Harvard University's Center for Population Studies holds that if technology and production inputs are epitomized, sufficient food can be produced to feed 38–48 billion people. These suppositions are based on a maximized land-technology situation and Revelle emphasized that poor distribution of arable land and the lack of technological interface with land resources greatly lower the world's potential. Eight billion is a more logical maximum world population that can be fed under present constraints.

A University of California (22) projection of population and food needs to 1985 (22) predicts a 35% world population increase in the 1975–1985 decade. Increases of 36–50% in need for rice, feed grains, oilseeds, and vegetables are indicated; and increases of 51–63% for sugar, pulse, nuts, fruit, and meat products are indicated. Most of the increased food must come from plant products because crops provide over 90% of the human need for calories. The cereals provide roughly 75% of the world's plant-derived calories; root crops, oilseeds, and sugar provide about 20%, and the remainder of the world's caloric needs are provided by vegetables and forage-fed animals. Marine sources provide only 3% and offer little promise for increase in the near term.

Increased production must come from higher yields from existing cropland or by production from expanded land areas. The limitations to increased yields on existing cropped areas can be partially overcome with greater production inputs such as fertilizer, water, or improved pest control. However, the major deterrent to consistently high yields is climatic variability. Crop varieties bred to provide a genetically controlled environmental "resiliency" can contribute markedly to yield stability.

Extension of cropping to marginal land areas will present many stress problems. Mineral ion deficiency or toxicity, drought, wind, salinity, and temperature extremes are factors other than terrain that will limit and delimit crop productivity in these environments. Tolerance and resistance to these stresses must be obtained if marginal lands are to make a significant contribution to the world food supply.

Limitations to Agricultural Productivity

It is the intent of the author to briefly document some of the world's major soil and climatic constraints to agricultural productivity. While it is recognized that imbalances in food supplies and population in various parts of the world cause severe social and political problems, that is not in the scope of the present discussion. The present effort is to explore the possibilities of increasing agricultural productivity on presently used land and adapting new land areas of perhaps marginal quality to agriculture through improved genetic adaptation to environmental extremes.

The food production limitations other than man-made stresses, such as air pollution, and pest losses are: (a) availability of productive soil, and (b) a climate supportive to crop growth. The former is a generally static condition and the latter is an uncontrollable variable.

Land Availability

It is estimated that there are from 1.6 to 3 billion hectares of land available in the world for producing reasonable crop returns (5). A 1967 survey (15) of potentially arable soil indicated that 24% of the ice-free land surface was suitable for cropping. A later survey by Buringh et al. (3), reported 25% of the ice-free land is arable and comprises some 3.4 billion hectares or about 0.88 hectares per capita. The world's major land problem is an unequal distribution of population on the land. Carter (4) projects a drop in arable land per person between 1970 and 1985, due to increased population and loss of land to urbanization. A University of California report (22), generally considered reliable, estimates that the world's land and water resources can feed 8 billion people while other reports (16) provide higher figures. The report points out that many problems would have to be met in bringing new land areas into production. Remoteness, lack of capital, poor transportation, and absence of processing facilities must be addressed. For example, the sparsely settled areas of the Central African Congo or the South American Amazon Basin offer as much as one billion hectares of potential agricultural land. Sanchez (18) holds that the tropical soils of the world will have to feed future population increases and cites soil management as the critical input for tropical agriculture.

About 70% of the unused tropical land has acidic soil belonging to groups called Ultisols and Oxisols. Soils of this type are also found in large areas of the temperate Eastern United States. Recent research data suggest that these soil areas can be moderately chemically amended to permit acid-tolerant crop selections to produce 80% of U.S. yields (18). By use of multicropping, annual production in tropical areas with adequate rainfall could be 150–200% of that in the temperate zones.

Tropical soil husbandry presents some unique contrasts to that of temperate soils. In temperate zones, cold temperature regulates both macro- and microbiology. Water supply may vary considerably but is generally a limiting factor during growing seasons; however, soil water supply recharges during winter. Loss of organic matter and nutrients is generally much less than in tropical soils. As a consequence, maintenance of organic matter is much more difficult except in some volcanic ash-derived and rainforest soils. Many tropical soils leach seriously and have little reserve holding capacity for either water or nutrients.

Temperature conditions of tropical soils are generally quite static. At root levels (50 cm), temperatures vary less than 5°C during the year and parallel average air temperatures. In general, temperatures are favorable for continuous support of microbial activity. Nitrogen fixation on the other hand is continuous in tropical soils and ceases in temperate zones during winter months.

Soil factors that reduce plant growth are generally predictable and may be modified or ameliorated by chemical and physical alterations to better fit plant requirements or conversely, plant genotypes can be selected that are adapted to soil conditions. The primary soil characteristics that alter plant productivity include (a) physical features (structure, organic matter, texture, profile, and slope) which control the flow and storage of water, nutrients, air movement, and heat exchange which in turn affect root penetration and water nutrient uptake, and (b) chemical make-up including chemical analysis, exchange capacity, and pH, all of which interact to mediate availability of nutrients to plant. Soil pH is one of the major factors that determines species occurrence or adaptation of domesticated crops. Acid soils generally have high (even toxic) levels of soluble aluminum, whereas the higher pH soils are deficient in available iron as will be discussed in a later chapter. Much alkaline soil is native to dry areas and is also often sodic. Irrigation sources in these areas are also often saline which further complicates the situation.

A survey of soils was initiated under FAO in 1961 (5) to map the land areas of the world (See Chapter 5). The maps provide knowledge on the properties and distribution of soils which can be related to potential productivity. Each map is accompanied by an explanatory volume. The task is near completion and when complete will be extremely valuable in more accurately assessing the total agriculturally useful soils of the world, and will give better insight into the inherent cropping problems of major soil types. The major deficiency of these soil maps is lack of a history of agricultural treatment which may greatly alter soil characteristics. Dudal (5) has provided a summary of the problems of the predominate soil types of the world. For example:

1 Ferralsols, Acrisols, Nitosols, Podzols, Podzoluvisols, and Andosols comprise 23% of the world land area and generally exhibit extremes of mineral stress.
2 Yermosols, Xerosols, Kastanozems, Solonchaks, Solentz Regosols, and Arenosols comprise 28% of world land area and generally are in drought stressed areas.
3 The Fluvisols, Gleysols, Vertisols, Plansols, and Histosols comprise 12% of the land area and occur in flooded or excess water areas.
4 The Lithosols, Rendzinas, Rankers, and Cambisols generally are very shallow soils comprising 24% of the world land area.
5 Chernozems, Creyzems, Phaeozems, and Luvisols occur in least stressed situations and constitute 10% of the land area.
6 Nonsoil areas (salt flats, rock debris, sands) and those affected by freezing (permafrost) comprise about 14%.

Thus of 14+ billion hectares of land in the world, only about 1.4 billion hectares of land area are classed by Dudal (5) as nonstressed or good crop land. An additional 2.9 billion hectares are limited in crop productivity by mineral stress, 3.7 by drought, 1.6 by excess water, 3.2 by shallowness, and 2.0 by permanent freezing.

Agriculture crops must continually cope with stress. Any increase in productivity will be dependent upon reducing the effect of climatic fluctuation and suboptimal soil conditions. As more marginal land areas are used for crop production, the stress problems will intensify. In like manner, as efforts increase to "fine tune" production in highly developed cropping systems, the sensitivity of the system to climatic perturbations will be amplified. A much greater emphasis will be placed on monitoring production systems. In many cases maximizing yields by escalating environmental amendments will not be economical.

Climatic Restrictions

It has been estimated that 60–80% of variability in crop production is a result of weather fluctuation (21). The major variable in climatic inputs to crop production is temperature and rainfall (water availability). The USDA-ERS 1974 report (20) on the world food situation states: "Although the effects of weather and climate on crops can be modified to some extent, most of the world's food supply still depends on the weather."

Climatic effect on crop production has two aspects; (a) long range climatic changes that may drastically alter agriculture over wide areas and (b) the day-to-day or weekly variability of rainfall and temperature from any other season.

Henry Wallace, corn breeder, Secretary of Agriculture, and U.S. Vice President under Franklin Roosevelt, conducted early studies on the interaction of weather and crop yield and reported results in his 1920 thesis (23). A later Iowa State researcher (Thompson) related 80% of crop yield variability to weather conditions; with others (11) he reported that an analysis of agricultural and climatological data indicates very strongly that production of grain in the United States has been favored by extremely good weather in recent years and that a part of the yield increase trend of 1940–1970 could be attributed to favorable weather rather than improved technology. By use of modeling techniques utilizing weather records and corn yields for the 1890–1972 period and equalizing all yearly yields to modern technology, McQuigg et al. (11), related weather patterns to yield. Their conclusions suggest that modern technology has little influence on the susceptibility of nonirrigated crops to weather.

Thompson and McQuigg's work stirred controversy with U.S. officials concerned with controlling food grain reserves. Thompson argued that present-day crops were endangered by weather as much as in the 1930s and that reserves should be maintained at a high level. Although Thompson's crop model is subject to error, as is any extrapolation of data into an unknown region, it does, however, point out that despite modern technology, weather patterns can seriously alter U.S. and world food production.

Other major food-producing areas of the world are also subject to adverse long- and short-term weather fluctuations. Canada experienced severe temperature conditions in 1974 that reduced wheat production by 10%. A part of the problem may be a cooling trend that commenced in about 1940, causing shorter growing seasons (10). Winstanley et al. (24), Canadian meteorologists, have analyzed and compared U.S. and USSR grain production climatic problems and noted, "Only 1.1% of Soviet Agriculture lies in rainfall areas of greater than 38 inches per year, compared with 60% of the United States—it is to be expected that Soviet grain production will continue to be highly vulnerable to rainfall fluctuations." Drought occurred in 1972 in the USSR and again in 1975. Added to USSR agricultural drought problems are temperature situations that can markedly reduce yields. Much of the Russian Wheat Belt can experience elevated temperatures and low humidity during the flowering period which is very damaging to yields.

Long-range alterations in climate have occurred in the past. Bryson (2) cites interesting pollen profile data to indicate the kinds of temperature and rainfall changes that have occurred in Minnesota over the past 13,000 years. Growing season, temperature, and moisture changes of wide magnitude are evident from the data. He cites evidence that rather drastic climatic changes have occurred in the United States in the past 1000 years. Temperature and rainfall shifts caused marked alterations in flora and fauna. Bryson (2) suggests that what has happened before can happen again. He points out that our climate patterns over the past 50 years have been more favorable to agriculture than any in the last 1000 years. During this time (1930–1978) most of our advances in crop productivity have been accomplished. Much of the advancement can be credited to crop adaptation to what Bryson considers an unusually favorable climate which has high statistical probability of changing very greatly in the near future. McQuigg et al. (11) voiced the warning:

> Any National Policy that does not take into consideration the fact that less favorable weather is far more likely than recent nearly optimum conditions, is likely to place us in most unfortunate circumstances.

Long-range trends in climate are almost impossible to predict. The excessive use of fossil fuel with resultant atmospheric increases in CO_2 and other gaseous contaminants is projected by some climatologists to alter the world climate. According to Schneider (19) an increase in atmospheric CO_2 to 400 ppm will result in a 1% increase in global temperature. Based on present CO_2 increase curves, that level may be reached by the year 2000. The implications of a 1.0% global change are large. It will constitute a greater change than any since the Little Ice Age of the 17th Century. Polar warming and ice cap melting may alter sea levels. Wind and rainfall patterns could be drastically altered.

Short-term fluctuations of temperature and rainfall are a certainty and their effect on agricultural productivity is likewise to be expected. The intense one-crop agriculture practiced in developed countries appears particularly vulnerable because the cultivars are developed for rigidly specific conditions. As agricultural production moves to more marginal land areas, crops with greater environmental resiliency will be necessary—perhaps at the expense of maximum yields.

Temperature is a primary climatic factor that delimits agricultural production areas. Most of the world's agriculture is restricted by temperature to the confines of 50° South to 50° North latitudes. Within these areas variable low temperature stresses exact considerable loss. Freeze injury to small grains with no snow cover in winter can drastically reduce grain supplies. Frost in Brazil upsets the world's coffee supply and price. Apple and peach growers routinely suffer frost damage at flowering time. A slight frost in Florida can greatly reduce citrus and vegetable supplies in the United States. Even in tropical areas chilling temperatures are a problem. Tropical fruits are often damaged at higher altitude and rice production is particularly reduced by chilling during grain development in India and China. Frost is a limiting influence on potato production at higher altitudes in Peru.

High temperature is a limiting factor in much of the lowland tropics. In temperate areas such as the U.S. Western Plains, heat limits corn and grain sorghum production often in combination with water stress. High soil temperatures are a production-limiting factor in West Africa where seed germination is inhibited (8).

Water availability to crops is the second primary climatic variable in limiting and delimiting crop production. Very little of the world's water is used by plants. Nace (13) estimated that only 2.5% of the world supply is nonoceanic or fresh—and that only 0.007% of all water is used by plants. It is impossible to estimate the impact of drought stress on world's agriculture. The major problem is departure from normal in any agricultural area. In rain-fed agriculture the problem of short-term (10–20 day) drought is common and exacts an intangible but real toll on productivity. Of course the incidence of short-term drought at critical periods of cropping cycles can be disastrous. Much of Europe, Russia, and eastern and central North America, and Africa are almost totally dependent upon rainfall for agricultural productivity. In modern operations in the United States, supplemental irrigation is used to some extent but may be marginally economic because of unpredictability of need and cost of equipment.

China leads in world irrigated agriculture with 40% of their land irrigated (14). Much of this is in rice. Irrigation of U.S. crops has increased at a rapid rate. Presently nearly 60 million acres are irrigated compared to 39 million in 1969. Irrigated acreage accounts for 10% of U.S. agricultural cropland but accounts for more than 25% of the crop production. Most of the cited U.S. irrigation is in the western states—nearly half in the Great Plains States where it is a requirement for economic crop production. In many of these areas water resources are rapidly being depleted—the Kansas, Oklahoma, Texas High Plains are good examples where over 3 million hectares of land are projected to revert to dryland culture due to exhaustion of ground water. Worldwide, irrigation is used on 201 million hectares or about 14% of cultivated land which produces almost one-half of the total world food.

One of the major constraints to expansion of agriculture in desert areas and in monsoon climates is lack of water. Some interesting data are presented in *Losses in Agriculture* (9) on the potential for increasing production through irrigation. Of the world's total arable land, 345 million hectares must be irrigated in order to produce a crop. Some 2.6 billion hectares will produce at least one crop without irrigation. If nonirrigated areas are multiply cropped, then crops could be produced equivalent to 4 billion hectares. If all cropland were irrigated as needed and multicropped, more than 16 billion hectares of crops could be produced annually. Unfortunately, this is impossible because the world's water resources are only sufficient to irrigate 470 million hectares of the arable land under presently used practices (3). Worldwide, billions of dollars have been spent to develop new irrigated areas with little attention paid to research on efficiency in water delivery systems or use by crop species. The present water use efficiency is about 35% in the United States and worldwide. Israel has achieved remarkable water use efficiencies ranging up to 80% through such methods as drip irrigation. Irrigation often causes serious salt accumulation problems and much excess water must be used with subterranean drainage to flush out salt accumulations. Kouda (7) estimated that 50–60% of the salt-loaded soils are a result of irrigation.

Although irrigation is extremely important to world food production, 85% of cultivated land is under rain-fed culture. Measures to husband and conserve water in soils is therefore extremely important to agriculture. In marginal rainfall areas, yield of crops can be doubled by such techniques at stubble mulching and minimum tillage. Adequate control of water runoff by maximizing soil infiltration is also essential to preventing soil erosion and increasing water harvesting. It is obvious that much of the world's agriculture will continue to depend on rainfall and production will fluctuate with rainfall variation.

Miscellaneous Environmental Factors

Several natural and man-made stresses are of localized and minor importance to crop culture and yield on a worldwide basis. Air movement, hail, light quantity, humidity, and air pollution are important examples. Wind is of major ecological importance in exclusion of trees on the plains of Western United States, Russia, and other like areas. It plays an extremely important part in water relations of plants, for example, physiological drought occurs in winter grains without snow cover. Wind also serves to transport insects, disease organisms, weed seeds, and pollen. High winds can limit fruit production, cause loss of grain by shattering, and coupled with low humidity, can exert drastic effects on flowering and fruiting of crops. Wind-soil erosion is a factor that prevents or modifies types of crop culture and limits yields. Air movement in the form of slope drainage is requisite for fruit culture in marginal temperature areas.

Atmospheric conditions can limit light levels and determine suitability of areas for crop culture. Fog may scatter or absorb selective bands of light; dust or industrial pollutants can also reduce light sufficiently to be limiting. Cloud cover or fog is a limiting factor in some minor areas.

Air pollution as a consequence of man's activities is of growing concern in the Western and Eastern United States, Western and Central Europe, and Japan. Ozone, SO_2, and NO_x compounds are of major concern. Other air contaminants such as aerosols, particulate matter, fluoride, copper, or other heavy metals can be very damaging in limited areas.

Preventing, Ameliorating, or Modifying the Impact of Environmental Stress on Crop Production.   Reducing environmental stress effects on plants either through prevention or postfacto therapy involves either alteration of the environment or the plant. Changing the environment may include such things as fertilization, pesticides, irrigations, land forming, deep tillage, and heating (smudge pots, etc.). Most environmental alteration is costly in labor and energy. Chemical alteration of the environment is increasingly being reviewed with concern. The alternatives are modification of the crop plant. Chemical modification of the plant to alter its resistance to environmental adversity is a possibility that is currently being researched. Plant growth regulators offer possibilities for stimulation of growth, timely induction and termination of flowering, induction of senescense, modification of internal water relations (8), and resistance to temperature extremes (17) or to air pollution (1). Genetic modification for crop adaptation is a practice as old as agriculture. Man has continually selected the most productive species and the best genotype. Since the beginning of directed efforts to breed better crops, the major goal has been to improve environmental adaptation. In most cases the stated breeding goal perhaps is improved yield and quality but the basic genetic alteration is improved genotype-environmental interaction, that is, plant adaptation.

Methods used by plant breeders have utilized large populations of diverse genotype from which selections are annually made for the desired trait. Such empirical methods are hit or miss in that the environment varies from year to year. The notable broad exception is in the area of disease resistance where directed efforts have been possible because of tangible specific selection criteria in the form of disease symptoms.

Soil scientists, chemists, plant physiologists, agronomists, and plant pathologists have over the past half century developed a considerable volume of information on plant response to abiotic and biotic stress. Much of the information concerns stress effects on basic plant functions such as water and ion uptake, photosynthesis, or chemical entities. These plant responses can serve as selection criteria for specific stresses and thereby be used to rapidly identify desirable genetic material. The following chapters detail much of the present knowledge of the physiology and genetics of the environmental factors that affect crop yield and how the information is being used to improve crop environmental resiliency.

REFERENCES

1   Bennett, J. H., E. H. Lee, and H. H. Heggestad. 1978. Apparent Photosynthesis and Leaf Stomatal Diffusion in EDU Treated Ozone-Sensitive Bean Plants. Proc. 5th Plant Growth Regulators Working Group. E. F. Sullivan, Ed.

2 Bryson, R. A. 1974. *Science* 184:753.

3 Buringh, P., H. D. J. van Heemst, and G. J. Staring. 1975. *Computation of the Absolute Maximum Food Production of the World*. Dept. of Tropical Soil Sci., Agric. Univ., Wageningen, The Netherlands.

4 Carter, H. D. 1975. *Perspectives on Prime Lands*. U.S. Dept. of Agriculture.

5 Dudal, R. 1976. Inventory of the major soils of the world with special reference to mineral stress. In *Plant Adaptation to Mineral Stress in Problem Soils,* M. J. Wright, Ed. Cornell Univ. Agric. Exp. Sta.

6 Food and Agriculture Organization. 1974. *Assessment of the World Food Situation Present and Future*. United Nations World Food Conference.

7 Kouda, V. A. 1971. *Soviet Geogr.* (Translated and revised.) 12:6–23.

8 Lal, R. 1974. *Plant and Soil* 40:129–143.

9 Losses in Agriculture. 1965. U.S. Department of Agriculture. *Agricultural Research Service Handbook* 291. U.S. Government Printing Office, Washington, D.C.

10 McKay, Gordon. 1974. World Food Supply in a Changing Climate. Sterling Forest Conference, George Kukla, Ed. Columbia Univ.

11 McQuigg, J. D., L. Thompson, S. LeDuc, M. Lockard, and G. McKay. 1973. *The Influence of Weather and Climate on the United States Grain Yields: Bumper Crops and Droughts*. Natl. Oceanic and Atmospheric Adm. U.S. Dept. Commerce, Washington, D.C., 30 pp.

12 Meyer, Jean. 1975. Agricultural productivity and world nutrition in crop productivity. Pp. 97–108 in *Research Imperatives,* A. W. A. Brown, T. C. Byerly, M. Gibbs, and A. San Pietro, Eds. Michigan Agric. Exp. Sta.

13 Nace, R. L. 1969. World water inventory and control. Pp. 31–42 in *Water, Earth and Man,* R. V. Chorley, Ed. London Methuen and Co., Ltd.

14 Peterson, D. F. 1974. *China Diary*. Arch. of Utah State Univ., Logan.

15 President's Science Advisory Committee. 1967. *The World Food Problem*. Vol. 2. U.S. Government Printing Office, Washington, D.C.

16 Revelle, R. 1974. *Sci. Am.* 231:168.

17 St. John, J. B., F. Rettig, E. Ashworth, and M. Christiansen. 1979. *Advances in Pesticides,* Part 2. Pergamon, New York.

18 Sanchez, P. A. 1976. *Properties and Management of Soils in the Tropics*. Wiley-Interscience, New York.

19 Schneider, S. H., and L. E. Mesirow. 1976. *The Genesis Strategy. Climate and Global Survival*. Plenum Press, New York.

20 The World Food Situation and Prospects to 1985. 1974. *Foreign Agriculture Economic Report No. 98*. U.S. Dept. of Agriculture.

21 Thompson, L. M. 1975. *Science* 188:535.

22 University of California Food Task Force. 1974. *A Hungry World: The Challenge to Agriculture*. University of California, Berkeley.

23 Wallace, H. A. 1920. *Monthly Weather Rev.* 48:439–446.

24 Winstanley, D., B. Emmett, and G. Winstanley. 1976. *Climatic Changes and the World Food Supply*. Environment Canada, Planning and Finance Report No. 5.

Chapter 2

THE PHYSIOLOGY OF TEMPERATURE EFFECTS ON PLANTS

ROBERT G. McDANIEL

Department of Plant Sciences, The University of Arizona, Tucson, Arizona

INTRODUCTION

Every plant is characterized by an optimum temperature for growth. This temperature response is governed by a complex interplay of genetic, developmental, and cultural factors which serve to condition the environmental response of the plant. Resistance of plants to environmental extremes, expectedly, may be related to the evolutionary pathway that resulted in the development of the species. Plants of tropical and subtropical origin, for example, possess markedly less tolerance to cold temperature than do montane and subarctic species.

As is so often the case in evolutionary sequences, the mechanisms by which plants withstand temperature stresses are varied. Ability to tolerate extremes in mean temperature often defines the range of plant species. Thus avoidance of temperature stress in colonization has been a major plant evolutionary strategy (1). It would appear that few interacting factors for high temperature stress tolerance have evolved, as plant physiological processes can be rather quickly modified to cope with temperature extremes approaching the point at which enzymes are thermally inactivated. Indeed, absolute upper and lower limits of growth would seem to have been established early in protist evolution: the upper limit of growth occurred at the point of protein denaturation, effectively shutting down the enzymatic machinery of the cell; the lower limit for growth at the point of intercellular ice formation. Ice would disrupt the continuity of the cytoplasmic gel and would effectively block the passage of hormone messages and synthetic

13

products necessary for cell growth and division. It is important to emphasize the distinction between cellular work, expressed as growth, and the maintenance energy required to keep plants alive under temperature stress. Plants and cultured plant cells can retain their viability under temperatures far in excess of those causing cessation of growth. The water relations of the tissues and cells become of paramount importance to plant survival under such conditions.

Plant response to temperature stress has been a major focus of agronomic and physiological research across a broad spectrum of agriculturally important species. Indeed, many scientists consider temperature, especially low temperature, as a prime limitation of where plant species may grow (25). The National Academy of Sciences published a research planning report in 1969 which emphasized the strong linkage of plant physiological limitations under temperature stress and potential crop productivity (98).

Bronk (21) has discussed specialized microorganisms that thrive under high temperatures. Some thermophilic bacteria seem uniquely suited to temperatures at which multicellular eukaryotes would exhibit immediate injury. Langridge and McWilliam presented some hypotheses to explain heat resistance in *Thermobiology* (96). Levitt's book, *Responses of Plants to Environmental Stresses,* has detailed the many physical and functional effects of temperature stress on plants (99). Lyons (104), Raison (158), and Lyons and Raison (105) have carefully reviewed the physiological effects of chilling injury to plants as well as evidence for the molecular mechanisms involved.

In a recent volume on plant environmental modification, Weiser et al. (189) reviewed symptoms and mechanisms of plant freezing injury, updating the previous review by Burke et al. (25) on the same topic. A considerable interest in freezing injury to plants is evidenced by the numerous reviews and books on aspects of cell and tissue freezing. Cryobiology, a rapidly emerging discipline, deals exclusively with responses of biological systems to freezing. Mazur's review of cryobiology (110) provides a good introduction to aspects of low temperature biology. Cryoprotectants—substances that afford freezing protection to plant tissues and cells—are a fascinating aspect of cryobiology (111).

GROWTH RESPONSES TO TEMPERATURE STRESS

High Temperature Injury

Plant responses to high temperature stress are closely linked with the water status of the plant (48, 1). Wilting, leaf burn, and leaf folding or abscission serve as early indicators of high-temperature-induced damage. Reduced soil moisture can potentiate these symptoms, especially when the high temperatures occur during periods of rapid leaf growth and expansion. In the southwestern United States, cotton grown under irrigation in sandy field areas wilts before adjacent rows in heavier soil. Retention of soil moisture is less in the sandy soil, and temperature rises more rapidly than in soil retaining more water. Afternoon air temperatures may exceed 45°C, causing leaves to wilt. The greater heat stress on cotton in sandy soil, associated with less water for transpirational leaf cooling, results in much reduced plant size. Paradoxically, this repeated, sublethal heat stress has likely produced cotton plants that are more thermostable, by reason of their slower growth rate and greater degree of tissue desiccation. Obviously, with continued tissue desiccation, plant growth will cease (100).

Chilling-Sensitive and Insensitive Plants

A large body of work has accumulated on effects of chilling on plants. Because of the great economic value of plants of tropical and subtropical origin, there has been considerable interest in their degree of susceptibility to chilling injury. Chilling injury may occur from the point of seed hydration (34, 144), through various stages of germination (27), throughout the growth and development of the crop (105), and during postharvest storage. Symptoms of injury due to exposure to temperatures in the range from 20 to 0°C include tissue necrosis, pitting and discoloration on crops such as cucurbits, citrus, sweet potatoes, and bananas (104). These crops are also subject to postharvest chilling injury. Apples and temperate stone fruits often may show discoloration during low temperature storage (104).

Chilling injury to germinating seedlings may be well illustrated with cotton, a crop of tropical origin. Christiansen (34) has described the symptoms elicited by chilling stress on cotton seedlings: radicle tip abortion caused by chilling during imbibition, and a root cortex decline which results from chilling during germination (144, 27). The necrotic and discolored cortical areas are subject to rapid invasion by damping off organisms and secondary soil pathogens (112). Damaging effects of chilling injury during germination were reflected in subsequent growth and yield reduction (35, 3).

In some instances the damaging effects of chilling stress could be alleviated or largely eliminated by preconditioning the seed (81, 182). The first obvious manifestation of chilling injury to imbibing seeds is solute leakage. Simon (166) proposed that as seeds imbibed moisture, cellular membranes were reorganized from a dehydrated, disorganized state of high permeability to a hydrated, selectively permeable state. If stress were imposed on the seed during the initial hydration process, then vital substrates might be leached out of cells to the detriment of the seedling. Christiansen et al. (38), Simon et al. (165), Simon and Wiebe (167), and Bramlage et al. (20) have confirmed the accelerated loss of cellular solutes due to chilling stress during imbibition.

There have been a number of reports of variability within chilling-sensitive species to chilling injury (104). Plants of temperate origin show little or no effects of chilling; and in fact seed of some species may require chilling treatment through stratification to optimize germination.

Freezing Resistance and Susceptibility

All plants are at risk to freezing injury. Below-freezing temperatures are responsible for enormous agricultural production losses. Physical damage to fruit, fruit drop, and outright killing of plants often result from freezing injury. These sorts of damage are often readily apparent in areas of the "sun belt" where citrus, avocado, and other crops of limited cold acclimation are found. In more northern climates, aside from the physical, structural damage of ice and snow, extremely low temperatures often result in extensive injury to deciduous trees and herbaceous perennials. The degree of injury sustained may not be obvious for several months. With the spring flush of growth, dead branches and trees, lack of regrowth by forages, and damage to shrubs and herbs become obvious. Winter cereals may have crowns killed, or may sustain root damage due to soil heaving during alternate freezing and thawing cycles (25, 189).

As has been discussed, many plants of tropical and subtropical origin are damaged by above-freezing temperature, and are killed by exposure to temperatures of 0°C, or a few degrees colder (104). Within species of tropical origin, many types of herbaceous perennials do exhibit considerable freezing tolerance, likely as a result of lignification associated with perennial habit, and adaptations to colder, higher elevation locations (189). With few exceptions, plants do not exhibit significant growth at below-freezing temperatures. Some solute translocation may still be occurring, and maturation of buds taking place, but metabolic processes have essentially halted and obvious growth has ceased (25, 189). Some workers believe photosynthesis may still be occurring, but at a greatly reduced rate making *in situ* measurements difficult.

Plants which grow in temperate areas exhibit four general mechanisms for resisting low temperature injury. Some plants avoid freezing stress by completing their life cycles in warm weather, persisting by over-winter survival of seeds (annual growth habit). Other plants with deep roots, crowns and/or rhizomes may become dormant and be somewhat protected from cold temperatures by snow, soil, and mulch cover. Plants that contain a high level of sugars and salts, and have concentrated cell cytoplasm can exhibit a freezing point depression of several degrees, effectively protecting the plants from intracellular freezing damage. Plants may avoid ice injury to tissues by supercooling, the phenomenon by which liquid water may exist at temperatures considerably below zero. This is made possible through temperature acclimation, which leads to the absence in plant cells of nucleating substances, which would act to "seed" ice crystals. Without heterogenous nuclei or disturbances to potentiate ice crystal formation, water may remain liquid to −38°C, the point of spontaneous ice formation (25). Following acclimation, some frost-sensitive plants can also supercool to −16°C (107). Plants such as winter-hardy tree species survive temperatures below −38°C by tolerating extracellular ice formation. When plants do survive ice formation, protoplasmic water has been effectively removed to extracellular areas enabling the cells to survive by becoming desiccated (104).

DEVELOPMENTAL TEMPERATURE TOLERANCE MECHANISMS

Seeds and Germination

Seeds of most plants are extremely tolerant of high and low temperature stresses. Seeds are often found to withstand the ravages of extreme temperatures by design, for survival of the seed is paramount to survival of the species. A large measure of the environmental stability of seeds is directly attributable to the desiccation of seed tissues. Much as low tissue water content provides a plant defense against temperature stress, low cell water content enables the seed to survive similar stresses. Coincident with a relatively dehydrated state, seed metabolism is also quiescent as the enzymatic machinery of the dry seed is at a near standstill.

Although the metabolic rate of mature, dry seeds is exceedingly slow, it can be shown that some gas exchange occurs. Seeds are living organisms: even through the respiratory "fires" characteristic of plant growth do not burn, the "coals" of the seed still glow. Gas exchange status of seeds is likely an important but little studied factor in their response to temperature stress. Certainly the oxygen status of imbibing seeds is critical in the ability of seeds to withstand temperature stress. Conditions which diminish the oxygen concentration, for example, excessive water added to the germination medium, may cause a dramatic reduction in seed viability. Low temperature stress potentiates this loss of germinability. Cocklebur seeds are an exception, where anaerobiosis, coupled with low temperature cycles, served to stimulate germination (61). This phenomenon has been termed *thermoperiodism,* although the absolute duration of the temperature cycles does not seem critical. Oxygen concentration has been shown to influence the formation of unsaturated fatty acids during the maturation of seeds. Unsaturated fatty acid synthesis increased as temperature decreased from 40 to 10°C. Doubling of the oxygen concentration increased synthesis of unsaturated fatty acids in sunflower and castor bean seeds (72). Degree of unsaturation of seed fatty acids has been shown to relate to the chilling sensitivity of seeds. Corn (23) and cotton (15) cultivars that showed increased seed fatty acid unsaturation, showed increased emergence at low temperatures.

Seed moisture content has proven to be a critical factor in response to chilling within chilling-sensitive species. Low moisture (less than 13%) soybean or sorghum seed was more severely injured by suboptimal germination temperatures than was seed of high moisture content (133, 143). Christiansen (37) has shown that preconditioning cottonseed by allowing the dry seed to imbibe water for several hours at 31°C followed by slow drying dramatically improves the ability of the seed to tolerate low temperatures when rehydration takes place. Germination response to chilling duration following preconditioning showed a genetic component in cotton (47), and probably reflected imbibitional rates as well. Lyons (104) has discussed the two periods of seed sensitivity to chilling injury, the first sensitive period during initial imbibition, the second occurring 18 to 36 hr following hydration of the seed (36). In view of the array of crop seeds susceptible to low temperature injury, understanding such stress responses is clearly important.

Growth and Maturation of Crop Plants

Cultivated horticultural and agronomic crops possess a remarkable tolerance to temperature stress. Crop plants face a wide gamut of environmental stress during their growth, development, and maturation. Man has consciously and unconsciously selected and molded plant genetic constitutions in such a manner that numerous crops of tropical or subtropical origin now are highly productive in temperate zones. Breeders are ever striving to extend the range of cereals, fruits, and other crops northward; still other plants are being bred for increased heat and disease tolerance to enable their cultivation in southern latitudes.

Cold stress is most damaging during seedling development in chilling-sensitive crops such as cotton or corn, when active expansion and division of embryonic tissues is taking place. Likewise, in tree crops, such as peaches (189), developing flower bud tissues are especially sensitive to cold injury.

During growth and development, plants may avoid high temperature injury by mechanisms such as dramatically increased transpiration rate, higher leaf reflectance, or angle of orientation to the sun. Leaf canopy may be designed so the more temperature resistant leaves serve to shade the more sensitive fruit. Protective layers of cork or thick bark may serve as a natural insulation to temperature extremes (41). Growth habit of the plant may be such that leaves are held at a distance from the soil surface, where the highest temperatures may occur. The elevated leaves of xerophytic vining cucurbits exemplify this adaptation.

Frost often serves as the environmental stimulus to hasten the ripening of fruit and to delimit the maturation of the crop. Unseasonal early frosts can prove extremely damaging, especially if they occur before the crop is mature and during periods of fruit or grain development prior to ripening. In general, the severity of crop damage due to temperature stresses can be closely correlated to the untimeliness of the stress. A majority of plants have the ability to gradually "harden" to resist high or low temperature stresses (1). The greatest crop damage may be expected when there has been little or no opportunity for gradual acclimation by plant response to steadily increasing or decreasing temperature over time.

Woody Species and Perennials

The gradual responses of plants to temperature fluctuations are especially vital to the development of temperature tolerance in woody species. Herbaceous species often show dramatic alterations in growth in response to lowering temperatures and coincident photoperiod changes. Phytochrome has been implicated in the initiation of cold acclimation in dogwood (120). Acclimation was linked to cessation of growth and to the initiation of terminal bud formation, with a strong influence of temperature.

In some crops, such as apples and grapes, improved cultivars may lack the degree of photoperiodic sensitivity necessary to escape injury from early fall temperature stress. The bred-in extension of growth period which results in increased productivity has, in fact, been achieved at the expense of the early cessation of growth, and subsequent winter hardening (189). The developmental continuum of freezing resistance has been studied in a number of crops (189, 82). In general, an initial increase in resistance to freezing injury may be seen by late August to early September. A dramatic increase in freezing resistance from around $-5°$ to $-35°C$ or lower is usually evident by late October to early November, and this maximal resistance is usually retained until late February or early March, when it is rapidly lost. Such environmental modifications of cold tolerance are closely associated with growth and metabolism of woody species and perennials. Maximum cold tolerance of a species may be lessened by an unseasonal warm spell in midwinter. The rate and degree of cold acclimation achieved will be expected to vary with the species and the climatic area (189, 1, 82). Table 1 illustrates the gradual spring deacclimation process in peach flowering, expressed as bud survival, when branches are challenged with low temperature stress.

Table 1 Apparent loss of winter cold hardiness (spring deacclimation) in peach (*Prunas persica* L.) flower buds averaged over 12 years. Data taken from Weiser et al. (189)

Date	Floral Developmental State	Temperature in °C at Which 50% of Buds are Killed
November to February	Full bud dormancy	−19.2
March 6	Initial bud swell	−15.2
March 20	Red calyx seen	− 9.1
March 28	First pink stage	− 6.0
April 2	First bloom seen	− 4.1
April 16	Bloom termination	− 3.0

PHYSIOLOGICAL RESPONSES TO TEMPERATURE STRESS

Temperature Effects on Proteins and Enzymes

Kinetic Properties of Temperature-Sensitive Enzymes. Few published investigations concern high temperature effects on plants. Investigators rapidly discovered that protoplasmic streaming was closely linked with high temperature stress, and some studies amplified this effect (1). With the advent of electrophoresis, which enabled the separation of enzymatic proteins based on molecular weight, net charge, and conformation, a number of thermal effects were related to enzymes, and to their alternative forms.

One cogent explanation for differential plant tolerance of high temperature stress is that heat-tolerant plants have more thermostable proteins and enzymes than do their heat-sensitive counterparts. Thermodynamically speaking, an enzyme with optimum kinetic activity at a high temperature must of necessity possess different properties than enzymes operational at lower, or suboptimum conditions. Why would this be expected? The ultimate function of enzymes as biological catalysts is to accelerate the speed with which substrate molecules are converted to product molecules. The multienzyme systems of higher plants and animals are extremely finely balanced catalytic pathways. Activation energy, the thermodynamic hurdle in making or breaking chemical bonds, is greater at low temperatures because the heat (enthalpy) in the system is less. Thus an enzyme whose rate function is ideal for the plant at 25°C may have unacceptably rapid activity at 35°C. Similarly, the converse is possible. How then can plants, as poikilotherms, maintain biosynthetic rate control across the temperature spectrum? (Poikilotherms are organisms whose metabolism is linked to environmental temperature. Cold-blooded animals are poikilotherms. Plants that are incapable of regulating temperature at levels significantly different from their environment are also considered poikilothermic organisms.)

There seem to have been two major evolutionary strategies for physiological heat resistance. One finds enzyme variants with altered kinetic properties which favor function at high temperatures. Low et al. (101) have elegantly demonstrated the temperature adaptation of enzymes in cold- and warm-blooded animal species as judged by the magnitudes of activation parameters. Somero (173, 175) has stated that activation energy parameters have no immediate effect on enzyme thermal adaptation, but have played an important role in evolution at the species level. Many enzyme systems showed no apparent changes in activation energies with temperature (1), consequently, some authors questioned the interpretation of the few published associations (122). One must consider, however, that it is necessary only for a single enzyme in a metabolic pathway to exhibit thermal inactivation, in order to drastically impair the function of the entire pathway. If one enzyme were more thermolabile than the others in a sequence then little evolutionary pressure for increased temperature stability would be brought to bear on other enzymes of the sequence. Selection of the appropriate enzyme system for study of temperature effects would then become important in determining the outcome of the experimental results. McMurchie et al. (121) showed an excellent correlation between activation energies for heart beat rate, membrane associated ATPases, and membrane phase changes between warm- and cold-blooded animals. More recent work has amplified the relation of membrane associated enzymes, lipid phase changes, and thermal sensitivity of both plants and animals (104, 105). It would thus appear that study of enzyme mechanisms will contribute significantly to our knowledge of plant thermal responses.

Protein Synthesis at High Temperature. Of the several factors associated with high temperature injury, effects on RNA metabolism and protein synthesis appear to be the most consequential. Nash and Grant (131) reported on the thermal stability of ribosomes from yeast species differing in heat sensitivity. Ribosomal properties of the more sensitive strain, including protein synthesis, were more drastically impaired by high temperature exposure (40°C for 5 min) than were those of the resistant strain. Interestingly, the ribosomal RNA of the heat-sensitive species contained significantly less adenine and significantly more guanine than was found in the base composition of rRNA of the heat-resistant strain. Baker and Jung (11, 12) also showed a reduction in total RNA content of perennial grasses at high (35°C) temperatures. Reduction in RNA and DNA content paralleled reported heat sensitivity in these grasses.

Interestingly enough, chilling stress nearly doubled rRNA synthesis in winter (cold-resistant) wheat, while no effect was seen in frost-sensitive spring wheat (54). Similarly, polyribosome content and RNA were maintained at high levels in potato leaves grown at low temperatures (184). Alterations in RNA metabolism would seem to result from both high and low temperature stress.

Isoenzymes and Temperature Stress. Somero (174) has reviewed the effects of temperature on the physical and chemical responses of enzymes. In general, the velocity of enzyme catalysis may be expected to double for every 10°C rise in temperature. Alterations in enzyme activity may be attributable to temperature effects at all levels of enzyme structure, and additionally, to hydrophobicity, solubility, salt linkages, hydrogen and disulfide bonding, and cooperative or allosteric effects (1, 24). Alterations in Michaelis constant (K_m), a measure of enzyme-substrate affinity, are considered to buffer against thermal effects on enzymes (174, 75). Such immediate enzyme activity changes could be elicited by pH alterations, enzyme cooperativity, and allosteric hindrance (1, 174, 24). Evolutionary changes in enzyme structure affecting K_m are believed to have come about as a means of optimizing enzymatic function at the level of substrate present throughout the organism's life cycle (75).

One method of accomplishing such structural evolutionary enzyme changes is by the development of the genetic capacity to synthesize variant enzymes either simultaneously or sequentially (75). Such an evolutionary strategy would enable a plant to react to environmental extremes through synthesis or activation of alternative enzymes which catalyzed the same reaction, but which differed in their catalytic properties. Thus, optimum metabolic function could be achieved across a range of environments. There is good evidence that maintenance of enzyme function at high temperatures depends on protein conformational stability (1, 41). Levitt (99) developed a theoretical model relating sulfhydryl disulfide bonding transitions to plant thermal stability, which helped tie together results from studies of thermal effects on enzymes. Sullivan and Kinbacher found that heat hardening of bean plants was reflected in the increased thermal stability of chloroplast fraction I protein. They attributed this increased stability to alterations in protein disulfide bonding during hardening (180).

Enzymes may exist as functionally or structurally variant forms called isoenzymes, which, by reason of differential kinetic properties, are believed to play an important role in heat tolerance. Much of the evidence comes from studies of cold- and warm-acclimated animals. Citrate synthase (74), an adenylate modulated enzyme, acetylcholinesterase (13), and other enzymes (75, 180) were found to differ in acclimated and nonacclimated animals. Significantly, the acetylcholinesterase existed as two electrophoretically separable forms, "a" and "b." Both isoenzymes were present in 12°C-acclimated animals. In 2°C-acclimated animals only the "a" form was present; in 17°C-acclimated animals, only the "b" form was present (13). Similar temperature-sensitive isoenzyme effects are seen in nematodes (isocitrate lyase; 159), *Pseudomonas* (histidine ammonialyase; 113), and in lily and tobacco (esterase; 137), among others. Some enzymes are reversibly heat inactivated, and should regain activity when cooled; others appear to be irreversibly denatured, and protein synthesis would be necessary to reestablish activity upon cooling.

An additional complexity in the interpretation of isoenzyme thermal responses is the intracellular distribution of isoenzymes. A good example is malate dehydrogenase, which is known to exist in several molecular forms, some of which are organelle specific (69). This enzyme has at least one isoenzyme (mitochondrial) which is membrane associated (116). Its *in situ* kinetic properties are distinctly different from those of the isolated, membrane-free isoenzyme (49). The contrasting effects of temperature on malate dehydrogenase isoenzymes from various organisms (7, 129, 76) exemplify the numerous variables which must be considered in establishing enzymatic cause-effect temperature relationships. Conditions under which enzyme assays are conducted can dramatically alter response of various isoenzymes to thermal stress, and can make *in vivo* interpretations difficult (18).

Hormonal Responses to Temperature Stress. Endogenous plant growth hormones are thought to play a significant role in plant temperature acclimation and response. Although a number of studies have established effects of growth retarding or growth regulating chemicals, little concrete evidence exists concerning the role of naturally occurring growth hormones in temperature stress. Cole and Wheeler (46) reported that treatment of cottonseed with gibberellic acid improved low temperature performance of the seed in laboratory experiments. Buxton et al. (28) also noted stimulatory effects of gibberellic acid treatments, but found the resultant seedlings were spindly and field emergence was poor. A number of reports have noted that gibberellic acid treatment increased tissue hydration. This was also found to be the case in cotton seedlings from gibberellic acid treated seed (McDaniel, unpublished). Increased tissue hydration elicited by hormone treatment could possibly contribute to lowered cold resistance under field conditions. Highly hydrated tissues would be expected to be tender.

Because of the association of photoperiod with the initiation of cold acclimation, phytochrome has been suggested as a hardiness mediator, possibly acting on the movement of abscissic acid through the plant, and its subsequent mediation of water relations (189). Nucleic acid derivatives can act to increase the cold tolerance of alfalfa, when applied as foliar sprays (69). Kinetin (a nucleic acid breakdown product) has been shown to affect plant membrane permeability (65). Several studies have documented the alleviation of thermodormancy in lettuce by cytokinin or kinetin treatments (85). Seed germination of some lettuce varieties was improved at high temperatures by treatment with gibberellic acid, ethephon (an ethylene source) and fusicoccin (a glucoside) in combination with kinetin (183). Treatment with growth hormones thus holds the potential of making seeds and seedlings more resistant to stress.

Probably the best evidence for growth hormone effects on plant temperature responses has come from studies of cytokinin response at high temperatures. Experimental evidence suggests that the cytokinin level in plant tissues falls with increasing temperature. At the same time abscissic acid increases, probably related to leaf abscission (85). Cole (48), using two pea cultivars that differed in sensitivity to high temperatures, observed that high root temperature inhibited either synthesis or transport of cytokinin to the aboveground portions of the plants, as judged by bioassays of cytokinin activity. The high temperature-tolerant variety maintained higher cytokinin activity under heat stress than did the sensitive variety. Nucleic acid levels and other parameters of growth also declined following high temperature exposure. Thermal injury at 40°C could be largely avoided if roots were maintained in soil at 20°C, as shown in Table 2. These data lend support to the role of high soil temperatures as a causal factor in reduced hormonal translocation and subsequent accelerated leaf senescence and sensitivity to temperature stress. Anderson and McNaughton (5) reported an effect of low

Table 2   Effect of root and shoot temperature on growth and cytokinin content of heat susceptible ('Alaska') and heat tolerant ('Wando') pea cultivars (*Pisum sativum*, L.). Nine-day-old plants were grown at the indicated temperatures in specially modified environmental chambers for nine days, after which measurements were made. Dry weight and protein analyses of shoots and roots showed the same trends as data presented here. Data of Cole (48)

Shoot-Root Temperature Maintained (°C)	Shoot Fresh Weight g/10 Plants		Shoot Cytokinin Activity Arbitrary Units/Plant	
	Alaska	Wando	Alaska	Wando
20°/20°	4.5	3.9	56	46
40°/20°	3.3	4.5	36	32
40°/40°	1.8	1.3	15	12
20°/40°	2.2	1.8	—	—

soil temperature on growth of a number of plant species. This effect was linked with possible impairment of cytokinin synthesis or transport, even though effects on photosynthesis and transpiration were not apparent at the stress levels imposed.

Chilling Injury and Membrane Functionality

Protoplasmic Streaming and Cytoplasmic Viscosity. The primary physiological response of plants to chilling temperatures has been the subject of many studies. Although a considerable body of evidence has been compiled which correlates cell membrane responses with plant chilling response, no unequivocal causal mechanisms have been established. Probably the most ubiquitous plant response to chilling temperatures is altered membrane permeability and related physiological effects.

A much studied and very dramatic indication of chilling damage is cytoplasmic streaming. It is generally accepted that cytoplasmic streaming is greatly reduced in chilling-sensitive plants at 10–12°C, whereas significant streaming can be observed in chilling-resistant plants down to 0°C (105). Protoplasmic streaming is an energy dependent process, and requires cell membrane integrity. Some reduction in protoplasmic streaming in plants which can cold-acclimate may be linked with simultaneous increases in cell protoplasmic viscosity. The thickening of cell sap as a means of initiating the tissue desiccation process so necessary for freezing survival has been termed *cytoplasmic augmentation* (145).

One interesting effect which has been suggested by spin label studies is the production of a high microviscosity in cells that are dense in membranous structures. In fact, the microviscosity of the cell sap might become high enough at low temperatures to make diffusive substrate transport more rapid along or within membrane limited organelles than through the cell sap itself (84). Implications of temperature related cytoplasmic viscosity effects and the function of the Golgi apparatus remain largely unstudied. Synthetic products of the Golgi, including lipids, proteins, and carbohydrates (190) which contribute to membrane content and turnover in cells, may be implicated in plant temperature response in the future. The contribution of the Golgi apparatus to plasma membrane structure deserves further study in this regard. The presence of microtubules and cytoplasmic fibrils influence the properties of the cytoplasmic gel, and most probably have some degree of influence on protoplasmic streaming, as well as cytoplasmic viscosity. Increased carbohydrate and protein synthesis (71, 50) which have been reported coincident with plant response to chilling temperatures would also serve to increase cytoplasmic viscosity when concentrated in cells. Olein (134) has proposed that presence of high-molecular-weight polysaccharides might be an intrinsic property of cold-hardy plants. These long chain sugars could dramatically alter cytoplasmic viscosity. Lipid synthesis over time may prove especially important in determining cytoplasmic viscosity at low temperatures.

Membrane Fluidity and Structural Changes. One of the earliest and most universal measures of plant temperature injury has been electrolyte leakage. Increased loss of electrolytes—presumably diffusible components of cell sap—has been documented across an array of tissues, species, and thermal stresses. Electrolyte leakage of sugars, amino acids, organic acids, proteins, and other solutes (169) characterizes cellular injury due to heat, chilling, or freezing stresses. The consensus is that electrolyte leakage reflects damage to cellular membranes. Disruption and damage to membranes would detrimentally alter their permeability, and result in loss of solutes. Nowhere is this effect more striking than in the hydration of dry seeds (169, 138). Simon (168) has discussed the relative disorganization of membranes of seeds in the dry state. Membrane phospholipids in nonhydrated tissues show a somewhat random distribution of the nonpolar, hydrophobic "tails" of the phospholipid molecules. This lack of orientation is believed to explain the greatly altered membrane permeability of dry seeds, as evidenced by loss of electrolytes from seed tissues upon hydration (166, 168, 171). The same altered permeability results when temperature stresses are imposed on cells.

What properties of membranes may account for altered permeability, and how then may these properties be related to the differential temperature resistance of various plant species? As Christiansen has aptly stated, the numerous changes at the tissue, cell, or molecular level reported to occur in plant tissues as a result of chilling stress are indicative of the complexity of stress effects on plants (41). Indeed, this complexity extends to freezing stress and high temperature effects also. Perhaps no simple answer exists.

Lyons and associates (104, 158) have developed what is probably the most correlatively tested theory for the causal mechanism of temperature injury to organisms. Stress, specifically cold stress, produces an alteration in the physical structure of cellular membranes; namely a phase transition from a "flexible" liquid-crystalline state in which permeability may be selectively controlled, to a "solid-gel" matrix which allows increased, unregulated permeability. According to Lyons and Raison (105), such a phase transition would occur at a specific temperature, the point of transition depending largely on the degree of unsaturation of the membrane lipids. Organisms sensitive to above-freezing temperature stress would exhibit a membrane shift. Insensitive, cold-tolerant organisms would not.

The plasma membrane, which delimits the bounds of the cell cytoplasm, is composed of phospholipids and intrinsic proteins. Additionally, many substances which possess a natural affinity for membranes may adsorb to the plasma membrane and also to cell organelle membranes. Proteins such as albumin, enzymes such as RNase and lipoxygenase (Goldstein, Anderson, and McDaniel, 1981. *Prep. Biochem.* 11:33), phenolic substances and growth regulators (McDaniel, unpublished) all may bind to membranes, with greater or lesser degrees of reversibility. Many of these associations have been simulated in the test tube, using aqueously dispersed fatty acid preparations or red cell membrane "ghosts."

Theory holds that the change in molecular ordering of membrane lipids from ordered to disordered states is the primary effect of low temperature stress. Alterations in membrane structure at some critical temperature were indicated by a considerable body of correlative experimental data accumulated in the early 1970s. Correlations between unsaturated lipid content of membranes and the temperature responses of organisms are numerous. Plants, animals, and microorganisms usually contain a high proportion of unsaturated fatty acids when grown at low temperatures (105, 1, 135). Altered activities of a number of enzymatic proteins occur simultaneously with the phase changes of the lipids isolated free of protein (105, 135). Only slight differences in hydrophobicity of proteins in hardy and nonhardy plants have been detected (33), which strengthens the pivotal role of membrane lipid composition in transition effects.

Fatty acid composition of membranes can also be altered by feeding or by incubating the organism in appropriate media (15). The fatty acid composition of the organism changes to approximate the fatty acid acyl chain composition of the supplement. Luck (102) found that the buoyant density of sucrose density gradient mitochondria changed with time of labeled choline uptake in a choline deficient *Neurospora* mutant. De la Roche et al. (52) have reported a uniform increase in fatty acid unsaturation of membrane lipids in wheat cultivars grown under cold stress. While these data (52, 92) did not support a role for fatty acid composition in relative cold hardiness of cultivars, other data (23, 67, 194) do show developmental and cultivar specific lipid differences in response to temperature stress. Clearly, the system studied, as related to the continuum of fatty acid biosynthesis and membrane turnover, will influence the outcome of experiments. Based on a considerable body of experimental results, de la Roche believes that increased membrane fatty acid unsaturation is a result of low temperature stress, rather than the cause of differential cultivar freezing tolerance (53). Using inhibitors of linolenic acid biosynthesis, Willemot (191) found that although linolenic acid accumulation was a necessary factor in tissue freezing resistance, levels of this fatty acid could not account for differences in the degree of freezing resistance among wheat cultivars. These results, taken together with the fact that dietary alterations or growth supplements can produce temporary changes in membrane lipid unsaturation, would indicate membrane lipid content to be an environmentally sensitive parameter and, therefore, under rather broad genetic control. Such control appears to be subtle, or very complex—by reason of enzymatic protein interactions (15).

Molecular Control of Membrane Properties. The myriad aspects of membrane structure will not be presented in depth here. A model of an archetypal cell membrane is shown in Figure 1, the elements of which are based upon presently accepted models in the literature (19). Singer and Nicolson's model (171) is especially relevant. Dynamic changes in the molecular structure of membranes include an alteration of the structural phase of the lipid acyl chains from a solid gel to an intermediate solid gel—liquid-crystalline state, to a fluid, flexible, liquid-crystalline state with increasing temperature (88).

In the study of temperature effects on kinetic function of cellular components, data are routinely presented as Arrhenius plots. These plots graphically present the logarithm of a reaction rate as a function of the reciprocal of the absolute temperature. Changes in activation energy with temperature are seen as abrupt alterations in the slope of the curve. Characteristically, a

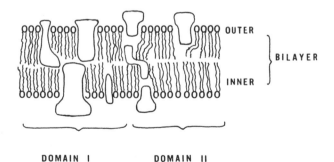

DOMAIN I   DOMAIN II

Figure 1 A conceptual model of cell membrane structure, adapted mainly from Singer and Nicolson (171), Kimelberg (88), Emmelot and Van Hoeven (60), and Esser and Souza (62). A lipid bilayer is illustrated, composed of polar head groups of phospholipid molecules (open circles) and fatty acyl chains (wavy lines), representing a fluid liquid-crystalline state characteristic of decreased packing and membrane thickness. Intrinsic and extrinsic proteins are represented by block forms.

phase change in a biological membrane or membrane lipid component will elicit an activation energy change which the Arrhenius plot will reflect.

The temperature(s) of phase changes depends mainly on degree of unsaturation of the component lipids (104, 60). Acyl chain length of lipids and the composition of the polar head groups also influence phase changes. Transition temperature and enthalpy can be influenced by the protein content (9) and protein-lipid ratio (51). As indicated in Figure 1, there is good evidence for numerous protein-lipid configurations. Some proteins would span the inner and outer membrane bilayer, with hydrophilic portions accommodating the transport of solutes across the membrane. Newly formed lipids may be inserted in a specific domain or may form specific associations before insertion (62). Proteins intrinsic to lipid domains probably differ. No definitive data exist on this point (88), although evidence for asymmetry of extrinsic protein placement on inner and outer membrane surfaces has been found (60). In fact, the suggestion has been made that lateral phase separations—with phase transitions signaling the initiation and termination of separations—are necessary to account for growth of organisms at various temperatures (62).

A phase separation theory would reconcile individual concepts such as alterations of fatty acid saturation and branching with growth temperature, phase transition data, responses of reconstituted phospholipids, and an uneven placement of proteins intercalated in the membrane. Such a mosaic model would have regulatory properties, enabling the system to respond to some deviation from growth temperature in a controlled manner, while at the same time possessing the property of dynamically reconstructing the membrane mosaic with time, in response to a new growth temperature. A membrane patchwork would enable the continued function of the membrane, albeit more slowly, with temperature changes. Because of the domain changes possible within portions of membranes, one can conceptualize a strong evolutionary role for this regulatory system. Ultimate genetic control of the flexibility of membrane response may reside in the protein synthesis machinery, which may function less accurately as growth temperatures vary from the optimum for that particular organism (62).

Martin et al. (108) and others (15) have reported that enzymatic desaturation of phospholipid fatty acids corresponded to the degree of membrane fluidity rather than to cell temperature. In *Tetrahymena,* growth following temperature shifts was not affected by fatty acid desaturation. Genetic control would appear to be at the transcriptional or translational level of fatty acid desaturase synthesis, with transient membrane fluidity changes responsible for the degree of environmental homeostasis expressed.

Freezing Injury of Membranes

Mazur (111) has discussed the paradoxical nature of freezing on physiological phenomena. The events following freezing are highly observable; their interactions are biologically complex. Hardy plants tolerate more freezing than do nonhardy plants (32). Kimball and Salisbury (87) have discussed alpine plants which can grow and reproduce under snow cover at near 0°C temperatures. They implicated photosynthesis as an important factor in low temperature plant growth. Some photosynthetic enzymes appear little affected by low temperatures (160), whereas chloroplast membranes and some kinetic photosynthetic functions appear closely regulated by temperature (64).

Steponkus and Wiest (176) utilized isolated protoplasts to study alterations in plasma membrane following freezing. They found protoplast surface area and osmotic potential of expansion and contraction were important and apparently independent factors in whether protoplasts lysed upon freezing. Palta and Li (136) also reported on cell membrane injury following freezing, and related increased transport of potassium across injured cell membranes to damage of membrane bound ATPases, enzymes important in energy linked ion transport.

Interestingly, reports of "antifreeze" glycoproteins in the blood serum of an antarctic fish may provide evidence of a rather unique evolutionary adaptation to minimize freezing temperature effects on active transport (59). These glycoproteins were found to depress the freezing point of serum well below the temperature at which sea water freezes (−1.9°C). Presence of these glycoproteins was correlated with increased serum sodium in these fishes. Effects on plasma cell ATPases and attendent ion transport may well be expected due to presence of a large quantity of glycoprotein in the serum. Studies of active transport may show a function of these proteins in preventing blood cell membrane injury. Heber (73) has described proteins capable of protecting chloroplast membranes against freezing.

BIOENERGETICS, MEMBRANES, AND TEMPERATURE

Plant Thermogenic Mechanisms

Perhaps much as fish developed an "antifreeze" protein, some plants have evolved mechanisms whereby heat is produced. The rationale for this is obvious—to accelerate germination or regrowth under marginally cold conditions, enabling the plant to get a head start of spring growth. A notable example is skunk cabbage (89) which has been shown to maintain a spadix (embryonic inflorescence) temperature as much as 35°C above ambient air temperature (−15°C to +15°C). Root starch supplies the respiratory substrate which fuels this temperature differential for two weeks or more. This respiratory heat production has been attributed to cyanide-insensitive respiration, as detailed in Solomos (172), and has been considered a mitochondrial function (16). Mitochondrial oxidative phosphorylation is thought to be capable of heat production even when well coupled (172, 170). It would appear that thermogenesis is developmentally controlled, judging from the changes in proportion of cyanide-insensitive respiration in various plant tissues (172, 130). Cyanide-insensitive respiration, mediated through an alternate oxidase (considered a mitochondrial electron transport component), has been demonstrated to occur in numerous crop seeds (193) and has been related to membrane effects in *Neurospora* (81).

Recent evidence has indicated caution in the interpretation of cyanide-insensitive respiration as a mitochondrial energy linked function. Parrish and Leopold have reported that lipoxygenase activity may account for the metabolic functions ascribed previously to a mitochondrial alternate oxidase (139). Lipoxygenase activity requires oxygen, is sensitive to salicylhydroxamate (SHAM) and propyl gallate (139, 8), and is associated preferentially with mitochondria isolated by conventional differential centrifugation (66). Significantly, further purification of wheat mitochondria on Percoll gradients effectively removed lipoxygenase activity (Goldstein, Anderson, and McDaniel, 1980. *Plant Physiol.* 66:488.) The resultant purified mitochondria were completely insensitive to SHAM and propyl gallate. Taken together, these data indicate that lipoxygenase activity could account for the effects ascribed to cyanide-insensitive mitochondrial respiration, at least in the systems studied to date.

Large amounts of lipoxygenase can be found in some seeds; one lipoxygenase substrate is linoleic acid (8). Cytokinin has been found to reduce lipoxygenase activity (70). Such effects serve to implicate lipoxygenase as an enzyme of significance in low temperature effects. Studies of temperature effects on the enzyme, and on mitochondria or membranes purified to be free of adhering lipoxygenase, will be necessary to confirm or disconfirm this suggestion.

Mitochondrial and Chloroplast Energetics

Cellular ATP levels and adenine nucleotide flux have been closely associated with low temperature stress. Chilling injury in chilling-sensitive species resulted in increased respiration (2, 192) and a decreased ATP level (192, 178). Decline in ATP level may be directly attributable to water stress, rather than to chilling temperature per se. Other workers have found ATPase activity linked with water relations in beans and corn (93, 14). Mory et al. (128) have shown that early synthesized proteins are prerequisites for subsequent DNA replication in germinating wheat embryos. Clay et al. (44) reported an association between chilling injury to germinating cottonseed, as evidenced by reduced germination, and seedling DNA synthesis following chilling. Reduction in high energy adenine phosphates in chilling-sensitive cultivars could markedly impair capacity for subsequent cell division, as judged by DNA replication. Cellular ATP content is a function of the adenylate energy charge of the cell. Production of ATP through coupled mitochondrial oxidative phosphorylation and electron transport is especially important during periods of intense metabolic activity, or when stored reserves are limited.

Mitchell (126) has discussed his theory for ATP synthesis—the chemiosmotic hypothesis—in a recent review on oxidative phosphorylation. The major theories extant are discussed in this same source (17). The Mitchell hypothesis is especially pertinent to temperature effects on plants because it more adequately reconciles the available data than do other explanations. This theory holds that respiratory and photosynthetic oxidation-reduction systems are membrane-located systems which perform work stoichiometrically with ATP synthesis or hydrolysis. Phosphorylation of ADP to produce ATP is carried out by proton flow across the membrane. This translocation of protons creates a potential difference across the membrane which effectively couples electron transport to phosphorylation. Mechanisms for solute and metabolite transport and an intrinsic ATPase function round out this hypothesis. Racker and Hinkle have shown that temperature alters the function of proton "pumps" studied in reconstituted vesicles (150). Additionally, evidence for transmission of protons (or of potential gradients) has been found (149). These electrochemical gradients probably function through gated pores or series of channels in the membrane (150, 177). Associations of specific ions, such as sodium and potassium, have been carefully studied with regard to permeability channels in nerve cells (177). More definitive data are necessary to transform theory into fact with regard to plant bioenergetics and temperature effects.

The best studied plant bioenergetic phenomena related to temperature stress have been mitochondrial respiratory function and oxidative phosphorylation, and implicit membrane effects. As discussed previously, temperature induced phase changes have been reported for a number of cellular components: enzymes, lipid components, membranes, and organelles (154, 185, 90). Biophysical techniques such as electron spin resonance using spin labels added to membranes confirmed the phase changes seen enzymatically (153). The majority of these studies have involved mitochondria. Essentially, the progression of experimental data from the 1960s to the present illustrates both the correlation of phase transitions with chilling sensitivity and the inability to distinguish degrees of chilling resistance within sensitive species (103, 119, 146, 157). The breaks in Arrhenius plots of mitochondria from wheat, a chilling-resistant plant (146), were considered to be the result of mitochondrial concentration differences, and aging effects after mitochondrial isolation (157). Similar State 3 respiration inhibition and aging effects have previously been reported with plant mitochondria (152, 115, 155). Another possible source of error could be anomalous breaks in Arrhenius plots if temperature-dependent increases in K_m are not compensated for by altering substrate concentrations (164). It would appear that interpretation of chilling responses of isolated organelles and their components requires a careful study and understanding of many potential artifacts. Results of typical mitochondrial respiratory measurements are illustrated in Figure 2. Arrhenius plots of chilling-sensitive cotton differing in chilling sensitivity show identical "breakpoints," or phase transitions. The observed differences in curves are attributable to different amounts of mitochondria per unit seedling tissue. Identical results have been achieved with a number of cotton cultivars and across a range of reaction and substrate concentrations.

Chapman and associates (31) have recently described a two-phase adaptive increase in mitochondrial lipid fluidity during tuber winter dormancy in Jerusalem artichoke. Nondormant tuber membranes showed phase transitions at 25° and 3°C; cold-adapted dormant tubers showed breakpoints at 9° and −5°C. Similar but less dramatic transition shifts were reported by Miller et al. (125) in wheat.

Chloroplast preparations showed temperature effects much like those observed with mitochondria. Photosynthetic physiology including chlorophyll fluorescence (124, 6), photosynthetic rate and unsaturation of chloroplast lipids (83, 141), and permeability (132) were subject to temperature effects. Phase transitions were seen in most parameters with chilling-sensitive plants.

Probably the most enticing correlation of all is that observed between Arrhenius phase transitions and growth rate of mung bean seedlings. Raison and Chapman (156) reported two growth phase transitions, one at 28°C and one at 15°C, as shown in Figure 3, which corresponded exactly to transitions observed in molecular ordering of mitochondrial and chloroplast membranes, and in succinate oxidation by isolated mitochondria. Taken together, the seasonal changes observed in phase transitions of artichoke membranes (31) and the mung bean growth measurements (156) provide compelling evidence for the role of membrane functionality in thermal responses of plants. More recently, Bagnell and Wolfe (10) have questioned whether the abrupt phase changes are in fact the best interpretation of Raison and Chapman's data; that a smooth curve may better describe the data. Until further statistical analyses of similar growth data are forthcoming, under identical experimental conditions, and using closely synchronized tissues (114), the correctness of one or the other of these interpretations cannot be judged.

ADP:O ratios of isolated mitochondria have been studied in conjunction with respiratory measurements. ADP:O ratios measure the phosphorylative efficiency of mitochondria—the number of ATP molecules produced per unit of carbon oxidized. In general, either slight or no effects of temperature on ADP:O's were seen in stress-resistant and susceptible animals (162). Wheat and rye cultivars (146), apple cultivars (119), and cucumber, tomato, sweet potato (103), and other cultivars showed no appreciable comparative differences in ADP:O ratios with temperature. In contrast, Table 3 illustrates the ADP:O ratios of mitochondria from seedlings of high elevation-adapted cottons are superior to those of low elevation varieties when assayed at low temperatures. The high elevation types possess a high degree of seedling cold tolerance; the low elevation types are resistant to high summer temperatures.

McDaniel (unpublished) did not find significant ADP:O differences between mitochondria from northern adapted alfalfa varieties and warmer cli-

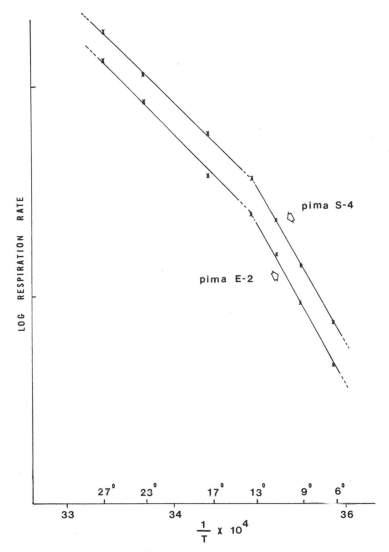

Figure 2 Effect of temperature upon State 3 respiration rate of mitochondria isolated from Pima Cotton (*Gossypium barbadense* L.) seedlings. 'S-4' is a cold-sensitive cultivar, 'E-2' is a relatively cold-tolerant type. Data presented as an Arrhenius plot. Substrate was α-ketoglutarate. Bovine serum albumin (350 mg) was added in two aliquots during centrifugation steps to maintain coupled oxidative phosphorylation. Temperature reciprocal scale is also expressed in equivalent degrees centigrade for clarity. Unpublished data of McDaniel.

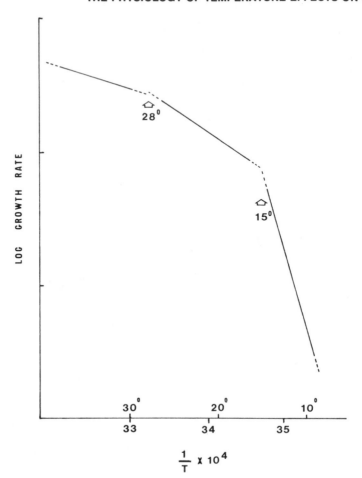

Figure 3 Growth rates of mung bean (*Vigna radiata* L.) cultivar 'Mungo' seedlings measured at 13 temperatures and presented as an Arrhenius plot. Temperature reciprocal scale is also expressed in equivalent degrees centigrade for clarity. Data of Raison and Chapman (156).

mate, nondormant clones. Reasons for the differential temperature response of plant mitochondria assayed under similar conditions remain unexplained.

BIOCHEMICAL MODIFICATIONS OF TEMPERATURE RESPONSE

Chemical Plant and Seed Treatments

Early workers (94) recognized that chemicals could alter plant temperature responses. Bonner reported that low temperature growth inhibition of *Cosmos* could be alleviated by addition of thiamine to the median. Peas and duckweed were reported to respond to addition of adenine and adenosine,

Table 3   Comparative mitochondrial efficiency of Pima cotton (*Gossypium barbadense* L.) cultivars of varying cold sensitivity. Mitochondria were isolated from cotton seedlings germinated at 22°C. Data are ADP:O ratio means calculated on three or more replicate mitochondrial extractions with α ketoglutarate as substrate. Unpublished data of R. G. McDaniel

Cultivar	Adaption and Seedling Cold Response	Mitochondrial Assay Temperature (°C)				
		9°	11°	17°	23°	27°
E-2	High elevation type (relatively cold-tolerant)	2.58	2.52	2.26	2.07	2.08
E-4	High elevation type (relatively cold-tolerant)	2.76	2.70	2.65	2.31	2.24
S-4	Low elevation type (relatively cold-sensitive)	2.10	2.11	2.12	2.15	2.12
S-5	Low elevation type (relatively cold-sensitive)	2.06	2.06	2.24	2.19	2.30

which improved the heat tolerance. Gibberellic acid was found to substitute for cold treatments (94). Jung and coworkers (80) reported that purine and pyrimidine treatments improved the development of predormancy cold tolerance in alfalfa. Purine nucleotide seed or seedling treatments may prove beneficial to a number of crop plants which are susceptible to cold injury during germination. Cole and Wheeler (46), as well as McDaniel and Taylor (118), have found that cotton seed treated with cyclic AMP (or AMP) germinated better under cold or salt stress than controls. Cohn and Obendorf (45) showed a 5- to 8-fold decline in corn embryo AMP content following a 48-hr cold hydration treatment. Anderson (4) reported that soybean axes soaked in adenine or adenosine synthesized large quantities of ATP. In potato tubers stored at low temperatures, tissue cyclic AMP levels declined from 30 pmol/g tissue to 7 pmol following 28 days of storage (78). As the germinating embryo must rely on stored reserves during initial germination stages, it seems reasonable that cold or other stresses would require extraordinary expenditures of seedling energy for growth. Exhaustion of seed ATP (and precursor molecules such as AMP) would be lethal if the seedlings were not yet autotrophic. Chemical seed treatments with ATP precursors could boost the seeds' ability to withstand low temperature stress (118). High temperature stress has likewise been alleviated by growth regulator-seed treatments of lettuce and other crops (22). Christiansen and Ashworth treated emerging cotton seedlings with antitranspirants to prevent chilling injury (42). Significantly, St. John and Christiansen also showed that treatment with pyridazanones altered unsaturated lipid content and resulted in loss of cold hardening and of chilling resistance in cotton (163). Free radical scavengers such as sodium benzoate and ethoxyquin served to protect sensitive vegetable fruit from chilling damage during storage (187). Cyclocel has been shown to elicit increases in chilling resistance (109) and cold hardiness (161). Most of the chemicals which increased chilling or cold resistance are classified as growth regulators; some are growth retardants (163).

Cryoprotectants afford protection to plant tissues during freezing. Farrant and Woogar (63) have proposed that cryoprotectants that enter cells, such as dimethyl sulfoxide (DMSO), and nonpenetrating cryoprotectants, including sucrose and hydroxyethyl starch, act by buffering the increase of other solutes as freezing occurs. In effect, cryoprotectants may act to alter the ratio of bound to unbound water in tissues (25, 189), thus potentiating more effective supercooling. Cryoprotectants, specifically DMSO, have been shown to protect cell organelles—mitochondria—from damage when frozen to liquid nitrogen temperatures (−196°C), and thawed under appropriate conditions (55, 56).

Mercaptoethanol, a thiol reagent, negated the effects of DMSO when cabbage leaf tissues were frozen (91). This was considered to be a sulfhydryl effect, presumably acting upon the disulfide linkages of proteins.

Chemical Modification of Plant Constituents

Treament of tomato seedlings with ethanolamine modified the membrane phospholipid composition (188). This amino alcohol elicited incorporation of substantial phosphatidylethanolamine, altering the acyl chain composition of phospholipids and resulting in an improvement in chilling resistance of cotyledon tissues (77).

The role of plant nutrient status in the amelioration of chilling injury has been little studied. Christiansen has implicated calcium as a significant factor in the chilling sensitivity of cotton seedlings. Addition of calcium or magnesium could eliminate chilling-induced solute leakage in cotton seedlings (39). Calcium has been related to membrane structure (195) and to the function of membrane associated enzymes such as ATPase (43). Zinc also may play a role in growth of chilling-sensitive species at low temperatures (79). Chapin (30) has described a decline of plant phosphate absorption capacity corresponding to the soil temperature of the latitude of origin. As mean July soil temperature of plant habitats increased, rate of phosphate absorption of roots of a number of species declined. Chemical treatments to improve temperature response of agronomic species will likely increase considerably as additional effective chemicals are documented.

PHYSIOLOGICAL CRITERIA FOR SELECTION OF TEMPERATURE TOLERANT PLANTS

Selection and Breeding Techniques

Chemical treatment to improve seed and seedling resiliency to temperature stress may be regarded as an interim solution that serves to bridge the gap between crop failure and eventual crop improvement through breeding. Evidence from a number of sources as discussed in previous sections suggests that conventional breeding and selection techniques may be effectively applied to temperature tolerance. Phatak (142) and Dickson (57) have shown that low temperature-tolerant strains of tomato and bean cultivars could be identified under controlled low temperature growing conditions. Patterson and associates (140) used leaf electrolyte leakage as a method of distinguishing chilling-resistant (low leakage rate) and chilling-sensitive (high leakage rate) *Passiflora* genotypes. Quamme (148) has recently detailed the use of freezing exotherm analysis to distinguish cold tolerance differences in apple cultivars. Inheritance of cold-hardiness appeared quantitative in apples.

Buxton and Sprenger (26) distinguished cotton cultivars differing in their ability to tolerate low temperature stress following planting. Maize inbred lines show similar responses (127). In cotton, there is some evidence for a maternal effect on seedling cold tolerance (106). Christiansen and Lewis (40) reported that the chilling sensitivity of F_1 cotton hybrids paralleled the response of the parent used as the female in the cross. Cytoplasmic inheritance of cold sensitivity in *Aspergillus* mutants has been recently reported (186). These mutants exhibited altered cytochrome spectra, and presumably were mitochondrial in nature. Again, a maternal component of chilling sensitivity was reported for corn seedlings (29). Differential winter hardiness responses of wheat, *Agropyron,* and their hybrids has been related to differential responses of mitochondria from seedlings of these lines (86). Considerable future work is indicated to confirm the genetic locus or loci of plant low temperature response. The high heritability for cold tolerance observed in most instances makes significant breeding progress possible.

Cell Culture Techniques

Price and Smith (147) have discussed the potential utility of tissue culture in selecting for improved cotton cultivars. Selection for altered plant response to temperatures could be effectively carried out by selecting for vigorous callus growth under controlled temperature stress. If tissue level reactions simulate whole plant response, then regeneration of the callus should result in an improved genotype. Dix and Street (58) have reported isolated cell lines of *Nicotiana* and *Capsicum* with enhanced survival under temperature stress, following mutagen treatment and chilling conditions. Mitochondria isolated from chilling-resistant and chilling-sensitive pepper cell lines exhibited much different Arrhenius temperature responses.

Studies on alterations of lipid content in plant tissue cultures as a function of temperature (157) and on artichoke callus cold acclimation (179) are illustrative of the utility of this approach to plant selection. New techniques for germplasm manipulation include fusion of protoplasts for production of interspecific and intergeneric hybrids (97). In this way a rapid incorporation of superior temperature tolerance of distantly related plants may be achieved. Genetic engineering, using plasmid vectors to transfer selected bits of DNA-containing genes for improved adaptive temperature responses, appears promising for future plant breeding.

Optimizing Plant Metabolism for Temperature Tolerance

Many of the physiological studies of plant temperature effects have been comparative in nature. The dissimilar response of two closely related organisms to a stimulus such as temperature can often shed light on the physiological mechanisms responsible for the difference. Hybrid plants are especially attractive for comparative studies.

Langridge (95) proposed a molecular basis for hybrid vigor based upon the greater stability of hybrid proteins at high temperatures. Data indicated that hybrids exhibited better environmental homeostasis, which would enhance hybrid vigor under environmental extremes. Griffing and Zsiros (68) feel that environmental stresses minimized heterotic effects. Other workers, however, found corn hybrids (123) and barley hybrids (117) to be superior to parental strains at high temperatures.

Recent work with alfalfa has demonstrated that physiological criteria are excellent plant breeding tools to select for temperature tolerant synthetics. When mitochondrial activity of an array of individual alfalfa genotypes within a variety was evaluated, a broad distribution of (a) mitochondrial phosphorylative efficiency (ADP: O ratios) and (b) mitochondrial respiration rates, per unit etiolated tissue, was observed. Individual plants showing combinations of efficiency levels and respiratory rates were grouped and used as parents in the production of insect-facilitated synthetic populations. Table 4 shows the forage yield of the four synthetics, selected solely on the basis of two mitochondrial parameters (and survival ability).

Under Arizona conditions, low elevation heat stress is the most severe limitation to forage crop growth. High temperatures cause reduced growth of alfalfa termed *summer slump*. The synthetic produced from parents with high ADP:O ratios and a low, controlled respiration rate (at 27°C) exhibited superior forage production compared with the other three synthetics. In fact, with no consideration of agronomic and morphological (plant type) characteristics, the high ADP:O, low respiration synthetic was numerically superior to the parent cultivar in productivity. These data emphasize the fact that use of multiple physiological criteria will improve breeding effectiveness. An interplay of two apparently unlinked physiological parameters is seen, where superior mitochondrial energy conversion efficiency is the critical factor relating to yield, with respiration rate of lesser, but nonetheless, significant importance in this example. Genotypes with exceedingly high or low respiratory rates failed to regrow following cutting and hence could be considered lethals under conditions of multiple harvesting. A high or uncontrolled rate of respiration was detrimental to the ability of the plants to with-

Table 4   Mitochondrial efficiency (ADP:O ratio with α-ketoglutarate as substrate, assayed at 27°C) and mitochondrial respiration rate of etiolated regrowth of three or more selected alfalfa clones (*Medicago Sativa* L.)[a]

Alfalfa Synthetic Population	Mean Parental ADP:O Ratio	Mean Parental Respiration Rate/ Unit Tissue	Three Year Forage Production 1976–1978 (kg Green Wt./Plot)
High ADP:O and low respiration	2.86	24	252.0
High ADP:O and high respiration	2.89	32	234.5
Low ADP:O and high respiration	2.07	38	217.0
Low ADP:O and low respiration	2.14	23	213.6
Mesa-Sirsa (check cultivar)	2.43	29	242.4

[a]Forage production of synthetic progeny populations derived from polycrosses within each group of clones was determined at Mesa, Arizona. Mesa Sirsa represents like-generation seed of the parent of the polycross source. Unpublished data of R. G. McDaniel, A. K. Dobrenz, and M. H. Schonhorst. Field data courtesy of R. K. Thompson.

stand summer temperature stress; whereas efficient coupling of oxidative phosphorylation and electron transport was well correlated with increased forage production. Similar approaches to physiological plant breeding both for disease resistance and yield capacity under stress are being undertaken with other important agronomic crops.

SUMMARY

Physiological and biochemical studies of plant temperature responses have proven to be fruitful areas of investigation. While the precise cause-effect relationships of processes important in climatic adaption and temperature tolerance of plants remain undiscovered, the weight of the findings to date have allowed the formulation of excellent working hypotheses to facilitate effective further experimentation.

Given the increased interest and participation of scientists in study of temperature related plant responses, one can reasonably look forward to the development and testing of unifying hypotheses of primary molecular effects of temperature on plants in the not too distant future. Certainly the better our understanding of plant temperature response at the physiological and molecular levels, the more effectively we can utilize this information in the production of temperature resistant, superior crops.

REFERENCES

1   Alexandrov, V. Y. and V. A. Bernstam, trans. 1977. *Cells, Molecules and Temperature: Conformational Flexibility of Macromolecules and Ecological Adaptation.* Springer-Verlag, Berlin.

2   Amin, J. V. 1969. *Physiol. Plant* 22:1184.

3   Amin, J. V. 1969. *Plant Soil* 31:365.

4   Anderson, J. E., and S. J. McNaughton. 1973. *Ecology* 54:1220.

5   Anderson, J. D. 1977. *Plant Physiol.* 60:689.

6   Armond, P. A., U. Schreiber, and O. Björkman. 1978. *Plant Physiol.* 61:411.

7   Atzpodien, W., J. M. Gancedo, W. Duntze, and H. Holzer. 1968. *European J. Biochem.* 7:58.

8   Axelrod, B., In Whitaker, Ed. 1974. *Food Related Enzymes.* Am. Chem. Soc., Washington, D.C., p. 324.

9   Bach, D., and I. R. Miller, 1976. *Biochem. Biophys. Acta.* 443:13.

10 Bagnall, D. J. and J. A. Wolfe. 1978. *J. Exp. Bot.* 29:1231.

11 Baker, B. S. and G. A. Jung. 1970. In *Proc. IX Int. Grassland Congr.* U. of Queensland Press, p. 499.

12 Baker, B. S., and G. A. Jung. 1972. *Bot. Gaz.* 133:120.

13 Baldwin, J., and P. W. Hochachka. 1970. *Biochem. J.* 116:883.

14 Barlow, E. W. R., T. M. Ching, and L. Boersma. 1976. *Crop Sci.* 16:405.

15 Bartkowski, E. J., D. R. Buxton, F. R. H. Katterman, and H. W. Kircher. 1977. *Agron. J.* 69:37.

16 Bendall, D. S., and W. D. Bonner, Jr. 1971. *Plant Physiol.* 47:236.

17 Boyer, P. D., B. Chance, L. Ernster, P. Mitchell, E. Racker, and E. C. Slater. 1977. *Annu. Rev. Biochem.* 46:955.

18 Brandts, J. F. 1967. In *Thermobiology,* A. H. Rose, Ed. Academic, New York, p. 25.

19 Bretscher, M. S. 1973. *Science* 181:622.

20 Bromlage, W. J., A. C. Leopold, and D. J. Parrish. 1978. *Plant Physiol.* 61:525.

21 Bronk, T. D. 1967. *Science* 158:1012.

22 Brown, J. W., and A. A. Khan. 1976. *J. Am. Soc. Hort. Sci.* 101:716.

23 Bubbels, G. H. 1974. *Can. J. Plant Sci.* 54:425.

24 Bull, H. B., and K. Breese. 1973. *Arch. Biochem. Biophys.* 158:684.

25 Burke, M. J. 1976. *Annu. Rev. Plant Physiol.* 27:507.

26 Buxton, D. R., and P. J. Sprenger. 1976. *Crop Sci.* 16:243.

27 Buxton, D. R., P. J. Sprenger, and E. J. Pegelow, Jr. 1976. *Crop Sci.* 16:471.

28 Buxton, D. R., P. J. Melick, L. L. Patterson, and C. A. Godinez. 1977. *Agron. J.* 69:672.

29 Cal, J. P., and R. L. Obendorf. 1972. *Crop Sci.* 12:369.

30 Chapin III, F. S. 1974. *Science* 183:521.

31 Chapman, E., L. C. Wright, and J. K. Raison. 1979. *Plant Physiol.* 63:363.

32 Chen, P. M., M. J. Burke, and P. H. Li. 1976. *Bot. Gaz.* 137:313.

33 Chou, J. C., and J. Levitt. 1972. *Cryobiology* 9:268.

34 Christiansen, M. N. 1963. *Plant Physiol.* 38:520.

35 Christiansen, M. N. 1964. *Crop Sci.* 4:584.

36 Christiansen, M. N. 1967. *Plant Physiol.* 42:431.

37 Christiansen, M. N. 1969. *Proc. Beltwide Cotton Prod. Res. Conf.*, p. 50.

38 Christiansen, M. N., H. R. Carns, and D. J. Slyter. 1970. *Plant Physiol.* 46:53.

39 Christiansen, M. N. 1971. *Proc. Beltwide Cotton Prod. Res. Conf.* p. 71.

40 Christiansen, M. N. and C. F. Lewis. 1973. *Crop Sci.,* 13:210.

41 Christiansen, M. N. 1978. In *Crop Tolerance to Suboptimal Land Conditions. Am. Soc. Agron.,* Madison, Wisconsin, p. 173.

42 Christiansen, M. N., and E. N. Ashworth. 1978. *Crop Sci.* 18:907.

43 Christiansen, M. N., and C. D. Foy. 1979. *Commun. in Soil Sci. Plant Anal.* 10:427.

44 Clay, W. F., D. R. Buxton, and F. R. H. Katterman. 1977. *Crop Sci.* 17:342.

45 Cohn, M. A., and R. L. Obendorf. 1976. *Crop Sci.* 16:449.

46 Cole, D. F., and J. E. Wheeler. 1974. *Crop Sci.* 14:451.

47 Cole, D. F., and M. N. Christiansen. 1975. *Crop Sci.* 15:410.

48 Cole, F. D. 1969. Ph. D. dissertation, University of Arizona.

49 Cole, F. D., and R. G. McDaniel. 1974. *Isoenzyme Bull.* 7:35.

50 Cothren, J. T., and G. Quinn. 1975. *Phyton* 33:131.

51 Davoust, J., B. M. Schoot, and P. F. Devaux. 1979. *Proc. Natl. Acad. Sci.* 76:2755.

52 De la Roche, I., M. K. Pomeroy, and C. J. Andrews. 1975. *Cryobiology* 12:506.

53 De la Roche, I. 1979. *Plant Physiol.* 63:5.

54 Devay, M., and E. Paldi. 1977. *Plant Sci. Letters* 8:191.

55 Dickinson, D. B., M. J. Misch, and R. E. Drury. 1967. *Science* 156:1738.

56 Dickinson, D. B., M. J. Misch, and R. E. Drury. 1970. *Plant Physiol.* 46:200.

57 Dickson, M. H. 1971. *Crop Sci.* 11:848.

58 Dix, P. J., and H. E. Street. 1976. *Ann. Bot.* 40:903.

59 Duman, J. G., and A. L. DeVries. 1974. *J. Exp. Zool.* 190:89.

60 Emmelot, P., and R. P. VanHoevan. 1975. *Chem. Phys. of Lipids* 14:236.

61 Esashi, Y., and Y. Tsukada. 1978. *Plant Physiol.* 61:437.

62 Esser, A. F., and K. A. Souza. 1974. *Proc. Natl. Acad. Sci.* 71:4111.

63 Farrant, J., and A. E. Woolgar. 1970. In *The Frozen Cell,* G. E. Wolstenholme and M. O'Conner, Eds. Churchill, London, p. 97.

64 Fehu, M., and M. Devay. 1975. *Biochem. Physiol. Pflanzen* 167:447.

65 Feng, K. A. 1973. *Plant Physiol.* 51:868.

66 Goldstein, A. H., J. O. Anderson, and R. G. McDaniel. 1979. *Plant Physiol. Suppl.* 63:812.

67 Grenier, G., and C. Willemot. 1974. *Cryobiology* 11:324.

68 Griffing, G., and E. Zsiros. 1971. *Genetics* 68:443.

69 Grimwood, B. G., and R. G. McDaniel. 1970. *Biochem. Biophys. Acta.* 220:410.

70 Grossman, S., and Y. Y. Leshem. 1978. *Physiol. Plant* 43:359.

71 Guinn, G. 1971. *Crop Sci.* 11:262.

72 Harris, P., and A. T. James. 1969. *Biochem. Biophys. Acta.* 187:13.

73 Heber, U. 1970. In *The Frozen Cell,* G. E. Wolstenholme and M. O'Conner, Eds. Churchill, London, p. 175.

74 Hochachka, P. W., and J. K. Lewis. 1970. *J. Biol. Chem.* 245:6567.

75 Hochachka, P. W., and G. N. Somero. 1973. *Strategies of Biochemical Adaptation.* W. B. Saunders Co., Philadelphia.

76 Haskins, M. A. H., and M. Aleksuik. 1973. *Comp. Biochem. Physiol.* 45B:343.

77 Ilker, R., A. J. Waring, and J. M. Lyons. 1976. *Protoplasma* 90:229.

78 Isherwood, F. A., and S. G. Ring. 1977. *Phytochemistry* 16:309.

79 Joham, N. E., and V. Rowe. 1975. *Agron. J.* 67:313.

80 Jung. G. A., S. C. Shih, and W. G. Martin. 1974. *Cryobiology* 11:269.

81 Juretic, D. 1976. *J. Bacteriol.* 126:542.

82 Kacperska-Palacz, A. 1978. In *Plant Cold Hardiness and Freezing Stress*, P. H. Li and A. Sakai, Eds. Academic, New York, p. 139.

83 Kaniuga, Z., and W. Michalski. 1978. *Planta* 140:126.

84 Keith, A. D., and W. Snipes. 1974. *Science* 183:666.

85 Khan, A. A., Ed. 1977. *The Physiology and Biochemistry of Seed Dormancy and Germination.* North Holland Publishing Co., Amsterdam.

86 Khristolyubova, N. B., V. V. Khvostova, V. T. Satonova, and T. K. Usova. 1974. *Theor. Appl. Genet.* 44:255.

87 Kimball, S. L., and F. B. Salisbury. 1974. *Bot. Gaz.* 135:147.

88 Kimelberg, H. K. 1978. *Cryobiology* 15:222.

89 Knutson, R. M. 1974. *Science* 186:746.

90 Kosiyachinda, S., and R. E. Young. 1977. *Plant Physiol.* 60:470.

91 Krull, E., and J. Levitt. 1972. *Physiol. Plant* 27:259.

92 Kuiper, P. J. C. 1970. *Plant Physiol.* 45:684.

93 Kuiper, P. J. C. 1972. *Physiol. Plant* 26:200.

94 Kurtz, E. B., Jr. 1958. *Science* 128:1115.

95 Langridge, J. 1962. *Am. Nat.* 116:5.

96 Langridge, J., and J. R. McWilliam. 1967. In *Thermobiology*, A. H. Rose, Ed. Academic, New York, p. 231.

97 Lao, K. N., F. Constabel, M. R. Michayluk, and O. L. Gamborg. 1974. *Planta* 120:215.

98 Lemon, E. R., Ed. 1969. *Physiological Limitations on Crop Production Under Temperature and Moisture Stress.* Natl. Acad. Sci., Washington, D.C., 36 p.

99 Levitt, J. 1972. *Responses of Plants to Environmental Stress.* Academic, New York.

100 Levitt, J. 1974. *Introduction to Plant Physiology,* 2nd ed. C. V. Mosby Co., St. Louis.

101 Low, P. S., J. L. Bada, and G. N. Somero. 1973. *Proc. Natl. Acad. Sci.* 70:430.

102 Luck, D. J. L. 1965. *Am. Nat.* 99:242.

103 Lyons, J. M., and J. K. Raison. 1970. *Plant Physiol.* 45:386.

104 Lyons, J. M. 1973. *Annu. Rev. Plant Physiol.* 24:445.

105 Lyons, J. M., J. K. Raison, and P. L. Steponkus. 1979. *Low Temperature Stress in Crop Plants.* Acad. Press, N.Y.

106 Marani, A., and A. Amirav. 1970. *Crop Sci.* 10:509.

107 Marcellos, H., and W. V. Single. 1979. *Cryobiology* 16:74.

108 Martin, C. E., K. Kiramitsu, Y. Nozawa, L. Skrivu, and G. A. Thompson. 1976. *Biochemistry* 15:5218.

109 Mayland, H. F., and J. W. Cary. 1970. *Adv. Agron.* 22:203.

110 Mazur, P. 1970. *Science* 168:939.

111 Mazur, P. 1969. *Annu. Rev. Plant Physiol.* 20:419.

112 McCarter, S. M., and R. W. Roncadori. 1971. *Phytopathology* 61:1426.

113 McClard, R. W., and H. M. Kolenbrander. 1974. *Experientia* 15:730.

114 McDaniel, R. G. 1969. *Plant Physiol. Suppl.* 44:39.

115 McDaniel, R. G., and I. V. Sarkissian. 1970. *Phytochemistry* 9:303.

116 McDaniel, R. G. 1971. In *Barley Genetics II*, R. A. Nilan, Ed. Wash. State U. Press, p. 323.

117 McDaniel, R. G. 1973. *Seed Sci. Tech.* 1:25.

118 McDaniel, R. G., and B. B. Taylor. 1979. *Agron. Abst.*, p. 115.

119 McGlasson, W. B., and J. K. Raison. 1973. *Plant Physiol.* 52:390.

120 McKenzie, J. S., C. J. Weiser, and M. J. Burke. 1974. *Plant Physiol.* 53:783.

121 McMurchie, E. J., J. K. Raison, and K. D. Cairncross. *Comp. Biochem. Physiol.* 44B:1017.

122 McNaughton, S. J. 1972. *Am. Natur.* 106:165.

123 McWilliams, J. R., and B. Griffing. 1965. *Aust. J. Biol. Sci.* 18:569.

124 Melcarek, R. K., and G. W. Brown. 1977. *Plant Physiol.* 60:822.

125 Miller, R. W., I. de la Roche, and M. K. Pomeroy. 1974. *Plant Physiol.* 53:426.

126 Mitchell, P. 1977. *Annu. Rev. Biochem.* 46:996.

127 Mock, J. J., and M. J. McNeill. 1979. *Crop Sci.* 19:239.

128 Mory, Y. Y., D. Chen, and S. Sarid. 1972. *Plant Physiol.* 49:20.

129 Mukirji, S. K., and I. P. Ting. 1969. *Arch. Biochem. Biophys.* 313:336.

130 Nakamura, K., and T. Osahi, 1976. *Arch. Biochem. Biophys.* 174:393.

131 Nash, G. H., and D. W. Grant. 1969. *Can. J. Microbiol.* 15:1116.

132 Nobel, P. S. 1974. *Planta* 115:369.

133 Obendorf, R. L., and P. R. Hobbs. 1970. *Crop Sci.* 10:563.

134 Olein, C. R. 1965. *Cryobiology* 2:47.

135 Overath, P., H. U. Schairer, and W. Stoffel. 1970. *Proc. Natl. Acad. Sci.* 67:606.

136 Palta, J. P., and P. H. Li. 1978. In *Plant Cold Hardiness and Freezing Stress*, P. H. Li and A. Sakai, Eds. Academic, New York, p. 93.

137 Pandy, K. K. 1972. *Nature, New Biol.* 239:27.

138 Parrish, D. J., and A. C. Leopold. 1977. *Plant Physiol.* 59:111.

139 Parrish, D. J., and A. C. Leopold. 1978. *Plant Physiol.* 62:470.

140 Patterson, B. D., T. Murata, and D. Graham. 1976. *Aust. J. Plant Physiol.* 3:435.

141 Peoples, T. R., D. W. Koch, and S. C. Smith. 1978. *Plant Physiol.* 61:472.

142 Phatak, S. C. 1970. *Hort. Res. Inst. Ont. Rep.*, p. 98.

143 Phillips, J. C., and V. E. Youngman. 1971. *Crop Sci.* 11:354.

144 Pollock, B. M., and V. K. Toole, 1966. *Plant Physiol.* 41:22.

145 Pomeroy, M. K., and D. Siminovitch. 1971. *Can. J. Bot.* 49:787.

146 Pomeroy, M. K., and C. V. Andrews. 1975. *Plant Physiol.* 56:703.

147 Price, H. J., and R. H. Smith. 1977. *Proc. Belt. Cotton Prod. Res. Conf.*, p. 51.

148 Quamme, H. A. 1978. In *Plant Cold Hardiness and Freezing Stress*, P. H. Li and A. Sakai, Eds. Academic, New York, p. 313.

149 Racker, E., and W. Stoeckenius. 1974. *J. Biol. Chem.* 249:662.

150 Racker, E., and P. C. Hinkle. 1974. *J. Membrane Biol.* 17:181.

151 Radwan, S. S., S. Grosse-Oetringhaus, and H. K. Mangold. 1978. *Chem. Phys. Lipids* 22:177.

152 Raison, J. K., and J. M. Lyons. 1970. *Plant Physiol.* 45:382.

153 Raison, J. K., J. M. Lyons, R. J. Mehlhorn, and A. D. Keith. 1971. *J. Biol. Chem.* 246:4036.

154 Raison, J. K., J. M. Lyons, and W. W. Thompson. 1971. *Arch. Biochem. Biophys.* 142:83.

155 Raison, J. K., J. M. Lyons, and L. C. Campbell. 1973. *Bioenergetics* 4:397.

156 Raison, J. K., and E. A. Chapman. 1976. *Aust. J. Plant Physiol.* 3:291.

157 Raison, J. K., E. A. Chapman, and P. Y. White. 1977. *Plant Physiol.* 59:623.

158 Raison, J. K. 1979. In *Encyclopedia of Plant Physiol,* P. K. Stumf, Ed.

159 Reiss, U., and M. Rothstein. 1974. *Biochem. Biophys. Res. Comm.* 61:1012.

160 Riov, J., and G. N. Brown. 1978. *Cryobiology* 15:80.

161 Roberts, C. W. A. 1970. *Can. J. Bot.* 49:705.

162 Roberts, J. C., R. M. Arine, R. H. Rochelle, and R. R. J. Chaffee. 1972. *Comp. Biochem. Physiol.* 41B:127.

163 St. John, J. B., and M. N. Christiansen. 1976. *Plant Physiol.* 57:257.

164 Silvius, J. R., B. D. Read, and R. N. McElhaney. 1978. *Science* 199:902.

165 Simon, E. W. 1972. *J. Exp. Bot.* 23:1076.

166 Simon, E. W. 1974. *New Phytol.* 73:377.

167 Simon, E. W., and H. H. Wiebe. 1975. *New Phytol.* 74:407.

168 Simon, E. W. 1978. *Pesticide Sci.* 9:169.

169 Simon, E. W. 1979. Proc. U.S./Australia Science Seminar.

170 Simon, R. G., C. E. Eykel, W. Galster, and P. Morrison. 1971. *Comp. Biochem. Physiol.* 40B:601.

171 Singer, S. J., and G. L. Nicolson. 1972. *Science* 175:720.

172 Solomos, T. 1977. *Annu. Rev. Plant Physiol.* 28:279.

173 Somero, G. N. 1969. *Am. Natur.* 103:517.

174 Somero, G. N. 1975. *J. Exp. Zool.* 194:175.

175 Somero, G. N. 1975. In *Isoenzymes II. Physiological Function,* E. Clement and L. Markert, Eds. Academic, New York, p. 221.

176 Steponkus, P. L., and S. C. Wiest. 1978. In *Plant Cold Hardiness and Freezing Stress,* P. H. Li and A. Sakai, Eds. Academic, New York, p. 75.

177 Stevens, C. F. 1977. *Nature* 270:391.

178 Stewart, J. McD., and G. Guinn. 1969. *Plant Physiol.* 44:605.

179 Sugawara, Y., and A. Sakai. 1978. In P. H. Li and A. Sakai, Eds. Plant Cold Hardiness and Freezing Stress, Academic, New York, p. 197.

180 Sullivan, C. Y., and E. J. Kinbacher. 1967. *Crop Sci.* 7:241.

181 Thomas, R. O., and M. N. Christiansen. 1969. *Proc. Beltwide Cotton Res. Conf.,* p. 51.

182 Thomas, R. O., and M. N. Christiansen. 1971. *Crop Sci.* 11:454.

183 Thomas, T. H. 1977. In *The Physiology and Biochemistry of Seed Dormancy and Germination,* A. A. Khan, Ed. North Holland Publishing Co., Amsterdam.

184 Vigne, J., P. H. Li, and C. R. Oslund. 1974. *Plant and Cell Physiol.* 15:1055.

185 Wade, N. L., R. W. Breidenbach, J. M. Lyons, and A. C. Keith. 1974. *Plant Physiol.* 54:320.

186 Waldron, C., and C. F. Roberts. 1978. *J. Gen. Microbiol.* 78:379.

187 Wang, C. Y., and J. E. Baker. 1978. *Plant Physiol. Suppl.* 61:172.

188 Waring, A. J., R. W. Breidenbach, and J. M. Lyons. 1976. *Biochem. Biophys. Acta.* 443:157.

189 Weiser, C. J., H. A. Quamme, E. O. Proebsting, M. J. Burke, and G. Yelenosky. 1979. In *Modification of the Aerial Environment of Crops,* B. J. Barfield and J. F. Gerber, Eds. Am. Soc. of Agric. Eng. Monogr., St. Joseph, Michigan, p. 55.

190 Whaley, W. G., M. Dauwalder, and T. P. Leffingwell. 1975. *In Current Topics in Developmental Biology*, Vol. 10. Academic, New York, p. 161.

191 Willemot, C. 1977. *Plant Physiol.* 60:1.

192 Wilson, J. M. 1976. *New Phytol.* 80:325.

193 Yentur, S., and A. C. Leopold. 1976. *Plant Physiol.* 57:274.

194 Yoshida, S., and A. Sakai. 1973. *Plant Cell Physiol.* 14:353.

195 Yoshida, S. 1979. *Plant Physiol.* 64:247.

Chapter 3

BREEDING FOR TOLERANCE TO HEAT AND COLD

H. G. MARSHALL

U.S. Department of Agriculture, Science and Education Administration, Agricultural Research, and the Pennsylvania State University, University Park, Pennsylvania

INTRODUCTION

Modern cultivars of the major crop species are well adapted to controlled cultural practices, but they generally are not highly tolerant of or resistant to extremes of uncontrolled environmental factors such as temperature. Temperature stress is the major factor affecting plant growth and development (38) and, therefore, crop yields. Since plant tolerance to temperature stress is heritable, selection and breeding can be used to improve the trait. Breeding cultivars with improved tolerance to temperature, however, is a difficult task. The development of effective selection and breeding technology has lagged behind that routinely used for many qualitative characters of the major crop plants. For cold tolerance, another limitation appears to be narrow ranges of genetic diversity in the gene pools of several crop species. Relatively little is known about genetic diversity for heat tolerance.

As indicated in the preceding chapter, many complex relationships are involved in the physiology of plant tolerance to temperature extremes. This already difficult situation is further complicated by the practical need to develop cultivars that have complex combinations of genes for tolerance of temperature extremes plus all the additional genes necessary for superior performance. The breeder must evaluate an enormous number of genotypes because the discovery of a plant with all the required genes is highly improbable. Allard (3) has provided an excellent basic discussion of theoretical expectations relative to the selective improvement of plants for quantitive characters.

Breeders have, in general, empirically selected for tolerance to temperature extremes under existing conditions in the field. Because of the diversity of the test environment and undependable occurrence of stress conditions, such selection efforts are usually futile. Although physiologists have demonstrated general correlations between temperature stress tolerance and many associated component traits, no precise and efficient selection tools for differentiation of narrow range genetic variability are available to the breeder. Unfortunately, most plant selection and progeny testing still are done in the natural field environment, and precision is achieved by the inefficient practice of testing in different environments as provided by several locations, years, or both. Perhaps the lack of more effective technology for the improvement of tolerance to temperature extremes reflects a lack of high priority, multidisciplinary team research.

Regardless of deficiencies in methodology, breeders have achieved valuable improvements in the temperature stress tolerance of major crop species. During recent years, however, there are few examples of significant increases in the upper levels of temperature tolerance. Additional major improvements in tolerance to temperature are needed to increase and stabilize crop yields under rigorous conditions in fringe areas of production and to adapt major crop species to new areas. My subsequent discussion will deal with some of the possibilities for the improvement of temperature stress tolerance in certain major crop species through breeding and selection. Space will not permit a systematic treatment of breeding prospects for each major crop species. Further, since breeding research with high temperature tolerance is limited, the major portion of my discussion will deal with breeding for tolerance to low-temperature stress.

BREEDING FOR TOLERANCE TO LOW TEMPERATURE

Progress from Breeding for Winter Hardiness

From the practical viewpoint, the breeder generally is interested in cold tolerance as it relates to winter hardiness in crop plants. Direct freezing injury is the usual cause of winterkilling (54, 39), but high freezing resistance does not necessarily guarantee superior winter hardiness. Other factors, such as heaving, smothering by ice sheets, desiccation, and disease, may contribute to or be the major cause of winterkilling of field crops in some locations or years. There are numerous and complex stresses associated with cultural practices, soil variation, and winter climatic variation that the plant must withstand in order to survive. The complexity of the breeding problem has caused Grafius (22) to suggest that present methods are inadequate for significant improvements in winter hardiness. Steponkus (67) recently concluded that breeding for winter hardiness is too broad an undertaking, and emphasized the need to determine the various components and to develop appropriate stress conditions for each. Unfortunately, it has not been possible to effectively apply knowledge about the physiology of freezing resistance to plant breeding methodology.

Regardless of complex genotype × environment interactions for winter hardiness and poor measurement techniques, early attempts to improve winter hardiness through breeding and selection led to some important successes. In wheat (*Triticum aestivum* L.), Hayes and Garber (27) developed 'Minhardi' wheat from the cross 'Odessa' × 'Turkey', and that cultivar has been widely used as a hardy parent in hard red winter wheat programs in the United States. Unfortunately, as Grafius (23) has pointed out, there is little evidence among modern cultivars of further significant increases in the upper level of winter hardiness of wheat in the United States. Apparently, the most winter-hardy domestic wheats known to man are old varieties like 'Albidum 11', 'Albidum 114', and 'Uljanovka' from programs in the USSR. They are unsatisfactory for modern commercial use and during a recent visit to the USSR, I found that Russian scientists generally are pessimistic about achieving significant increases in the upper level of winter-hardiness based on transgressive segregation from crosses involving these varieties. Several Russian scientists stated that no increase in the upper level of winter hardiness in wheat has been achieved during the last 50 years.

Although the upper level of winter hardiness in wheat apparently has plateaued in modern times, breeders have developed many varieties that combine excellent winter hardiness with improvements in various traits like lodging resistance, disease resistance, grain quality, and yield. In fact, improvements in traits like these may alter the economics of the risk of loss by winterkilling to the extent that increased use of an overwintering crop may result even though there has been no increase in the inherent level of winter hardiness. Improved management practices coupled with the use of such varieties may raise the yield potential to the point where it more than offsets the periodic losses due to winterkilling.

Similarly, most of the improvement in winter hardiness in barley (*Hordeum vulgare* L.) occurred during the early years of organized breeding effort. The old variety 'Tennessee Winter' apparently was the first known introduction of barley into the United States (25) and has served as a long-time standard for measuring progress in breeding for winter hardiness. Wiebe and Reid (78) reviewed the comparative winter hardiness of barley varieties grown in the Uniform Barley Winter Hardiness Nursery from 1937 to 1956 and concluded that improved winter hardiness had resulted from selection following hybridization. At the end of that 20 year period, the variety 'Kearney' was the most winter-hardy, followed closely by 'Dicktoo'. 'Kearney' was selected from a composite cross ('C.I. 5530') of 13 varieties and apparently resulted from transgressive segregation because only moderately hardy to nonhardy varieties were parents in the composite cross. In 242 tests at various stations in the United States and Canada from 1949 to 1956 'Kearney' averaged 82% survival compared to 46% for 'Tennessee Winter' ('C.I. 6034'). Unfortunately, the parents in the composite cannot be compared directly to 'Kearney', but 'Wisconsin Winter' and 'C.I. 2159' (the most winter-hardy parents in the composite) averaged 66% survival compared to 57% for 'Tennessee Winter' check for 1937–1948. While such a major increase in winter hardiness gives cause for optimism, there apparently have been no further significant increases since 'Kearney' was introduced in 1949. There have been several varieties in the 'Kearney'-'Dicktoo' hardiness class, and good progress has been made in combining the upper level of winter hardiness with improvements for lodging resistance, disease resistance, and other useful traits.

The area of winter oat adaptation also has moved northward as a result of selection for winter hardiness in old varieties or following hybridization. Coffman (11) reviewed progress in breeding for improved winter hardiness based on survival comparisons in the Uniform Winter Oat Hardiness Nursery from 1926 through 1956. Progress was measured relative to 'Winter Turf' check, generally considered the most winter-hardy variety in 1926. 'Wintok', entered in the test in 1938, proved to have a major improvement in winter hardiness, and that cultivar still is one of the most winter-hardy oats. The variety 'Hickory' was released for parental use in 1959 (52) and probably is the most winter-hardy oat known. Several modern cultivars have winter hardiness slightly above that of 'Wintok', but there has been no major advance in the upper level of winter hardiness in oats.

Among the forage crops, the major breeding effort for cold tolerance has been with alfalfa (*Medicago* spp.). The initial alfalfa introductions in North America were *M. sativa* L. strains with little cold tolerance. Adaptation of alfalfa to the northern United States and Canada resulted from the development of the variegated alfalfas (*M. media*) from hybrids of *M. sativa* × *M. falcata* L. The evolution of 'Grimm' cultivar on the Minnesota farm of the German immigrant Wendelin Grimm over a 35-year period is a classic example of natural selection (7). 'Teton' is an example of a more recent variety with improved cold tolerance derived from *M. falcata* (1).

Genetics Considerations

Heritability. Theoretically, because of the complexity of the genotype × environment interaction, heritability for winter hardiness should be low. This has not always been true in the relatively few studies that have been reported. In a diallel cross series involving 18 winter barley varieties, Rhode and Pulham (58) found that heritabilities for freezing resistance were highest from winter-hardy × winter-tender crosses. Amirshahi and Patterson (5) reported heritabilities for freezing resistance ranging from 18 to 93% for F_3 lines from 14 winter × spring oat crosses. Working with 18 different oat crosses, Muehlbauer et al. (48) found that heritability estimates for winter hardiness ranged from 62 to 89% following a year with a consistent, differential winterkill. The heritability estimates for freezing resistance of naturally hardened plants in the same populations ranged from 56 to 87%. These heritability estimates may be unduly encouraging since they reflect plant responses to only a limited portion of the broad spectrum of stresses that may occur in the field during different winters.

Transgressive Segregation. Another important consideration is that the level of winter hardiness needed to move overwintering crops into more northerly areas generally is dependent on the creation and detection (selection) of genotypes that transgress available domestic parents. Theoretically, transgressive genotypes will occur only rarely and large populations should be produced and selected to provide a reasonable chance of finding them, but the occurrence of transgressive segregation for winter hardiness is well documented in the cereal crops. As I mentioned earlier, 'Kearney' barley resulted from transgressive segregation. Dantuma (16) and Reid (57), however, found no transgressive segregation among progeny from intercrosses of several winter and spring cultivars of barley.

The increased winter hardiness of 'Minhardi' and 'Minturki' wheat cultivars resulted from transgressive segregation (54), but this apparently has been an infrequent occurrence in wheat. I found no reports in the literature of transgressive segregation from winter by spring wheat crosses. During a visit to the USSR in 1977, I found that although Russian scientists are optimistic about finding transgressive segregation in progeny from crosses between moderately hardy varieties, they are pessimistic about increasing the upper level of winter hardiness through transgressive segregation in populations from crosses among the old varieties with elite winter hardiness. Their primary effort involves crosses among moderately hardy varieties that are adapted for intensive management, and they hope for some minor improvements in winter hardiness through transgressive segregation.

In the cereal crops, transgressive segregation for winter hardiness has been reported most often in oats. The hardiness of 'Wintok', mentioned earlier, exceeds that of its parents (12). 'Hickory' (52) resulted from a cross between two hardy varieties. 'Dubois' (8), 'LeConte' (61), and 'Norline' (14) are varieties derived from winter × spring crosses with winter hardiness resulting from transgressive segregation. 'Dade' (52) and 'Pennwin' (42) are varieties with complex pedigrees involving at least one spring parent, and their hardiness exceeds that of the best winter parent. Finkner (20), Jenkins (32) and Muehlbauer et al. (48) have reported transgressive segregation for winter hardiness in genetic studies with oats.

There is little question but what transgressive segregation for winter hardiness occurs in the cereal crops, and the breeder should be optimistic that at least small increments of further improvement based on this genetic phenomenon are possible. There is little evidence, however, of breeding programs in which the first priority is the use of parental materials and breeding systems designed to provide the maximum opportunity for transgressive segregation beyond the upper level of winter hardiness and in which primary selection is for winter hardiness regardless of other characteristics.

Mode of Gene Action.   Knowledge about the mode of gene action and undesirable genetic linkages should be useful for designing effective breeding systems for cold tolerance. There is only limited indisputable genetic information available, however, for cold tolerance in crop plants, and most of that must be gleaned from studies of the inheritance of winter hardiness. Winter hardiness generally appears to be under polygenic control, with a major component of variance attributable to additive effects. The detectable gene action may vary, however, depending on the environmental conditions. Rhode and Pulham (58), for example, found that the genetic control of winter hardiness in barley may be either dominant or recessive, depending on test location and conditions. With wheat, several workers have reported dominance of winter hardiness under mild stress conditions and a lack of dominance under more severe conditions (28, 45, 53, 59, 81). Working with populations from a diallel set of crosses involving divergent oat species and cultivars, Jenkins (32) found that certain parents had high general combining ability for freezing resistance. Muehlbauer et al. (48) calculated the estimated variance components for winter survival of F_3 lines derived from a diallel set of hardy-winter × spring oat crosses. Specific combining ability of parents and differences among F_3 lines within crosses were important sources of variation; general combining ability, maternal effects, and reciprocal effects over populations were relatively unimportant. The most important sources of variation for winter survival were differences between lines within crosses. They suggested that it may be possible to evaluate crosses (populations) at only one location to identify those with elite selection potential, but lines within crosses probably should be tested at more than one location.

In alfalfa, the relative importance of general and specific combining ability for cold tolerance has varied depending upon the parents involved (76). From genetic studies with varieties of *M. sativa* that are widely different for cold tolerance, Daday and Greenham (15) concluded that there were no indications of dominance and gene action was largely additive.

Based on current knowledge, there is little question but what cold tolerance and winter hardiness are complex, polygenic traits. Breeders should strive to simplify their task by reducing those traits into the smallest possible units of genetic control. When we can critically define and measure the effects of these components of genetic control, major improvements in cold tolerance are likely to follow.

Availability of Genetic Diversity.   The prospects for improvement of the cold tolerance of the major crop plants are dependent on the availability of exploitable genetic diversity. Breeding efforts with some crops have been primarily at the intraspecific levels while others have crossed species or even genera boundaries. Harlan and deWet (26) have classified cultivated crops and their wild relatives into primary, secondary, and tertiary gene pools. Briefly, the primary gene pool includes races that can be crossed within a crop to give reasonably fertile hybrids. Species within the secondary gene pool also can be crossed but the gene flow tends to be restricted by factors such as sterility and poor chromosome pairing that restricts recombination. The tertiary gene pool includes species that can be crossed but techniques such as embryo culture and tissue culture are necessary to effect gene transfer. Cultivated species and varieties usually constitute the primary gene pool with the wild relatives in the secondary and tertiary gene pools. However, the accessibility of these gene pools for gene transfer varies with the crops. In the cereals, for example, there are several wild hexaploid species in the oat primary gene pool. *Hordeum spontaneum* L. is in the cultivated barley gene pool. Hexaploid wheat (*Triticum aestivum*) has no wild relatives in its primary gene pool.

Based on recent progress, some breeders fear that further increases in the upper levels of cold tolerance may be restricted by a lack of readily exploitable genetic diversity. In most cases, current levels of cultivar hardiness are near the apparent upper level of the known range in the primary gene pool. Grafius (23), for example, believes that future progress based on intraspecific crosses will be slow at best. Past achievements certainly support this contention, and the required genetic diversity for major breakthroughs (like wheat with the cold tolerance of rye) probably does not exist in the primary gene pools of the cereal crops. Because of the intangible nature of genetic potential for transgressive segregation, however, this does not preclude achievement of small increases in cold tolerance that could result in major changes in crop adaptation and use. Historically, the gradual infringement of over-wintering crops on the traditional territories of summer annual crops has resulted from subtle improvements for winter hardiness. The tendency for pessimism regarding further improvement based on intraspecific genetic diversity may reflect our inability to systematically ferret out and use available genetic diversity within the primary gene pools. There are substantial germplasm collections in most major crops. The problem is how can a breeder identify parents with elite combining ability for cold tolerance and winter hardiness? Diallel crosses have been used in certain over-wintering crops (58, 76, 32) to evaluate differences in combining ability among small groups of parents, but there are no practical tests to screen our extensive germplasm collections for parental combinations with elite potential for transgressive recombination. This should be a high priority area of research.

In the cereal grain crops, domestic germplasm with a spring growth habit may be a valuable source of genetic diversity for cold tolerance (11, 51). Examples of cultivars with transgressive winter hardiness derived from winter × spring crosses were cited earlier, but breeders have evaluated only a few crosses out of the enormous number of possible combinations. Again there is a need for efficient, accurate tests to identify those spring parents that have the greatest potential for use in the breeding program. Since there is little evidence of transgressive segregation from hardy × hardy crosses in the winter cereals, it may be feasible to use a representative hardy parent as a tester parent and use the mean F_2 bulk performances in controlled freezing or field survival tests to identify elite parents. Information is needed about the relationship of F_2 bulk cold tolerance means from such an array to potential population variances in subsequent generations.

Perhaps the most promising sources of genetic diversity for large increases in the upper levels of cold tolerance are to be found in wild relatives of the major crop species. In cultivated hexaploid oats, the primary gene pool includes all the known wild hexaploid relatives. Suneson and Marshall (74) described several different freezing resistant lines of *Avena fatua* L. that may be useful for improving the winter hardiness of oats. I have used these in crosses to several hardy oat cultivars, and developed a bulk composite that has been grown under severe natural selection pressure near University Park, Pennsylvania. The population has mean winter hardiness approaching that of elite cultivars, but lines have not been extracted to test for transgressive segregation. Accessions of *A. sterilis* and *A. sterilis* var. *ludoviciana* also represent potential sources of useful genetic diversity for freezing resistance and winter-hardiness. I also have developed bulk composites involving these wild hexaploids, and a few derived lines have had slightly transgressive freezing resistance in crown freezing tests.

Grafius (23) recently discussed the possibility of interspecific gene transfer to increase winter hardiness in barley. Interspecific hybridization within the *Hordeum* secondary gene pool is difficult, with the possible exception of *H. vulgare* × *H. bulbosum* L., and there is little opportunity for interspecific recombination because essentially no chromosome association occurs in the hybrids. Grafius is optimistic about using callus culture of *H. vulgare* × *H. jubatum* L. hybrids to find plants with partial fertility that might be used to effect transfer of cold tolerance genes from *H. jubatum* into domestic barley. He also reports evidence of increased resistance to nonequilibrium freezing in progeny from crosses between *H. vulgare* and an amphiploid of *H. brachyantherum* Nevski × *H. bogandii* Wil. Additional work is underway with this material and with crosses involving *H. vulgare* and *Agropyron trachycaulum* (Link) Malte.

Modern procedures for tissue culture and so-called genetic engineering may lead to gene transfers or to chromosome additions or substitutions that previously were not possible within the secondary and tertiary gene pools, but many barriers to success remain. These endeavors should be in addition to, and not at the expense of, proven breeding approaches that are likely to produce small but economically valuable increments of improvement in cold tolerance.

Apparently there are no wild species of *Triticum* with outstanding cold hardiness, but certain related genera have high levels of winter hardiness. Certain *Agropyron* species can be crossed with *T. aestivum* and have long been cited as promising sources of genetic diversity to improve the cold tolerance of wheat (64). To date, however, the cold tolerance of *Agropyron* has not been combined with all the *T. aestivum* traits necessary to create usable, meiotically stable wheat cultivars. In 1977, I visited the Scientific Research Institute of Non-black Soils, Moscow, USSR, where there is an active project to develop wheat cultivars with increased winter hardiness derived from *Agropyron*. Dr. Lapchenko has worked with four *Agropyron* species that cross readily with *T. aestivum*. Of these, derivatives from crosses with *A. glaucum (intermedium)* have been the most promising, and I saw $2n = 42$ chromosome, wheat-like plant types from lines that are under test. These lines reputedly have good fertility and winter hardiness exceeding that of commonly grown Russian cultivars. Some of the wheatlike lines were $2n = 56$ chromosome types with high winter hardiness and high protein content but small seeds and low yield.

Triticale, a genomic alloploid resulting from hybridization of wheat with rye (*Secale cereale* L.), has attracted much attention as a potential cereal crop for areas too cold for wheat culture (31, 37). To date, however, triticales do not have cold tolerance or winter hardiness exceeding that of the most hardy wheat cultivars. Based on cold tolerance studies with octaploid triticales, Dvorak and Fowler (19) recently concluded that rye is an unsuitable source of genes for the improvement of cold tolerance in hexaploid wheat. They suggest that the polyploid relatives of wheat represent a more promising gene pool because the genetic variability for cold tolerance has evolved in a polyploid system. Rajhathy (55), however, is more optimistic about combining the superior grain quality of wheat with the superior winter hardiness of rye because of a recently expanded germplasm base, a shift from the octaploid to primary and then secondary hexaploid types, and new genetic knowledge that has altered the effectiveness of selection for fertility.

Prospects for Improving Selection Techniques

Laboratory selection techniques have not been accurate or efficient, and the breeder's best choice has been to select under stress conditions in the field. I came across no reports of cultivars tracing to selections made with laboratory techniques. Physiologists have demonstrated that various physiological adaptations occur during cold hardening of crop plants and sometimes have suggested that breeders should select on the basis of such relationships. In general, this is not possible because many of the cause and effect relationships of plant reponse to low temperature stress remain unclear. For example, Christiansen (10) has suggested that many of the recently observed responses to temperature, particularly the lipid changes, can be used as selection tools to increase the limits of tolerance. The application of lipid changes to selection, however, is not obvious, because fatty acid unsaturation apparently is not correlated with genotype differences in cold tolerance (17, 60, 63, 79, 80). Most breeders have not had the equipment or training required to test the breeding application of the various physiological adaptations of plants in response to low temperature. The sharing of laboratory facilities and cooperative research by breeders and physiologists are essential for the evolution of improved selection techniques for cold tolerance at the tissue or cellular level.

Utilization of Natural Selection. Because cold tolerance is a major component of winter survival in heterogeneous populations of crop plants, natural selection should increase the frequency of genes for that characteristic. Suneson (72) has summarized multigeneration results with a spring barley composite cross and suggested an "evolutionary method of plant breeding" to take advantage of natural selection. There are few examples, however, of long-term endeavors to improve the cold hardiness of populations by utilizing natural selection.

Sprague (66) briefly discussed the results of six cycles of natural selection for winter-hardiness in a barley composite cross in which simulated random mating was achieved by the use of genetic male sterility. The history and results with this composite, designated 'ms CC18^2', are not well documented. The following discussion is based on several sources including personal communication with D. A. Reid. Based on winter survival tests at six locations in 1942, G. A. Wiebe selected 18 elite winter barleys of geographically diverse origin (from the United States, China, Korea, and Caucasus) and crossed them individually to a genetic male-sterile. After four generations of backcrossing to male-sterile segregates, a diallel of the 153 possible crosses was completed. According to Reid's unpublished account in the 1958 Barley Winter Hardiness Nursery Report, the F_1 plants were grown at Sacaton, Arizona in 1951. In 1952, F_2 progeny rows were grown at five locations including Lincoln, Nebraska, and Urbana, Illinois. The F_2 populations grown at Lincoln were used for genetic studies of winter-hardiness as reported by Rhode and Pulham (58). Severe winterkilling occurred in the populations at Urbana and the seeds from surviving male-sterile plants were harvested. The next generation, constituted of F_1's between surviving plants, was grown at Sacaton, and harvested in bulk to constitute the initial composites. This composite was grown at Madison, Wisconsin, in 1954 and winterkilling again was severe. As before, seeds from only male-sterile plants were harvested, and the new F_1 population was grown under the mild conditions in Arizona. This natural selection for winter hardiness was continued for six cycles. In 1962, Suneson and Wiebe (73) suggested that this breeding method should be useful for improving other characteristics in composite cross populations, and the method has since been termed "male-sterile facilitated recurrent selection" by R. F. Eslick (T. A. Ramage, 1974 Barley Newsletter, p. 63).

The effectiveness of natural selection was estimated by comparing the winter hardiness of the sequential composite populations to that of a bulk constituted from equal amounts of seed (by weight) from the 18 parents. These comparisons were made in the 1958 (cycles 1 through 4) and 1963 (cycles 4 through 6) Barley Winter Hardiness Nursery grown at numerous

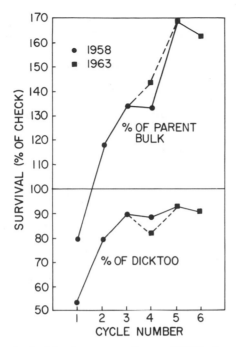

Figure 1 Winter survival of barley composite ms CC18^2 after successive cycles of natural selection relative to the parental bulk population and the 'Dicktoo' check variety.

locations in the United States and Canada. Sprague (66) tabulated those results from data compiled by Dr. Reid, but no comparison was made to 'Dicktoo', the most hardy parent included in the composite. Figure 1 shows the results I obtained from resummarizing the data from the unpublished nursery reports and only using data from locations where at least some winterkilling occurred in all of the composite populations. Based on the 1958 test, comparing the first four cycles to the parent bulk, winter-hardiness of the composite actually was below that of the parent bulk after the first cycle. At the end of three cycles, the winter hardiness of 'ms CC18^2' was 134% of the bulk parent population and 90% of 'Dicktoo'. There was no change during the fourth cycle. Based on the 1963 data, however, the cold hardiness of the composite increased through the fifth cycle relative to the parent bulk. The relationship to 'Dicktoo' plateaued after the third cycle at about the 90% level. The relationship to 'Kearney', a check variety with hardiness similar to that of 'Dicktoo', was almost identical. Why did the cold hardiness appear to level off sooner relative to the two hardy check varieties? One clue is that the cold hardiness of the parent bulk was 67% of 'Dicktoo' under the average test conditions of 1958 but only 55% of 'Dicktoo' in 1963. Evidently, genetic changes occurred during cycle 5 in 'ms CC18^2', providing an additional advantage over the parent bulk but not over the two hardy check varieties because they, too, had the superior level of tolerance to critical stress factors.

Grafius (23) recently reported that the average nonequilibrium freezing resistance (i.e., crowns at high moisture level) of 'ms CC18^2' was slightly above that of 'Dicktoo'. Two cycles of recurrent selection for nonequilibrium freezing stress did not cause any further increase in the cold tolerance. He concluded that natural selection had exhausted the genetic variability of the population—at least for nonequilibrium freezing resistance.

Unfortunately, there is little documentation about the controlled selection effort applied to 'ms CC18^2'. Selections from the Wyoming, Illinois, Wisconsin, and Harrow, Canada, programs have periodically been entered in the Barley Winter Hardiness Nursery, and a few lines were at least equal to 'Dicktoo' for winter hardiness over a 3- or 4-year period. Apparently, no significant transgressive gains for cold tolerance occurred, but there is no question but that natural selection rapidly increased the frequency of genes for cold tolerance to a point at which the mean tolerance was near that of the most hardy parent involved. Several more cycles of simulated random mating and natural selection in that diverse population should have been enlightening regarding the long-term potential value of natural selection for cold tolerance.

The effectiveness of natural selection for low temperature resistance varies with the frequency, uniformity, and severity of freezing stress in the field. Finkner (21) reported that natural selection was not efficient in sorting out the most winter-hardy plants from solid-planted, bulk hybrid oat populations during a 6-year period. In my work with oats in Pennsylvania, however, I found that severe winter stress caused large increases for freezing resistance in populations with initially low mean hardiness and generally small or nonsignificant changes in hardy populations (41, 43). Populations that did not change had an average winterkill in the 30–60% range, and it was apparent that spotty winterkilling in unfavorable environmental niches within plots caused no consistent changes in freezing resistance. Surviving patches generally showed little evidence of plant elimination by winterkilling. The pattern of winterkilling was similar to that observed in hardy check varieties.

Warnes and Johnson (77) also found that the greatest genetic gains for winter hardiness in barley occurred in populations from crosses between hardy × nonhardy and hardy × moderately hardy varieties. There were no significant changes in populations from hardy × hardy crosses.

There have been reports of increases in the winter hardiness of alfalfa (*Medicago* spp) populations as a result of natural selection. One of the earliest reports was by Brand (7) regarding adaptive changes in alfalfa grown under Minnesota conditions. Smith (65) found that the winter hardiness of 'Narragansett' and 'Vernal' cultivars increased in winter hardiness when grown north of their area of adaptation for a single generation and decreased if they were grown south of their area of adaptation.

Working with alfalfa under severe winter stress in Alaska, Klebesadel (35) developed 'A-syn. B' synthetic of variegated alfalfa (*M. media*) from plants that survived in various strains and varieties over about a 20-year period. In field tests from 1968 to 1970 'A-syn. B' was more winter-hardy than 33 strains from Canada, northern Europe, and the conterminous United States. Klebesadel suggested that 'A-syn. B' is constituted of genotypes better able to attain a high level of hardiness under the unique photoperiod-temperature interrelationships that prevail during late summer and autumn in Alaska.

Although it is apparent that natural selection may cause increases in winter hardiness in populations with wide genetic variability for that trait and low mean survival relative to that required for consistent survival, it is questionable whether stress in the natural winter environment is frequent and consistent enough to differentiate among genotypes in the elite part of the distribution range. There is little evidence that breeders have attempted to manipulate cultural practices or environment to favor the development of more uniform stress; more needs to be done.

Natural selection for cold tolerance should be more precise if the breeder can manipulate environmental components to provide more uniform and consistent freezing stress. This is difficult to accomplish but an important first step is to grow the populations at locations where the probability of temperature near the critical level is high. Homogeneity of topography, soil type, drainage, fertilizer treatment, land preparation, weed control, and other cultural practices should be the goal. Even when there is essentially complete winterkilling in a heterogenous population, surviving plants tend to occur in little patches of a few plants. Slight ridges or depressions, compaction in wheel tracks, and more subtle differences obviously have large effects on the level of stress. In fact, plant breeders commonly are frustrated to find that even scattered single plant survivors are escapes. For example, in 1959, I developed lines from 98 scattered plants that were the sole survivors in a 0.8-ha nursery of hardy oat bulk populations. In subsequent tests, none of these lines was superior to the best parents in the source populations. Uncontrolled natural selection stress apparently is not a critical selection tool when the range of genetic variability is narrow.

Other manipulations of the environment may improve the precision of natural selection. With modern techniques for soil fumigation, for example, it may be possible to restrict or remove the soilborne disease component. An even more extreme approach would be to completely replace the top soil with a homogeneous mixture carefully leveled over a coarse base that would provide uniform drainage. Soil moisture level in the fall could be controlled by irrigation. Another approach to improve plot uniformity would be to grow the plants in areas free of wheel tracks after preparation with tobacco bedding machines or similar equipment. Growing plants on small ridges also might provide more uniform stress for cold tolerance. In my preliminary studies with an oat bulk composite, this practice increased the amount of winterkilling (compared to a level planting surface) and, based on controlled freezing tests, the rate of genetic gain for cold hardiness.

Another difficulty with natural selection stress is the undesirable genetic changes that may occur for other important traits. I recently sampled 12 winter oat bulk populations (43) after 4 years of natural selection pressure for winter survival and found that 5 out of 12 populations had improved for freezing resistance in crown freezing tests. On the negative side, five populations made undesirable changes toward tall plant height, and all populations became later in maturity. Such undesirable changes probably should be ignored in all-out efforts to increase the upper level of cold tolerance, but if directly useful cultivars are the primary goal, controlled selection pressure for traits such as plant height and maturity may be necessary during one or more early generations and periodically thereafter. With the evolution of mechanical planters that facilitate the planting of large numbers of progeny rows, and water-soluble seed tape and other space planting systems, alternate cycles of controlled and natural selection are more manageable.

Suneson (71) has pointed out the potential value of using genetic male sterility to facilitate prolonged heterozygosity in populations in natural selection breeding systems. Chemical male gametocides, even if they are not absolute, also could be used to effect periodic cycles of cross pollination in those crops in which genetic male sterility is unknown. A degree of assortive mating, however, is likely with either approach because of differences in time of flowering.

Individual Plant Selection. The breeder usually cannot effectively select for cold tolerance based on individual plant responses for reasons implicit in the above discussion, and this has been a major barrier to improvement of the trait. Because of inability to assess cold hardiness (or its components) on an individual plant basis (without destroying the plant), selection generally is based on some form of progeny testing. Pure individual plant selection for cold and winter hardiness probably is practiced only in the form of mass selection when surviving plants are bulked together to form a new population. As indicated previously, the probability of genetic gain from such selection is low since surviving plants probably escaped lethal injury because of fortuitous location within the plot. About the only immediate prospects for improving individual plant selection in the field reside is manipulation and control of the plant environment to reduce winter survival into components under simpler genetic control.

Selection Based on Progeny Tests. Selection for cold tolerance usually is based on progeny performance for winter survival following genotype extraction from a population that may or may not have been under severe winter stress. Initial tests will be on a local basis, but promising lines may be progressively evaluated in cooperative state, regional, national, or international tests. Obviously, the number of selections a breeder can evaluate with fixed resources is affected by the number of generations and tests required to ascertain cold hardiness.

Regardless of the crop, every effort should be made to systematically mechanize all steps of the testing process. The best breeding approach with current technology has been to work with large numbers of genotypes in the most uniform plot environment achievable and at a location where differential winter stress is likely during most years. In the cereal crops, a short (0.5 to 1.5 m), single-row plot containing the progeny from a single head or panicle selection commonly is used to do preliminary screening for winter hardiness. The pitfalls are apparent from my previous discussions, and the occurrence of "differential" winterkilling does not necessarily provide for accurate selection of superior genotypes. The cereal breeder, however, can improve his chances of success by using modern mechanical planters to increase sample size. Using a planter equipped with a seed tray indexer, two workers can plant about 3000 single-row, 0.5-m-long plots per hour. Obviously, threshing, seed packeting, record keeping, and other related operations also must be mechanized to exploit these planting capabilities. In order to sample large numbers of genotypes, maximum mechanization should be applied to winter hardiness breeding programs in all crops.

Short, single-row plots also have several advantages over multiple-row plots for additional testing of preliminary selections. More replication is possible because of a lower seed requirement with short rows. Perhaps of more importance, smaller test areas should include less environmental diversity and result in reduced experimental error relative to that in larger areas.

The ultimate test for winter-hardiness in winter cereal crops is performance over several environments in regional and national uniform winter hardiness nurseries. These tests are assembled by the U.S. Department of Agriculture and grown by various cooperators in the United States and Canada. They provide critical winter hardiness data on relatively advanced generation varieties that may be useful as commercial cultivars or as parents because of unique gene combinations for winter hardiness. These tests do not serve as a primary selection tool for the breeder, however, because of the amount of seed required and restrictions on the number of entries per year. In any event, such cooperative tests are of proven value, and breeders working on crops that do not have them should seriously consider the feasibility of initiating cooperative tests.

Selection with Controlled Freezing Tests. Although disease or other factors may be the critical stresses that cause winter injury in certain years or locations, freezing resistance is generally recognized as the principal component of winter hardiness. This association has stimulated extensive research efforts during the past 60 years to develop controlled freezing tests for the measurement of freezing resistance of divergent genotypes in breeding programs. Precise and efficient freezing tests would circumvent the problems associated with field testing and would be an invaluable breeding tool for: (*a*) identification of genetic diversity in germplasm collections, (*b*) extraction of elite genotypes from heterogeneous populations, (*c*) progeny testing, (*d*) application of recurrent cycles of mass selections or other breeding systems, (*e*) evaluation of progress under various breeding systems, and (*f*) genetic studies.

Efforts to develop freezing technology have led to many innovations and generated numerous optimistic reports regarding the positive correlation between winter hardiness and freezing resistance in the various field crops. Dexter (18) has reviewed early work with controlled freezing tests. Larson and Smith (36) studied the reliability of various plant constituents and freezing tests for determining winter hardiness in alfalfa. Newer or modified techniques have evolved periodically (4, 6, 24, 30, 33, 34, 62). Subsequent practical experience, however, has demonstrated that all techniques reported have limited application in breeding programs because of inefficiency and high experimental error. Although there has been considerable use of freezing tests for basic studies of the physiological and genetical bases of cold tolerance and to evaluate breeding methodology for winter hardiness, there is little evidence of effective application of freezing tests in breeding systems. There are several reasons for this: (*a*) high genotype × environment interaction associated with direct freezing of plants in pots or flats of soil, (*b*) techniques are not efficient for screening large numbers of genotypes, (*c*) techniques are not precise enough for individual plant selection, and (*d*) poor control over several distinct forms of possible freezing stress.

Regardless of the problems of freezing plants intact in the growing medium, the technique is still used. During my visit to the USSR in 1977, I observed that freezing plants in wooden flats is the prevalent procedure used to test cereal crop varieties for freezing resistance. The plants are hardened outdoors and periodically frozen in controlled temperature cabinets during the fall and winter to determine LD_{50} values (temperature at which 50% of the plants are killed). The Russian scientists conceded that these tests are only accurate enough to place varieties and cultivars into broad classes and do not serve as primary selection tools. The potential of controlled freezing tests for more precise selection work is recognized, however, and large phytotrons for use in cold tolerance research are under construction at several locations in the USSR.

In cereal crop plants, the crown contains the meristematic tissues critical for freeze survival. The precision in freezing tests recently has been improved by applying freezing stress to crowns after removal from the growing medium (34, 40, 49). Further precision and specificity in freezing tests may be achieved by innovations to control the uniformity of crown moisture to characterize plant responses to equilibrium or nonequilibrium freezing stresses or both (24, 50, 44). Chen and Gusta (9) recently reviewed the role of water in cold hardiness of winter cereals and concluded that water content of the crown is crucial under conditions of fast freezing and rapid removal of latent heat. Grafius (23) also stressed the importane of differences in crown moisture in relation to freezing injury to winter cereals in the field and attributed the rapid loss in winter hardiness that may occur during "January thaws" to increased crown water content. Thus, measurement of plant responses to specific stresses associated with freezing conditions and water status may be useful for accurate selection of genotypes that are likely to have superior cold or winter hardiness under the usual soil moisture levels that occur during periods of natural winter stress. Techniques for selection of genotypes in small grain crops based on different forms of freezing stress recently have been outlined by Marshall et al. (44). Although selection based on more than one form of freezing stress may be desirable to predict broad range winter hardiness, this will not be feasible in many breeding programs because of limited resources and facilities for freeze testing. If there is generally good agreement between freezing resistance in a single, standardized test and the average winter hardiness of adapted varieties, the measurement of one form of stress may be sufficient for routine progeny tests. In fact, where selection is to be based on individual plant survival, only one freezing test is possible with current technology. Of course, cloning is possible in certain crop species, but this greatly reduces the number of genotypes that can be tested with a fixed amount of effort. If the soil tends to be dry during critical periods of winter stress, first priority probably should be placed on the measurement of equilibrium freezing stress (low crown moisture). Where the soil moisture tends to be high during winter stress, like in the eastern United States, measurement of nonequilibrium freezing resistance (high crown moisture) may be more important. Michigan workers (24) have collected data supporting the hypothesis that differences in the ability of wheat genotypes to withstand equilibrium and nonequilibrium freezing are heritable.

In any event, the value of a freezing test for selection purposes should be assessed on the basis of correlation with average winter survival of a differential set of varieties over a number of environments as provided by locations and years within the problem area. Field survival data from only one or two test environments may be atypical and independent of freezing resistance.

Although the development of modern equipment and use of techniques like crown freezing of cereal crop plants have improved the accuracy and efficiency of freezing tests, there still is need for further improvement if the breeder is to use such tests effectively as a selection aid. Differentiation of genotypes within a narrow genetic range, and this is essential, still requires an inefficient level of replication because of uncontrolled experimental error. To be of ultimate value to the breeder, freezing tests must facilitate individual plant selection and progeny testing based on a few plants. To achieve this, seed for all freezing tests should be grown under uniform, stress-free conditions and be free of seedborne pathogens. Absolute environmental control seemingly is impossible, but this must be the goal during the plant growth, hardening, freezing, and recovery stages of freezing tests. Theoretically, under absolutely uniform experimental conditions, the expression of freezing resistance and degree of freezing injury should be uniform among all plants in a homogeneous population because plant responses should be caused by genetic variation and not by environmental variation or by interactions between these two sources of variation.

In my current research with the application of freezing technology to selection and breeding systems for winter hardiness in oats, plants are grown and hardened in a growth chamber in individual plastic tubes in a semi-nutriculture system. Individual crowns are frozen in vials on a turntable as detailed in an earlier report (44) and surviving crowns are transplanted and grown to maturity. While it is apparent that nongenetic variation still causes selection errors, performance of extracted lines in progeny freezing tests and in field tests indicates useful genetic gains. In fact, a few lines grown from occasional survivors (5% or less) in heterogeneous populations appear to have transgressive freezing resistance (unpublished results). I am applying cycles of recurrent selection for crown freezing resistance, with and without sib-mating of survivors, to both domestic and domestic × wild oat bulk populations.

PROSPECTS FOR IMPROVING BREEDING SYSTEMS

Because there has been little or no improvement in the upper level of cold tolerance in the major crop species during recent decades, one might conclude that continued use of current breeding systems will not result in further improvements for that trait. In barley and other cereals for example, Grafius (22, 23) is pessimistic about achieving major increases in cold tolerance by use of current methods and wonders why breeding methods successful for other traits have failed for winter hardiness. His discussion, however, deals primarily with the breeder's inability to effectively select for cold tolerance and an apparent lack of useful genetic variability for the trait. Perhaps the breeding systems, per se, have not been the primary cause for failure. In fact, in the absence of effective measurement techniques for cold tolerance, it is somewhat meaningless to discuss the improvement of breeding systems.

Assuming that some genetic gain can be achieved with modern controlled freezing tests or by manipulation of the natural environment to improve the uniformity of freezing stress, certain breeding systems may be theoretically superior for achieving cold tolerance. Based on prior discussion, the system should facilitate increasing the frequency of genes for the components of cold tolerance, and in most crops must provide the maximum opportunity for transgressive segregation.

As indicated earlier, the experiences with bulk composite breeding in cereal crops have been discouraging in that no cultivars with major increases in the upper level of cold tolerance have resulted to date. It seems apparent, however, that natural selection does result in gradual elimination of the least hardy genotypes from populations and that these populations will plateau near the level of cold tolerance required for winter survival in an area. The failure to find genotypes with highly transgressive cold tolerance in populations like the barley composite 'ms CC18^2' may result from insufficient genetic diversity for the evolution of the required new polygenic systems or, if the necessary genetic diversity is present, insufficient time and opportunity for the assemblage of the required new polygenic systems.

Although it seems likely that natural, evolutionary breeding systems can be improved upon and may eventually lead to improvements in cold hardiness, a mass selection system involving relatively controlled sib-crosses between selected genotypes within a genetically diverse population may be more effective in the short run. It may be possible to identify elite populations for mass selection by controlled freezing tests of F_2 bulk populations from test crosses. After a population has been chosen, controlled freezing stress could be applied to the F_2 generation at a level that will eliminate most individuals. The survivors can be sib-crossed in as many combinations as feasible. Seed from the resulting F_1 hybrids can be bulked to form the F_2 bulk population for the next cycle of selection. Remnant seed from each cycle could be increased for use in the breeding program and to assess genetic gain for cold tolerance. Assuming adequate genetic variability, success will be highly dependent on uniform, selective freezing stress. Because of recent improvement of freezing tests for cereal crops, the application of sib-mating systems may be a feasible and effective approach for improving cold tolerance.

For the cereal crops, Grafius (22, 23) has suggested a scheme of recurrent selection that employs a form of panmictic mating accompanied by mild inbreeding and selection. His scheme is dependent on selection based on crown freezing of progeny from genetic male-sterile plants in the population and mild inbreeding is effected by restricting the number of lines consolidated into each cycle. This system is being used in the Michigan program for cold hardiness in barley and progress will be evaluated after three cycles of recurrent selection.

This discussion has provided few suggestions for improving breeding systems for cold tolerance. Based on genetic theory, however, some form of mass selection with minimum inbreeding should be most effective for the evolution of new genotypes with transgressive cold tolerance. The delineation of effective breeding methodology for cold tolerance is dependent upon the development of precise measurement and selection techniques for that trait or closely associated components, and this is dependent upon close team work among breeders and physiologists.

BREEDING FOR HEAT TOLERANCE

Although high temperature stress may be a major cause of yield loss in major crop plants, relatively little specific effort has been made to improve high temperature tolerance through breeding and selection. Perhaps one reason for this is that injury and yield losses associated with high temperature stress tend to be more subtle than those associated with winterkilling or major disease or insect infestations. Also, the breeding task is especially difficult for at least the following reasons: (a) limited understanding of the genetic and physiological bases of heat tolerance in plants, (b) confounding of heat and drought stress in field and laboratory tests, (c) inability to control temperature in the natural environment, (d) lack of efficient and accurate laboratory tests to select for the component physiological mechanisms that give elite heat tolerance, (e) limited understanding of what stages of plant growth or what tissues can be used as selection criteria, and (f) limited information about the range of genetic diversity for heat tolerance. It is not surprising that little breeding effort has been devoted to the improvement of tolerance to heat stress.

Genetic Diversity

Although there is little specific information about the range in genetic diversity for heat tolerance in the gene pools of the major crops, the existence of heritable variation has been established in several crops. Heritable differences in heat tolerance have been reported in sorghum, *Sorghum bicolor* (Linn.) Moench. (56, 68, 69); corn, *Zea mays* L. (29, 75); soybeans, *Glycine max* (L.) Merr. (47); oats (13); and a number of other crops. In some cases this variation in heat tolerance has been related to yield levels under high temperature stress, but more information is needed in order to determine the range of useful genetic diversity.

For most crops, only a small segment of the primary germplasm pool has been evaluated for heat tolerance. Future improvements of heat tolerance in crop plants is highly dependent on the characterization of that trait in our germplasm collections. Obviously, the collection of such data is dependent on the development and use of efficient, accurate, and meaningful (i.e., relative to product yield or quality) laboratory and field tests.

Selection and Testing

Application of Laboratory Tests.   Because of our inability to critically control temperature and to separate heat and drought stress in the natural environment, plant selection and line testing for heat tolerance in field tests are likely to be futile efforts. Sullivan and Ross (70) have suggested that the basic concepts of heat stress resistance involving tolerance and avoidance factors are sufficiently sound to facilitate the development of systematic selection techniques for plant breeding applications. They have conducted studies for a number of years with a technique (68, 70) that measures electrolyte leakage from leaf discs following a shock-type heat stress treatment. The test is simple and rapid and has been positively correlated with response of other plant processes to the heat stress, for example, soluble protein and enzyme resistance to denaturation, heat stability of photosynthesis in intact leaf tissue, and with responses of intact plants subjected to heat stress. The responses obtained with this technique have correlated well with known field responses.

Martineau et al. (46) recently used the leaf disc technique to characterize the heat tolerance of several soybean cultivars. Variation among plants was large and cultivar separation was dependent on the use of bulked leaflets from several plants. Based on the work with sorghum and soybeans, the leaf disc technique appears to have promise for screening germplasm for elite parental material and for evaluation of lines. Apparently, the technique is not precise enough for individual plant selection or to apply selection pressure to segregating populations.

Alexandrov (2) reviewed several methods of assessing heat resistance in plants and concluded that measurements of depression of protoplasmic streaming and retardation of photosynthesis were among the most sensitive. Methods of measurement, however, have been too inefficient to permit practical application to breeding and selection research. Sullivan et al. (69) have developed a portable unit for rapid measurement of photosynthesis in plants growing in the field or laboratory. The method is nondestructive and apparently has considerable promise for selection for heat tolerance.

Perhaps, as with measurement of low temperature tolerance, a major need is to develop techniques that reduce plant-to-plant variation caused by environmental diversity. The development of genetically homogeneous lines (perhaps with modern tissue culture techniques) may be necessary in order to accomplish this. Regardless of the trait to be evaluated, accurate and efficient individual plant selection for heat tolerance surely will not be possible so long as nongenetic variation is an important cause of phenotypic difference.

Selection Based on Field Tests. Assuming the presence of significant genetic variability and at least periodic high temperature stress, the breeder may occasionally achieve genetic gains by empirical selection for what appears to be heat tolerance under field conditions. There are few documented cases, however, of successful selection for heat tolerance in the major crop plants. Sullivan and Ross (70) reported that a sorghum population selected under conditions of heat and drought stress in Arizona and Nebraska had significantly higher heat tolerance than another population derived from the same base population but apparently not under effective selection pressure for heat tolerance. In addition to occasional genetic gains from conscious empirical selection for heat tolerance under rigorous stress conditions, breeders undoubtedly have made more or less automatic gains for heat tolerance by selection for superior performance traits, like high yield or plump grain, that may be associated with superior heat tolerance under subtle stress conditions.

Selection for heat tolerance under field conditions could be improved by the development and application of techniques which would provide consistent, uniform stress conditions. Effective selection based on individual plant responses may not be feasible, but some improvement in replicated progeny testing should be possible. Since the effects of heat stress may be confused with yield losses or other damage caused by factors such as drought, nutritional problems, and disease infestation, the goal should be to eliminate or minimize the latter causes of stress.

To increase the frequency of differential test years in the field, the breeder may have to use locations outside of his region where high temperatures are likely to occur at the proper time to provide stress at the critical stage of plant growth. Cooperative regional, national, and even international tests may be useful to identify potentially useful germplasm for parental or cultivar use. Irrigation may be necessary during the periods of high temperature so as to minimize the confounding of heat and drought stress. Modern fungicides may be used to control foliar diseases on many crop plants and soil fumigation can be used to control soilborne pathogens.

REFERENCES

1 Adams, M. W., and G. Semeniuk. 1958. *South Dakota Agric. Exp. Sta. Bull. 469.*

2 Alexandrov, V. Ya. 1964. *Quart. Rev. Biol.* 39:35.

3 Allard, R. W. 1960. *Principles of Plant Breeding.* First ed. Wiley, New York.

4 Amirshahi, M. C., and F. L. Patterson. 1956. *Agron. J.* 48:181.

5 Amirshahi, M. C., and F. L. Patterson. 1956. *Agron. J.* 48:184.

6 Andrews, J. E. 1958. *Can. J. Plant Sci.* 38:1.

7 Brand, C. J. 1911. *U.S. Dept. Agric. Bull.* 209.

8 Caldwell, R. M., J. F. Schafer, L. E. Compton, F. L. Patterson, and J. F. Newman. 1957. *Purdue Univ. Agric. Res. Bull.* 642.

9 Chen, P., and L. V. Gusta. 1978. "The Role of Water in Cold Hardiness of Winter Cereals." In *Plant Cold Hardiness and Freezing Stress*, P. H. Li and A. Sakai, Eds. Academic, New York, N.Y.

10 Christiansen, M. N. 1978. "The Physiology of Plant tolerance to Temperature Extremes." In *Crop Tolerance to Suboptimal Land Conditions*, G. A. Jung, Ed. Am. Soc. Agron., Madison, Wisconsin.

11 Coffman, F. A. 1937. *Am. Soc. Agron.* 29:79.

12 Coffman, F. A. 1957. *Agron. J.* 49:187.

13 Coffman, F. A. 1957. *Agron. J.* 49:368.

14 Compton, L. E., R. M. Caldwell, J. F. Schafer, and F. L. Patterson. 1965. *Purdue Univ. Res. Bull.* 799.

15 Daday, H., and C. G. Greenham. 1960. *J. Hered.* 51:249.

16 Dantuma, G. 1958. *Euphytica* 7:189.

17 De la Roche, I., M. K. Pomeroy, and C. J. Andrews. 1975. *Cryobiology* 12:506.

18 Dexter, S. T. 1956. The Evaluation of Crop Plants for Winter Hardiness. In *Advances in Agronomy*, Vol. 8, A. G. Norman, Ed. Academic, New York.

19 Dvorak, J., and D. B. Fowler. 1978. *Crop. Sci.* 18:477.

20 Finkner, V. C. 1964. *Crop Sci.* 4:465.

21 Finkner, V. C. 1966. *Crop Sci.* 6:297.

22 Grafius, J. E. 1974. *Michigan Agric. Exp. Sta. Res. Rep.* 247:16.

23 Grafius, J. E. 1981. "Breeding for Winter Hardiness." In *Analysis and Improvement of Plant Cold Hardiness*, C. R. Olien, Ed. CRC Press, West Palm Beach, Florida.

24 Gullord, M., C. R. Olien, and E. H. Everson. 1975. *Crop Sci.* 15:153.

25 Harlan, H. V., and M. L. Martini. 1936. *U.S. Dept. Agric. Yearbook*, p. 303.

26 Harlan, J. R., and J. M. J. deWet. 1971. *Taxonomy* 20:509.

27 Hayes, H. K., and R. J. Garber. 1919. *Minnesota Agric. Exp. Sta. Bull.* 182.

28 Hayes, H. K., and O. S. Aamodt. 1927. *J. Agric. Res.* 35:223.

29 Heyne, E. G., and A. M. Brunson. 1940. *J. Am. Soc. Agron.* 32:803.

30 Hodges, H. F., L. V. Svec, and A. L. Barta. 1970. *Crop Sci.* 10:318.

31 Jenkins, B. C. 1963. *Proc. 2nd Int. Wheat Genet. Symp.*, p. 301.

32 Jenkins, G. 1969. *J. Agric. Sci., Camb.* 73:477.

33 Jenkins, G., and A. P. Roffey. 1974. *J. Agric. Sci., Camb.* 83:87.

34 Kretschmer, G. 1960. *Der Zuchter* 30:251.

35 Klebesadel, L. J. 1971. *Crop Sci.* 11:609.

36 Larson, K. I.., and D. Smith. 1964. *Crop Sci.* 4:413.

37 Larter, E. N. 1973. "Progress in the Development of Triticale in Canada." In *Triticale, Proc. Int. Symp.* 69. R. G. Anderson, Ed.

38 Leopold, A. C. 1964. *Plant Growth and Development.* First ed., McGraw-Hill, New York.

39 Levitt, J. 1972. *Responses of Plants to Environmental Stress.* First ed., Academic, New York.

40 Marshall, H. G. 1965. *Crop Sci.* 5:83.

41 Marshall, H. G. 1966. *Crop Sci.* 6:173.

42 Marshall, H. G. 1973. *Crop Sci.* 13:581.

43 Marshall, H. G. 1976. *Crop Sci.* 16:9.

44 Marshall, H. G., C. R. Olien, and E. R. Everson. 1981. "Techniques for Selection of Cold Hardiness in Cereals." In *Analysis and Improvement of Plant Cold Hardiness*, C. R. Olien, Ed. CRC Press, West Palm Beach, Florida.

45 Martin, J. H. 1927. *J. Agric. Res.* 35:493.

46 Martineau, J. R., J. H. Williams, and J. E. Specht. 1979. *Crop Sci.* 19:75.

47 Martineau, J. R., J. H. Williams, and J. E. Specht. 1979. *Crop Sci.* 19:79.

48 Muehlbauer, F. J., H. G. Marshall, and R. R. Hill. 1970. *Crop Sci.* 10:646.

49 Olien, C. R. 1967. *Annu. Rev. Plant Physiol.* 18:387.

50 Olien, C. R. 1977. *U.S. Dept. Agric. Tech. Bull.* 1558.

51 Poehlman, J. M. 1952. *Econ. Bot.* 6:176.

52 Poehlman, J. M., and F. A. Coffman. 1969. *Crop Sci.* 9:396.

53 Quisenberry, K. S. 1931. *U.S. Dept. Agric. Tech. Bull.* 218.

54 Quisenberry, K. S. 1938. *J. Am. Soc. Agron.* 30:399.

55 Rajhathy, T. 1977. *Can. J. Genet. Cytol.* 19:595.

56 Rao, G. N., and B. R. Murty. 1963. *Indian J. Agric. Sci.* 33:155.

57 Reid, D. A. 1965. *Crop Sci.* 5:263.

58 Rhode, C. R., and D. F. Pulham. 1960. *Nebraska Agric. Exp. Sta. Res. Bull.* 193.

59 Rosenquist, C. E. 1933. *J. Am. Soc. Agron.* 25:528.

60 St. John, J. B., M. N. Christiansen, E. N. Ashworth, and W. A. Gentner. 1979. *Crop. Sci.* 19:65.

61 Shelby, T. O. 1959. *Tennessee Agric. Exp. Sta. Bull.* 293.

62 Siminovitch, D. H., H. Therrien, F. Gfeller, and B. Rheaume. 1964. *Can. J. Bot.* 42:637.

63 Singh, J., I. A. de la Roche, and D. Siminovitch. 1977. *Cryobiology* 14:620.

64 Smith, D. C. 1942. *J. Agric. Res.* 64:33.

65 Smith, D. 1958. *Agron. J.* 50:226.

66 Sprague, G. F. 1966. "Quantitative genetics in crop improvement." In *Plant Breeding,* K. J. Frey, Ed. Iowa State Univ. Press, Ames.

67 Steponkus, P. L. 1978. "Cold hardiness and freezing injury in plants." In *Advance in Agronomy,* Vol. 30. N. C. Brady, Ed. Academic, New York.

68 Sullivan, C. Y. 1972. "Mechanisms of heat and drought resistance in grain sorghum and methods of measurement." In *Sorghum in Seventies*, N. G. P. Rao and L. R. House, Eds. Oxford and Indian Book House, New Delhi, p. 247.

69 Sullivan, C. Y., N. V. Norcio, and J. D. Eastin. 1977. "Plant Responses to High Temperatures," In *Genetic Diversity in Plants,* A. Muhammed, R. Aksel, and R. C. von Borstel, Eds. Plenum Publishing Corp., New York, p. 301.

70 Sullivan, C. Y., and W. M. Ross. 1977. *Proc. Int. Conf. on Stress Physiol. in Crop Plants.* Boyce Thompson Institute.

71 Suneson, C. A. 1945. *Agron. J.* 37:72.

72 Suneson, C. A. 1956. *Agron. J.* 41:459.

73 Suneson, C. A., and G. A. Wiebe. 1962. *Crop Sci.* 2:347.

74 Suneson, C. A., and H. G. Marshall. 1967. *Crop Sci.* 7:667.

75 Tatum, L. A. 1954. *Proc. Ninth Annu. Hybrid Corn Ind. Res. Conf.* 9:22.

76 Theuer, J. C., and L. J. Elling. 1963. *Crop Sci.* 9:245.

77 Warnes, D. D., and V. A. Johnson. 1972. *Crop Sci.* 12:403.

78 Wiebe, G. A., and D. A. Reid. 1958. *U.S. Dept. Agric. Tech. Bull.* 1176.

79 Willemot, C., H. J. Hope, R. J. Williams, and R. Michaud. 1977. *Cryobiology* 14:87.

80 Willemot, C. 1977. *Plant Physiol.* 60:1.

81 Worzella, W. W. 1935. *J. Agric. Res.* 50:625.

Chapter 4

PLANT RESPONSE TO MINERAL ELEMENT TOXICITY AND DEFICIENCY

RALPH B. CLARK

U.S. Department of Agriculture, Science and Education Administration, Agricultural Research, and the University of Nebraska, Lincoln, Nebraska

Mineral stresses that occur in soils throughout the world are among the important factors that restrict growth and production of crop plants. Mineral stresses vary among soils and areas, but nearly one-fourth of the world's soils are considered to have some kind of mineral stress (60). If plants are to grow on these soils economically, adjustments must be made to compensate for many of these mineral stresses. In some soils, it may be a matter of adding a mineral element that is needed for normal plant growth; in other soils, a mineral element may be in excess or toxic. Sometimes mineral deficiencies and excesses occur in the same soil. These problems are complex and each needs to be examined carefully in order to achieve a balance to maintain good plant growth.

The purposes of this chapter are to identify some of the mineral stresses found in soils throughout the world, to discuss some of the problems of mineral element use, to detail some of the effects of element deficiencies and excesses on plants, and to consider how plants might be adapted to specific mineral stresses.

MINERAL STRESS PROBLEMS OF THE WORLD

Mineral stress problems in soils are often related to the kind of parent material that makes up a particular soil. Because of this, many mineral stress

Contribution of the U. S. Department of Agriculture, Science and Education Administration, Agricultural Research and Department of Agronomy, University of Nebraska, Lincoln, Nebraska 68583.

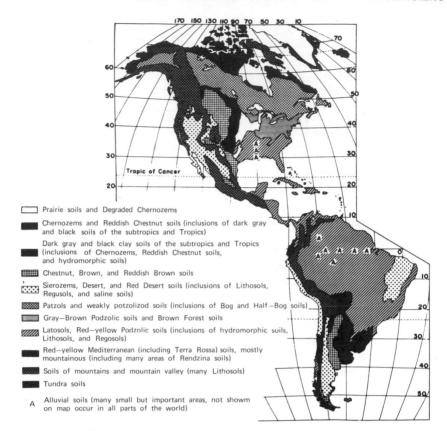

Prairie soils and Degraded Chernozems

Chernozems and Reddish Chestnut soils (inclusions of dark gray and black soils of the subtropics and Tropics)

Dark gray and black clay soils of the subtropics and Tropics (inclusions of Chernozems, Reddish Chestnut soils, and hydromorphic soils)

Chestnut, Brown, and Reddish Brown soils

Sierozems, Desert, and Red Desert soils (inclusions of Lithosols, Regusols, and saline soils)

Patzols and weakly potzolizod soils (inclusions of Bog and Half—Bog soils)

Gray—Brown Podzolic soils and Brown Forest soils

Latosols, Red—yellow Podzolic soils (inclusions of hydromorphic soils, Lithosols, and Regosols)

Red—yellow Mediterranean (including Terra Rossa) soils, mostly mountainous (including many areas of Rendzina soils)

Soils of mountains and mountain valley (many Lithosols)

Tundra soils

A   Alluvial soils (many small but important areas, not showm on map occur in all parts of the world)

Figure 1   A small scale soil map of the world (107).

features may be predicted from the substratum of the soil (60). Examples of this are the Andosols, Arenosols, Thionic Fluvisols, Histosols, Nitosols, Rendzinas, and Vertisols. For other soils, mineral stresses may be related to the type of soil-forming processes and these cause specific properties for some soil groups. Examples of these are Acrisols, Ferralsols, Chernozems, Kastonozems, Phaeozems, Planosols, Podzols, Solonchaks, Solonetz, Xerosols, and Yermosols. In still other soils, mineral stresses may not be consistent with the soil group or classification because of the variability of the soil. Examples of these are Cambisols, Fluvisols, Lithosols, Luvisols, Podzoluvisols, Rankers, and Regosols.

Dudal (60) outlined some of the mineral stress features related to parent materials and to soil-forming processes (Table 1). More complete descriptions of these soils and their terminology are found in the Soil Taxonomy Manual of the Soil Survey Staff (160) and the soil map of the world series of the Food and Agriculture Organization of the United Nations (76). Other

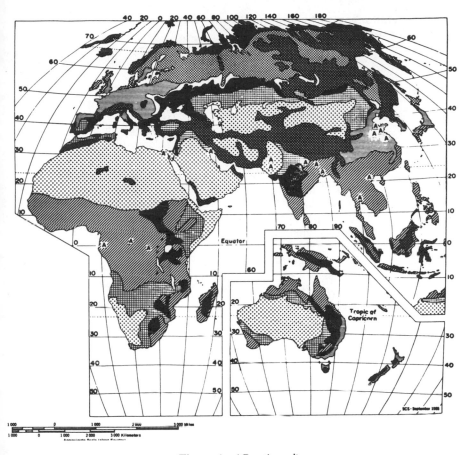

Figure 1 (*Continued*)

mineral stresses may occur that are related to management, water availability, erosion, topography, climate, and so forth.

Distribution of Soils in the World

The distribution of mineral stresses throughout the world can be assessed to a fair extent from the distribution of the soil groups. The amounts of land mass making up the different soil units are shown in Table 2. The geographical distribution of these soil units can be noted by relating them to the distribution of the Great Soils Group (Figure 1). Table 3 shows the distribution of these soil groups and their approximate relationship to the new soil classification system. Further descriptions of these soil groups are given by Kellogg and Orvedal (107). Specific mineral and other stresses may overlap. A general division of the various stresses for soils of the world is shown in

Table 1 Relationships of soil groups of the world and mineral stresses (from 60)

Soil group	Properties of soils	Element Deficiency	Element Toxicity
Acrisol	Tropical and subtropical, very acid, subsurface accumulation of clay, low base status, low content of weatherable minerals, strong leaching, low cation exchange capacity.	N, P, and most other elements	Al, Mn, Fe
Andosol	Volcanic ash, weakly developed, high proportion of weatherable minerals, amorphous hydrated oxides.	P, B, Mo, Ca (more Mg than Ca)	Al
Arenosol	Sandy, weak horizon differentiation, easily leached, low water and cation holding capacities.	K, Zn, Cu, Fe	
Chernozem	Cold and subhumid, high surface horizon organic matter, high base status, lime accumulation at shallow depth, high pH, low $CaCO_3$, may be water stress.	Zn, Mn, Fe	
Ferralsol	Tropical, strongly weathered, mainly kaolinite, quartz, and hydrated oxides, low base exchange capacity, high phosphate fixation.	P, Ca, Mg, Mo	Al, Mn, Fe, sometimes Cr, Co, Ni
Fluvisol	Alluvial, highly variable, no special mineral stresses unless it is a thionic Fluvisol (swamps and marshes). If so, then low pH and Al, Mn, and Fe toxicities.		Al, Mn, Fe
Gleysol	Excess water, low oxidation potential reduces nitrification	Mn	Fe, Mo
Histosol	Prolonged water saturation, thick surface horizon of decomposed organic matter, may have low oxidation potential like Gleysols.	Si, Cu	
Kastanozem	Subarid, organic matter in surface horizon, high base status, often calcareous, usually water stress.	P, Mn, Cu, Zn, K	Na
Nitosol	Tropical, strongly weathered, deep accumulation of clay, similar to Ferralsol but not as acute, low base saturation.	P	Mn (acid)

Table 1 *Continued*

Soil group	Properties of soils	Element Deficiency	Toxicity
Phaeozem	Humid temperate forest and high grass prairie, high organic matter in surface horizon, high to medium base status, deep $CaCO_3$ leaching, no special mineral problems.		Mo (if poorly drained)
Podzol	Acid, iron or humus accumulation in subsurface horizon, coarse textured, mineral stresses associated with excessive leaching and formation of organic matter and metal compounds.	N, K, P, Ca (micronutrients also)	Al
Planosol	In level or depressed topography seasonal surface waterlogging, low cation exchange capacity in surface horizon, strongly leached.	Most nutrients	Al
Rendzina	Shallow, high organic matter surface horizon, highly calcareous, mineral stresses related to high free $CaCO_3$.	Mn, Zn, Fe, P	
Solonchak	Salty, high water stress, hindrance of normal nutrient uptake by salt, different anion or cation that accumulates in soil may be toxic.		B, Na, Cl
Solonetz	Subsurface clay accumulation, high Na saturation, poor physical conditions, unfavorable drainage, and limited available moisture.	N, P, K, Zn, Cu, Mn, Fe	Na
Vertisol	Swelling clay soils with deep and wide cracks when dry, base saturation is high with Ca and Mg, sometimes high pH, low organic matter.	P, N	S (sulfide)
Xerosol	Semiarid, strongly saline and Na-saturated, drought stress, high $CaCO_3$, may have high gypsum.	P, Fe, Zn, Mg, K	Na (salt)
Yermosol	Arid, high water stress, if water applied mineral stresses similar to Xerosol.	P, Fe, Zn, Mg, K	

Table 2   Distribution of the major soils of the world in land masses (60)

Soil associations dominated by	1000 Hectares	Percent of total
Fluvisols	316,450	2.40
Gleysols	622,670	4.73
Regosols and Arenosols	1,330,400	10.10
Andosols	100,640	0.76
Vertisols	311,460	2.36
Solonchaks and Solonetz	268,010	2.03
Yermosols	1,175,980	8.93
Xerosols and Kastanozems	895,550	6.79
Chernozems, Greyzems, and Phaeozems	407,760	3.08
Cambisols	924,870	7.02
Luvisols	922,360	7.00
Podzoluvisols	264,120	2.00
Podzols	477,700	3.63
Planosols	119,890	0.91
Acrisols and Nitosols	1,049,890	7.97
Ferralsols	1,068,450	8.11
Lithosols, Rendzinas, and Rankers	2,263,760	17.17
Histosols	240,200	1.82
Miscellaneous land units (icefields, salt flats, rock debris, shifting sands, etc.)	420,230	3.19
World land area	13,180,390	100.00

Table 4. About 1950 million hectares, or 14.8% of the world land area, are affected by freeze stresses and permafrost which render them limited in use for food or crop production. Other stresses may exist because of textural and sandy surfaces which may limit the use of many soils for plant production.

Acid and Alkaline Soils

Mineral stresses can often be associated with the pH of the soil. The relationships of pH and the relative availability of mineral elements in organic and mineral soils are shown in Figure 2. The pH for greatest availability of most elements in organic soils is at about 5.5 and in mineral soils at about 6.5. Aluminum, Fe, and Ca interfere with P. Thus, at low and high pH values, insoluble P compounds form.

Acid Soils.   An acid soil normally forms when the basic ions supplied to the soil or from the parent material are less than their losses by leaching. The geographic distribution of acid soils throughout the world is shown in Figure 3; the soils are classed into four major categories: Oxisols, Utisols, Alfisols, and Andepts (Inceptisols) (165). Each of these acid soil groups is defined and

Table 3 Estimated area of potentially arable soils in the world both cultivated and not cultivated (107)

Great soils group	Total area of map unit (million hectares)	Percent of total
Prairie soils and degraded Chernozems (Mollisols, Alfisols)	122.3	0.9
Chernozems and reddish chestnut soils (inclusions of dark gray and black soils of the subtropics and tropics) (Mollisols, Alfisols)	381.5	2.9
Dark gray and black clay soils of the subtropics and tropics (inclusions of Chernozems, reddish chestnut soils and hydromorphic soils) (Vertisols)	500.0	3.8
Chestnut, brown and reddish brown soils (Mollisols, Alfisols, Aridisols)	1,203.8	9.2
Sierozems, desert, and red desert soils (inclusions of Lithosols, Regosols and saline soils) (Aridisols, Entisols)	2,798.2	21.3
Podzols and weakly podzolized soils (inclusions of bog and half-bog soils) (Spodosols, Inceptisols, Alfisols, Histosols)	1,294.5	9.8
Gray-brown podzolic soils and brown forest soils (Alfisols, Inceptisols)	605.2	4.6
Latosols, red-yellow podzolic soils (inclusions of hydromorphic soils, Lithosols and Regosols) (Oxisols, Utisols, Inceptisols, Entisols)	3,214.0	24.4
Red-yellow Mediterranean (including Terra Rossa) soils mostly mountainous (including many areas of Rendzina soils) (Alfisols, Mollisols)	111.8	0.9
Soils of mountains and mountain valleys (many Lithosols)	2,465.4	18.7
Tundra soils (Inceptisols)	459.1	3.5
	13,155.8	100.0

classified according to the amount of weathering (measured as the cation exchange capacity of the clay fraction), leaching (measured as the base saturation of the exchange capacity), and illuviation (presence of a clay horizon) that has occurred in them. Oxisols are usually highly weathered, severely leached, and have low cation retention. Their ion adsorption is dependent on acid-base reactions with organic matter, Al and Fe oxides, and kaolinite (1:1 layer clay). Alfisols have not been strongly leached of cations and the base saturation is greater than 35%. Andepts are highly leached with

Table 4 General breakdown of specific stresses in soils of the world (from 60)

Stress	Soil groups		World land area	
			Million hectares	Percent
Mineral	Ferralsols Acrisols Nitosols	Podzols Podzoluvisols Andosols	2961	22.5
Water	Yermosols Xerosols Kastanozems Solonchaks	Solonetz Regosols Arenosols	3670	27.8
Excess Water	Fluvisols Gleysols Vertisols	Planosols Histosols	1611	12.2
Shallow	Lithosols Rendzinas	Rankers Cambisols	3189	24.2
Least Stress	Chernozems Greyzems	Phaeozems Luvisols	1133	10.1
"Nonsoils"	Icefields Salt Flats	Rock Debris Shifting Sands	420	3.2

less weathering and illuviation and are common in volcanic mountainous areas of high rainfall. Oxisols and Utisols are found mostly in warm climates on old landscapes. Alfisols lie outside the tropic latitudes. Oxisols, Utisols, and Inceptisols of the tropics occupy about one billion hectares; leached soils of the temperate regions total about 325 million hectares, and Utisols equal about 130 million hectares. In terms of potential arable land that does not require irrigation, these soils represent 33, 10, and 4%, respectively, of that area. In other words, nearly half of the nonirrigated arable lands are acid soils.

Acid soils are frequently low in available P, low in base exchange capacity, and high in leaching capacity. Thus, nearly all nutrients have to be added in order to maintain fertility in these soils. Acidity increases the availability of Fe, Mn, and Al and these elements are often toxic to plants. Liming is usually needed to raise the soil pH to overcome toxicity problems. Selection for plant tolerance of these elements is an alternative method for overcoming some elemental toxicity problems.

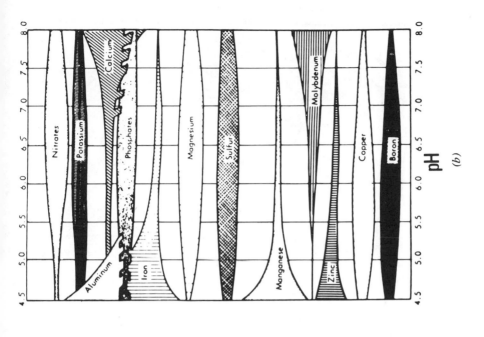

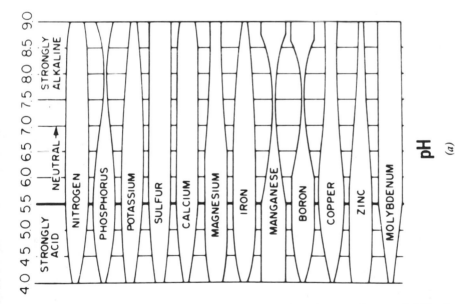

Figure 2 The relationship between soil pH and relative plant nutrient availability for organic (*a*) and mineral (*b*) soils (59). The wider the bar, the greater the availability. Where Al, Fe, and Ca interlock with P, these elements combine to form insoluble compounds.

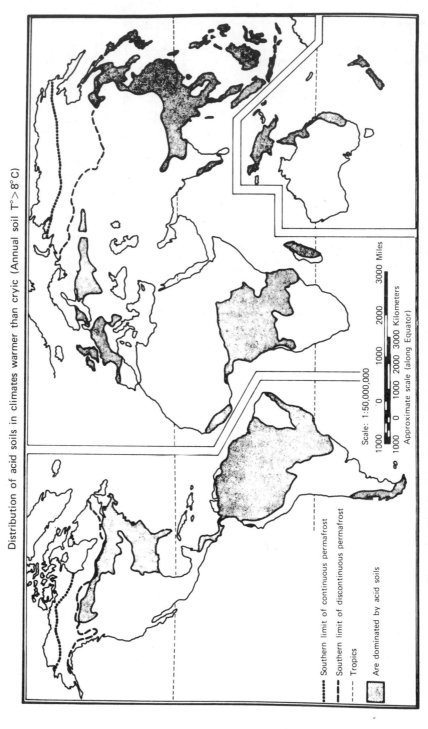

Distribution of acid soils in climates warmer than cryic (Annual soil T° > 8°C)

Scale: 1:50,000,000

1000   0   1000   2000   3000 Miles

1000   0   1000   2000   3000 Kilometers

Approximate scale (along Equator)

•••••• Southern limit of continuous permafrost

— — Southern limit of discontinuous permafrost

– – – Tropics

Are dominated by acid soils

Figure 3 Distribution of acid soils in the world in climates warmer than cryic (165).

Alkaline Soils.   It is difficult to make a sharp separation between acid and alkaline soils. A main distinction between acid and alkaline soils is the amount of precipitation compared to the evapotranspiration. If precipitation exceeds evapotranspiration in most years, the soils usually are leached and acid. However, if precipitation exceeds evapotranspiration in only a few years, the soils are usually neutral or alkaline. The general boundary between acid and alkaline soils appears near the natural boundary between areas where forests and prairies predominate (74). In a broad sense, the boundary between acid and alkaline soils in the United States is the boundary between where maize and small grains are grown as the predominant crops. Alkaline soils are usually considered to be in the subhumid, semiarid, and arid climates.

Alkaline soils frequently contain soluble salts in the profile. In dry soils, high amounts or proportions of Na and K are retained. If ground water is not close to the surface or if irrigation has been properly managed, these salts are usually leached from the surface soil and are not a detriment to crop production. However, if the ground water is close to the soil surface or irrigation has not been properly managed, these salts may accumulate in the soil surface and may interfere with crop production. Alkaline and neutral soils tend to have abundant 2:1 layer clay minerals (especially montmorillonite) and are usually very fertile soils. If these soils are mismanaged or become high in Na, they can become saline.

Alkaline soils are usually adequately supplied with Ca and Mg. In some places, Mg may be high and cause Ca to be limiting. Many of these soils contain adequate K for crop production, but if vermiculite clays are present, K may be fixed. Nitrogen and P are not usually sufficiently high to maintain high yields without additions. Sulfur deficiencies have been known to occur on many alkaline soils and B toxicities may be common. Molybdenum and F may be high in alkaline soils, and even though they may not be toxic to plants, they may be toxic to animals. Most alkaline soils contain Fe that could supply plants almost indefinitely, but the Fe in the soil is not available for plant use. Many plants require Fe added in available forms to remain green.

Assessments of Mineral Problems in Soils

Many countries, international centers, and scientists have and are making surveys to identify specific mineral element problems in soils. Attempts can then be made to alleviate or find solutions to the problems so that the land may be used effectively for food production.

Asia.   One of the more extensive surveys made (and being continued) has been the survey of problem soils associated with the production of rice in south and southeast Asia [International Rice Research Institute (IRRI), reference 102]. Nearly 100 million hectares were identified as having mineral

Table 5   Current and potential soils for rice production in south and southeast Asia (102)

Kind of soil	Million hectares
Saline	62.5
Alkali	2.2
Acid sulfate	9.8
Organic	29.0

problems for rice (*Oryza sativa* L.) production (Table 5). Of those identified, saline and organic soils made up about 90% of the mineral stress problems. Countries that showed the greatest saline problems were India, Indonesia, Pakistan, Malaysia, Vietnam, and Bangladesh (Figure 4, upper). Acid sulfate soil problems in this area of the world were also noted (Figure 4, lower). The survey by IRRI (102) also showed that Fe toxicities occurred on strongly acid soils, P deficiencies on Oxisols, Utisols, and Vertisols, Zn deficiencies on alkaline and organic soils, and that wet soils limited rice yields on yet another 50 million hectares.

South America, India, and Australia. As shown in Figure 3, most of the soils in South America, especially those east of the Andes Mountains and northeast of Argentina, are acid. Brazil contains a large amount of the acid soils of South America and these soils have extensive Al toxicity and P, N, and Zn deficiency problems. Some of the worst P (Figure 5) and Zn deficiencies observed by this author were seen on plants grown on the Cerrado soils of Brazil. Low fertility and Al, Mn, and Fe toxicities in Colombian soils have been recognized for many years (161). Deficiencies of Zn and toxicities of Mo, B, Se, and sodicity have been recognized in Northwest India (141). Low supplies of P and micronutrients have been recognized in Australia (116, 117) and New Zealand (41) for many years. Many other mineral stresses prevail throughout the world (75, 105, 179).

United States. In the United States, deficiencies of N, P, and K have been reported to be the most extensive deficiency mineral element problems. Beeson (15) and Berger (17) reported widespread distribution of Mg, Cu, Fe, Zn, B, and Mn deficiencies. Deficiencies were noted in 41 states for B, 30 for Zn, 25 for Fe and Cu, 21 for Mo, and 13 for Cu. A large number of plants were affected. Surveys like these usually lead to additional investigations to determine similar problems in nearby areas or states. Upon close inspection, nearly all deficiencies can be found in particular plants. For example, if apple orchards were to be grown in every state, B deficiencies would most likely be reported from every state. Or if sorghum were to be grown in every state, more Fe deficiencies would likely be reported. In addition, as deficiencies of specific minerals are identified and corrected, sometimes with excess amounts of the once-deficient mineral, other mineral element problems arise or are recognized.

The distribution of micronutrient problems throughout the United States have been reviewed by Kubota and Alloway (111) and are summarized as follows:

Manganese.   Manganese deficiency is relatively widespread on soils in both the arid or semiarid West and the humid East. Manganese deficiencies are most often observed on naturally wet soils that have been drained to bring into crop production. Most of these soils are dark and have high organic matter contents. Manganese deficiencies often appear on soils with calcareous subsoils or on limed soils followed by improved drainage. Manganese toxicities may be a problem in acid soils and have been reported in New York, Wisconsin, and North Carolina.

Iron.   Iron deficiencies have been reported on both acid and alkaline soils, but most of the iron deficiencies appear in alkaline, particularly calcareous, soils. Some plants show Fe deficiency even on acid soils. The list of plants showing Fe deficiency on calcareous soils is extensive and about 48 plants have been classified for their tolerance or susceptibility to Fe deficiency (126). Iron toxicities have not been common.

Boron.   Boron deficiencies have been reported in most states and the distribution of B deficiency may be more closely related to the production areas of susceptible plants than to the differences in available B in soils. Boron deficiencies appear more in dry years. Liming of acid soils may induce temporary B deficiencies in susceptible crops like alfalfa (*Medicago sativa* L.) and apples (*Malus* spp.). Boron toxicity in plants has been reported for the western states and this problem appears to be associated with the use of high B waters in irrigation.

Copper.   Copper deficiencies have been reported for the acid organic soils of the East coast, acid sands of Florida, Podzolic soils of Wisconsin, and very sandy alkaline soils of the Pacific coast. Copper toxicities have been reported where Cu fungicides have been used excessively and in soils affected by mine wastes. Even though Cu deficiencies have been the least common of the deficiencies, whenever organic soils are converted to agricultural use, Cu deficiencies could arise.

Zinc.   Zinc deficiencies have been reported to be widespread throughout the United States and Zn may potentially become deficient in nearly all intensively cropped soils. Zinc deficiency is most likely to occur in plants growing on calcareous, leached, scraped or scalped, acid, and sandy soils, as well as on soils high in available P. Zinc deficiencies are largely found to occur on

84

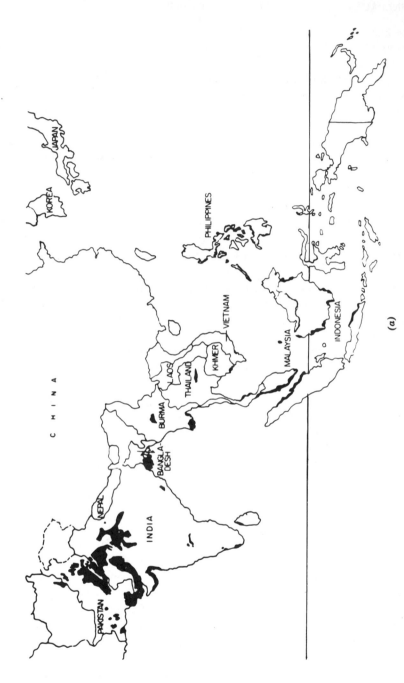

Figure 4 Saline (*a*) and acid sulfate.

(*a*)

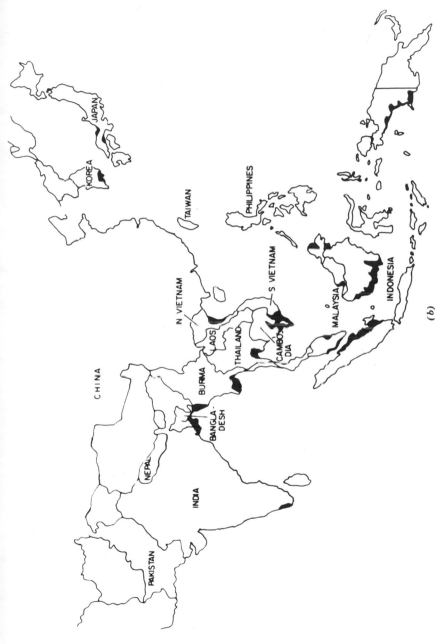

Figure 4 (*Continued*) Saline (*b*) soils of south and southeast Asia (102).

Figure 5   Effects of added P on sorghum grown in a Cerrado soil, Brazil (Clark, personal observation).

small spots of soils rather than generally throughout fields. Seldom have Zn toxicities been noted, but some cases have been reported on organic soils in New York.

Molybdenum.   Areas for most frequently reported Mo deficiencies have been the Eastern seaboard, the Great Lakes states, and the Pacific coast states in alfalfa, cauliflower (*Brassica oleracea* var. *botrytis* L.), broccoli (*Brassica oleracea* var. *botrytis* L.), soybeans (*Glycine max* (L.) Merr.), clover (*Trifolium* spp.), truck crops, and citrus. Molybdenum deficiencies appear to be the result of growing specific crops on acid soils. Molybdenum toxicity has not been reported for plants, but high Mo in plants has induced problems in animals, especially in the western states.

Other Elements. *Cobalt* is required for nitrogen fixation by microorganisms in a symbiotic relationship with legumes, but the requirement is low. Cobalt in plants has caused problems in animals. Areas of the United States showing these problems have been Florida, the Eastern seaboard, the northeastern, and northern mid-west states. *Iodine* has not been shown to be essential to plant growth and is usually low in plants. Its absence has been related to the incidence of goiter in humans. The highest frequency of goiter cases (male draftees during World War I) were noted in the Northwest, northern mid-west, and northern Great Plains states. Some plants of the Great Plains and Rocky Mountain states contain sufficiently high amounts of *selenium* to be toxic to animals that feed on them. *Sodium* has become a problem in the western states where ground waters have been close to the surface and where irrigation waters have not been properly managed. Most of these saline soils are spotty throughout this area, but may be extensive in some areas. Soils near oceans may also become saline. *Aluminum* can be toxic in acid soils. Incidents of Al toxicity have been reported over many areas of the eastern and southeastern states. *Silicon* may become limiting in rice although no Si toxicities have been reported in any of the states. Toxicities of other elements like *lead, cadmium, mercury,* and *nickel* have been reported in plants, but these have been man-made or associated with local contaminations (like around buildings painted with Pb paints). *Sulfur* and *fluoride* toxicities in plants have been reported and these have usually been the result of mining and industrial operations and to a lesser extent of occurrence in parent material. High *magnesium* soils have been reported in California with the serpentine soils. High *calcium* is well known throughout the Great Plains and western states. These calcareous soils induce considerable problems with Fe deficiency and to a lesser extent with Zn and Mn deficiencies.

PROBLEMS OF MINERAL ELEMENT USE

Once mineral element problems have been identified and recognized, fertilizers or amendments can be added to alleviate deficiency or excess problems. Once these have been added, other factors come into play and need to be understood before use may be made of the elements by plants. These factors may not be just a matter of adding the elements to soils, but how and when to add these elements and how much to apply. How and when to add elements to soils are beyond the scope of this article, but these are dependent on the soil, the plant to be grown, the desired yield, the climate, and the management practices, to mention a few. The amount of element added could cause interactions with other elements and thus cause additional problems. Innumerable interactions of one element with another are known (96, 97, 133, 159) and these interactions could be of major significance.

The addition of P is one of the prime examples of interactions with other elements. As noted in Figure 2, P can be bound by other compounds dependent on pH, thus rendering it relatively unavailable. The reaction of P with soils is such that P does not readily move in soils. Thus, plant roots have to grow to the P that has been placed in the soil and, as such, relatively small amounts of P are actually used by the plant. In addition, P may interact with many other elements, especially the micronutrients (Mn, Fe, Cu, Zn, and Mo), to cause deficiencies of these. Phosphorus may also bind other elements that could be toxic (Al, Fe, Mn). A proper balance of the various elements is needed to sustain good plant growth. Other examples of elemental interaction are the balance of cations in plants. Most plants maintain a relative constancy of Ca, Mg, and K (14, 168). If one element is unbalanced, one of the other elements usually compensates. Excessive lime can also cause imbalances, especially if an element is near its critical level for plant needs in the soil. Many examples of mineral imbalances could be cited, but the important factor is that a proper balance among elements be maintained.

Additional mineral element problems arise from the source of N used. Ammonium or nitrate cause soil pH values to change and these pH changes must be considered. Toxicities of some elements have arisen because of the application of pesticides. For example, Cu toxicities appeared in citrus because of the Cu contained in fungicides (111). Sprays have been used which contained trace minerals that have built up and caused interactions with other elements or caused toxicities themselves. Additional use of fertilizers usually means better yields and more luxurious plants, but with this, new problems with insects and diseases usually arise.

Of more recent vintage have been the concerns for land, water, and air pollution or contamination from excess mineral elements. These have been brought about mostly from man-made activities such as mining, manufacturing, disposing of industrial and municipal wastes, and applying fertilizers. Excess N, particularly nitrate, can leach through soils to ground waters or move with water on soil surfaces to places other than the site of application. Surface erosion and the leaching of P and other elements are also of concern. These problems are real and must be controlled, but they are beyond the scope of this discussion.

Fertilizer Efficiency

One of the current problems in mineral element nutrition is the efficiency with which the elements can be used by plants. It is important that minerals added to soils are used effectively and efficiently. The resources and methods appear to be readily available to produce the fertilizers that will be needed for the near future, but costs and distribution may be limiting factors. Costs are increasing at greater rates than ever before and these are of concern, especially in countries where capital is not readily available. Thus, the efficient use of fertilizer that is applied becomes all the more important in order to keep food production at a level sufficient to provide that needed by the peoples of the world.

Definition of Mineral Efficiency

The terms "mineral efficient" or "fertilizer efficient" have been defined in many ways. Usually efficiency is defined as output divided by input. A plant that produces the highest yields per given amount of element added can be considered to be an efficient plant. This may not be a plant that produces the highest total output at higher or lower levels of nutrients. A "mineral efficient" plant may also be one that produces the greatest dry matter per unit of element taken into the plant. Another definition is a plant that produces the most harvestable produce (usually grain or fruit) per unit element added or taken up. Other definitions for mineral element efficiency have also been suggested.

"Mineral efficiency" in plants may be better understood if mineral requirements for plants are better understood. Loneragan and colleagues (118–121) suggested that element requirements of plants fall into three categories: "solution requirement, functional requirement, and critical concentration."

Two aspects of "mineral efficiency" have been considered here: (a) the efficiency with which plants recover fertilizer elements that are applied to the soil for use by the plant, and (b) the efficiency with which plants use elements after they have been absorbed by the plant. The first view considers the amount of element recovered by the plant that is being grown immediately or by plants grown successively. It also considers the amount of element lost from the soil or the amount that becomes unavailable (fixed) to the plant in the soil. The second view considers the use and dry matter production by plants with a given amount of element inside the plant.

Soil Factors in Element Efficiency

To use mineral elements, plant roots must make contact with the elements in the growth medium, or the element must be sprayed on leaves, but somehow elements must come in contact with plant parts. Because soil additions of mineral elements are the most common method of application, this aspect will be considered. In soils, the roots must grow through the soil to make contact with the elements or the elements must move to the roots. Most fertilizer elements are absorbed by roots growing in a soil and roots contact only about 1 or 2% of the soil volume (11). Thus, movement of elements to the roots in soils is important. The mechanisms for movement of elements to roots in soils are mass flow and diffusion (9). Mass flow is movement of elements through the soil to the root in the flow of water. If water flow does not supply the root requirements, element absorption by the root at or near its surface will reduce the existing concentration. This forms a concentration gradient with the soil volume away from the root. Ions may then move by diffusion in response to this concentration gradient. Inasmuch as N, P, and K are needed in the greatest quantities and are added frequently (usually annually and sometimes more often), these elements will be considered in the discussion.

Mass flow supplies a large part of the N used by plants because nitrate is readily moved with the flow of water. Phosphorus and K are not moved as readily by mass flow and availability at root surfaces is dependent on diffusion (Figure 6). Although diffusion values vary with the soil and water in the soil, diffusion distance values for N, P, and K have been calculated to be about 1 cm, 0.2 cm, and 0.02 cm, respectively (11). The mean distance between maize roots in the top 15 cm of soil is about 0.7 cm (124) so some molecules would need to move half this distance (0.35 cm) before they would become available for absorption by the roots.

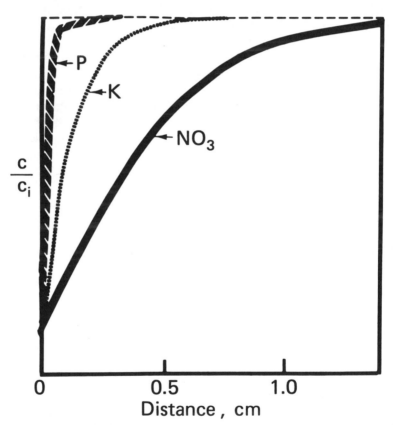

Figure 6 The distribution of NO_3, P, and K radially from a plant root growing in silt loam soil, after uptake for five days, where diffusion controlled the nutrient flux to the root (11). (C is the resulting concentration as a fraction of C_i, the initial concentration of available nutrients.)

Recovery of added N, P, and K depend on many factors, but estimates for the amount recovered are 50% for N, 10% for P, and 20 to 40% for K (11). Improvement for the use of each element could be in the order of two or more, but many factors determine whether this can be accomplished. Continued research is needed to understand the many factors involved. A few examples of N efficiency studies are discussed briefly to note types of approaches taken and the results obtained. Results from these studies should undoubtedly lead to better practices and the improvement of mineral element efficiency.

Recovery of Applied Elements

The uptake of N by maize (*Zea mays* L.) as affected by rates of N fertilizer and water applications in the field ranged from 30 to 68% (Figure 7). Maximum N uptake occurred at fertilizer levels that produced maximum grain yields. When N applications were above this level, the amount of N in the plant remained essentially unchanged, but accumulated in the soil. Unless N rates were in excess of plant needs, very little nitrate derived from the fertilizer escaped from the root profile. Greater efficiency in N uptake was noted when higher rates of water were applied (Figure 7). Increased amounts of water also increased the utilization of N inside the plant (Table 6).

Nitrogen uptake by maize was also studied for the effects of N application rate, frequency of water application, but with the same amount of water, and time of N application (54, 134). When fertilizer was applied, grain yields over 3 years benefited more from lighter, more frequent irrigations (5 cm biweekly) than from heavier, less frequent irrigations (10 cm monthly) (Figure 8). In addition, efficiency of N uptake was higher for lighter, more frequent irrigations regardless of N application (Table 7). This was probably due to less denitrification and/or leaching. Greater uptake efficiency of the N applied was noted when N was added as a side-dressing when plants were about 60 cm tall than when N was applied at the time of planting (Figure 8). The greater efficiency of N from side-dressing could be partially explained by the higher grain/forage ratio of N. That is, less N was tied up in the leaves and was readily translocated to the grain.

Table 6 Grain yields per unit N for maize grown under irrigation for three seasons (23)

| Irrigation (cm) | Grain yield (kg grain/kg N) | | | |
	1973	1974	1975	$\bar{x}$
20	41.64	42.73	45.48	43.28
60	45.48	42.90	46.43	44.94
100	45.53	42.85	50.02	46.13

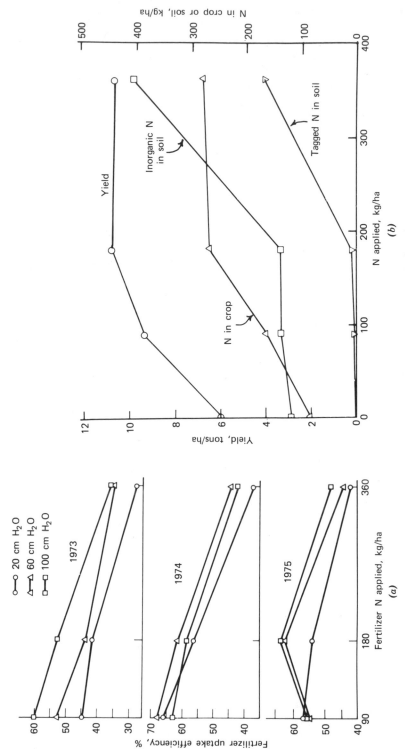

Figure 7 (a) Fertilizer N utilization efficiency in irrigated maize during three years; and (b) Yield, N uptake, and residual soil inorganic N after three years, 60 cm irrigation, California 1973–1975 (23).

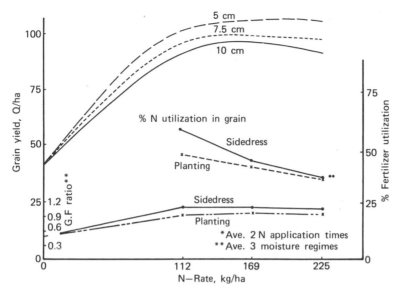

Figure 8   Grain yield response and N use efficiency of irrigated maize with varied irrigation regimes and times and rates of fertilizer N application, Nebraska 1973–1976 (134). (Same total quantity of water applied for each irrigation regime but in varied increments of 5 cm approximately biweekly, 7.5 cm every 3 weeks, and 10 cm monthly.)

Considerably more residual N remained in the upper part of the root profile with the side-dress application than with the planting time application. Similar results were obtained for the lighter, more frequent irrigation than with the heavier, more infrequent irrigations. Essentially no nitrate leached to the deep portions of the soil (300 cm) when N rates were at the levels that gave highest yield (112 kg N/ha) compared to higher N application rates. In addition, the N in the upper profiles was readily available for plant use in a subsequent year when no N was applied. Regardless of irrigation level or N rate, yields were higher for side-dress applications than for planting time applications. Irrigated maize was capable of taking up more residual N from the deep profiles in the soil (180 cm or deeper) if previous N applications were not high (85 compared to 250 kg N/ha). The N taken up by the plants from the deeper portions in the soil had a greater distribution in the grain compared to the stover if the residual N was low. The N taken up late in the growing season was translocated more directly to the grain without extensive vegetative entrapment that occurs with N taken up earlier in the season (134).

Barber (11) also suggested that fertilizer placement might improve nutrient uptake. Mixing P or K with all the soil might result in the fixation of the portion added to the soil. Restricting soil contact by banding fertilizer in the soil should greatly restrict the amount of roots making contact with the

Table 7   Efficiency of N uptake by maize irrigated with different increments of water (same total amount of water) and at three different N application rates (avg. of 3 years) (from 134)

N Rate (kg/ha)	Percent recovery for increment of water[a]		
	5	7.5	10
112	54	54	50
169	43	42	40
225	37	34	32

[a]5 = 5 cm biweekly; 7.5 = 7.5 cm every three weeks; and 10 = 10 cm monthly.

added element. However, greater efficiency in uptake should occur when the added element is distributed somewhat uniformly about the roots rather than concentrated about a small fraction of the root system; but this is contrary to restriction of nutrient absorption by the soil. An intermediate approach to the two extremes of broadcasting and banding might be beneficial. A strip of K fertilizer was placed on the soil surface before plowing so that after plowing, the fertilizer would mix with 10–20% of the soil. Over 5 years, average yield increases of 13% above that of the band treatment and 8% above that of the broadcast treatment were obtained (10). The concentration of K in the ear leaf of plants grown by this "strip-plow" method was 25% higher than the other treatments which were equal.

Maize planted earlier in the season had higher amounts of dry matter in the grain than if it was planted later in the season (Table 8). The differences in maize yield between treatments with or without K were greater for earlier plantings than for later plantings. Higher maize yields were also obtained with a given amount of N when it was applied as a side-dress compared to a preplant treatment.

Table 8   Effect of planting date on yield of maize and proportion of dry matter appearing as grain (11)

Planting date	Dry matter (10^3 kg/ha)		Dry matter in grain (%)
	Grain	Total	
April 26	8.06	15.5	52
May 9	8.46	15.8	54
May 22	6.94	14.7	47
June 3	6.16	14.4	43

Efficiency of Elements After Uptake

Several cultivars and hybrids of crop plants, grown in various areas of the United States, yielded differently and showed differences in utilization of N, K, and P (Table 9). Plants with the highest yields did not necessarily have the highest efficiency for the utilization of the elements. Maize had a higher efficiency for the utilization of N and K than any other plant reported. Many of the plants noted took up more N and K than maize, but they did not produce as much dry matter per unit N or K. Of the other plants reported, sorghum [*Sorghum bicolor* (L.) Moench] was next highest to maize in N efficiency and reed canarygrass (*Phalaris arundinacea* L.) was the lowest. Coastal

Table 9 Dry matter yields and dry matter produced per unit element absorbed for various plants grown throughout the United States (128)

Plant	State grown	Total dry matter yield (kg/ha)	Dry matter / unit element (kg/kg)		
			N	K	P
Maize	Colorado	26.8	81.0	—	
	Illinois	25.8	89.6	91.0	
	New Jersey	23.7	99.6	85.8	
	Florida	19.2	81.8	81.8	
		17.7	83.6	63.2	
	Michigan	19.0	80.4	82.4	
		18.4	82.6	74.1	
	Nebraska	17.6	78.1	75.9	
	Mean	21.0	84.6	79.2	
Alfalfa	Indiana	18.4	40.0	31.6	—
Tempo	Indiana	17.5	31.8	38.2	240
DuPuits	Indiana	17.4	43.4	39.9	245
	New Jersey	16.6	35.7	34.5	361
Scout	Indiana	16.5	33.8	36.0	250
Vernal	Indiana	15.7	29.8	42.2	246
	Mean	17.0	35.8	37.1	268
Bermudagrass, coastal	Georgia	21.5	31.8	53.8	
Sorghum-Sudan		17.6	49.2	40.4	
Orchard	Indiana	14.8	30.2	31.4	
Soybeans	North Carolina	13.9	30.1	59.9	
Sorghum	Kansas	13.5	65.0	42.0	
Canarygrass, reed	Manitoba, Canada	11.2	26.3	31.8	
	Mean	15.4	38.8	43.2	

bermudagrass (*Cynodon dactylon* L.) and soybeans had markedly higher dry matter yields per unit K than did orchardgrass (*Dactylis glomerata* L.) and reed canarygrass.

Sixty-six strains of snapbeans (*Phaseolus vulgaris* L.) and numerous tomato (*Lycopersicon esculentum* Mill.) genotypes were exposed to an artificial selection pressure of a growth-limiting supply of K, P, and N (87). Dry matter yields and the efficiency ratios (mg dry weight per mg element in the plant) for plants considered to be efficient and inefficient are given in Table 10. Differences between the efficient and inefficient strains were sufficient that success should be achieved in increasing plant efficiency for mineral element utilization. Yields of efficient and inefficient strains differed in the utilization of the elements by as much as 44% for N, 72% for P, and 100% for K.

Differences as high as 143% were noted in dry matter yields between maize inbreds showing differential Ca efficiency (Table 11). When these inbreds were grown at near optimum levels of Ca, their dry matter yield differences were only 14% in favor of the more efficient inbred. Differences in their efficiency factor (g dry weight/g Ca) were even more pronounced. At low levels of Ca, 'Oh43' produced 5.3-fold more dry matter/unit Ca than did 'A251'. Regardless of Ca level, 'Oh43' had a higher efficiency factor than did 'A251'. Somewhat similar results were noted for maize inbreds that showed differential Mg efficiency (Table 11). At the low levels of Mg, 'B57' produced more dry matter and had higher efficiency factors for Mg utilization than 'Oh40B'. However, at high levels of Mg, 'Oh40B' produced more dry matter and had a higher efficiency factor for Mg than 'B57'. Differential dry matter

Table 10   Dry matter yields and efficiency ratios (mg dry wt/mg element in plant) of strains of snapbeans and tomato representing the extremes in efficiency of N, P, and K (87)

Species	Stress element	Level of stress element (mg/plant)	Strain[a]	Dry matter yields (g/plant)	Efficiency ratio
Snapbean	K	11.3	I	6.00	157
			E	8.83	294
Tomato	K	5	I	0.95	173
			E	1.97	358
Snapbean	P	2	I	0.87	562
			E	1.50	671
Tomato	N		I	2.51	83
			I	2.71	88
			E	3.51	110
			E	3.62	118

[a]I = inefficient, E = efficient.

yields and efficiency factors for two maize inbreds grown at varied levels of Zn were also noted (Table 11). 'H84' produced more dry matter and had a higher efficiency factor for Zn utilization than 'A635' at low Zn levels. However, the Zn-efficient inbred at low Zn application became Zn inefficient at high Zn application.

When grown with limited P, sorghum genotypes showed marked differences in dry matter yields, P contents, efficiency factors for P (mg dry weight/mg P), and distribution of P between roots and tops and within upper and lower leaves (Table 12). Plants that produced the highest dry matter were not necessarily the plants that had the highest efficiency factor for P. Differences among genotypes for P efficiency might also be explained in terms of differences in translocation of P within leaves, since some genotypes retained more P in the roots than others rather than translocating it to leaves. Some genotypes also retained more P in lower leaves than others, not moving it to the newer upper leaves. Differences of up to 100% in rates of P uptake were noted among snapbean genotypes (87) and these differences indicated that roots had different capacities to absorb P. Specific sorghum parents showed different N efficiency traits when N partitioning patterns and dry matter produced per unit N were used for the selection of N efficiency (123). The most N-efficient hybrids were not the highest yielders of grain.

Additional factors that can affect mineral element efficiency are root numbers, lengths, densities, types, sizes, and surface activities. Shoot/root ratios could also be important since lower shoot/root ratios might mean fewer leaves or other tissue dependent on a given volume of roots. In N uptake, a large number of small fibrous roots with large numbers of long root hairs may not be critical, but in P uptake, this may be very important. Micorrhizal associations with plant roots appear to be important in P efficiency (11, 12, 117). The translocation and re-export of elements to keep pace with growth rates could also be important. Plants unable to translocate elements to keep pace with growth rates might be expected to show deficiencies. Some plants appear to be more efficient for some elements because of their slower growth (117).

These and other factors play significant roles in mineral element efficiency and plant responses to mineral elements. They need to be put in perspective and considered when studying mineral elements in plants.

PLANT RESPONSES TO MINERAL DEFICIENCIES

During seed germination and early growth, seedlings depend on the seed for its mineral elements; upon depletion of this source, plants are dependent on the external medium for minerals. If these required minerals are not supplied in sufficient amounts, visual deficiency symptoms appear. If the deficiencies are severe, plants are delayed in flowering and maturity or they may die.

Table 11 Dry matter yields and efficiency factors of maize inbred tops and roots grown at varied levels of Ca, Mg, and Zn (46, 47)

Element varied	Solution level	Dry matter yields (g/plant)				Efficiency factor (g dry wt/g Ca and Mg)			
		Tops		Roots		Tops		Roots	
	(mM)	Oh43	A251	Oh43	A251	Oh43	A251	Oh43	A251
Ca	0	0.28	0.14	0.08	0.05	88	35	24	11
	0.037	1.15	0.40	0.53	0.29	479	91	196	94
	0.15	2.56	1.12	1.04	0.41	800	287	433	121
	0.62	2.92	2.51	1.07	1.00	562	380	282	227
	2.5	3.05	2.17	1.23	0.77	363	252	228	92
	10.2	2.45	1.16	0.77	0.33	227	94	92	23
	25.4	1.03	0.70	0.21	0.14	52	43	19	13

Mg	B57	Oh40B	B57	Oh40B	B57	Oh40B	B57	Oh40B
	0.70	0.24	0.10	0.03	2593	1263	370	73
0.006	1.44	0.52	0.30	0.08	3600	1576	1034	160
0.031	2.53	1.00	0.94	0.21	4081	2222	1741	288
0.12	2.37	2.05	0.88	0.61	1185	2181	1257	718
0.62	2.52	2.23	0.96	0.77	884	1784	914	856
1.2	2.22	2.68	0.81	1.13	752	1624	648	983
2.1	2.18	2.77	0.82	1.09	581	1385	469	752
3.1	2.02	3.68	0.72	1.27	569	1636	300	635
4.1								

Zn	H84	A635	H84	A635	H84	A635 (g dry wt/mg Zn)	H84	A635
(µM)								
0	0.71	0.23	0.33	0.14	23.7	11.1	10.1	2.8
0.017	0.96	0.25	0.42	0.16	29.8	11.9	10.5	2.5
0.17	1.18	0.52	0.43	0.21	35.9	16.3	9.3	3.5
1.7	1.60	0.84	0.51	0.39	33.4	21.5	8.6	5.4
16.8	1.69	1.39	0.61	0.59	21.7	28.3	7.4	6.1
50.5	1.61	1.59	0.56	0.75	11.4	28.9	6.7	4.6
168	2.00	1.96	0.56	0.87	12.0	15.5	3.0	3.8
505	1.91	2.33	0.65	0.99	8.8	11.5	2.4	4.5

Table 12　Means and ranges of high and low dry matter yields, P contents, P efficiency factors, and distributions of P for 42 sorghum genotypes grown with limited P (50)

Parameter	Plant part	High[a] $\bar{x}$	High[a] Range	Low[a] $\bar{x}$	Low[a] Range
mg dry wt/plant	Tops	1283	(977–1492)	855	(637–1006)
	Roots	902	(689–1040)	521	(346–666)
	Whole plant	2052	(1579–2332)	1437	(1000–1852)
	Top/root	1.97	(1.63–2.31)	1.13	(1.09–1.15)
μg P/plant	Upper leaves	1269	(863–1775)	771	(530–938)
	Lower leaves[b]	298	(189–378)	85	(27–154)
	Tops	1492	(1003–1959)	950	(708–1262)
	Roots	1202	(658–2179)	612	(424–762)
	Whole plant	2468	(1555–3781)	1636	(1126–2046)
mg dry wt/mg P	Tops	1173	(1012–1376)	690	(603–818)
	Roots	1185	(1123–1292)	681	(451–836)
	Whole plant	1139	(1000–1334)	706	(592–838)
μg P ratio in plant parts	Upper/lower	21.2	(7.9–46.2)	3.3	(2.2–4.6)
	Tops/roots	2.04	(1.74–2.28)	1.10	(0.76–1.37)

[a]Averages and ranges of five experiments in which the highest and lowest values for the genotypes in each experiment were used (total of 42 genotypes).
[b]Lowest four leaves on the plant.

Mineral-deficient plants are usually weakened extensively and are more susceptible to disease, insect, and physical damage. Plant metabolism is changed under mineral deficiencies because each element has specific functions. The organic constituents that change under mineral deficiencies of the various elements are enormous and are the subject of an extensive review by Hewitt (96). These changes are too extensive to be considered in this chapter. Inasmuch as plants generally tend to have a relative constancy of cations and anions (14, 97, 140, 159, 168), when a plant has one cation or anion that becomes low or limiting, other cations or anions are taken up to compensate for the loss of the ion that has become low. However, under relatively severe mineral deficiency conditions, plants tend to increase concentrations of many of the mineral elements because of the concentrating effects of the reduced growth (Figure 9).

Visual Deficiency Symptoms of the Macronutrients

Visual deficiency symptoms from mineral elements in plants vary among plant species, but basic principles hold for most elements. Symptoms of particular plants need to be observed if correct diagnoses of deficiencies are to be made for certain plants. Visual mineral deficiency symptoms for many

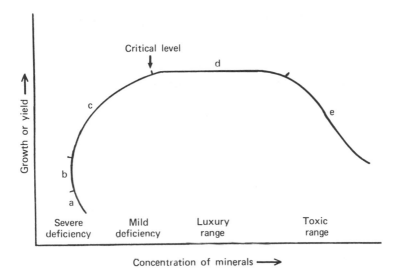

Figure 9 Relationship of mineral composition in plant tissue to growth (159).

plant species have been documented (42, 43, 96, 162, 177). With few exceptions, the symptoms described here are for young sorghum plants.

Nitrogen. Nitrogen is required in many compounds, particularly in amino
acids, amides, peptides, protein, nucleic acids, nucleotides, and coenzymes.
It is concentrated in actively growing cells and organelles because N is contained in the nucleic acids, the basic genetic material of cells. Nitrogen is
essentially ubiquitous throughout the plant. Except for C, H, and O, which
are derived from the air or water, N is the element most needed by plants.
Except for legumes, which have the ability to fix their own N, N must be
supplied to plants for growth. It is usually added as a fertilizer and is required for all types of soils.

Nitrogen is mobile to a large extent in plants and moves from older tissues to newly developing tissues and is accumulated extensively in seeds.
Deficiency symptoms appeared on the older or lower leaves first. Nitrogen-
deficient leaves turned pale yellow and the color was relatively uniform over
the entire leaf. Roots were not affected to any great extent, although fibrousness appeared to decrease.

Phosphorus. Like N, P is required in numerous compounds in cells and
organelles. Phosphorus is associated with many structural components of
the plant. These compounds are associated with numerous components of
metabolism (sugar phosphates, nucleic acids, nucleotides, coenzymes,
phospholipids) and are closely associated with energy transfer (triphos-
phonucleotides) and the genetic controlling materials (nucleic acids). Plants

must have P for growth and development and P is a major element in many fertilizers.

Phosphorus is mobile in plants and moves from older or lower leaves and tissues to newer or upper leaves and tissues. Extensive amounts of P are stored in seeds. Deficiency symptoms appeared first on older leaves and were characterized by a dark green appearance with reddish-purple streaks along the veins. In more severe P deficiency, the streaks coalesced so that uniform coloring over the leaf appeared. Phosphorus-deficient roots were long, spindly, with little fibrousness and turned dark (reddish-purple to dark red) with increased severity of deficiency.

Potassium.   Potassium is required in relatively high amounts in most plant tissues. It acts as an activator of many enzymes (reference 73) and has been associated with the osmotic regulation of cells. It is the main cation associated with the balance of anions (including organic acids) (97), and an extensive amount of K accumulates in seeds. Potassium must be available for plant growth and is usually found in adequate amounts in alkaline or neutral (arid or semiarid) soils. In acid soils, K has usually been leached to the extent that additional K must be added as a fertilizer.

Potassium is mobile and retranslocates from older to newer tissue, thus deficiency symptoms appear first on the lower or older leaves. Potassium deficiency symptoms were characterized by bronze and yellowish-brown color with necrotic specks appearing uniformly over the leaf. Roots of K-deficient plants appeared to be relatively normal, except they were somewhat shorter.

Magnesium.   Magnesium is required in lower amounts than the previously mentioned nutrients. About 10% of the Mg is a component of chlorophyll and Mg is required for the catalysis of many enzyme reactions, especially those requiring P transfer. Extensive amounts of Mg translocate to seeds. Soils usually contain adequate Mg; when it is required, however, it is usually added as a fertilizer or with lime amendments. Acid soils usually require the addition of Mg. Dolomitic limestone contains relatively high amounts of Mg, and most Mg needs have been met by lime additions.

Magnesium is a mobile element and deficiency symptoms appear first in the older leaves and tissues. Symptoms of Mg deficiency were characterized by a light coloring between the veins. Leaves became brittle and margins curled. Dark necrotic spots appeared and leaves usually turned reddish-purple with severe deficiency. Magnesium-deficient roots turned dark red, were shorter, and had little growth.

Calcium.   Calcium is required as a structural component of cells (Ca pectate in the middle lamella) and is needed for the integrity of cells. Cells without Ca tend to lose their semipermeability and collapse. Calcium has been found to catalyze some enzymes involved in the hydrolysis of adenosine triphosphate (ATP) and phospholipids and to partially replace Mg in some reactions. Calcium is also involved with binding of ribonucleic acid (RNA) to protein in chromosomes and has been found to ameliorate toxicities of heavy metals (168). Very little Ca accumulates in seeds, but Ca is required in fruits to alleviate "bitter bit" or "blossom end rot" symptoms. It has also been suggested that Ca is needed only in micronutrient quantities if heavy metals are low (170). Calcium is usually abundant in alkaline and neutral soils. Calcium usually has to be added to acid soils and this practice (liming) supplies the necessary Ca for plant growth.

Calcium is immobile in plants and Ca deficiencies appear first in the newly emerging leaves or forming tissues. Even with Ca-deficient new leaves, the older or lower leaves may contain extensive amounts of Ca. Calcium deficiency was characterized by dead and tightly curled tips which were usually bent over and sticky or gummy to the touch. Calcium-deficient leaves commonly guttated and had small, sticky liquid vesicles on leaf margins and surfaces. Less severely affected leaves were speckly yellow and somewhat brittle and frequently showed serrated breaks. Lower leaves were usually darker in color than were normal leaves (Figure 10a). Tips of old and new roots of Ca-deficient plants were stubby, darker brown (dead), and decreased slightly in fibrousness.

Sulfur. Sulfur is required in the amino acids cystine, cysteine, and methionine, and thus in proteins. It is also a component of lipoic acid, coenzyme A, glutathione, biotin, and several other compounds. Many of the S radicals form "-S-S-" bonds to give specific configurations in protein, and these bonds have also been associated with metallic binding of Fe with Mo in ferredoxin, nitrogenase, and nitrate reductase (132). Extensive amounts accumulate in seeds or fruits. Sulfur may be needed for both acid or alkaline soils. If S is insufficient, it is usually added as a fertilizer. Many lime materials contain extensive amounts of S.

Sulfur is relatively immobile and deficiency symptoms are found first in newly emerging leaves. Sulfur-deficient leaves became lighter in color and turned yellow with near uniformity over the leaf. Sulfur deficiency appeared in the newly emerging leaves. Roots of S-deficient plants appeared to be relatively normal with some growth reduction.

These six elements are considered to be the macronutrients; they are required in relatively large amounts for plant growth and are usually the nutrient elements most frequently added to soils. The major elements in fertilizer formulations are N, P, and K with additions of other elements as needed for particular locales. These macronutrients are no more important than the micronutrients (Mn, Fe, B, Cu, Zn, Mo) in plant growth, but they are needed in greater amounts. Micronutrients are usually added less frequently and may be needed only in particular soils, for particular plants, or for particular locations. Special problems may arise for certain micronutrients in some plants. For example, Fe deficiency is a major problem for sorghum and

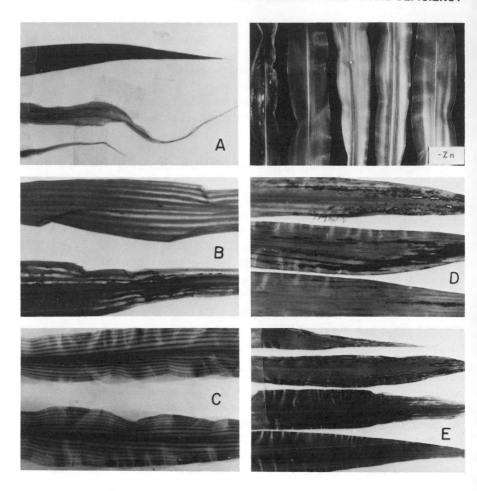

Figure 10 Symptoms for Ca (*a*), Mn (*b*), Fe (*c*), and Zn (*Zn*) deficiency and NO_3-N (*d*) and P (*e*) excess on sorghum leaves (Clark, personal observation). (Zinc deficiency is on maize leaves.)

several other plants grown in many of the Great Plains states (Nebraska, Kansas, Oklahoma, Texas, New Mexico, and Colorado) of the United States.

Visual Deficiency Symptoms of the Micronutrients

Manganese. Manganese has been known to substitute for Mg in many ATP-dependent enzymes (carboxylases and kinases) and is required for oxidases, peroxidases, arginase, phosphotransferase, and several other enzymes. Manganese also appears to participate in the oxygen-evolving system of photosynthesis. Manganese is usually added by spraying the foliage or by applying it to the soil.

Manganese is somewhat mobile and deficiency symptoms first appear in the lower or older leaves. Manganese deficiency symptoms are characterized by a lightening of the interveinal tissue ("Christmas tree" design in trees and broadleaves) or a long dead necrotic streak in the interveinal tissue of cereal and grass leaves (Figure 10b). The remainder of the leaf remains relatively dark green. Roots appeared to be normal with somewhat darker color.

Iron. Iron is found in nonhaem proteins (ferredoxin) and is involved in key metabolic reactions such as nitrogen fixation, photosynthesis, electron transfer, and hormone biosynthesis. Iron is also found in haem proteins (cytochromes) which are important as respiratory electron carriers. Iron has also been found to be associated with chlorophyll synthesis, with nitrate, nitrite, and sulfite reductases, and with the redox metabolism of plants. Both acid and alkaline soils have relatively high amounts of Fe, but for many plants, the Fe is not available for use. Iron must be added to plants which are susceptible to Fe deficiency. Except for some Fe chelates, added Fe has little or no residual effect for plants to be grown the next season or growth period. Iron sprayed on foliage must be added several times during the growing season to keep susceptible plants green. Costs frequently prohibit soil additions of Fe chelates unless the plant is of high value or esthetically important. In many cases, the Fe deficiency problem is alleviated by selecting or breeding plants that are not susceptible to Fe deficiency.

Iron is immobile in plants and the new leaves show deficiency symptoms first. Iron deficiency symptoms were characterized by an intense yellowing of the interveinal leaf tissue and the bright green veins (Figure 10c). Roots were not seriously affected by Fe deficiency.

Boron. The role of B in plants has not been clearly determined, but B has been closely associated with sugar and growth regulator translocation, metabolism, and responses; stimulation and inhibition of some enzymes; nucleic and phenolic acid biosynthesis; cell differentiation, lignification, and several other metabolic processes in plants (62). Dicotyledonous plants, especially clovers, alfalfa, and apples appear to be more susceptible to B deficiency than monocotoledonous plants. Adequate B is usually found in the water or parent material and may be more of a toxicity problem than a deficiency problem. The range of B concentration between deficiency and toxicity is relatively narrow. Boron deficiencies appear in new leaves or newly forming tissue. When B is needed, it is usually added with fertilizers. Boron is routinely added to fertilizers intended for use on certain crops. Boron deficiencies are characterized by brittleness of leaf petioles, by brittle and crinkled new leaves, and by the appearance of dark or bronze hollow

Figure 11   Deficiency symptoms of B (left) on tomato, Cu (upper right) on sorghum, and Mo (lower right) on maize leaves (32, 34, 36).

stems or cracked and decayed stems. Figure 11 (left) shows B-deficient tomato leaves.

Copper.   Copper is a component of cytochrome, ascorbate oxidases, laccase, tyrosinase, and urease. It is associated with the reduction of oxygen to water and of plastocyanin, a protein associated with photosynthetic electron transfer (132). Copper has also been associated with Ca metabolism and is required for wheat (*Triticum aestivum* L.) and several other plants in the reproductive phase (35). Deficiencies of Cu have been associated with fruit and citrus trees more than with many other plants; however, Cu deficiencies have been reported in cereals, vegetables, grasses, and forage legumes. Copper has been added as a spray (usually as Cu fungicides before the nutritive value of Cu was appreciated) and as a soil amendment.

Copper is not readily retranslocated and deficiencies appear first in newer leaves and are characterized by yellowing of new leaves with bronzing. In cereals, Cu deficiency appears in new leaf tips similar to Ca deficiency and may indeed be a Ca deficiency (Figure 11, upper right). Roots of plants grown without Cu appeared normal.

Zinc.   Zinc is required as an activator or structural component of several enzymes like glutamic, alcohol, and lactic dehydrogenase, carbonic anhydrolase, carboxypeptidase, alkaline phosphatase, and DNA polymerase I; these are important metabolic enzymes (132). Zinc is usually added as a soil amendment in fertilizers or may be sprayed on plants. Sometimes it is added around trees, shrubs, and high-value plants as a Zn chelate or inorganic salt.

Zinc is relatively immobile and deficiency symptoms usually appear first in new leaves. Deficiency symptoms appear as smaller, yellow new leaves, and stem elongation is usually inhibited (rosettes). In cereals, Zn deficiency appears as faded or bleached streaks or patches in the centers of newly emerging leaves (Figure 10*d Zn*). Leaf margins many times turn dark red. Sorghum or maize roots showed no abnormal symptoms under Zn deficiency.

Molybdenum. Molybdenum is a component of at least five enzymes that catalyze diverse and unrelated reactions; these include nitrogenase; nitrate reductase, and xanthine, aldehyde, and sulfite oxidases. The usual manner to overcome Mo deficiencies is by the addition of lime which increases the availability of native soil Mo. *Brassicas,* lettuce (*Lactuca sativa* L.), tomato, sugarbeet (*Beta vulgaris* L.), clovers, and alfalfa are relatively susceptible to Mo deficiency and large-seeded legumes, grasses, cereals, and potatoes (*Solanum tuberosum* L.) are susceptible to a lesser extent.

Molybdenum deficiency symptoms appear first in older leaves, but may also be apparent in new leaves. It is characterized by mottling, puffing, and curling of leaves with lighter color in the interveinal tissue. In cereals, Mo deficiency appears in young leaves as die-back and curling of the emerging leaves similar to that of Ca deficiency (Figure 11, lower right). No abnormalities were noted on roots of plants grown without Mo.

Other Elements. In addition to the above mineral elements which are essential to plant growth, other elements are also known to be essential or functional in plant metabolism. Deficiency symptoms are not easily obtained for these elements unless specific or "indicator" plants are used. Evidence has been reported that indicates the essentiality of Cl, Na, and Si. *Chlorine* has been associated with photosynthetic reactions involved with oxygen evolution and is an important ion in cation-anion balance in plants (22). *Sodium* has been found to be essential in *Atriplex vesicaria* and *Anabaena cylindrica* (132). Partial substitution or sparing effect of Na for K has been noted (21). *Silicon* has been shown to be required by rice (21, 115). Silicon does not seem to be indispensible for vegetative growth but may be necessary for the normal development of the plants. Without Si, rice plants were found to have higher P, Fe, Mn, and B; each of these may become toxic.

Several other elements may be classed as functional nutrients in some plants (21). *Iodine, fluorine,* and *selenium* have been classed in this category because they have been found to be metabolized and incorporated into defined organic compounds (21, 156). Classed similarly are *bromine, rubidium,* and *strontium* because they exert a sparing effect on the utilization of Cl, K, and Ca, respectively (21). *Cobalt* probably also falls into this classification. Cobalt is required by microorganisms in their symbiotic association with legume roots for nitrogen fixation. Cobalt is found as cyanocobalamin (Vitamin B_{12}) and is associated with group transfer reactions. Evidence has also been reported for the biosynthesis of cobalt-containing proteins in higher and lower plants (132). Cobalt deficiencies have been noted in nonnodulating subterranean clover grown with nitrate-N or urea and in wheat grown with nitrate-N. In each case, plants were grown under aseptic conditions. *Vanadium* has been proposed as being capable of substituting for Mo in nitrogen fixation and has been shown to replace Mo in nitrogenase, but with greatly lower activity than with Mo (132). Information is not sufficient about the mechanisms for the beneficial effects observed for *aluminum, beryllium,* and *barium.* Sparing effects or partial replacement of an essential element by these functional elements can be envisaged. A number of mechanisms were proposed for the beneficial effect of Al (81). Evidence has not been convincing that *nickel* is essential for plants, but some workers argue that it behaves like an essential nutrient in that it is ubiquitous in the biosphere, shows *in vitro* biological activity affecting certain enzymes, has a low molecular weight and has two interchangeable valencies, and is nontoxic at low levels. In studies with animals, nickel has also been found to replace Co (164).

Mineral Deficiency Effects on Growth

As would be expected, mineral deficiencies reduce plant growth. However, top or root growth are not affected equally when elements become deficient (Table 13). Deficiencies of P, Ca, S, and Zn affected top growth more than roots, deficiencies of NO_3^--N, Mg, Mn, and Cu affected roots more than tops, and deficiencies of NH_4^+-N, K, and Fe affected tops and roots about equally. These effects could be important when considering the physical ability of roots to explore a soil for nutrients. For example, Loneragan (117) and Barber (11) pointed out that a large surface area is imporatnt for the absorption of P from soils. Some plants have responded to low available P in soils by decreasing the size of their tops relative to their roots. Thus, the effectiveness of the roots to meet the plant's requirements for P should be enhanced. In some cases, mild P deficiency enhanced root growth while depressing top growth (93). On the other hand, Mg deficiency showed an opposite effect on root growth as that of P. The implications of the inability of a Mg-deficient plant to explore the soil for Mg are apparent. A Mg-deficient plant would have a much poorer chance to compete for nutrients than other plants because of its greatly restricted root growth.

Mineral Deficiency Effects on Other Elements

With mineral deficiencies, other elements tend to increase in plants because of the concentrating effect of reduced growth (140, 159, 168). Higher concentrations of elements under a mineral deficiency condition are depicted by

Table 13 Top/root ratios of maize grown with deficient levels of mineral elements (44)

Strength of nutrient in solution[a]	Mineral deficiency										
	NH_4-N	NO_3-N	K	P	Ca	Mg	S	Fe	Mn	Zn	Cu
Full	1.81	2.26	2.83	2.42	3.21	2.94	2.69	2.36	2.59	2.31	2.28
1/10	1.96	3.00	3.25	1.34	3.09	3.92	2.41	2.53	2.88	2.48	2.46
1/50	1.91	4.03	3.12	1.08	2.25	4.00	2.11	2.56	3.24	1.82	2.88
1/200	1.98	4.32	3.07	1.03	1.46	5.67	2.00	2.66	3.36	1.58	2.87
0	1.85	5.50	2.92	1.16	2.27	6.58	2.20	2.37	3.37	1.71	2.75

[a]The salt concentrations (mmoles/liter) used in the full-strength solutions were: $CaCl_2$ (1.30), $Ca(NO_3)_2$ (3.20), $(NH_4)_2SO_4$ (0.76), KH_2PO_4 (0.72), KCl (2.21), $MgSO_4$ (1.25) and Fe Citrate (0.27). The composition of micronutrients (μmoles/liter) were H_3BO_3 (23.12), $MnCl_2$ (4.57), $ZnSO_4$ (0.76), $CuSO_4$ (0.32) and H_2MoO_4 (0.14).

the bottom part of the "C curve" as shown in Figure 9. Plants that remained for a prolonged time in a mineral-deficient condition accumulated the element in deficiency even though the deficiency may not have been corrected (104).

Patterns for mineral element changes with mineral deficiencies are shown for citrus leaves in Table 14. Some of these modifications were relatively large and others were small. Some elements were affected by nearly all element deficiencies and others were affected by only a few.

Studies on the effects of varied N, P, K, Ca, Mg, S, Mn, Fe, B, Cu, and Zn in nutrient solutions on maize plants showed decreases in dry matter yields with each deficiency, except B (Table 15). Boron deficiencies could not be obtained for maize in nutrient solutions. Even though some growth depression was noted for the minus Cu treatment, no Cu deficiency symptoms were observed. Some element deficiencies had dramatic effects on dry matter yields and concentrations of other mineral elements. Table 16 shows the patterns of these modifications for the elements. Nearly every element changed with the deficiency of another element.

PLANT RESPONSES TO MINERAL EXCESSES

Like the responses of plants to mineral deficiencies, mineral excesses, if sufficiently great, will inhibit plant growth and show visual symptoms. These excess or toxicity symptoms are relatively specific, but are not as well defined as are mineral deficiency symptoms. In many cases, mineral toxicity symptoms resemble many mineral deficiency symptoms because of element interactions. In fact, many excess symptoms may very well be deficiency

Table 14   General effect of an applied element on the mineral composition of citrus leaves[a] (159)

Element added	Elements measured in leaves									Total number
	N	P	K	Ca	Mg	Cu	Zn	Mn	B	
N	+	−	−	+	+	−	−	−	0	8
P	0	+	−	+	−	−	+	+	0	7
K	−	0	+	−	−	0	0	0	−	5
Ca	0	0	−	+	0	?	?	?	?	2
Mg	0	0	−	−	+	−	+	+	0	6
Cu	0	0	+	0	0	+	−	−	0	4
Zn	0	0	+	−	−	−	+	−	0	6
Mn	0	0	+	0	−	−	0	+	0	4
B	0	−	+	−	−	0	0	−	+	6
Total No.	2	3	9	7	7	6	5	7	2	48

[a]Increased concentration in leaf is indicated by (+), decrease by (−).

Table 15 Mineral element concentrations of maize leaves of plants grown at low levels of mineral elements in nutrient solution (from 45)

Element deficiency	Dry matter yields (% of control)	Element concentration										
		%						µg/g				
		N	P	K	Ca	Mg	S	Mn	Fe	B	Cu	Zn
None[a]	100	3.35	1.26	4.95	0.62	0.37	0.30	39	118	20	14	62
NH_4^+-N	76	2.79	0.66	7.63	0.83	0.35	0.31	72	134	14	10	43
NO_3^--N	61	2.04	1.02	4.52	0.32	0.18	0.38	19	309	16	10	21
P	34	3.20	0.44	4.71	1.05	0.47	0.37	38	1968	37	21	72
K	20	4.06	2.85	1.25	2.24	0.77	0.49	92	585	51	15	92
Ca	30	3.94	3.18	6.84	0.05	0.32	0.32	81	1029	27	24	89
Mg	29	2.95	1.73	5.91	0.79	0.07	0.19	86	252	30	20	84
S	38	4.88	1.47	3.70	0.52	0.39	0.05	36	184	28	17	43
Mn	27	4.27	1.74	6.54	0.76	0.35	0.39	11	218	28	22	68
Fe	63	4.12	1.48	7.36	1.08	0.32	0.40	97	63	14	14	67
B	102	3.34	0.92	5.05	1.22	0.50	0.25	78	147	8	13	39
Cu	78	3.87	0.82	5.97	1.02	0.39	0.30	63	238	14	7	87
Zn	57	4.05	2.46	6.99	1.23	0.41	0.37	115	1813	33	21	19

[a]Separate nondeficient controls were grown with each element deficiency, because five levels of each nutrient deficiency were used. The values for the nondeficient leaves are those of the control for the element which was in deficiency; N values are for the NO_3^--N control.

Table 16 Patterns of mineral element changes in maize leaves as nutrients become deficient[a] (44)

Nutrient in deficiency	Nutrient in leaf										
	N	K	P	Ca	Mg	S	Mn	Fe	B	Cu	Zn
NH₄-N		+ + +				–				–	
NO₃-N	– – –	– – –	– – –	– – –	– – –		–		– –	– – –	– – –
K		– – –	+ +	+ + +	+ +		+		+ +		+ +
P		– –	– –	– –	+ +				+ +	+	+ + +
Ca			+ + +	– –	– –	+ + +	– – –	+ + +	+	+ + +	+ + +
Mg	– –	+	+ + +	– –	– –	– – –	– – –	– – –	+	+ + +	+ + +
S	+ +		+ + +	– –	– –	– – –				+ +	
Mn			+	– –	– –				– –		
Fe		+	+ +	– –	–	+ +	+		+	+	+ + +
B								–	–		+
Cu				– – –		+ +				– –	
Zn		+ +	+ +					+ +	+ + +		– –

[a]Significance at the .10, .05 and .01 level for – or +, – – or + +, and – – – or + + +, respectively. The minus marks (–) refer to decreases and the plus marks (+) refer to increases in nutrient content compared to plants grown in nondeficient nutrient solutions.

symptoms. Symptoms are mostly plant specific, but certain principles for mineral excesses are apparent.

Visual Symptoms of Excess Levels of Macronutrients

Visual symptoms of excess amounts of mineral elements on young sorghum plants grown in nutrient solutions have been noted (51, 52). Sorghum is a somewhat unusual plant in that under many stress conditions leaves tend to turn reddish-purple. Even under normal growth conditions, many sorghum genotypes have a high degree of red coloration compared to other plants.

Ammonium-Nitrogen. Leaves of plants given excess NH_4^+-N (NH_4NO_3) showed dark red lesions with greater intensities near the margins and fewer near the midrib. Leaves were also lighter green in color. Under severe conditions, the margins turned brown and died. Symptoms were relatively uniform over the leaf. No adverse effects were noted on roots of these plants.

Nitrate-Nitrogen. Excess nitrate [$Ca(NO_3)_2$] caused leaves to be lighter in color with reddish-purple lesions extending throughout the leaf. Interveinal tissue turned yellow-brown. Severely affected leaves had necrotic spots intermingled with the red and yellow-brown areas (Figure 10d). Excess NO_3^--N caused roots to decrease in growth, especially the secondary roots, and fibrousness near the tips increased (Figure 12a).

Phosphorus. Excess P (KH_2PO_4) caused lower leaves to have extensive dark red lesions and necrotic spots or red-speckling. The intensity of this red

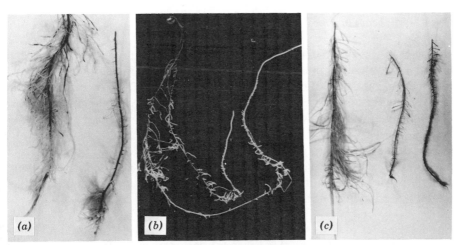

Figure 12 Symptoms of excess NO_3-N (a) (normal root on left and excess NO_3-N on right), Al (b), Cd and Ba (c) (normal root on left and excess Cd in middle and excess Ba on right) on sorghum roots (52; R. Clark, personal observation).

speckling has been found to increase with increased levels of P (83). Iron deficiency was severe in the upper leaves and appeared to some extent in the base sections of the lower leaves (Figure 10e). Roots of plants with excess P appeared normal, but were decreased in growth relative to the control.

Potassium. Leaves of plants given excess K (KCl) showed "firing" and browning of tip sections which progressed somewhat uniformly toward the leaf base. Leaves wilted or "dampened off" (Figure 13a). Severely affected leaves shrivelled, browned, and died. Symptoms of plants grown with K_2SO_4 had less severe burning, but had more reddening than plants grown with KCl. Roots of plants grown with KCl or K_2SO_4 were slightly darker and began to die.

Calcium. Leaves of plants given excess Ca ($CaCl_2$) turned blackish-brown and had light red streaks along the veins. Darker red spots also appeared. Leaves showed wilting or "dampening off" effects. Roots turned darker and were slimy. Plants grown with excess $CaSO_4$ had less severe symptoms than those observed with excess $CaCl_2$ and more reddening appeared.

Magnesium. Leaf tip sections of plants given excess Mg ($MgCl_2$) were more seriously affected than the base section of leaves. Marginal tissue turned brown and died. Interveinal tissue became lighter in color from tip to base. Roots turned dark brown and were slimy.

Sulfur. Excess S (Na_2SO_4) symptoms were not distinguishable from NH_4^+-N excess symptoms. Roots turned darker red, were slimy and considerable stubbling appeared.

Visual Symptoms of Excess Levels of Micronutrients

Manganese. Leaves of plants given excess Mn were darker green than normal leaves and small dark reddish-purple spots appeared extensively over the leaf. Margins of the leaf tip sections turned light brown and died. In some plants, excess Mn caused Fe deficiency to appear in the upper leaves. No abnormal symptoms were noted on roots.

Iron. Leaves of plants given excess Fe turned blackish-straw color from margin toward midrib before death. Less severely affected leaves were sometimes dark green with only the marginal tissue showing symptoms (Figure 13b). Roots were dark, slimy, and very red.

Boron. Leaves of plants given excess B turned straw color at the margins. The boundary between the dead and the less severely affected leaf tissue was distinct and abrupt. Otherwise, leaf tissue was dark green (Figure 13c). No abnormal effects of excess B on roots were noted.

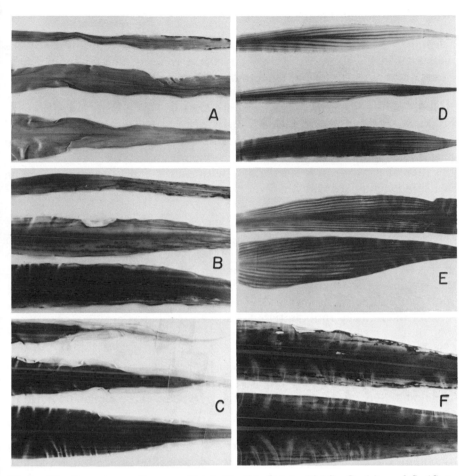

Figure 13 Symptoms of excess K (*a*), Fe (*b*), B (*c*), Al (*d*), Co (*e*), and Sr (*f*) on sorghum leaves (Clark, personal observation).

Copper. Leaves of plants given excess Cu became slightly light streaked similar to Fe deficiency, with red streaks along the margins. Roots were slightly darker red and slimy.

Zinc. Leaves with Zn excess were lighter in color with some necrotic lesions in the interveinal tissue. The lighter color spread uniformly over most of the leaf. Some "dampening off" appeared at tips and some Fe deficiency symptoms appeared in leaf material emerging from the whorl. Roots showed no deleterious effects.

Molybdenum. Symptoms of excess Mo in leaves were indistinguishable from excess Se. Roots showed no abnormal effects.

Visual Symptoms of Excess Levels of Other Elements

Selenium.   Leaves of plants given excess Se showed symptoms that were indistinguishable from Mo excess and these were similar to P deficiency. Roots showed no abnormal symptoms with excess Se.

Sodium and/or Chloride.   Leaves of plants given excess NaCl showed blackish-straw colored margins with a distinct border between affected and unaffected tissue. Leaves became lighter from the margin toward the midrib in a relatively uniform manner. Symptoms were more severe near the tips than at the base of leaves. Roots with excess NaCl turned light brown and some stubbiness was observed.

Aluminum.   Excess Al symptoms in leaves of this sorghum genotype resembled relatively severe Fe deficiency (Figure 13*d*). Other genotypes given excess Al have shown more typical P deficiency symptoms. Roots given excess Al showed marked reductions in root elongation, were thicker, darker, and very stubby (Figure 12*b*). The secondary and new auxiliary roots were affected most. Under relatively severe Al toxicity, a greater number of auxiliary roots initiated, but were unable to grow extensively.

Cobalt.   Symptoms of excess Co were similar to Fe deficiency, except that the symptoms appeared only in the leaf just emerged from the whorl or on the sheath next to the whorl and not in the leaf tip sections (Figure 13*e*). The symptoms were more diffuse than those typical of Fe deficiency. Roots showed some stubbiness.

Nickel.   Symptoms of excess Ni were also similar to Fe deficiency. These symptoms did not extend as far out toward the leaf tip as is noted for typical Fe deficiency. Roots were stubbier and showed symptoms resembling those of excess Al, but not as severe.

Barium.   Excess Ba caused dark red lesions to appear with lighter color near the margins progressing inward toward the midrib. Symptoms were more severe from the whorl toward the tip. Roots of plants given excess Ba had no secondary root lengthening and were dark in color (Figure 12*c*).

Strontium.   Leaves of plants given excess Sr became necrotic in a splotchy pattern at the margins with a lighter color appearing in the margin progressing toward the midrib (Figure 13*f*). Roots were dark red, coarse, stubby, and somewhat slimy.

Chromium.   Leaves with excess Cr turned light reddish-brown from tip toward base and from margin to midrib. Some leaves had somewhat dark reddening on tips and margins. Roots were darker and stubbier and growth was inhibited extensively.

Lead.   Leaves of plants given excess Pb were reddish-brown and had necrotic dead spots with red around them. Leaf tips were affected more than the leaf base and symptom severity progressed from margin to midrib. Roots were stubbier and had fewer auxiliary roots, but were more normal than roots grown with Cd.

Cadmium.   Leaves with excess Cd turned a fiery red from margin to midrib. Severely affected leaves became bright red over the entire leaf. Roots were dark red, small, and had no growth on secondary roots (Figure 12, right).

Mercury.   Mercury turned leaves blackish-brown with dark and necrotic lesions. Leaves became water-soaked, were leathery, and curled extensively. Leaves did not turn lighter in color. Roots were somewhat inhibited in growth, but otherwise were relatively normal.

Mineral Excess Effects on Growth

Once requirements of mineral elements for plant growth have been met, higher concentrations of these elements may be toxic and cause reductions in growth. Concentrations of the macronutrients may be relatively high before growth reductions occur. However, in the case of the micronutrients and the nonessential elements, relatively low levels can cause growth reductions. At very low levels, some of the nonessential elements have been noted to benefit plant growth (21, 81, 115). The level at which growth reduction occurs is dependent on the element and the plant species. Some plants can tolerate much higher levels of heavy metals than others. The effects of varied levels of the micronutrients and several heavy or trace metals on the growth of bush beans (*Phaseolus vulgaris* L.) are shown in Table 17. Except for FeEDDHA [Fe ethylenediamine di(o-hydroxyphenylacetate)] and Li, element levels above about $10^{-4} M$ were detrimental to plant growth. At the $10^{-5} M$ level, some elements caused greater growth reductions than others. Growth reductions in cotton (*Gossypium hirsutum* L. and *G. barbadense* L.) (143–148) were similar to those of bush beans (Table 18), but soil levels of the heavy metals showed somewhat different results from those in nutrient solutions. In the case of some elements, soil had the capacity to alter the toxicity limits. The effects of Ca and chelating agents have been shown to have pronounced buffering effects on heavy metal toxicities (see A. Wallace and colleagues 143–148, 169–176). In the case of chrysanthemum (*Chrysanthemum coronarium* L.) (135), growth reductions appeared for five of the six elements studied at $10^{-6} M$ (Figure 14). Of the elements studied, Zn gave the least effect on growth reduction and Cr, Cu, and Cd had the greatest effects.

The threshold values for toxicity of several heavy metals on lettuce seedling root elongation are shown in Table 19 (19). A wide range of threshold values was noted for the various elements tested; Cd had the low-

Table 17 Relative dry matter yields of bush bean plants grown with varied levels of micronutrients and heavy metals in nutrient solution (calculated from data of 169, 171, 173–175)

Concentration (M)	Micronutrient						Heavy metal							
	Mn	FeEDDHA	FeSO$_4$	B	Zn	Mo	Cd	Sn	Ti	Cr	V	Co	Ag	Li
	% of control (normal plants)													
Control[a]						100	100	100	100	100	100	100	100	100
10^{-6}		101			106		76	96	84	39	107	78	42	
10^{-5}	100	100	100	100	100		65	118	75	9	88	77	15	
10^{-4}	83	107	110	107	70	69	39	126	24		50			
2.0×10^{-4}			90		33	75								
2.5×10^{-4}					25									
5×10^{-4}			36	55										
10^{-3}	37	98		39		13		19						81
2×10^{-3}														75
5×10^{-3}														55
10^{-2}	17													26

[a]No heavy metals were added.

Table 18 Relative dry matter yields of cotton plants grown with varied levels of heavy metals in nutrient solution and soil (calculated from data of 143–148)

Concentration	Heavy metal (%)							
	Cu	Zn	Co	Mn	Cr	Li	Ni	Cd
Nutrient solution (M)								
Control[a]	100	100	100	100	100	100	100	100
10^{-6}	89		117		69	83		
5×10^{-6}		94						
10^{-5}	87		111	96	72	80	112	46
5×10^{-5}		140						
10^{-4}	29		25	82	20	76	7	9
5×10^{-4}		21						
10^{-3}				77				
Soil (µg/g)								
Control	100	100	100	100	100	100	100	100
25					130	114		
50					136	109		
100	104	85	31		129	23	51	35
200	93	96	15				22	15
400	60	41	8					
500				91				
1000				39				
2000				14				

[a]The control solution contained $0.97 \times 10^{-6} M$ Zn and $4.6 \times 10^{-6} M$ Mn. The other elements were not added to the control solution.

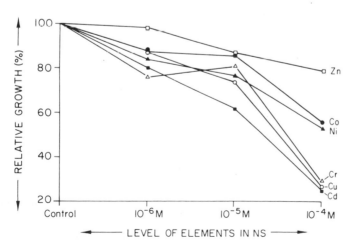

Figure 14 Relative growth in relation to different levels of Zn, Co, Ni, Cr, Cu, and Cd in nutrient solution (135).

Table 19   Toxicity levels of several heavy metals on lettuce seedling root growth (19)

Element	Threshold toxicity	50% Growth toxicity (μeq/liter)	Lethal toxicity
Cd	1.6	7	90
V	8	19	590
Co	13	62	1,500
Cu	15	16	22
Ni	19	29	5,000
Cr	42	67	2,000
Zn	63	105	2,500
Mn	3,200	5,600	27,000

est threshold toxicity value and Mn had the highest. Even though the threshold toxicity level may have been relatively low, the lethal toxicity levels for the elements were not necessarily the same magnitude of difference from the threshold levels. For example, the threshold toxicity level for Cu was 15 μeq/l, but the lethal toxicity level was only 22 μeq/l, a 1.5-fold difference. The threshold toxicity level of Ni was 19 μeq/l, but the lethal toxicity level was 5000, more than a 260-fold difference.

Plant species differ in the manner in which high levels of heavy metals affect growth. The tops of cotton decreased more than the roots at the lower levels of the heavy metals. This was reflected in decreased top/root ratios for most elements (Table 20). However, at the highest element levels added, abrupt changes in top/root ratios were noted for several of the elements. Close analysis of the dry weight data showed that at the highest element level (usually $10^{-4} M$ or higher), root weights decreased considerably more than top weights. The effects of higher levels of P on growth were opposite those noted for the heavy metals, however. Root weights of bush beans decreased extensively as P increased in solution from near 10^{-4} to $10^{-2} M$ while top weights remained nearly the same and in some instances increased (176). On the other hand, chrysanthemum showed markedly different effects from those of cotton, bush beans, and many of the other plants noted. Top (stem)/root ratios of chrysanthemum increased as Cu, Co, Cd, Ni, and Cr increased in solution from none added to $10^{-5} M$ (Table 20) (135). Root weights were markedly inhibited by these heavy metals compared to the top weights. However, at the highest level of the elements ($10^{-4} M$), both top and root weights were severely affected, with roots being affected more than tops for most of the elements. In the case of Zn, top/root ratios decreased as Zn in solution increased, indicating more inhibition of top than root growth. At $10^{-4} M$ Zn, the top/root ratio was near that of the control, thus both top and root weights decreased about equally at the highest Zn level.

Table 20   Top/root ratios of cotton and chrysanthemum plants grown with varied levels of heavy metals in nutrient solutions (135, 143, 145, 147)

Concentration (M)	Heavy metal (top/root ratio)							
	Cu	Zn	Co	Mn	Cr	Li	Ni	Cd
Cotton								
Control	4.6	4.6	4.6	4.6	5.6	5.5	6.2	6.2
10^{-6}	4.4		4.8		3.8	5.0	5.3	5.4
5×10^{-6}		4.3						
10^{-5}	3.8		3.8	5.0	4.5	4.5	4.5	3.8
5×10^{-5}		3.2						
10^{-4}	5.5		3.8	4.6	4.3	5.0	5.8	7.6
5×10^{-4}		4.6						
10^{-3}				4.2				
Chrysanthemum								
Control	4.1	1.6	6.7	0.6			3.7	—
10^{-6}	8.1	1.0	74.7	>262			12.0	24.0
10^{-5}	18.5	0.9	>10.0	>350			17.6	25.0
10^{-4}	8.6	1.3	> 1.2	> 25			>10.2	3.0

Mineral Excess Effects on Other Elements

As noted for mineral deficiencies, element excesses change the composition of other elements in the leaves and roots of plants. The effects of 21 elements in excess on the patterns of P, K, Ca, Mg, Mn, Fe, Cu, and Zn composition in sorghum leaves and roots are shown in Table 21. Excess levels of the elements altered many of the mineral elements essential to plant growth. Phosphorus changed for 11 of the elements in excess; Mn for 10; Ca and Cu for 9; Mg, Mn, and Fe for 7; and so forth for the leaves. Manganese changed for 17 of the elements in excess, K for 15, Cu for 13, P for 12, and so forth for the roots. More changes were noted for roots than for leaves. Since some elements are not translocated from roots to shoots as readily as other elements (172), the element in excess might not have as great an effect on those elements that are not translocated to the leaves as they would in the roots. It was of interest to note that P and Fe increased in almost all cases with excess amounts of the mineral elements and the other elements usually decreased.

GENETIC ADAPTATION OF PLANTS TO MINERAL STRESS

Many factors influence the adaptability of plants to mineral stress conditions. In the past, the common approach has been to change the soil to fit the

Table 21 Patterns of mineral element changes in sorghum leaves and roots when elements were in excess in nutrient solutions (from 51, 52)

Element in excess	Element affected by element excess[a]							
	P	K	Ca	Mg	Mn	Fe	Cu	Zn
Leaves								
P	++	+		+				– –
K	++	++	–	– –	–	+		
Ca	++		++		–	+		–
Mg	+	–		++	– –	+		
Mn			–		++	+		
Fe		– –			+	++	–	
B					–			
Cu			– –	–	– –		++	–
Zn				–				++
Mo	–						– –	–
Na							–	
Al			–	–	–		– –	–
Se	–						–	– –
Ba	–		–					
Sr		–	–				–	
Pb			–		–		–	
Ni	+							
Cd	+				–		–	
Co	++							
Cr	++		++			++		
Hg			++	+		++		
No. of elements to change	11	5	9	7	10	7	9	7
Roots								
P	++	+		+	– –	+	++	++
K	++	++	–	– –	–		–	
Ca	++	– –	++	– –	– –		–	
Mg	++	–	– –	++	– –		–	
Mn			–	–	++	–		
Fe		– –	– –	– –		++	– –	
B		+			– –		–	
Cu					– –		++	
Zn		+		–	–			++
Mo	+	–			++	++	– –	–
Na		– –			–			–
Al	++	–	–	–	–	++	– –	
Se						++		–
Ba		–	–		–		– –	–

122

Table 21 *Continued*

Element in excess	Element affected by element excess[a]							
	P	K	Ca	Mg	Mn	Fe	Cu	Zn
Sr	+ +	−	−			+	+	+
Pb	+	−			−			
Ni	+ +				−	+ +		
Cd	+ +	−			− −	+ +		+ +
Co					−			
Cr	+ +		+		−	− −		−
Hg	−	−	−	− −	− −	+ +	− −	− −
No. of elements to change	12	15	10	10	17	10	13	9

[a] + (increase) and − (decrease) were significant at .05 (+ or −) or .01 (+ + or − −).

needs of the plant. Epstein (70) pointed out that the usual approach to solve many problems of plants has been a combined effort to improve the environment in which the plant is to be grown and to breed or select resistant or tolerant plants. In the case of agricultural scientists interested in mineral stress problems, including salinity, the approach has relied almost exclusively on improving the environment with little attention given to the genetic approach. Alternate strategies for coping with mineral stress problems deserve attention. Selection and genetic manipulation of plants appears to hold considerable promise.

For a variety of reasons, the genetic approach of developing plants that are more mineral efficient and tolerant to low and high levels of elements and to high salinity is beginning to receive added attention. More evidence is becoming available to indicate that plants contain sufficient genetic variability for mineral element responses to make selection and breeding not only a possibility, but a reality.

Differential Mineral Element Responses in Plants

An early report suggested ". . . the possibility of breeding directly for efficiency in the use of manure and for the raising of varieties partially suited for soil types known to be deficient in some essential constituent" (94). Differential responses of maize and tomato genotypes to nitrogen and potassium were found to be inherited and the traits for the more efficient utilization of these elements were found to be partially dominant (95). Marked differences among maize inbreds for the accumulation of several elements in leaves were noted (154). Since these early reports, a number of reviews and special articles have been written on the subject of genetic control and differential responses of plant genotypes to mineral elements (3, 25, 32, 49, 68, 69, 71, 78, 81, 86, 105, 110, 113, 125, 127, 129, 166, 179).

Considerable effort has been directed toward the development or identification of plants that may overcome mineral stress problems and to note some of the properties, heritability, and genetics of plants that show these differential responses to mineral elements. Some of the early work was initiated to find and develop plants that could exclude ^{85}Sr, an element resulting from nuclear reactions toxic to plants. The early research results showed that the uptake and accumulation of many of the mineral elements in plants was genetically controlled (7, 8, 88–90, 109, 136, 158, 180). Scientists have sought to understand physiological and chemical differences in the responses of plants to various elements and to explain the abilities of plants to survive with low or to tolerate high levels of minerals (24, 25, 27–30, 32, 77–82). Salt tolerance in plants has also been a major concern for a long time, and differences among plant species and within plant species have been noted for many plants by scientists (4, 5, 16, 18, 56, 61, 66, 67, 72, 92, 99, 101–103, 106, 114, 142, 151, 153, 163). Many scientists are attempting to solve some of the mineral stress problems by selecting and developing more mineral-efficient and tolerant plants.

Heritability of Mineral Elements

Studies have shown that responses to specific mineral stress characters are heritable. However, the degree of heritability and the gene action involved for the various elements have not been widely investigated. Studies on the inheritance of N, P, and K utilization by efficient and inefficient strains of tomatoes and snapbeans differed by 44% for N, 72% for P, and 100% for K (85, 87). The levels of segregation in seedling progeny for these elements were found to be adequate for selection in breeding programs. Inheritance for Sr-Ca, Mg, K, P, Zn, Cu, B, Al-Fe, and Mn accumulations in maize leaves was associated with two or three genes acting in an additive manner (90). Later studies showed that the inheritance of P, Ca, and Mn in maize was associated with chromosome 9 (131). Copper efficiency in rye (*Secale cereale* L.) was associated with one arm of a single chromosome (91). Single gene control for the utilization of B in celery and tomato and Fe in soybean and maize was established relatively early (13, 138, 167, 178). Reciprocal approach grafts of Fe-efficient and Fe-inefficient tomatoes and soybeans located the source of Fe inefficiency to be in the rootstocks (33, 37). Numerous differential responses of plants to the various mineral elements have been noted and inheritance and genetic mechanisms involved with many of these differential responses in plants should only await investigation and reporting.

Considerable research has also been conducted with metal toxicities in plants. Aluminum tolerance in certain barley (*Hordeum vulgare* L.) and wheat has been found to be controlled by single major, dominant genes (108,

149, 150), but studies with other wheat cultivars indicated that the genes controlling Al tolerance were more complex (112). The site of Al tolerance in "Atlas 66" was found to be on chromosome 5D (139). Manganese tolerance in alfalfa was attributed to additive gene action with little or no dominance (63). The inheritance of Mn tolerance in lettuce varied from one to four genes in five genotypes studied (65). Evidence for the genetic control of Mn toxicity in plants is not as extensive as it is for Al, but evidence is sufficient for genotypic differences to Mn that genetic control of Mn accumulation should be found (55, 150). Ecotypes provide useful plants for solving heavy metal toxicity problems, particularly on mine spoil areas, but the selection within ecotypes was not as advantageous as the selection for superior ecotypes (98). Sufficient evidence is available for cereals and other plants to suggest that genetic improvement for Al and Mn tolerance as well as tolerance to low pH conditions should be achieved (55, 149, 150).

Research Approaches to Mineral Stress

Other research results suggest that mineral element uptake and use are genetically controlled in plants and that the traits are manipulable. Thus, they could be valuable for improving plants to better adapt to mineral stress conditions. A unique procedure was used to transfer a desirable Cu efficiency character from an efficient plant to an inefficient plant (91). The Cu efficiency character from rye (Cu-efficient) was transferred to triticale (*Triticum × Secale*), and in turn, the relevant rye chromatin was transferred (by translocation) to wheat (Cu-inefficient). Tissue cultures were used to produce cell lines and regenerated plants with enhanced tolerance to Al and Na (58, 130). Salt tolerance in cultures of tobacco (*Nicotiana tobacum* L.) were increased from near 800 ppm NaCl to 8200 ppm (130). Attempts to incorporate desirable traits from wild or exotic plant species, such as salt tolerance in tomato, into domesticated crop species have been investigated (70, 152). In addition, barley genotypes fed undiluted sea water survived and set seed. Three of the best genotypes studied had yields that averaged 21% of those considered good on nonsaline irrigated soils (70). Using plants with roots that can restrict the translocation of some elements to the tops and not others, and plants that can readily accumulate certain elements and not others, might be used to develop tolerant plants to toxic elements (20, 172).

Extensive studies have been conducted on differential Fe efficiency in plants. Many Fe efficient plants have been found to respond more favorably to an Fe deficiency to alleviate the stress than do Fe-inefficient plants. Plants considered to be Fe-efficient released more H-ions from their roots, released more Fe-reducing compounds from their roots, reduced more Fe at their root surfaces, and produced more organic acids (particularly citric) in their roots (30). In addition, Fe-efficient genotypes had a greater ability than inefficient genotypes to absorb more Fe from the growth medium, especially in competition with a high number of inefficient plants (1), to reduce and use more Fe from highly stable Fe^{3+}-chelates (31) and insoluble Fe^{3+}-phosphates (26), to have greater tolerances for heavy metals (38), and to have higher nitrate reductase activities (39). Many studies have shown that inefficient genotypes tend to take up and accumulate higher levels of interacting elements which could cause deficiency problems (24, 36, 40, 48). A specific example for this was for oat cultivars, one of which became Fe deficient when grown in the field, and another did not (Figure 15). The only difference between the cultivars in mineral element composition was that the one that became Fe deficient contained about two times more Ca than the other cultivar. A sorghum genotype that grew well on an acid soil contained about twice the P as a genotype that grew poorly on the same soil. When these same genotypes were grown on an alkaline soil, the genotype that had the higher P became Fe deficient whereas other genotypes grew normally and had no Fe deficiency (Figure 16). The sorghum genotype that grew well on the acid soil was also more Al-tolerant than the other genotype (84). Iron-inefficient genotypes of maize and soybean contained higher amounts of the elements like P, Ca, Zn, and Mn, elements which are known to interact with Fe (133), compared to the Fe-efficient genotypes (24, 48). These principles have been useful to help identify and understand plants that adapt to Fe stress conditions. Similar approaches could be used to identify and characterize plants that might be used to adapt to other mineral stress conditions.

International Research on Plant Adaptation to Mineral Stress

Numerous scientists have been motivated to exploit differences among plant genotypes for their adaptation to mineral stress in order to help increase food production using mineral-stressed soils. Interest in this kind of approach

Figure 15 Iron-deficient 'TAM-0312' (left) and green 'Coker 227' (right) oat cultivars grown in the field (40). (Photo courtesy of M. E. McDaniel.)

Figure 16 'SC369-3-1Jb' (left), 'PI-405107' (center), and 'NK212' (right) sorghum genotypes grown on alkaline (upper) and acid (lower) soils (36).

may be, in part, because capital in many countries is and has not been available to buy fertilizers and soil amendments which can be added at rates used in the developed countries. Many people must be fed and soils must be used in much the way they exist, with only small amounts of fertilizer and capital available to them.

Advances have been made in Brazil for the adaptation of wheat to their acid soils (157). In this program, breeders selected plants for Al tolerance and adaptation to low pH soils. In 1940, on a newly cleared acid soil, a yield trial began with three Brazilian wheat varieties and 22 of the best wheat varieties from Uruguay and Argentina. Only the Brazilian varieties yielded anything; the others were all dead one month after sowing. It is normal practice in Brazil to describe the Al tolerance for all new wheat varieties released for use by farmers. Many of the improved wheat varieties developed in other parts of the world did not perform well on Brazilian soils. Breeders of maize and sorghum in Brazil are also concerned about tolerance of genotypes to Al (6, 137, 155) and wide differences among genotypes have been observed. When grown on those acid soils, some sorghum genotypes died as seedlings while others matured and produced grain (Figure 17). Some maize hybrids were considerably more tolerant to Al than other hybrids and produced relatively good yields under the acid conditions (Figure 18). In fact, one hybrid (H-4) responded very little to added P or lime and others (H-10 and H-9) did not respond to added P if 2 tons lime/ha had been added. Aluminum made up one-third of the base saturation in the 0–20 cm soil profile and over half in the 20–40 cm soil profile (Magnavaca, personal communication). Progress has been good for the development of Al-tolerant sorghum and maize hy-

Figure 17 Differential responses of sorghum genotypes to Al tolerance (acid soil) grown on Cerrado soil, Brazil. (Photo courtesy of R. E. Schaffert, Sete Lagoas, Brazil.)

brids. Many other crop plants are being screened for tolerance to low P and Zn and high Al (122; Clark, personal observation).

Probably the greatest amount of effort for screening and breeding plant genotypes for tolerance or efficiency under mineral stresses and adaptation to problem soils is being conducted at IRRI in the Philippines (101–103). Such a program has been in progress since 1965. Rice genotypes have been screened for relative responses to deficiencies of Zn, Fe, and P, and tolerances to excessive salt, alkali, Fe, Al, and Mn, and tolerances to organic and acid sulfate soils. Several thousand rice genotypes have been screened for some of these mineral stresses. For example, as of late 1978, over 16,000 rice genotypes had been screened for their ability to grow under various mineral stresses (Ponnamperuma, personal correspondence). Nearly 10,000 rice genotypes had been screened in greenhouses for salt tolerance, nearly 8000 for alkali soil tolerance, over 9000 for zinc deficiency, and hundreds of

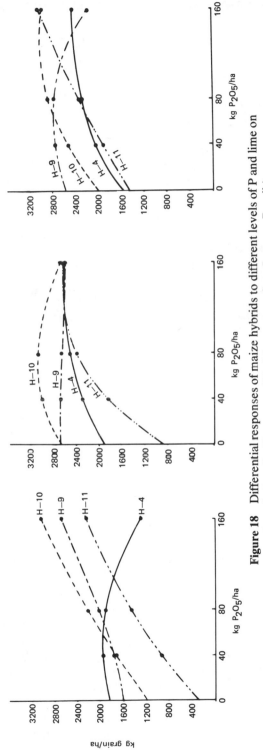

Figure 18 Differential responses of maize hybrids to different levels of P and lime on Cerrado soil, Brazil. (Courtesy of R. Magnavaca, Sete Lagoas, Brazil.)

Figure 19   Differential responses of rice genotypes to salt (103).

others for the other mineral stresses. For each mineral stress, efficient or tolerant genotypes were found. Figure 19 shows some of the differential responses of rice to salt. Few genotypes were selected compared to the number of materials tested, but these few may be the germplasm needed to overcome specific problems. Of the 13,500 rice genotypes screened for mineral stresses in 1977, only 259 showed efficiency or tolerance. Differences in relative yields of some of the plants were large, as illustrated by 18 rice genotypes in salt-screening experiments (Table 22). Yield differences of nearly 3-fold in 1977 and nearly 5-fold in 1978 were noted among rice genotypes grown on a coastal saline soil in the Philippines (Table 23). In many cases, the tolerances were found not only in initial breeding and world collection stocks, but also in advanced breeding materials. For some mineral stresses, the tolerances found by screening exceeded the tolerances of genotypes naturally or normally grown on the problem soils.

Studies have been conducted at IRRI to determine the inheritance in rice for these mineral stress responses (101–103). Although the genetic behavior of salt tolerance could not be estimated decisively, the heritability was quantitative. The recovery of salt-tolerant progeny was about 5–20% regardless of the source of tolerance or the number of topcrosses or double crosses involved. Table 24 gives data on the distribution among salt tolerance ratings of selected crosses for a particular rice genotype tested four different times. Several of the progeny from a number of crosses with tolerant genotypes had good survival rates, indicating that with systematic screening the tolerance

Table 22   Salt injury symptoms and relative grain yield of 18 rice genotypes grown on a saline soil in the Philippines (102)

Genotype	Salt injury ratings[a]	Relative grain yield (%)[b]
IR28	8.3	7
IR32	2.7	75
IR2061-465-1-5-5	4.3	38
IR2071-88-8-10	3.0	29
IR2153-26-3-5	1.7	65
IR2681-163-5-2-2-2	4.3	35
IR4432-103-6	3.3	41
Banih Kuning	2.3	76
Kalarata 1-24	1.0	98
Kuatik Putih	1.3	68
Kuatik Serai Rendah	3.0	61
Kuatik Serai	1.0	70
Lal Buchi	7.7	4
Mala Kuta	8.9	10
Merak	2.0	78
Mi Pajang	7.7	8
Mosiha Bide	8.0	5
Pulot Daeng Maradka	1.0	92

[a] 1 = normal, no adverse symptoms; 9 = almost all plants dying. Ratings made 10 weeks after transplanting to a saline soil.

[b] $\dfrac{\text{Yield on saline soil}}{\text{Yield on normal soil}} \times 100$

traits from the tolerant genotype could be introduced into other genotypes with good agronomic backgrounds. Rice genotypes with good tolerance to alkalinity and other mineral stress problems have been distributed to other locations and some of these locations have reported good tolerance to these conditions. Progress has been good for the improvement and adaptation of plants to other mineral stresses in soils and continued progress should be made with the efforts and concerns given.

Scientists in Colombia (Centro Internacional de Agricultura Tropical–CIAT) have also initiated screening and breeding programs to determine tolerances and adaptations of plants to acid, infertile, low agriculturally productive soils of the humid and subhumid American tropics (55, 161). Early trials indicated that the major limitations to cropping in these areas were low phosphorus and soil acidity factors. They found that variability among genotypes for tolerance to acidity was low for some species and high for others. Rice genotypes derived from outside of Colombia produced equally with or without lime (Figure 20, left). Cassava (*Manihot esculenta* Crantz), which has been reported to yield well under low fertility conditions, was well adapted to acid conditions and responded both negatively and positively to

Table 23 Yields of selected rice genotypes grown on coastal saline soils in the Philippines (103)[a]

	1977 Wet season			1978 Wet season	
Genotype	Yield (t/ha)	Relative to control (%)	Genotype	Yield (t/ha)	Relative to control (%)
Pokkali	2.87	100	IR9884-54-3	3.46	132
IR2071-104-5-4	2.78	97	IR2071-586-6-3	2.71	103
IR4630-22-3-3-2	2.62	91	IR4595-4-1-15	2.69	102
C4-63 G	2.58	90	BPI-76	2.69	102
IR2823-399-5-6	2.49	87	Pokkali	2.63	100
IR2153-26-3-5-2	2.38	83	C4-63 G	2.42	92
BPI-76 (NS)	2.36	82	IR4432-28-5	2.41	92
IR2071-586-5-6-3-4	2.31	80	IR4630-22-2-5-1	2.38	90
IR2071-88-8-10	2.09	73	IR9859-87-2	2.26	86
IR2153-96-1-5-3	1.86	65	IR10199-108-1	2.24	85
IR2070-820-2-3	1.63	57	IR26	2.15	82
IR30	1.33	46	IR4673-13-1-12	2.09	79
IR26	1.05	37	IR9849-80-1	2.07	79
IR28	1.00	35	IR10206-18-3	1.81	69
			IR9849-72-3	1.73	66
			M1-48	0.70	27

[a]1978 data used by permission of the International Rice Research Institute, Los Banos, Philippines.

Table 24 Distribution of salt tolerance ratings of selected IR8073 crosses tested four times (103)

Test number	Number of selections tested	Selections at the different ratings[a] (% of total tested)					
		1	2	3	5	7	9
1	67	2	3	4	24	31	36
2	164	18	3	26	19	22	12
3	116	8	1	23	39	23	5
4	262	1	1	10	15	23	50
Mean		7	2	16	24	25	26

[a]1 = almost normal plants; 9 = almost all plants dying.

added lime (Figure 20, right); yields were highest with low or no lime treatments. However, cassava was found not to tolerate high pH (above 7.8 to 8.2) and had high external phosphorus requirements (64). Apparently cassava could use phosphorus that was relatively unavailable in the soil.

Screening of plants, especially rice, has recently begun in Nigeria and Liberia (International Institute of Tropical Agriculture—IITA) for Fe and Mn toxicity and for low N, P, K, S, and Zn fertility. Striking differences among cultivars were obtained in growth, yields, and visual symptoms (100). Rice cultivars from South America and the Philippines were among the genotypes evaluated. Preliminary data indicated that genotype screening "under different low nutrient levels would be useful for African conditions and should be carried out as a priority." Scientists in India have been

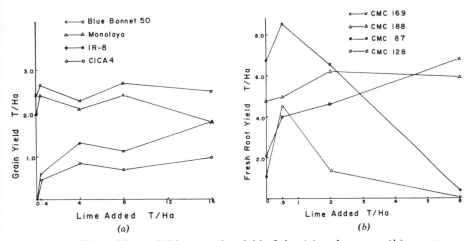

Figure 20 Effect of lime additions on the yield of rice (*a*) and cassava (*b*) genotypes in Colombia (161).

screening plants for their response to Zn deficiency, their tolerance to toxicity of Mo, B, and Se, and for sodicity (141; Seetharama, personal communication). Identifying and understanding plant responses and the ability of plants to grow with low supplies of phosphorus and micronutrients have been the objectives of scientists in Australia (116, 117) and New Zealand (41) for many years.

Other international groups are and have been studying means to improve the responses of plants to mineral stresses, but those cited above illustrate the problems of mineral stress and the belief that selection of plants can overcome many of the problems. Multidisciplinary groups of scientists are making an effort to develop plants that use mineral elements efficiently and that adapt to mineral stresses or other suboptimal soil conditions. The rising costs of energy and fertilizers and the concerns for conservation and ecology make this approach attractive and essential. This approach to solving mineral stress problems was emphasized by scientists of the stress group at the Michigan-Kettering conference entitled "Crop Productivity—Research Imperatives" (57). Three major imperatives were listed: "1) Manipulate crops or their environments in ways which avoid or reduce stress injury and increase productivity; 2) Exploit genetic potential for developing new varieties of crops resistant to environmental stress, and 3) Elucidate the basic principles of stress injury and resistance in plants, and evaluate the scope and nature of stress damage to crops." Over the past few years, increased emphasis has been made to this approach and this subject has been the object of a number of symposia and workshops (2, 105, 179).

CONCLUSION

Mineral stress problems exist in many soils throughout the world. Many of these mineral-stressed soils can be used for the production of crop plants. The usual method of making these soils more productive and useful has been to change the soil to fit the needs of the plant. With the extensive amounts of genetic variability in plants for efficiency or tolerance to unfavorable levels of mineral elements, plants should be developed to grow better on mineral-stressed soils. The solution to overcoming some of the mineral stress problems should be somewhere between the extremes of the two approaches. Mineral nutrients need to be added for plant growth or mineral toxicities need to be alleviated. Plants that are more efficient for uptake and use of nutrients, that require fewer nutrients to perform their normal functions, and that are more tolerant to toxic levels of elements could help solve some of the mineral stress problems that have previously been solved almost exclusively by trying to modify the soil. This potential appears to be a practical and feasible approach. The plant breeder, soil scientist, agronomist, and plant physiologist need to work together to achieve these goals. Hopefully the end product will be a proper balance between soil inputs and plant adaptability so that fewer soil inputs and greater utilization of the soil inputs can be implemented for the production of adequate plant resources to maintain and satisfy mankind's needs and desires.

REFERENCES

1 Ambler, J. E., and J. C. Brown. 1972. *Plant Physiol.* 50:675–680.
2 American Society of Agronomy. 1975. Symp. Am. Soc. Agron. Mtg., Knoxville, Tennessee.
3 Antonovics, J., A. D. Bradshaw, and R. G. Turner. 1971. *Adv. Ecol. Res.* 7:1–85.
4 Ayers, A. D., C. H. Wadleigh, and L. Bernstein. 1951. *Proc. Am. Soc. Hort. Sci.* 5:237–242.
5 Ayoub, A. T. 1974. *J. Agric. Sci.* 83:539–543.
6 Bahia, A. F. C., G. E. de Franca, G. V. E. Pitta, R. Magnavaca, J. F. Mendes, F. G. F. T. C. Bahia, and P. Pereira. 1976. Evaluation of maize lines and populations under high acidity conditions. Pp. 51–58 in E. Paterniani, Ed. Proc. 11th Annu. Brazilian Maize Sorghum Rev. Piracicaba, Sao Paulo, Brazil (in Portuguese).
7 Baker, D. E., R. R. Bradford, and W. I. Thomas. 1967. Accumulation of Ca, Sr, Mg, P, and Zn by genotypes of corn (*Zea mays* L.) under different soil fertility levels. Pp. 465–467 in *Symp. on Use of Isotopes in Plant Nutrition and Physiology.* Int. Atom. Eng. Ag., Vienna, Austria.
8 Baker, D. E., W. I. Thomas, and G. W. Gorsline. 1964. *Agron. J.* 56:352–355.
9 Barber, S. A. 1962. *Soil Sci.* 93:39–49.
10 Barber, S. A. 1974. *Fert. Solutions* 18:24–25.
11 Barber, S. A. 1976. Efficient fertilizer use. Pp. 13–29 in *Agronomic Research for Food*, F. L. Patterson, Ed. Am. Soc. Agron. Sp. Publ. 26, Am. Soc. Agron., Madison, Wisconsin.
12 Barrow, J. 1978. Problems of efficient fertilizer use. Pp. 37–52 in *Plant Nutrition 1978*, A. R. Ferguson, R. L. Bieleski, and I. B. Ferguson, Eds. Proc. 8th Int. Colloq. Plant Anal. Fert. Prob., Auckland, New Zealand.
13 Beadle, G. 1929. *Am. Nat.* 63:189–192.
14 Bear, F. E., and A. L. Prince. 1945. *Agron. J.* 37:210–222.
15 Beeson, K. C. 1945. *Soil Sci.* 60:9–13.
16 Berg, C. Van den. 1950. *Fourth Int. Cong. Soil Sci. Trans.* 1:411–413.
17 Berger, K. C. 1962. *J. Agric. Food Chem.* 10:178–181.
18 Bernstein, L., and A. D. Ayers. 1953. *Proc. Am. Soc. Hort. Sci.* 61:360–366.
19 Berry, W. L. 1978. Comparative toxicity of VO_3^{2-}, CrO_4^{2-}, Mn^{2+}, Co^{2+}, Ni^{2+}, Zn^{2+}, and Cd^{2+} to lettuce seedlings. Pp. 582–589 in *Environmental Chemistry and Cycling Processes*, D. C. Adriano and I. L. Brisbin, Jr., Eds. U.S. Dept. of Energy, Tech. Inf. Ctr., Washington, D.C.
20 Berry, W. L., and O. R. Lunt. 1978. Tolerance to trace elements in plants. In *Handbook of Nutrition and Food.* CRC Press, Cleveland, Ohio.
21 Bollard, E. G., and G. W. Butler. 1966. *Annu. Rev. Plant Physiol.* 17:77–112.
22 Bove, J. M., C. Bove, F. R. Whatley, and D. I. Arnon. 1963. *Z. Naturforsch.* 18:683–688.
23 Broadbent, F. E., and A. B. Carlton. 1978. Field trials with isotopically labeled nitrogen fertilizer. Pp. 1–41 in *Nitrogen in the Environment, Vol. I.* D. R. Nielsen and J. G. McDonald, Eds. Academic Press, New York.

24   Brown, J. C. 1961. *Adv. Agron.* 13:329–369.

25   Brown, J. C. 1963. *Annu. Rev. Plant Physiol.* 14:93–106.

26   Brown, J. C. 1972. *Agron. J.* 64:240–243.

27   Brown, J. C. 1976. Genetic potentials for solving problems of mineral stress: Iron deficiency and boron toxicity in alkaline soils. Pp. 83–94 in *Plant Adaptation to Mineral Stress in Problem Soils,* M. J. Wright, Ed. Cornell Univ. Agr. Exp. Sta., Ithaca, New York.

28   Brown, J. C. 1976. Screening plants for iron efficiency. Pp. 355–357 in *Plant Adaptation to Mineral Stress in Problem Soils,* M. J. Wright, Ed. Cornell Univ. Agr. Exp. Sta., Ithaca, New York.

29   Brown, J. C. 1977. Genetically controlled chemical factors involved in absorption and transport of iron by plants. Pp. 93–103 in *Advances in Chemistry Series, No. 162. Bioinorganic Chemistry*—II, K. N. Raymond, Ed. Am. Chem. Soc., Washington, D.C.

30   Brown, J. C. 1978. Physiology of plant tolerance to alkaline soils. Pp. 257–276 in *Crop Tolerance to Suboptimal Land Conditions,* G. A. Jung, Ed. Am. Soc. Agron., Madison, Wisconsin.

31   Brown, J. C., and J. E. Ambler. 1974. *Physiol. Plant.* 31:221–224.

32   Brown, J. C., J. E. Ambler, R. L. Chaney, and C. D. Foy. 1972. Differential responses of plant genotypes to micronutrients. Pp. 389–418 in *Micronutrients in Agriculture,* J. J. Mortvedt, P. M. Giordano, and W. L. Lindsay, Eds. Soil Sci. Soc. Am., Madison, Wisconsin.

33   Brown, J. C., R. L. Chaney, and J. E. Ambler. 1971. *Physiol. Plant.* 25:48–53.

34   Brown, J. C., and R. B. Clark. 1974. *Soil Sci. Soc. Am. Proc.* 38:331–333.

35   Brown, J. C., and R. B. Clark. 1977. *Plant Soil* 48:509–523.

36   Brown, J. C., R. B. Clark, and W. E. Jones. 1977. *Soil Sci. Soc. Am. J.* 41:747–750.

37   Brown, J. C., R. S. Holmes, and L. O. Tiffin. 1958. *Soil Sci.* 86:75–82.

38   Brown, J. C., and W. E. Jones. 1975. *Commun. Soil Sci. Plant Anal.* 6:421–438.

39   Brown, J. C., and W. E. Jones. 1976. *Physiol. Plant.* 38:273–277.

40   Brown, J. C., and M. E. McDaniel. 1978. *Crop Sci.* 18:551–556.

41   Butler, G. W. 1978. The importance of fertilizer and mineral nutrition research in New Zealand. Pp. 61–72 in *Plant Nutrition 1978,* A. R. Ferguson, R. L. Bielski, and I. B. Ferguson, Eds. Proc. 8th Int. Colloq., Plant Anal. Fert. Prob., Auckland, New Zealand.

42   Chapman, D. H. (Ed.) 1966. *Diagnostic Criteria for Plants and Soils.* University of California, Div. Agric. Sci., Riverside, California.

43   Childers, N. F. (Ed.). 1966. *Fruit Nutrition.* Horticultural Publications, New Brunswick, New Jersey.

44   Clark, R. B. 1970. *Agrichem. Age* 13:4, 8.

45   Clark, R. B. 1970. *Ohio Agric. Res. Dev. Center Res. Circ.* 181.

46   Clark, R. B. 1976. Plant efficiencies in the use of calcium, magnesium, and molybdenum. Pp. 175–191 in *Plant Adaptation to Mineral Stress in Problem Soils,* M. J. Wright, Ed. Cornell Univ. Agric. Exp. Sta., Ithaca, New York.

47   Clark, R. B. 1978. *Agron. J.* 70:1057–1060.

48   Clark, R. B., and J. C. Brown. 1974. *Plant Soil* 40:669–677.

49   Clark, R. B., and J. C. Brown. 1980. Role of the plant in mineral nutrition as related to breeding and genetics, in *Moving up the Yield Plateau: Obstacles and Advances,* L. S. Murphy, E. C. Doll, and L. F. Welch, Eds. Soil Sci. Soc. Am., Madison, Wisconsin. Pp. 45–70.

50   Clark, R. B., J. W. Maranville, and H. J. Gorz. 1978. Phosphorus efficiency of sorghum

grown with limited phosphorus. Pp. 93–99 in *Plant Nutrition 1978*, A. R. Ferguson, R. L. Bieleski, and I. B. Ferguson, Eds. Proc. 8th Int. Colloq., Plant Anal. Fert. Prob., Auckland, New Zealand.

51 Clark, R. B., J. W. Maranville, and P. A. Pier. 1979a. Mineral element deficiency and toxicity symptoms of sorghum. P. 28 in *Proc. 11th Grain Sorghum Res. Util. Conf.* Wichita, Kansas.

52 Clark, R. B., P. A. Pier, D. Knudson, and J. W. Maranville. 1981. Effects of trace element deficiencies and excesses on mineral nutrients in sorghum, in *Proc. Symp. Trace Metal Stress in Plants*, A. Wallace, W. L. Berry, and E. M. Romney, Eds. Los Angeles, California. *J. Plant Nutrition* 3:357–373.

53 Cock, J. H., and R. H. Howeler. 1978. The ability of cassava to grow on poor soils. Pp. 145–154 in *Crop Tolerance to Suboptimal Land Conditions*, G. A. Jung, Ed. Am. Soc. Agron., Madison, Wisconsin.

54 Deibert, E. J., R. A. Olson, and G. Rehm. 1976. *Nitrogen Balance in the Soil-Plant System under Intensive Irrigation Agriculture.* Annu. Tenn. Valley Auth. Proj. Report.

55 Devine, T. E. 1976. Genetic potentials for solving problems of soil mineral stress: Aluminum and manganese toxicities in legumes. Pp. 65–72 in *Plant Adaptation to Mineral Stress in Problem Soils*, M. J. Wright, Ed. Cornell Univ. Agric. Exp. Sta., Ithaca, New York.

56 Dewey, D. R. 1962. *Crop Sci.* 2:403–407.

57 Dilley, D. R., H. E. Heggestad, W. L. Powers, and C. J. Weiser. 1975. Environmental stress. Pp. 309–355 in *Crop Productivity—Research Imperatives, Book 1*, A. W. A. Brown, T. C. Byerly, M. Gibbs, and A. San Pietro, Eds. Mich. Agric. Exp. Sta., East Lansing, Mich. and C. F. Kettering Found., Yellow Springs, Ohio.

58 Dix, P. J. and H. E. Street. 1975. *Plant Sci. Letters* 5:231–237.

59 Donahue, R. L., R. W. Miller, and J. C. Shickluma. 1977. *An Introduction to Soils and Plant Growth.* Prentice-Hall, Englewood Cliffs, New Jersey.

60 Dudal, R. 1976. Inventory of the major soils of the world with special reference to mineral stress hazards. Pp. 3–13 in *Plant Adaptation to Mineral Stress in Problem Soils*, M. J. Wright, Ed. Cornell Univ. Agric. Exp. Sta., Ithaca, New York.

61 Dudley, J. W., and L. Powers. 1960. *J. Am. Soc. Sugar Beet Technol.* 11:97–127.

62 Dugger, W. M. 1973. Functional aspects of boron in plants. Pp. 112–129 in *Trace Elements in the Environment*, E. L. Kothny, Ed. Adv. Chem. Ser. 123. Am. Chem. Soc., Washington, D.C.

63 Dussureaux, L. 1959. *Euphytica* 8:260–265.

64 Edwards, D. G., C. J. Asher, and G. L. Wilson. 1977. Mineral nutrition of cassava and adaptation to low fertility conditions. Pp. 124–130 in *Proc. IV Symp. Int. Soc. Trop. Root Crops*, J. Cock, R. McIntyre, and M. Graham, Eds. Cali, Colombia.

65 Eenink, A. H., and G. Garretsen. 1977. *Euphytica* 26:47–53.

66 Ehlig, C. F. 1960. *Proc. Am. Soc. Hort. Sci.* 76:323–331.

67 Elzam, O. E., and E. Epstein. 1969. *Agrochimica* 13:187–195.

68 Epstein, E. 1963. Selective ion transport in plants and its genetic control. Pp. 284–298 in *Desalination Research Conference*, National Academy of Science, National Research Council Publ. 942. Natl. Acad. Sci., Natl. Res. Coun., Washington, D.C.

69 Epstein, E. 1972. Physiological genetics of plant nutrition. Pp. 325–344 in *Mineral Nutrition of Plants: Principles and Perspectives*, E. Epstein. Wiley, New York.

70 Epstein, E. 1976. Genetic potentials for solving problems of soil mineral stress: Adaptation of crops to salinity. Pp. 73–82 in *Plant Adaptation to Mineral Stress in Problem Soils*, M. J. Wright, Ed. Cornell Univ. Agr. Exp. Sta., Ithaca, New York.

71 Epstein, E., and R. L. Jefferies. 1964. *Annu. Rev. Plant Physiol.* 15:169–184.

72 Epstein, E., and J. D. Noryln. 1977. *Science* 197:249–251.

73 Evans, H. J., and G. J. Sorger. 1966. *Annu. Rev. Plant Physiol.* 17:47–76.

74 Flach, K. W. 1976. Formation, distribution and consequences of alkaline soils in agricultural development. Pp. 25–30 in *Plant Adaptation to Mineral Stress in Problem Soils*, M. J. Wright, Ed. Cornell Univ. Agric. Exp. Sta., Ithaca, New York.

75 Ferguson, A. R., R. L. Bieleski, I. B. Ferguson (Eds.). *Plant Nutrition*, 1978. Proc. 8th Int. Colloq. Plant Anal. Fert. Prob., Auckland, New Zealand.

76 Food and Agriculture Organization of the United Nations. 1974. FAO-UNESCO soil map of the world. Vol. I, Legend. UNESCO, Paris, France.

77 Foy, C. D. 1973. Manganese and plants. Pp. 51–76 in *Manganese*. National Academy of Sciences, National Research Council, Natl. Acad. Sci., Natl. Res. Coun., Washington, D.C.

78 Foy, C. D. 1974. Effects of aluminum on plant growth. Pp. 601–642 in *The Plant Root and Its Environment*, E. W. Carson, Ed. Univ. Press Virginia, Charlottesville, Virginia.

79 Foy, C. D. 1974. Effects of soil calcium availability on plant growth. Pp. 565–600 in *The Plant Root and Its Environment*, E. W. Carson, Ed. Univ. Press Virginia, Charlottesville, Virginia.

80 Foy, C. D. 1976. General principles involved in screening plants for aluminum and manganese tolerance. Pp. 255–267 in *Plant Adaptation to Mineral Stress in Problem Soils*, M. J. Wright, Ed. Cornell Univ. Agric. Exp. Sta., Ithaca, New York.

81 Foy, C. D., R. L. Chaney, and M. C. White. 1978. *Annu. Rev. Plant Physiol.* 29:511–566.

82 Foy, C. D., and A. L. Fleming. 1978. The physiology of plant tolerance to excess available aluminum and manganese in acid soils. Pp. 301–328 in *Crop Tolerance to Suboptimal Land Conditions*, G. A. Jung, Ed. Am. Soc. Agron., Madison, Wisconsin.

83 Furlani, A. M., R. B. Clark, and C. Y. Sullivan. 1978. *Agron. Abstr.* 1978:153.

84 Furlani, P. R., and R. B. Clark. 1978. *Agron. Abstr.* 1978:153.

85 Gabelman, W. H. 1976. Genetic potentials in nitrogen, phosphorus, and potassium efficiency. Pp. 205–212 in *Plant Adaptation to Mineral Stress in Problem Soils*, M. J. Wright, Ed. Cornell Univ. Agr. Exp. Sta., Ithaca, New York.

86 Gerloff, G. C. 1963. *Annu. Rev. Plant Physiol.* 14:107–124.

87 Gerloff, G. C. 1976. Plant efficiencies in the use of nitrogen, phosphorus, and potassium. Pp. 161–173 in *Plant Adaptation to Mineral Stress in Problem Soils*, M. J. Wright, Ed. Cornell Univ. Agric. Exp. Sta., Ithaca, New York.

88 Gorsline, G. W., D. E. Baker, and W. I. Thomas. 1965. *Pennsylvania State Univ. Agric. Exp. Sta. Bull.* 725.

89 Gorsline, G. W., W. I. Thomas, and D. E. Baker. 1964. *Crop Sci.* 4:207–210.

90 Gorsline, G. W., W. I. Thomas, and D. E. Baker. 1968. *Pennsylvania State Univ. Agric. Exp. Sta. Bull.* 746.

91 Graham, R. D. 1978. Plant breeding for nutritional objectives. Pp. 165–170 in *Plant Nutrition 1978*, A. R. Ferguson, R. L. Bieleski, and I. B. Ferguson, Eds. Proc. 8th Int. Colloq., Plant Anal. Fert. Prob., Auckland, New Zealand.

92 Greenway, H. 1962. *Aust. J. Biol. Sci.* 15:16–38.

93 Greenway, H., and A. Gunn. 1966. *Planta* 71:43–67.

94 Gregory, F. G., and F. Crowther. 1928. *Ann. Bot.* 42:757–770.

95 Harvey, P. H. 1939. *Genetics* 24:437–461.

96 Hewitt, E. J. 1963. The essential nutrient elements: Requirements and interactions in plants. Pp. 137–360 in *Plant Physiology, a Treatise, Vol. 3: Inorganic Nutrition of Plants*, F. C. Steward, Ed. Academic, New York.

97   Hiatt, A. J., and J. E. Leggett. 1974. Ionic interactions and antagonisms in plants. Pp. 101–134 in *The Plant Root and Its Environment,* E. W. Carson, Ed. Univ. Press Virginia, Charlottesville, Virginia.

98   Humphreys, M. O., and A. D. Bradshaw. 1976. Genetic potentials for solving problems of soil mineral stress: Heavy metal toxicities. Pp. 95–109 in *Plant Adaptation to Mineral Stress in Problem Soils,* M. J. Wright, Ed. Cornell Univ. Agr. Sta., Ithaca, New York.

99   Ikehashi, H. 1978. *Agron. Abstr.* 1978:155–156.

100 International Institute of Tropical Agriculture. 1976. Pp. 17–19 in *Int. Inst. Trop. Agric. Annu. Rep.,* Ibadan, Nigeria.

101 International Rice Research Institute. 1975. Pp. 164–170 in *Int. Rice Res. Inst. Annu. Rep.,* Los Banos, Laguna, Philippines.

102 International Rice Research Institute. 1976. Pp. 97–104, 222–226 in *Int. Rice Res. Inst. Annu. Rep.,* Los Banos, Laguna, Philippines.

103 International Rice Research Institute. 1977. Pp. 113–129 in *Int. Rice Res. Inst. Annu. Rep.,* Los Banos, Laguna, Philippines.

104 Jones, J. B., Jr. 1967. Interpretation of plant analysis for several agronomic crops. Pp. 49–58 in *Soil Testing and Plant Analysis, Part II,* G. W. Hardy, Ed. Soil Sci. Soc. Am., Madison, Wisconsin.

105 Jung, G. A. (Ed.) 1978. *Crop Tolerance to Suboptimal Land Conditions.* Am. Soc. Agron. Madison, Wisconsin.

106 Kannan, S. 1975. *Commun. Soil Sci. Plant Anal.* 6:63–69.

107 Kellogg, C. E., and A. C. Orvedal. 1969. *Adv. Agron.* 21:109–170.

108 Kerridge, P. C., and W. E. Kronstad. 1968. *Agron. J.* 60:710–711.

109 Kleese, R. A., D. C. Rasmusson, and L. H. Smith. 1968. *Crop Sci.* 8:591–593.

110 Kruckeberg, A. R. 1959. Ecological and genetic aspects of metallic ion uptake by plants and their possible relation to wood preservation. Pp. 526–536 in *Marine Boring and Fouling Organisms,* D. L. Ray, Ed. Univ. Washington Press, Seattle, Washington.

111 Kubota, J., and W. H. Alloway. 1972. Geographic distribution of trace element problems. Pp. 525–554 in *Micronutrients in agriculture,* J. J. Mortvedt, P. M. Giordano, and W. L. Lindsay, Eds. Soil Sci. Soc. Am., Madison, Wisconsin.

112 Lafever, H. M., L. G. Campbell, and C. D. Foy. 1977. *Agron. J.* 69:563–568.

113 Läuchli, A. 1976. Genotypic variation in transport. Pp. 372–393 in *Encyclopedia of Plant Physiology,* New Series, Vol. 2, Part B, U. Lüttge and M. G. Pitman, Eds. Springer-Verlag, New York.

114 Läuchli, A., and J. Wieneke. 1978. Salt relations of soybean mutants differing in salt tolerance: Distribution of ions and localization by x-ray microanalysis. Pp. 275–282 in *Plant Nutrition 1978,* A. R. Ferguson, R. L. Bieleski, and I. B. Ferguson, Eds. Proc. 8th Int. Colloq., Plant Anal. Fert. Prob., Auckland, New Zealand.

115 Lewin, J., and E. F. Bernhard. 1969. *Annu. Rev. Plant Physiol.* 20:289–304.

116 Loneragan, J. F. 1976. Plant efficiencies in the use of B, Co, Cu, Mn, and Zn. Pp. 193–203 in *Plant Adaptation to Mineral Stress in Problem Soils,* M. J. Wright, Ed. Cornell Univ. Agr. Exp. Sta., Ithaca, New York.

117 Loneragan, J. F. 1978. The physiology of plant tolerance to low phosphorus availability. Pp. 329–343 in *Crop Tolerance to Suboptimal Land Conditions,* G. A. Jung, Ed. Am. Soc. Agron., Madison, Wisconsin.

118 Loneragan, J. E., J. S. Gladstones, and W. J. Simmons. 1968. *Aust. J. Agric. Res.* 19:353–364.

119 Loneragan, J. E., and K. Snowball. 1969. *Aust. J. Agric. Res.* 20:465–478.

120 Loneragan, J. E., and K. Snowball. 1969. *Aust. J. Agric. Res.* 20:479–490.

121 Loneragan, J. E., K. Snowball, and W. J. Simmons. 1968. *Aust. J. Agric. Res.* 19:845–857.

122 Malavolta, E., and F. A. L. Amaral. 1978. Nutritional efficiency of 104 bean varieties (*Phaseolus vulgaris* L.). Pp. 313–317 in *Plant Nutrition 1978*, A. R. Ferguson, R. L. Bieleski, and I. B. Ferguson, Eds. Proc. 8th Int. Colloq., Plant Anal. Fert. Prob., Auckland, New Zealand.

123 Maranville, J. W., R. B. Clark, and W. M. Ross. 1979. Nitrogen efficiency in grain sorghum. P. 59 in *Proc. 11th Grain Sorghum Res. Util. Conf.*, Wichita, Kansas.

124 Mengel, D. B., and S. A. Barber. 1974. *Agron. J.* 66:399–402.

125 Millikin, C. R. 1961. *J. Aust. Inst. Agr. Sci.* 27:220–233.

126 Mortvedt, J. J. 1975. *Crop Soils* 27 (Aug.-Sept.):10–13.

127 Munson, R. D. 1970. Plant analysis: Varietal and other considerations. Pp. 84–104 in *Proceedings from a Symposium on Plant Analysis*. F. Greer, Ed. Int. Min. Chem. Corp., Skokie, Illinois.

128 Munson, R. D. 1974. Plant breeding and nutrient concentration or uptake: A perspective. Mimeo of talk presented at Am. Soc. Agron. Natl. Mtgs., 1974, Chicago, Illinois.

129 Myers, W. M. 1960. Genetic control of physiological processes: Consideration of differential ion uptake by plants. Pp. 201–226 in *Radioisotopes in the Biosphere*, R. S. Caldecott and C. A. Snyder, Eds. Univ. of Minnesota, Minneapolis, Minnesota.

130 Nabors, M. W., A. Daniels, L. Nadolny, and C. Brown. 1975. *Plant Sci. Lett.* 4:155–159.

131 Naismith, R. W., M. W. Johnson, and W. I. Thomas. 1974. *Crop Sci.* 14:845–849.

132 Nicholas, D. J. D. 1975. The functions of trace elements in plants. Pp. 181–198 in *Trace Elements in Soil-Plant-Animal Systems*, D. J. D. Nicholas and A. R. Egan, Eds. Academic, New York.

133 Olsen, S. R. 1972. Micronutrient interactions. Pp. 243–264 in *Micronutrients in Agriculture*, J. J. Mortvedt, P. M. Giordano, and W. L. Lindsay, Eds. Soil Sci. Soc. Am., Madison, Wisconsin.

134 Olson, R. A. 1979. Isotope studies on soil and fertilizer nitrogen, in *Proc. FAO/IAEA International Symposium on the Use of Isotopes and Radiation in Research on Soil-Plant Relationships*. Int. Atomic Energy Ag., Vienna, Austria. Pp. 3–32.

135 Patel, P. M., A. Wallace, and R. T. Mueller. 1976. *J. Am. Soc. Hort. Sci.* 101:553–556.

136 Pinkas, L. L. H., and L. H. Smith. 1966. *Plant Physiol.* 41:1471–1475.

137 Pitta, G. V. E., W. L. Trevisan, R. E. Schaffert, G. E. de Franca, and A. F. C. Bahia. 1976. Evaluation of sorghum lines under high acidity conditions—preliminary note. Pp. 553–557 in *Proc. 11th Annu. Brazilian Maize Sorghum Rev. Piracicaba*, Sao Paulo, Brazil (in Portuguese).

138 Pope, D. T., and H. M. Munger. 1953. *Am. Soc. Hort. Sci. Proc.* 61:481–486.

139 Prestes, A. M., C. F. Konzak, and J. W. Hendrix. 1975. *Agron. Abstr.* 1975:60.

140 Prevot, P., and M. Ollagnier. 1961. Law of the minimum and balanced mineral nutrition. Pp. 257–277 in *Plant Anal. Fert. Prob.*, W. Reuther, Ed. Am. Inst. Biol. Sci., Publ. 8, Washington, D.C.

141 Randhawa, N. S., and P. N. Takkar. 1976. Screening of crop varieties with respect to micronutrient stresses in India. Pp. 393–400 in *Plant Adaptation to Mineral Stress in Problem Soils*, M. J. Wright, Ed. Cornell Univ. Agric. Exp. Sta., Ithaca, New York.

142 Ratanadilok, N., V. Marcarian, and C. Schmalzel. 1978. *Agron. Abstr.* 1978:160.

143 Rehab, F. I., and A. Wallace. 1978. *Commun. Soil Sci. Plant Anal.* 9:507–518.

144 Rehab, F. I., and A. Wallace. 1978. *Commun. Soil Sci. Plant Anal.* 9:519–527.

145 Rehab, F. I., and A. Wallace. 1978. *Commun. Soil Sci. Plant Anal.* 9:637–644.

146 Rehab, F. I., and A. Wallace. 1978. *Commun. Soil Sci. Plant Anal.* 9:645–651.

147 Rehab, F. I., and A. Wallace. 1978. *Commun. Soil Sci. Plant Anal.* 9:771–778.

148 Rehab, F. I., and A. Wallace. 1978. *Commun. Soil Sci. Plant Anal.* 9:779–784.

149 Reid, D. A. 1971. Genetic control of reaction to aluminum in winter barley. Pp. 409–413 in *Barley Genetics II. Proc. 2nd Int. Barley Genetics Symposium,* R. A. Nilan, Ed. Washington State Univ. Press, Pullman, Washington.

150 Reid, D. A. 1976. Genetic potentials for solving problems of soil mineral stress: Aluminum and manganese toxicities in the cereal grains. Pp. 55–64 in *Plant Adaptation to Mineral Stress in Problem Soils,* M. J. Wright, Ed. Cornell Univ. Agric. Exp. Sta., Ithaca, New York.

151 Richards, L. A. (Ed.). 1954. *U.S. Dept. Agric. Handbook 60.* U.S. Govt. Print. Office, Washington, D.C.

152 Rick, C. M. 1972. Potential genetic resources in tomato species: Clues from observations in native habitats. Pp. 255–269 in *Genes, Enzymes and Populations,* A. M. Srb, Ed. Plenum, New York.

153 Rush, D. W., and E. Epstein. 1976. *Plant Physiol.* 75:162–166.

154 Sayre, J. D. 1955. Mineral nutrition of corn. Pp. 293–314 in *Corn and Corn Improvement,* G. F. Sprague, Ed. Academic, New York.

155 Schaffert, R. E., A. J. McCrate, W. L. Trevisan, A. Bueno, J. L. Meira, and C. L. Rhykerd. 1975. Genetic variation in *Sorghum bicolor* (L.) Moench for tolerance to high levels of exchangeable aluminum in acid soils of Brazil. Pp. 151–160 in *International Sorghum Workshop.* Univ. Puerto Rico, Mayaguez, Puerto Rico.

156 Shrift, A. 1969. *Annu. Rev. Plant Physiol.* 20:475–494.

157 Silva, A. R. da. 1976. Application of the genetic approach to wheat culture in Brazil. Pp. 223–231 in *Plant Adaptation to Mineral Stress in Problem Soils,* M. J. Wright, Ed. Cornell Univ. Agric. Exp. Sta., Ithaca, New York.

158 Smith, L. H., D. C. Rasmusson, and W. M. Myers. 1963. *Crop Sci.* 3:386–389.

159 Smith, P. F. 1962. *Annu. Rev. Plant Physiol.* 13:81–108.

160 Soil Survey Staff, Soil Conservation Service, U.S. Dept. of Agriculture. 1975. *Soil Taxonomy: A Basic System of Soil Classification for Making and Interpreting Soil Surveys.* U.S. Dept. Agric. Handbook No. 436. U.S. Govt. Printing Office, Washington, D.C.

161 Spain, J. M. 1976. Field studies on tolerance of plant species and cultivars to acid soil conditions in Colombia. Pp. 213–222 in *Plant Adaptation to Mineral Stress in Problem Soils,* M. J. Wright, Ed. Cornell Univ. Agric. Exp. Sta., Ithaca, New York.

162 Sprague, H. B. (Ed.). 1964. *Hunger Signs in Crops.* Third Edition. David McKay Co., New York.

163 Taylor, R. M., E. F. Young, Jr., and R. L. Rivera. 1975. *Crop Sci.* 15:734–735.

164 Underwood, E. J. 1971. *Trace Elements in Human and Animal Nutrition.* Academic, New York.

165 Van Wambeke, A. 1976. Formation, distribution and consequences of acid soils in agricultural development. Pp. 15–24 in *Plant Adaptation to Mineral Stress in Problem Soils,* M. J. Wright, Ed. Cornell Univ. Agric. Exp. Sta., Ithaca, New York.

166 Vose, P. B. 1963. *Herbage Abstr.* 33:1–13.

167 Wall, J. R., and C. F. Andrus. 1962. *Am. J. Bot.* 49:758–762.

168 Wallace, A. (Ed.). 1963. *Solute Uptake by Intact Plants.* Edwards Bros., Ann Arbor, Michigan.

169 Wallace, A., G. V. Alexander, and F. M. Chaudhry. 1977. *Commun. Soil Sci. Plant Anal.* 8:751–756.

170 Wallace, A., and E. F. Frolich. 1966. *Nature* 209:634.

171 Wallace, A., and E. M. Romney. 1977. *Commun. Soil Sci. Plant Anal.* 8:699–707.

172 Wallace, A., and E. M. Romney. 1977. Roots of higher plants as a barrier to translocation of some metals to shoots of plants. Pp. 370–379 in *Biological Implications of Metals in the Environment*. Hanford Life Sci. Symp., Energy Res. Dev. Admin. Ser. No. 42, Washington, D.C.

173 Wallace, A., E. M. Romney, G. V. Alexander, and J. Kinnear. 1977. *Commun. Soil Sci. Plant Anal.* 8:741–750.

174 Wallace, A., E. M. Romney, G. V. Alexander, S. M. Soufi, and P. M. Patel. 1977. *Agron. J.* 69:18–20.

175 Wallace, A., E. M. Romney, J. W. Cha, and F. M. Chaudhry, 1977. *Commun. Soil Sci. Plant Anal.* 8:773–780.

176 Wallace, A., R. T. Mueller, and G. V. Alexander. 1978. *Soil Sci.* 126:336–341.

177 Wallace, T. 1951. *The Diagnosis of Mineral Deficiencies in Plants by Visual Symptoms*. Chem. Publ. Co., New York.

178 Weiss, M. G. 1943. *Genetics* 28:253–268.

179 Wright, M. J. (Ed.). 1976. *Plant Adaptation to Mineral Stress in Problem Soils*. Cornell Univ. Agric. Exp. Sta., Ithaca, New York.

180 Young, W. I., and D. C. Rasmusson. 1966. *Agron. J.* 58:481–483.

Chapter 5

GENETIC FITTING OF CROPS TO PROBLEM SOILS

T. E. DEVINE

U.S. Department of Agriculture, Science and Education Administration, Agricultural Research, Cell Culture and Nitrogen Fixation Laboratory, Beltsville, Maryland

INTRODUCTION

Historical Perspective

Man depends on continuing supplies of fresh water for his own physiological functions. As a result, human populations have concentrated in areas of ample rainfall or river valleys that provide irrigation, such as the Nile. Consequently, domestication of crops has emphasized species adapted to soils found in such areas: that is, soils leached of excess soluble salts and of moderate or low pH. To utilize other environments for crop production there are three alternatives available to man: (*a*) alter the environment by irrigation or amend the soil with lime or fertilizers, (*b*) domesticate wild species adapted to adverse environments, or (*c*) genetically adapt crop species to adverse environments.

In many instances the first alternative is not economically feasible and it is those circumstances that concern us here. The second alternative is now being pursued with increasing vigor (137), as more intense efforts are made to obtain specialized chemicals from plant sources for industrial use in our increasingly complex society and for therapeutic use in our increasingly sophisticated medicine. The effort to domesticate jojoba, *Simmondsia* spp., is an example of an attempt to domesticate a wild species (137), and the survey of the plant kingdom to obtain products with chemotherapeutic activity against cancer is an example of the search for therapeutic plant substances (119).

The third alternative, namely adaptation of crop species to stress environments, is currently receiving considerable attention. Most species now used as crops are suited for agricultural use by virtue of their ability to accumulate significant quantities of starch, protein, or oil in reproductive or

143

vegetative tissues. Such accumulations are the result of the development, through natural selection, of growth patterns programmed as a response or preparation for environmental stress periods, such as a dry season or cold season, which occur in the life cycle of the species. Indeed then, crop species may be ideally suited for use as crops precisely because they respond to stress by the accumulation of desirable products in harvestable plant tissues.

Presumably the early domesticators of crop plants concentrated their attention on species adapted by natural selection to their locality. With the advance of trade and human migration, crop selections were tested in new environments. Those performing satisfactorily were retained and those not adapted were discarded. Until very recently, this has been the procedure used in adapting crops to mineral stress. Research has recently begun to define the specific nature of soil toxicities and mineral nutrient deficiencies. Cooperative research between plant breeders, plant physiologists, and soil scientists is beginning to identify and characterize plant responses to mineral toxicities and deficiencies, and the precise genetic control mechanisms governing these responses. It is now possible to characterize crop cultivars for mineral response and predict their adaptation to specific soils. In the past such adaptation was determined on the basis of trial and error.

At present, the immediate practical application of our knowledge of crop responses to mineral stress is in the choice of crop cultivar for planting. Examples are the choice of Fe-efficient soybean cultivars for planting on calcareous soils in the midwestern United States to avoid Fe deficiency chlorosis symptoms, and the choice of Al-tolerant wheat cultivars for the Al-toxic Cerrado soils of Brazil to reduce the effect of Al toxicity on the wheat crop.

For the future, research on genetic control mechanisms and the heritability of plant responses to soil stress should lead to the development of new crop cultivars specifically bred for adaptation to problem soils.

In this chapter I will consider the principles pertinent to the judgment of whether a breeding program is an appropriate solution to the utilization of a problem soil, the development of breeding programs, the genetic control systems that influence the choice of breeding strategies, the relationship of stress tolerance to vigor and yield, symbiotic relationships, and the progress made in breeding for edaphic tolerance. I have chosen to organize this chapter not by grouping information under crop species nor by grouping under responses to specific edaphic stresses (such as tolerance to excess Mn). Rather, I have organized this chapter on the basis of genetic systems and mechanisms with the objective of clarifying the genetic principles involved in the evolution of edaphic stress tolerance and in the development of breeding strategies for the synthesis of cultivars adapted to definable complexes of stresses. Within topics an effort was made to arrange information primarily on the basis of increasing complexity of the genetic system and increasing precision of the genetic analysis. It is hoped that this form of organization will also demonstrate more clearly the gaps in our present knowledge and provide guidelines for the research needed to develop a more comprehensive understanding of biological adaptation to edaphic stress.

Prerequisites to Development of a Breeding Program

Several assumptions or decisions are prerequisite to undertaking a breeding program for specific edaphic adaption. First, the specific soil problem or problems must be identified for the region for which the breeder is attempting to develop cultivars. Second, the problem must be sufficiently severe and sufficiently widespread to merit the investment of the breeders' resources. Third, the problem must not be amenable to alleviation more economically by other means such as fertilization or liming. Fourth, the problem must be adjudged amenable to solution through breeding. This fourth assumption requires (a) that techniques are available to assay plant response to the particular edaphic stresses, (b) that there is useful genetic variation for the plant characteristics needed either in agronomically suited cultivars or in noncultivated forms of the crop species or in related species, (c) that the character be heritable, and finally (d) that the estimated degree of improvement in adaptation (determined from the range of variation and heritability) is sufficient to be of applied use.

Thus, from the very inception of a program to develop crop cultivars to fit specific edaphic stress environments, the expertise of both soil science and plant breeding are needed for problem definition and goal development. This combination of expertise is usually developed in research teams or in individuals with strong training in both disciplines.

Development of the Breeding Program

The opportunity to capitalize on crop adaptation to defined soil stresses may become available at any stage in a breeding program. Ideally, however, such an opportunity would be most valuable at the initiation of the program. I will consider the latter case.

After the nature of the soil stress condition (nutrient deficiency, Al or Mn toxicity, saline toxicity, etc.) is defined, the breeder will test a diverse range of germplasm for response to the stress. If adequate tolerance or nutrient uptake efficiency can be found in otherwise agronomically adapted cultivars, the problem is resolved by choosing proper cultivars for planting and by the use of such germplasm in breeding cultivars for the future.

If the level of tolerance in the agronomically improved cultivars is not deemed adequate, the germplasm collection of the crop species would be screened for sources of the desired variability. If the intraspecific variation is not adequate, the more arduous task of locating sources of the desired genes in related species may be undertaken. Such genes will then have to be transferred by interspecific hybridization to crop cultivars. Usually undesirable alleles must be eliminated in this process. Mutagenesis may be employed in an effort to produce the desired variability in cases where such variation is not readily available.

In searching for desired variants in natural populations, it would be most helpful to have information on the geographical distribution of problem soils. Such information would be used to identify ecotypes that may have evolved a level of tolerance through natural selection. Unfortunately, detailed information on specific soil characteristics is often lacking, and indeed, information on the origin of plant introductions maintained in germplasm collections is sometimes so general that little can be surmised as to the edaphic adaptation of individual plant introductions. Hopefully, in the future, germplasm collection programs will provide more detailed information.

The choice of breeding methods to be used for the characteristics in question will depend upon several considerations: (a) the method of reproduction (self-pollinated vs. cross-pollinated, vegetative vs. sexual, tolerance to inbreeding, seed increase methods, etc.); (b) mode of gene action (multigenic vs. monogenic, dominant vs. recessive, heterosis, epistasis, etc.); (c) sources of tolerance available (crop cultivars, noncultivated forms of the crop species, related species, etc.); and (d) priority assigned to goal in relation to other agronomic traits (disease, insect, nematode resistance, quality factors, etc.).

GENETIC CONTROL MECHANISMS

Information on genetic control systems for plant responses to edaphic stresses is still limited and fragmentary. Several reviews are available on tolerance to edaphic stresses and the efficiency of nutrient uptake and utilization (4, 5, 11, 13, 43–48, 56, 61, 80, 88, 94, 110, 139, 148).

Interspecific Variation

Knowledge of species differences in edaphic adaptation was implicit in the selection of crop species by farmers through the centuries. With methods for more precise characterization of edaphic stress, the differences among species, both cultivated and wild, have been studied.

McCormick and Steiner (100) determined the Al tolerance of 11 tree species in solution culture. A hybrid poplar and autumn-olive (*Elaeagnus umbellata Thunb.*) were very sensitive to low Al concentrations (10–40 ppm), but European black alder [*Alnus glutinosa* (L.) Gaertn.], *Betula* spp., *Pinus* spp., and *Quercus* spp. were tolerant to higher concentrations (80–160 ppm).

Foy and Wheeler (53) surveyed 69 species of ornamental plants for Al tolerance in a greenhouse pot test using an acid Tatum subsoil adjusted to different pH levels by liming. Marked difference in tolerance were observed.

Gerloff et al. (57) surveyed native plants in Wisconsin, the United States, for selective absorption of mineral elements. Potassium content of some species was 7.0% when the average concentration of other species in the same area was only 2.0%. *Cornus canadensis* appeared to selectively exclude Mn; for example, a sample from a bog of pH 4.0 contained 149 ppm Mn vs. an average of 1061 ppm in other species from the same area. *Nemopanthus mucronata* accumulated Zn, with samples containing 300–700 ppm Zn vs. less than 50 ppm for other species on the same sites.

Nyborg (115) studied Mn deficiency of wheat, oat, and barley cultivars. Oats were most sensitive to Mn deficiency, wheat less sensitive, and barley (cultivars 'Olli' and 'Parkland') tolerant. The greater sensitivity of oats was attributed to poor ability to take up Mn rather than a higher Mn requirement.

Spain (135) reported many forage grasses and legumes have evolved in the allic soil ecosystem and have the ability to tolerate extreme soil acidity and low fertility and thus have minimal fertilizer requirements. Among the species suited to allic soils he lists *Hyparrhenia rufa*, *Melinis minutiflora*, and *Brachiara decumbens*. *Panicum maximum* also appears to tolerate extreme soil acidity but requires higher fertility than is usually found in allic soils.

McDaniel and Dunphy (101) reported that field trials at Beeville, Texas, indicated that oats were more sensitive to Fe deficiency chlorosis than the lines of wheat, rye, triticale, and barley tested. Marked differences in oat cultivars were also described.

Intraspecific Variation

Several reviews describe research on intraspecific variation for tolerance to edaphic stresses, and evidence continues to accumulate as research in the area accelerates. Some examples are considered.

Gallagher and Walsh (55) determined differential cultivar response to Mn deficiency on Irish soils for wheat, barley, rye, and oats, and reported that the effects of deficiency of available manganese in the soil may be largely overcome by choice of suitable cereal varieties. Early work by Neenan (112) with wheat cultivars grown in culture with solutions containing several levels of Al or Mn indicated cultivar differences for the individual factors in the soil acidity complex. At the higher levels of Mn all cultivars contained approximately the same level of Mn (i.e., 3000–4000 ppm), suggesting that the tolerance of the four cultivars tested was due to their ability to tolerate Mn rather than to prevent uptake in toxic proportions.

Mesdag and Slootmaker (105) tested 300 cultivars of spring and winter wheat for tolerance to highly acid soil conditions in a greenhouse test. They reported that Brazilian cultivars had the greatest tolerance to low pH and that Scandinavian, particularly Finnish, cultivars were more tolerant than Dutch cultivars. Variation in tolerance was broader among the German and French cultivars than among other European cultivars.

Hill et al. (71) found differences in wheat cultivars in susceptibility to Cu deficiency. The lower grain yield of 'Argentine IX' wheat in a greenhouse experiment under Cu deficiency stress was attributed to the ability of this cultivar to form many tillers which competed with the developing grain for Cu retranslocated from other plant parts.

Shukla and Raj (131) reported variations among eight maize genotypes in the expression of Zn deficiency symptoms when grown on a Zn-deficient soil. Genotype differences in the concentration of Zn in plant tissue were associated with shoot dry matter production.

Downton (41) demonstrated genetic variation for salt tolerance in rootstocks of avocado. He grafted a common scion 'Fuerte' to the rootstocks of three genotypes: Mexican, Guatemalan, and Zutano. Scions on the Mexican and Zutano rootstocks were less tolerant of salinity than the scions on the Guatemalan stock as measured by growth of stem diameter and biomass production. Chloride accumulated in the leaves over a 65-day period in all plants, but at a slower rate with the Guatemalan stock. Sodium accumulated in leaves of the Mexican and Zutano rootstock plants, at high rates of salt treatment, but not in the plants of Guatemalan rootstock. Flowering on the Mexican rootstock was decreased by the highest NaCl treatment (20 mM), but the Guatemalan rootstock showed no depression.

White et al. (146) described soybean cultivar variation for tolerance to Zn toxicity. The same researchers (145) made reciprocal root/shoot grafts of the Zn-tolerant cultivar 'Wye' and the Zn-susceptible cultivar 'York'. Grafts of the tolerant shoots on the susceptible rootstocks displayed a tolerant reaction. Shoot grafts of the susceptible cultivar on the tolerant root stocks behaved as susceptible plants. The leaf levels of Zn were similar in both the Wye/York and reciprocal combinations. Soybean tolerance to toxic levels of soil Zn was attributed to shoot genotype, and a factor or factors other than foliar Zn content were indicated as responsible for cultivar variation in tolerance. In a similar manner, Heenan and Carter (67) and Brown and Jones (21) used reciprocal grafts of soybean shoots and rootstocks to establish that the genotype of the rootstock had little or no control over tolerance to Mn. Tolerance was controlled by the genotype of the plant tops.

Carter et al. (28) surveyed 30 soybean genotypes for response to Mn toxicity and reported that the cultivars 'Hill', 'Dare', 'Davis', 'Delmar', and 'Bragg' were very sensitive and 'Hawkeye', 'Custer', 'Lee', and 'Amredo' were very tolerant. Their study of the parentage of the cultivars assayed did not reveal any clear parental source of Mn sensitivity. Armiger et al. (6) reported significant variation among 48 soybean cultivars from 10 maturity groups for tolerance to Al toxicity in an acid Bladen soil. Sartain and Kamprath (128) also reported differences in Al tolerance among 29 cultivars tested.

Able and MacKenzie (2) reported wide variation in salt tolerance in 50 soybean lines adapted to soils of low salinity. Detailed studies with 6 lines indicated that there were no varietal differences in chloride accumulation in the roots, but that the chloride content of leaves of intermediate and salt-susceptible lines were approximately 10- and 15-fold higher, respectively, than the tolerant lines. Salt tolerance during germination was unrelated to tolerance during later growth stages.

Ouellette and Genereux (116) reported differential response to Mn toxicity among the six potato (*Solanum tuberosum*) cultivars studied. Keser et al. (83) found variation for Al tolerance among and within 116 sugar beet populations in solution culture tests on seedling plants. More detailed study of 5 tolerant and 2 susceptible cultivars suggested that although the probability of finding tolerant genotypes is greater in certain cultivars, breeders should not disregard any agronomically desirable cultivar as a source of resistant genotypes.

Kotze (86) reported differences in Al toxicity among several apple rootstocks and suggested that rootstocks might be selected that are best adapted to growth under acid soil conditions. Foy et al. (52) reported that *Eragrostis* species and accessions within species varied significantly in tolerance to a calcareous soil (pH 7.3) as indicated by susceptibility to an Fe-related chlorosis and top growth yield.

Murray and Foy (109) tested 35 Kentucky bluegrass cultivars, 15 fineleaf fescue cultivars and 6 tall fescue cultivars for Al tolerance in an acid Al-toxic Tatum soil in a greenhouse pot experiment. They used the ratio of performance at two pH levels (4.6 and 5.7) to determine tolerance. Variation among cultivars was found in each of the three groups. For example, in the bluegrasses the top yield under Al stress ranged from 8% of growth at pH 5.7 for 'Windsor' to 91% for 'Fylking'.

Hutton et al. (76) measured Mn tolerance of several lines of the forage legume, *Macroptilium atropurpureum* in sand culture. Plant top growth showed considerable differences in Mn tolerance among lines. These investigators suggested that tolerance operated at three levels: (a) an inherently vigorous line usually grows better than a less vigorous line in the presence of Mn stress, (b) some lines resist the uptake of Mn, and (c) genotypes vary in the ability to tolerate a given Mn concentration in their tops.

Hill and Guss (70) reviewed several studies on variation for mineral accumulation in forage species.

Natural Selection for Edaphic Adaptation

Several studies have indicated that natural selection or inadvertent selection by man may alter the edaphic response of a population. Gregory and Bradshaw (62) compared populations of *Agrostis tenuis* from pastures and different mine workings with levels of metal (Cu, Ni, Zn, Pb) contamination toxic to most higher plants. In measuring tolerance by the rooting of tillers, the mine-derived populations displayed remarkable tolerance to the particular metals present in high quantities in the soils of their original habitats. The pasture populations did not show this tolerance. Tolerance was specific to the metal characterizing the site, except in the case of Ni and Zn where tolerance to one metal conferred tolerance to the other. Individual tolerances did occur together and were related to the occurrence of the two metals together in toxic quantities at the collection site. The specificity of tolerance to individual metal stresses has been challenged by Cox and Hutchinson (31).

Davies and Snaydon (33) measured the Al tolerances of populations of the grass *Anthoxanthum odoratum* derived from either high or low pH plots of the "Park Grass Experiment," at Rothamsted, and natural populations from acid soil locations. The populations from acid soil sites were more tolerant of high concentrations of Al than populations from calcareous soils. Differential liming treatments were first imposed on the plots 65 years prior to testing.

Wild and Bradshaw (147) reported that a larger number of species occupy the metalliferous and other anomalous soils of South Central Africa than comparable sites in Northern Europe. It was suggested that this difference may be due to the longer period for the operation of evolutionary processes to occur in Southern Africa, which did not experience the Pleistocene Ice Ages that eliminated whole floras in Northern Europe.

Foy et al. (51) reported that wheat cultivars developed in Ohio tended to be more tolerant to Al toxicity than cultivars developed in Indiana. Lafever and Campbell (92), after an extensive survey of wheat cultivars developed in Ohio and Indiana, concluded that Al tolerance was important in the adaptation of the older Ohio cultivars. Although some Indiana cultivars have a small percentage of Al-tolerant lines in their parentage, they lack Al tolerance. Many recently developed lines in the Ohio breeding program are derived from the crosses of the older Al-tolerant cultivars and Al-sensitive lines from Indiana and elsewhere. These new lines have been selected on limed soils where the effects of Al have been reduced and are found to lack the high degree of tolerance of the older Ohio cultivars.

Gene Action and Heritability

Determining the genetic mechanism (monogenic vs. multigenic, major gene vs. minor genes, qualitative vs. quantitative, and dominant vs. recessive) controlling a particular edaphic response is important to formulating the breeding strategy. If a trait is controlled principally by a single gene with discrete qualitative effect, it may often be readily transferred to the pertinent cultivars by backcrossing systems. If the trait is controlled by many genes, resulting in quantitative inheritance, other breeding systems may prove more efficient. In the case of a quantitatively inherited trait, the degree of heritability will be important in predicting success in breeding and in developing the suitable breeding procedures. Heritability in a broad sense refers to that portion of the phenotypically expressed variation which is due to genetic variation rather than environmental variation. In a more narrow sense, it is a measure of the degree to which a trait can be modified by selection. Estimates of the heritabilities of plant responses to edaphic stresses are being developed in a number of breeding programs.

Ratanadilok et al. (124) studied the inheritance of salt tolerance during germination and seedling growth in sorghum, *Sorghum bicolor* (L.) Moench. They concluded that data on parents, midparents, and F_2 mean values, indicated that salt tolerance was controlled by complementary gene action, incomplete dominance, and dominance or additive effects of several genes.

Cacco et al. (26) observed differences in sulfate uptake efficiency among three maize inbreds. They reported heterosis for sulfate uptake in the hybrids.

Gorsline et al. (58) studied the inheritance of strontium and calcium in corn and concluded that both Ca and Sr accumulation levels were controlled largely by genes acting in an additive manner and that breeding efforts would be successful in changing Sr and Ca contents of corn selections if deemed desirable. Gorsline et al. (59) studied the inheritance of the concentrations of P, K, Mg, Cu, B, Zn, Mn, Al, and Fe in maize leaves and grain. They reported additive gene action for ear concentrations of P, K, Mg, Cu, B, Zn, Mn, Al, and Fe. Nonadditive gene action was indicated for ear-leaf concentrations of some elements and for grain concentration of K. Ear-leaf concentrations of P, K, Mg, Cu, Zn, Mn, and Fe were not phenotypically associated with grain concentration of the same elements, suggesting that different genetic factors were involved, which act during distribution and/or deposition.

Nielsen and Barber (113) surveyed 12 corn inbred lines in solution culture and noted a 1.8- to 3.3-fold variation in maximal P influx, root weight, root length per unit of plant weight. Comparison of the inbreds and F_1 hybrids grown in the field suggested genotypic control of P uptake.

Shuman et al. (132) summarized research conducted at the Pennsylvania State University on Zn in maize. They reported that genetic control of Zn accumulation was evident in several single cross hybrids over a wide variety of environments. They concluded that hybrids could be selected for soils where high availability of P might induce plant Zn deficiency.

Dessureaux (34) studied manganese tolerance in alfalfa, *Medicago sativa* L., and concluded that tolerance was heritable and controlled by additive genes with little or no dominance. He reported (35) that reciprocal differences for manganese tolerance occurred in crosses of alfalfa plants. These differences were evident in the number of trifoliate leaves developed, dry matter production and unifoliate leaf area. Differences were generally associated with seed weight. However, reciprocal differences were not detected in visual symptom ratings. He concluded that maternal influences were unimportant in assessing the reaction of alfalfa seedlings to manganese toxicity, but that seed size may sometimes affect characters used to measure tolerance.

Kemp and t'Hart (82) and Butler (24) suggested that the ratio K:(Ca +

Mg) was important in the incidence of grass tetany hypomagnesemia. Prompted by this suggestion, Sleper et al. (133) used two diallel mating sets with six parents each to study the heritability of the accumulation of Mg, Ca, K, P, and the ratio $K:(Ca + Mg)$ in tall fescue (*Festuca arundinacia* Schreb.). They reported differences in parents and progenies for all minerals. General combining ability was more important than specific combining ability. Heritabilities indicated that breeding for higher mineral levels should be effective. The high heritability of the $K:(Ca + Mg)$ ratio suggested that progress could be made in breeding a tall fescue with low hypomagnesemia potential.

Hovin et al. (74) studied the heritability of mineral concentration in reed canarygrass (*Phalaris arundinacea* L.) using twelve parent clones and 30 two-clone crosses for 3 harvest years at two locations. All characters: N, P, K, Ca, Mg, Zn, Cu, Mn, and the ratio $K:(Ca + Mg)$ showed general combining ability, but only Mn displayed specific combining ability. Additive genetic varience was found for all elements but N, and nonadditive genetic variance was observed only for Mn concentration.

Halevy (64) studied the growth of two cotton cultivars differing in tolerance to K deficiency. He reported that the cultivar 'Acala 1517-C' had difficulty taking up K from the soil, because of its smaller root system and lack of sufficient fine roots, when K demand for developing flowers and fruits was high. The total K uptake by 'Acala 1517-C' was 164 kg/ha compared with 185 kg/ha in 'Acala 4-42'. Yield of lint and seed however was higher for 'Acala 1517-C'. The author suggested that weather conditions can interact with K metabolism. When the environment is favorable for early and rapid flowering and nonshedding of bolls, there is a heavy demand for K within a short time interval. In K-deficient soil the demand will not be satisfied and leaf browning will occur, bolls will not be filled, and yield will be low. If conditions favor more gradual and prolonged boll-setting there will not be a heavy demand for K for boll development during a short interval. Leaf browning will be delayed, and the bolls will be fully developed and yield will be good in spite of leaf browning.

Peedin and McCants (118) described the differential development of symptoms of Ca deficiency in the tobacco cultivars 'Coker 254' (susceptible) and 'North Carolina 2326' (slightly susceptible). Calcium concentrations of the apical tissues in 'C-254' were substantially lower than those in 'NC 2326'. In field studies, symptoms were less pronounced at locations subjected to moisture stress during vegetative growth, indicating that Ca deficiency is more apt to develop on immature plants that undergo periods of rapid growth.

Foy and Gerloff (50) selected a strain of the alga *Chlorella pyrenoidosa* for tolerance to Al by subjecting the original strain to increasing Al stress in culture. Even though the original strain was killed by an Al concentration of 12 ppm, the selected strain showed no decrease in yield when cultured with 48 ppm Al at a pH 4.6.

Meredith (103) demonstrated that differences in cultivar tolerance to Al in tomato were expressed in callus cultures with 200 and 400 μM Al. These results indicated that the expression of tolerance was not exclusively a whole-plant phenomenon. Meredith (104) also selected Al-resistant variants from cultured tomato cells by subjecting callus and plated suspension cultures to 200 μM Al for several months. Growth of the variants was vigorous in the presence of Al but not as vigorous as in the absence of Al. In the absence of continued selection pressure for Al tolerance, most of the variants retained tolerance but several lost tolerance. The origin of the variants was attributed to genetic mutation, but the possibility of an epigenetic basis was not ruled out.

Croughan et al. (32) selected a salt-tolerant line of alfalfa cells by repeated subculturing of cells on 1% NaCl media. The selected line grew better than the unselected control culture in media with high levels of NaCl. The selected line performed poorly in the absence of NaCl and appeared to require a substantial amount of NaCl for its growth. Salt tolerance appeared to be stable under maintained selection pressure in culture. In this, as in many other programs, regeneration of plants will be necessary to determine the stability and heritability of tolerance through seed propagation.

Cell and tissue culture techniques are potentially powerful selection techniques which may prove valuable when integrated into long term crop improvement programs.

As indicated by these reports, general statements regarding the heritability of response to edaphic stresses would be hazardous. Evidence to date indicates that each case, whether it be tolerance to a specific mineral toxicity or deficiency or nutrient utilization, should be assessed without preconceptions.

Genes with Major Effect

Several genes with major effect on edaphic responses have been reported. In a soybean introduction from northeastern China, Weiss (144) found a recessive gene, designated *fe*, which in homozygous mode conditions low efficiency in iron utilization. These plants are deficient in the ability to reduce Fe^{3+} to Fe^{2+} at the root surface. Bernard and Howell (10) determined the existence of a major gene pair, designated *Np np*, controlling sensitivity to excess phosphorus in soybeans. Three phenotypes were distinguishable: *Np Np* plants with no symptoms to slight brown foliar splotching, *Np np* with slight to moderate splotching, and *np np* with severe splotching, stunting, and chlorosis. Abel (2) established the existence of a major gene pair in soybeans, *Ncl ncl*, controlling chloride exclusion from plant tops. Chloride exclusion was completely dominant. Chloride-"including" genotypes contained over 7000 ppm chloride whereas chloride-excluding genotypes usually contained less than 1000 ppm chloride. Bell et al. (9) reported that the yellow-stripe mutant in corn, ys_1, was unable to efficiently utilize ferric iron supplied to the roots.

Rhue et al. (126) reported the F_1 hybrids of crosses of maize inbreds differing in Al tolerance were equal to or superior to the more tolerant parent. F_2 and backcrosses segregated into two distinct classes with 3:1 and 1:1 ratios, respectively, for tolerant:sensitive plants. Selfed progeny from a composite also segregated into distinct classes. The wide range in tolerance among genotypes and the segregation into two distinct classes suggested that Al tolerance was controlled at a single locus by a multiple allelic series. Screening of reciprocal crosses and inbred lines transferred to various cytoplasms showed no evidence of cytoplasmic inheritance of Al tolerance in corn.

Reid (125) identified a dominant gene, designated Alp, conferring tolerance to Al toxicity in winter barley. Both the cultivars 'Dayton' and 'Smooth Awn 86' carry this gene.

Prestes et al. (123) identified an Al tolerance factor in 'Atlas 66' wheat and determined, through the use of chromosome substitution lines, that this factor was located on chromosome 5D. However, Lafever and Campbell (92) reported that while their genetic data indicated that Al sensitivity was conditioned by a single recessive gene and selection for Al sensitivity was effective, that selection for tolerance or intermediate response was less effective, suggesting that inheritance of tolerance was more complex than a single gene. Lafever et al. (93) suggested that the cultivars 'Atlas 66', 'Blueboy', and 'Pennoll' represent alternative germplasm pools and perhaps different genes for Al tolerance. Henning (68) attributed the differential Al tolerance of the roots of wheat cultivars to the molecular construction of the plasmalemma of the root tip cells rather than the structural features of the root tip. He reported that the cultivar 'Atlas 66' required 100 to 200 times as much Al as the cultivar 'Brevor' before Al could penetrate the plasmalemma of meristematic cells. Once inside the cell, Al was equally harmful to either the most sensitive or most tolerant wheat cultivar.

As in the case of heritability, the influence of genes with major effect in control of edaphic responses needs to be determined for each case individually. Further, it should be borne in mind that often, even after major genes with major effect have been identified, modifying genes of lesser, but important, effect are detected. Thus the intensity of expression of a trait may be increased by selecting for modifying genes or genetic background effects to complement the genes with major effect.

Organization and Interrelation of Genetic Control Elements Governing Edaphic Responses

The organization, in the chromosome complement or cytoplasm, of the genetic material controlling edaphic response is important in predicting the behavior of these genes in segregating populations (90). For example, if a breeder desires to separate a gene for sensitivity to a particular mineral stress from genes controlling salinity tolerance, the existence of genetic linkage and the tightness of the linkage will determine the size of the population of progeny which must be generated to recover the desired recombinants.

Information is now beginning to develop on how the genes controlling tolerance to mineral stress are organized in the chromosome complement.

Mugwira et al. (107) studied several lines of triticale, the wheat-rye hybrid, along with wheat and rye cultivars for response to lime and nutrient uptake. They concluded that the Ca, Mg, Al, Mn, and P compositions of the lines suggested that the genetic control of nutrient uptake by triticale was inherited from wheat, while plant growth response suggested that tolerance to soil acidity was induced by a rye genome in triticale.

Mesdag et al. (106) reported a positive correlation (.39) of higher protein kernel content with higher tolerance to soil acidity in a study of 121 F_4 wheat lines. They concluded that, in the material studied, the two characteristics are genetically different but linked to some extent.

Slootmaker (134) tested several wheat-related species and triticale for acid soil tolerance using a 3:1 mixture of acid soil and peat, amended with sulfuric acid to create two levels of stress for differentiating tolerance. He concluded that the A genome of *Triticum* species contributes to tolerance discernable at the lower stress level and that the D genome carries one or more genes that contribute to tolerance to high soil acidity in hexaploid wheats. He suggested that the increased geographic adaptation of the hexaploid *Triticum aestivum* over the lower ploidy level wheats is partially due to increased tolerance to high soil acidity. The even wider adaptation of triticale was attributed to the addition of the R genome from rye.

Lopez-Benitez (97) in a study of Al tolerance of several triticale crosses and their parents in solution culture, reported that the wheat parent exercised the dominant influence on the response of the derived triticale. Crosses between a sensitive wheat and a sensitive rye produced amphidiploids with a level of tolerance similar to that of the least sensitive parent. In the tolerant wheat × sensitive rye cross, the amphidiploid was equal in Al tolerance to the parental wheat. The amphidiploid derivative of the sensitive wheat × tolerant rye cross was less tolerant than the rye parent and equal to tolerant wheat lines. Crosses of tolerant wheat × tolerant rye produced amphidiploids twice as tolerant as the tolerant wheat parent but not as tolerant as the rye parent. He concluded that doubling the chromosomes of F_1 hybrids had no effect on the reaction of Al of amphidiploids with respect to their parent and F_1 hybrids.

It is of interest that there are several reports of clustering or grouping on the same chromosome of genes controlling edaphic response. Novick and Roth (114) found that penicillinase plasmids, a series of extrachromosomal resistance factors in the bacteria *Staphylococcus aureus,* carried determinants of resistance to a series of inorganic ions in addition to drug resistance. Separate genetic loci for resistance to arsenite, lead, cadmium, mercuric, and bismuth ions were demonstrated. Resistance to antimony and zinc were also found but were not separated genetically from resistance to arsenite and cadmium, respectively. The ion resistance markers were located in a cluster on the plasmid.

Eenink and Garretsen (42) studied the inheritance of tolerance to high exchangeable Mn in steam-sterilized soils in the lettuce species *Lactuca sativa* and *L. serriola*. From intercrosses of the sensitive cultivars 'Neptune' and the insensitive cultivars 'Plenos', 'Troppo' and 'Celtuce', of *L. sativa* and an insensitive accession of *L. serriola*, they concluded that different numbers of genes for insensitivity were present in the parents; one gene in 'Plenos' and 'Troppo', two genes in *L. serriola*, and possibly four genes in 'Celtuce'. Three of the loci were linked, and a genetic map was developed for these genes.

Barber (8) reported that studies with reciprocal translocations in maize indicated that accumulation potentials for many elements were affected by loci on chromosome 9. In further research Naismith et al. (111) used marker genes and translocations with supernumerary chromosomes to study the location on the chromosomes of genetic loci controlling accumulation of Ca, P, and Mn in the middle leaves of maize. Data for chromosomes 1 and 8 indicated the absence of loci controlling these three elements. Chromosome 9, however, carried loci influencing accumulation of all three elements. Loci affecting Ca and P accumulation were located on both the short and long arms of chromosome 9. Loci influencing Mn accumulation appeared to be located distal on the long arm of chromosome 9. Linkage intensity units were calculated.

Little information is available on the effect of polyploidy per se on response to soil stresses. Cacco et al. (25) determined the efficiency of root uptake of sulfate and potassium at different ploidy levels in wheat ($2n$, $4n$, $6n$), sugar beet ($2n$, $3n$, $4n$) and tomato ($2n$, $4n$). In wheat and sugar beet, efficiency increased with increasing ploidy level, but in tomato, efficiency decreased with increased ploidy.

Brown and Devine (16) found differences between reciprocal crosses of soybean lines with contrasting tolerances to excess Mn and suggested that cytoplasmic inheritance, in addition to chromosomal inheritance, may play a role in Mn tolerance.

Correlated Response

When a plant characteristic is modified in a population by shifting gene frequencies through selection or in a breeding line by the introduction of new genetic traits, there is a possibility of unintentionally altering other characteristics affecting crop performance. Such changes may result from pleiotropic responses, genetic linkage, or coherence. A pleiotropic response occurs when a single gene produces two or more phenotypic characteristics. Genetic linkage results in two or more genetic loci on the same chromosome being passed to progeny together more frequently than would be expected if they were associated on different chromosomes. The coherence phenomenon, which has been reported to occur in progeny populations of crosses between distantly related taxa, is expressed in the polarization of the phenotypes of the segregating progeny toward the parental types, rather than fitting the expectations of random segregation. Fragmentary evidence of the interrelationship of edaphic responses is available, but much more research is needed to clarify the operation of these genetic control systems.

An apparent example of a pleiotropic effect occurs in tomato. Andrus (3) found a sublethal mutant in the tomato stock T3238 which produced a "brittle-stem" phenotype. This gene was assigned the gene symbol *btl*. Later Wall and Andrus (141) determined that the brittle stem phenotype resulted from the inefficient transport of boron from the roots to above ground portions of the plant. The threshold concentration of boron required for expression of the brittle stem phenotype varied with light duration and intensity, temperature and soil fertility level. Wann and Hills (142, 143) determined that the gene *fer* in tomatoes which produces an Fe-inefficient response, was not linked to the brittle stem gene.

Brown et al. (14) characterized the Fe inefficient response of the *fer* mutant of tomato. Reciprocal approach grafts of the Fe-efficient and inefficient scions and rootstocks indicated that the Fe response mechanism was controlled by the genotype of the rootstock. Both Fe-efficient and Fe-inefficient tops grafted to Fe-inefficient rootstocks developed Fe deficiency chlorosis symptoms. On the rootstocks of Fe-efficient plants, both the Fe-efficient and inefficient tops were free of chlorosis symptoms. The normal tomato genotype *(Fer)* responded to Fe deficiency stress by releasing hydrogen ions from its roots, increasing reduction of Fe^{3+} to Fe^{2+} at its roots, and increasing the citrate concentrations in its roots. In contrast, the Fe-inefficient mutant *(fer)* displayed little response to Fe stress. Brown and Ambler (12) reported that Fe deficiency stress prompted development of lateral roots in susceptible tomatoes. The association and physiological relationship of these responses suggest pleiotropic control.

Brown and McDaniel (22) reported an interaction of response to Fe, Zn, Cu, Ca, and P in two cultivars of the oat, *Avena byzantina* C. Koch. The Fe-efficient cultivar 'Coker 227' reduced Fe^{3+} to Fe^{2+} at the root and was lower in Ca concentration than the Fe-inefficient cultivar 'TAM 0-312'. These authors suggested that Ca competes with or inactivates Fe in 'TAM 0-312', causing Fe chlorosis.

Brown and Jones (17) studied four Mn-inefficient and four Mn-efficient oat cultivars (*Avena sativa* L.). The Ca concentration in the tops of the Mn-efficient cultivars was consistently greater than that in the Mn-inefficient cultivars. They suggested that Ca may physically substitute for Mn at nonspecific sites in the plant thus making more Mn available for chemical reactions specifically requiring Mn.

In a study of differential maize susceptibility to Zn deficiency (127), susceptibility to Zn deficiency was not related to Zn absorption, but to the P and Fe absorption and transport mechanism. Brown et al. (15) reported that under Fe deficiency stress conditions the most P-efficient sorghum genotypes developed Fe chlorosis, whereas the most P-inefficient plants remained green.

Lafever and Campbell (92) pointed out that the Al response gene of 'Atlas 66' wheat, located on chromosome 5D, is on the same chromosome reported to be responsible for the association between leaf rust resistance and high protein in 'Atlas 66'.

Kudelova et al. (89) suggested that genetic resistance to plant pathogens may affect mineral uptake by plants. After inoculation with bacterial wilt, *Corynebacterium insidiosum* (McCull) H. L. Jens., the wilt-resistant alfalfa, *Medicago sativa* L., cultivar 'Hodoninka' was only 11% lower than the uninoculated control in ^{54}Mn uptake whereas the susceptible cultivar 'Prerovka' was 93% lower. The uptake of ^{65}Zn was 53% lower in the resistant cultivar but 17% higher in the susceptible variety.

Vose and Randall (140) reported that in a survey of ryegrass cultivars, resistance to toxic levels of both Al and Mn were associated with low cation exchange capacity of the roots. They suggested that this association may be an extension of the tendency for plants of low fertility requirements to have lower cation-exchange capacity.

Hutton et al. (76) reported that studies of *Macroptilium atropurpureum* indicated that high Mn tolerance was found in some selections made only for characters associated with yield.

Hill et al. (71) found that wheat cultivars differed in susceptibility to Cu deficiency. The lower grain yield of Argentine IX wheat in a greenhouse experiment under Cu deficiency stress was attributed to the ability of this cultivar to form many tillars which competed with the developing grain for retranslocated Cu.

Bruetsch and Estes (23) studied 12 corn genotypes of varying maturity for variation in uptake N, P, K, Ca, Fe, Mg, and Zn in field tests in New Hampshire, USA. Significant variation was found for dry matter yield and accumulation of P, K, Ca, Mg, Fe, and Zn. Foliage P (concentration and content) was positively correlated with percent dry matter, indicating that earlier maturing genotypes had higher P levels. The efficiency of dry matter production per unit of P absorbed was greater in the later maturing hybrids. Genetic differences were also found for dry weight production per gram of N, P, K, Ca, and Mg, absorbed.

Tan and Rajaratnam (136) studied genetic variation in oil palm, *Elaeis guineensis,* leaves for N, P, K, Mg, Ca, B, and Cu and suggested on the basis of phenotypic and genotypic correlations that selection for increased N concentration would also increase P and Cu concentrations.

Hovin et al. (74) studied forage yield and N, P, K, Ca, Mg, Zn, Cu, and Mn concentrations in reed canarygrass. Concentrations of major elements were negatively correlated with forage yield in the parent plants, and N and P were negatively correlated with yield in the progeny. Phenotypic correlations between micronutrients were small, but some correlations between macronutrients, and macronutrients with micronutrients were significant.

Foy et al. (49) reported that 'Atlas 66' wheat (*Triticum aestivum* L.) was more tolerant to excess Al, but less tolerant to excess Mn than 'Monon' wheat. They concluded that tolerance to one acid soil factor in a given genotype does not necessarily mean tolerance to another. However, Cox and Hutchinson (31) reported that a population of *Deschampsia cespitosa* (L.) Beeauv, derived from the metal-contaminated barren lands near the Coniston smelter in Ontario, Canada, was tolerant not only to Ni, Cu, and Al which were present in elevated amounts at the collection site, but also to Pb and Zn even though these metals were not present at elevated levels in the soil from which the population was derived. They suggested the existence of some common physiological mechanism of metal tolerance by which the tolerance mechanism for one metal may confer tolerance to one or more other metals not exerting a selection pressure for tolerance in the environment of the plant ecotype.

PRINCIPLES FOR SELECTION TECHNIQUES

Fundamental to the development of crop cultivars for tolerance to edaphic stresses is the development of techniques and procedures for assaying plant response to the pertinent stress factor. Ideally the assay procedures should have several characteristics.

The assay should correctly simulate the appropriate stress. The area in which the crop is to be grown should be accurately characterized for the significant stress factor or factors. For example, is the problem one of pH per se, or Al toxicity or Mn toxicity, or a combination of these? Because of the very complex interaction of pH, mineral solubility, and the complexing of mineral constituents in nutrient media, care must be taken to assure that the intended stress is actually being imposed.

The assay procedure should provide values which are both accurate and precise. Fluctuations in values due to environmental effects should be minimized. The assay should characterize the genotype of the zygote being tested rather than the residual maternal effects due to the genotype of seed bearing parents or the seed production environment. This will be particularly important in early seedling test.

Devine (36) tested the effect of seed source on the expression of soybean seedling tolerance to Al toxicity in an acid Al-toxic Tatum soil. The comparison of seedlings from seed of the cultivars 'Kent' and 'Dare' from sources in Maryland, Virginia, Oklahoma, and Arkansas indicated that the environmental effects on seed production did not mask genetic expression of tolerance by the seedling. However, in analyzing plant introductions where variation in seed size is much greater than that encountered in commercial lots of U.S. cultivars, Devine et al. (39) found seed weight was significantly correlated with seedling tolerance. Hanson and Kamprath (66) reported that one cycle of selection for Al tolerance resulted in a 21% increase in seed weight and heavier roots.

The test environment should simulate the stress environment in which it is ultimately intended to grow the crop. There is evidence that prolonged selection under the artificial conditions of the greenhouse or laboratory may result in loss of fitness for field adaptation. This may result from the relaxation of selection for tolerance to disease, insect or nematode damage, or a change in the selection pressure for physiological responses such as winter-hardiness, date of flowering, rate of regrowth after cutting, and so forth.

It is desirable to use an assay technique that provides the maximum expression of genetic variation without undesirable genotype by environment interactions. For example, the level of Ca in a nutrient solution designed to induce Al toxicity stress will markedly affect the severity of root stunting symptoms. It is important to know whether the genotypes selected for tolerance at the level of Ca which permits maximum expression of genetic variation for tolerance would be essentially the same genotypes selected at other Ca levels, and if there is an important difference, which set of genotypes would express the desired tolerance in the crop production environment.

The assay procedure should be rapid and inexpensive in order to permit a large number of genotypes to be tested. This would be especially important in the case of the cross-pollinated polyploid crop species, such as the perennial forage species, which are especially subject to inbreeding depression. In these cases, if the gene frequency in the populations for resistance is low, several thousand or tens of thousands of individuals may have to be screened to obtain sufficient resistant plants for intercrossing in a phenotypic recurrent mass selection program. The cost of various alternative techniques may vary considerably from program to program. Complex laboratory equipment may be difficult to obtain and operate in some locations, and research assistants to perform detailed plant measurements may be more readily available in some locations than others.

Usually, for many characteristics, the seedling stage is the most efficient stage for screening very large numbers of plants. Care must be taken to assure that the seedling response is also predictive of mature plant response for the character in question.

The development of screening techniques for determining plant response to soil stresses is an active area of research. Techniques vary with the species under study and with the particular stress or complex of stresses being imposed. Current literature provides information on techniques now in use. Some examples are useful in illustrating important principles.

Sartain and Kamprath (128) studied Al tolerance of several soybean cultivars in short-term (48-hour exposure to Al) solution culture and long-term (growth to full bloom) soil culture in the greenhouse. In solution culture studies the cultivars 'Lee', 'Lee 68', 'York', 'Dare', and 'Ogden' were Al tolerant. The cultivars 'Lee', 'Bragg', 'Picket 71', and 'York' were more tolerant in the soil studies. These investigators concluded that the short-term solution culture studies accounted only for the effects of Al on cell elongation and cell division, while the soil growth studies reflected the continued effect of Al on top growth, root growth, and nutrient uptake.

Howeler and Cadavid (73) used the relative rootlength of plants grown in nutrient solution at 30 and 3 ppm Al to assay for Al tolerance in rice (*Oryza sativa* L.) cultivars. For the 240 cultivars studied, a good relationship was indicated between the laboratory assay and field performance by the correlation of relative rootlength values and grain yield in the field with low lime. Al-tolerant cultivars had higher levels of P and Ca and lower levels of Al in their shoots than Al-susceptible cultivars. Hypomagnesemic tetany or grass tetany occurs in grazing ruminants pasturing on cool season grass pastures. It is often associated with low forage Mg concentration and high K concentration. Haaland et al. (63) suggested that if plants with heritability for high Mg and low K could be identified, tall fescue could be bred for low grass tetany potential. They reported screening seedlings of tall fescue for Mg and K concentrations at 2 and 21% rhizosphere O_2 in sand culture. Eight genotypes selected to represent a range in Mg and K levels at low O_2 were evaluated as vegetative propagules in soil for Mg and K concentrations at 2 and 21% rhizosphere O_2. Low Mg genotypes remained low in Mg, but K levels of all selections were high in soil and the seedling evaluation of K was not, therefore, considered a reliable indicator of mature plant response.

McLean and Gilbert (102) used the hematoxylin stain technique to determine that Al absorbed by corn roots was accumulated in the cortex, mainly in the protoplasm and concentrated in the nuclei. Polle et al. (120, 121) described laboratory techniques for screening wheat and maize for Al tolerance. These methods require an assay of rootlength in stress or nonstress solution culture, or exposing the seedlings to the stress in solution culture and then staining with hematoxylin. The degree of staining indicated Al susceptibility. The staining pattern correlated with Al tolerance expressed as root elongation and performance in field trials under Al stress (122). Seedlings evaluated by staining remain viable and may be transplanted and grown to maturity permitting this method to be used for screening segregating populations.

SELECTION FOR DIFFERENTIAL SOIL FERTILITY LEVEL

Several studies have been conducted on the influence of the level of soil fertility on the heritability and the effectiveness of selection for yield and other agronomic traits. Gotoh and Osanai (60) studied the progress in selection in progeny from the cross of two winter wheat cultivars under one-half, standard, and double the standard level of applied fertilizer. After two generations of selection under the three fertility levels, the selected F_4 lines were evaluated under each fertility level. Results indicated that superior yielding lines were obtained more frequently from the population selected at the lowest fertility level. The lines selected at the lowest fertility level had the widest range of adaptation over the three fertility environments. Heritability (calculated by regression of F_3 lines on F_2 plants) was highest with selection under the lowest fertility management.

In contrast to the results of Gotah and Osanai (60), Krull et al. (87)

analyzed data from locations in the Near East, Mexico, and Columbia for 25 spring wheat cultivars. They reported that in the four most productive and four least productive environments the highest yielding cultivars were the same. They concluded that in evaluating wheat lines for grain yield, test sites with optimum management and high fertility should be used because the highest yielding cultivars were selected under this management.

Frey (54) initiated selection in a population of 1200 F_2 oat plants, derived from four crosses, under fertility stress and nonstress environments. The stress area was an unfertile droughty outcrop resulting from erosion of the topsoil from a knoll. The nonstress area, 15 m (50 f) away, at the foot of the knoll, had good moisture retention and high fertility. Selection for vigor, proper maturity and height, and desirable appearance was practiced in the F_2 and F_3 generation. The F_4, F_5, and F_6 generations were selected for yield. During the early cycles of selection fewer lines were eliminated from the stressed population because of the inability of the environment to elicit as good a differential response from the oat plants and lines as the nonstress environment. The heritability for grain yield was higher in the nonstress environment than in the stress environment. Selection in the nonstress environment resulted in the retention of oat strains with a wide adaptation reaction whereas selection in the stress environment did not.

Johnson and Frey (79) tested 27 oat cultivars at three levels of soil nitrogen and four levels of soil phosphorus. Decreasing the fertility of the test environment caused greater genotypic differentiation among oat lines, but environmental variance also increased, and there was no consistent increase or decrease in heritability with lessened environmental stress. Heritability was measured for heading date, height, plant weight, panicles per plot, spikelets per panicle, weight per 100 seed, and grain yield. Heritabilities increased for all characters with phosphorus application,. With higher nitrogen application, they increased for number of panicles per plot, responded erratically for plant height, and decreased for all other characters.

Vela-Cardenas and Frey (138) studied the heritability of plant height, spikelets per 5 panicles, 100-seed weight, and grain yield with 240 oat lines at 10 environments varying in fertility, date of planting, and density of stand. The low fertility environments had the highest heritability value for 100-seed weight. For all other characters, the heritabilities at low fertility were below the mean of environments with high fertility.

Johnson (78) compared the effectiveness of index selection versus selection for yield alone in oats at four rates of soil P. His results suggested that selection for yield per se would be more efficient at high levels of productivity, but at low productivity levels, panicle number might be an effective aid to selection for yield. The predicted selection advance increased for all indices with increased P, and predicted advance based on yield alone also increased with increased P.

These studies relate only to differential fertility levels and moisture stress and not to specific toxicity characteristics such as soil pH or excess Al or Mn or soil salinity. In addition the research has concentrated on the cereal grains, therefore care should be exercised in extrapolating these results to other species or other stresses.

YIELD AND STRESS TOLERANCE

The association, noted in some plant ecotypes, between adaptation to metal-contaminated mine spoils and lower biomass production, has given rise to the suggestion that adaptation to edaphic stress will result in lower biomass yield.

Hickey and McNeilly (69) studied the relative performance of metal-tolerant ecotypes from a Welsh mine site and homologous nontolerant ecotypes from the same vicinity of the species *Agrostis tenuis* Sibth., *Anthosanthum odoratum* L., *Plantago lanceolata* L., and *Rumex acetosa* L. Comparing the field performance of clonal propagules of the tolerant and nontolerant ecotypes in competition with swards of *Lolium perenne* L., cultivars 'S23' and 'S24', indicated that in all four species the tolerant ecotypes were weaker competitors relative to the nontolerant ecotypes, both in dry matter production and in plant survival. However, there were distinct differences among the species in the relative fitness of their tolerant populations in the order: *Agrostis = Rumex > Plantago > Anthoxanthum*.

Cook et al. (30) compared metal-tolerant ecotypes of *A. tenius, A. odoratum,* and *P. lanceolata* (derived from mine sites) with nontolerant ecotypes. Tests were conducted in pot experiments in a nontoxic potting mixture without competition, under interecotypic competition, and competition with 'S23' ryegrass. Dry weight data indicated a significant advantage for the nontolerant ecotypes over the tolerant ecotypes in all cases except that of *P. lanceolata* in mixed stand with ryegrass. In general, the advantage for the nontolerant ecotypes was greater in the competition treatments than in the pure stand treatments.

Parsons (117) cited several research reports which indicated that ecotypes from infertile soil have inherently slower growth rates than ecotypes from more fertile soils. He suggested that slow growing species would be at a selective advantage on infertile soils since they would not outgrow the nutrient supply available.

In many cases the acid metal toxic mine spoils sites are also low in nutrient fertility. Ecotypes that have evolved on these sites may have been selected not only for metal tolerances but also for adaptation to low fertility. The slower growth rates reported for some ecotypes may be a mechanism for such adaptation. If this is the case, metal tolerance need not necessarily condition slow growth.

In considering the relationship between stress tolerance and growth rates, the important distinction between biomass production and economic yield should be borne in mind. If the product of economic value is a plant chemical or specific plant component, reduced plant biomass may not necessarily be detrimental to economic yield.

While research on stress tolerance and growth rates have clarified the significant questions in this area, the issue is not settled. The development of isogenic lines differing in edaphic stress tolerance factors would provide valuable biological tools for critical assessment.

SYMBIOTIC ASSOCIATIONS

Symbiotic nitrogen fixation permits the culture of properly nodulated legume crops without application of synthetic nitrogen fertilizer. Selection and use of strains of *Rhizobium* adapted to particular edaphic stresses and compatible with the host crop genotype is important in achieving good nitrogen fixation.

Loneragan (96) reported wide variation for acid tolerance among rhizobial species and strains within species. Ham et al. (65) studied the distribution of the soybean microsymbiont, *Rhizobium japonicum* (Kirchner) Buchanan, over 75 locations in Iowa. Strains of *Rhizobium* belonging to the serological group 123 were the dominant group on soils with pH below 7.5, whereas on soils with pH above 7.8 serogroup 135 was most common. Soil pH accounted for 81% of the variation in the occurrence of serogroup 135.

Dobereiner (40) reported that in acid soil strains 413 and 411 of *Rhizobium phaseoli* showed differential response to the addition of toxic levels of Mn in both nodulation and the efficiency of fixation.

Keyser and Munns (84) reported that all three strains of cowpea *Rhizobium* studied in a defined liquid media were limited in maximum population density by low P. Low Ca limited maximum population density in one strain. A survey of several cowpea and soybean *Rhizobium* indicated high Mn did not stop the growth of rhizobia, while low Ca stopped the growth of only a few strains. Some strains were slowed in growth by high Mn and low Ca. All strains tolerant of Al were also tolerant of high Mn and low Ca. Keyser and Munns (85) also reported that low P limited total attainable population density and slowed the growth of some strains. Acidity slowed the growth of most strains and stopped growth of 50% of them. Tolerance of acidity did not necessarily confer Al tolerance. High Al slowed the growth of almost all strains tolerant of low pH. In general, the cowpea rhizobia were more tolerant of Al than the soybean rhizobia. These authors reported that a rapid test of rhizobial strains could be based on ability to attain visible turbidity in liquid culture under acid or Al stress when initial rhizobial density is low ($<<10^5$ cell/ml).

Munns et al. (108) assessed the tolerance of 40 strains of *Rhizobium* to soil acidity on two cultivars of mung bean (*Vigna radiata* L.). Some strains combined high acid soil tolerance with high N_2 fixation effectiveness. A strain's acid tolerance could not be predicted from the abundance or effectiveness with which it nodulated at a favorable pH, nor from its growth rate or acid production in yeast mannitol medium. Some strains varied in sensitivity with the host genotype used, implying that acid tolerances of rhizobial strains and legumes cannot be compared with validity in trials using only a single host genotype.

The danger of loss of symbiotic competence by *Rhizobium* through selection for adaptation to stresses associated with acid soils was suggested by Lie's observation (95) that ineffective strains of *Rhizobium* are predominant in acid waterlogged soils in Britain and the Netherlands. In a study of Irish soils, Masterson (98) found a significant positive association between the effectiveness of rhizobial isolates with the commercial white clover cultivar 'S100' and the pH of the soil at the site from which the isolate was obtained.

Holding and Lowe (72) subjected a strain, designated 8 Al, of *Rhizobium trifolii* Dangeard to 10 successive subcultures with 0 or 900 ppm of added Mn. After each subculturing, the effectiveness of the culture was tested on 'S184' white clover. No change occurred with 0 ppm Mn, but a substantial loss in effectiveness occurred with subculturing at 900 ppm Mn. In contrast, Sherwood (130) reported only a minimal change in rhizobial effectiveness on 'S100' white clover after eight transfers on media containing subtoxic levels of Mn. Six strains increased slightly in effectiveness, one was reduced, and one was unchanged.

Masterson and Sherwood (99) have suggested that the native ecotypes of clover occurring on wet, acid soil are adapted to such conditions and differ from bred cultivars of clover in both ecological adaptation and symbiotic affinity. In their test, isolates of *Rhizobium* from such soils were often very ineffective with bred clover cultivars, but very effective with indigenous clovers from the sites from which they were isolated. If coevolution of host and microsymbiont is an important factor affecting the function of symbiosis as suggested by Devine and Breithaupt (38), the source of the rhizobial strain will be an important consideration in breeding *Rhizobium* for edaphic stress.

The complexity of the host × *Rhizobium* × soil environment interaction is further complicated by the role of mycorrhiza. These fungi colonize plant roots and greatly increase the uptake of nutrients by extending mycelium beyond the immediate root zone. Increased P uptake through mycorrhizal association has been reported to result in improved nitrogen fixation by alfalfa in soil with low available P (7).

Investigation of genotypic differences among mycorrhiza for adaptation to soil stresses, efficiency of nutrient uptake and transfer, and affinity and compatibility with host species, should prove a challenge to researchers in the future.

The development of systems and techniques for controlled genetic exchange in economically important *Rhizobium* such as *Rhizobium japonicum* is in a very early stage. Such techniques are a prerequisite for directed breeding of highly effective genotypes of *Rhizobium* adapted to important edaphic stresses.

PROGRESS IN BREEDING

Brown and Jones (18–20) made an important contribution to crop breeding programs by a series of tests conducted to characterize the germplasm of soybeans, cotton, and sorghum that is being widely used in breeding programs to develop the cultivars of the future. To make this assessment they used a series of soils selected to produce deficiencies or toxicities of Fe, Zn, Cu, Al, or Mn. Differential cultivar response was found for many of the stress conditions, thus clearly indicating the value of selecting the proper parental germplasm for breeding programs to develop cultivars adapted to important soil groups. For example a high yielding crop of a Mn-tolerant cotton cultivar on a Richland (Grenada) soil could be followed by a crop failure with Forrest or Bragg soybeans because of the sensitivity of these cultivars to Mn toxicity. Liming Richland soil to grow Forrest soybeans could result in B and Zn deficiency stresses in a subsequent cotton crop. Alternatively, choosing soybean breeding lines with Mn tolerance as parental lines to develop cultivars for this soil should provide control of the problem. Similar characterizations of germplasm are needed in other crops.

Examples of crop cultivars particularly bred for tolerance to soil stresses are very few at present. However breeding programs now underway should produce such cultivars in future years.

Acid Soils

As yet no crop variety has been released that is a product of a selective breeding program specifically for adaptation to acid soil factors. However, researchers from the Botany Department of the University of Liverpool, England, have collected naturally evolved ecotypes of the grass *Agrostis tenuis*, identified their specific mineral tolerances, and released several as cultivars (75). These include 'Goginan' which is intended for use on neutral and acid Pb/Zn-contaminated wastes, and 'Parys' intended for use on neutral and acid Cu-contaminated wastes.

In 1978 the United States Department of Agriculture and the Arizona Agricultural Experiment Station released Barley Composite Cross *XXXIV* as a source of germplasm for selection of aluminum- and/or acid soil-tolerant lines or parents. The composite was derived from crosses involving all available barleys reported to be tolerant to acid soils and/or aluminum in the soil solution.

Devine et al. (37) established that recurrent phenotypic selection was effective in altering the tolerance of alfalfa to an acid, Al-toxic soil. A broad base of alfalfa germplasm was subjected to two cycles of selection in Bladen soil. From this common base one population was selected for tolerance and another for susceptibility. The tolerant and susceptible selections were tested on an acid Al-toxic Tatum soil and found to be significantly different in tolerance.

The Ohio Agricultural Research and Development Center developed and released the soft red winter wheat cultivar 'Titan' in 1978 (91). 'Titan' was described as outstanding in grain yield in Ohio trials and moderately tolerant to acid soil conditions. Research in Ohio has shown that the most severe Al toxicity problems occur in the eastern third of the state, and although mild toxicities may occur in the extreme southwest, Al toxicity is not known to be a problem in western Ohio (27). For this reason a breeding program has been designed to develop wheat cultivars for the eastern part of the state utilizing Al-tolerant Brazilian and European lines in crosses with adapted wheats with high yield potential. Progeny are being screened in the seedling stage for Al tolerance.

The International Rice Research Institute initiated research in 1976 to screen rice germplasm for tolerance to acid sulfate soils, which are estimated to occupy 15 million hectares of land in the tropics and subtropics that are climatically and physiographically suited to rice culture (77). Screening of several hundred breeding lines for tolerance to Fe toxicity has identified several lines of rice which have been used in crosses designed to improve tolerance. IRRI has also identified the rice cultivars 'Palawan', 'AZ uclna', 'Os 6', and 'MI-48' as tolerant to Al and Mn toxicities.

Centro Internacional de Agricultura Tropical (CIAT) has initiated screening of the cassava germplasm collection to identify genotypes tolerant to acid soils low in P and K (29). Cultivars which produce good yields with little or no liming on Al toxic soils have been identified. CIAT programs directed at characterizing the tolerance of forage grass and legume species to acid Al-toxic soils are also underway.

Alkaline Soils

An ecotype of *Festuca rubra*, 'Merlin', which is tolerant of Pb and Zn, was released for use on neutral and basic contaminated land sites by the Botany Department of the University of Liverpool (75).

In 1979, the Iowa Agricultural and Home Economics Experiment Station in cooperation with the Experiment Stations at Puerto Rico, Michigan, Minnesota, South Dakota and the USDA released the soybean cultivar 'Weber' which was described as having good resistance to iron deficiency chlorosis on calcareous soils. A significant criteria in the selection of this cultivar was the evaluation of progeny of the F_5 generation for iron deficiency chlorosis on calcareous soil.

IRRI has screened several thousand rice genotypes for tolerance to alkalinity and identified several hundred with tolerance. Alkali-tolerant cultivars with high yield potential are now available for use in areas where alkalinity is an important problem. Tolerant cultivars have been used as parents in breeding programs and two breeding lines, IR 4427-28-3-2 and IR 4227-10 4-3-1 have shown outstanding alkali tolerance.

Salinity

IRRI is engaged in a vigorous program of breeding rice for tolerance to salinity. Cultivars which have been grown in coastal saline soils are salt tolerant, but are low in yield due to insect and disease susceptibility and plant morphology. Several hundred salt-tolerant rice genotypes have been identified from screening over 5000 lines. Lines with high salinity tolerance have been developed from crosses of tolerant germplasm and should prove valuable in future cultivar development.

Researchers at the University of California at Davis are breeding barley for culture with seawater irrigation (45). Lines have been developed which survive and set seed (yields in the order of 1188 kg/ha) under irrigation with undiluted sea water. Similar breeding is underway with wheat.

The same researchers screened for salt tolerance in tomato cultivars with little success. However, a wild tomato, *Lycopersicon cheesmanii*, collected from sites close to the shore on the Galapagos Islands was able to survive in saline cultures equivalent in salinity to seawater. While the fruit of the wild species is too small for commercial use, F_2 progeny of crosses of the wild species and commercial cultivars include segregates with acceptable fruit size (similar to cherry tomatoes) and tolerance to a salinity equivalent to one-third that of seawater (45).

Shannon (129) screened 32 accessions of tall wheatgrass, *Agropyron elongatum* (Host) Blauv., a forage grass used on the western rangelands of the United States, for salt tolerance. He classified plants by ability to recover from salt stress and identified seven tolerant genotypes from diverse geographic origins for continued selection.

A long-term program of evaluation and selection of avocado rootstocks for tolerance of salinity has resulted in the development of very successful avocado orchards in regions with saline irrigation water in Israel (81).

Nutritional Tolerances

The IRRI has conducted an extensive screening program (testing more than 6000 lines) to locate rice lines tolerant to Zn deficiency, and identified over 100 tolerant lines. IRRI has also reported marked differences among rice lines for tolerance of P and Fe deficiency.

CIAT is screening bean germplasm for tolerance to low P. Their researchers report that several cultivars have been identified which yield as well or better at a low P level than with high rates of P application. CIAT researchers have also characterized species and ecotypic difference for tolerance of low P among forage species. Cassava cultivars are also the subject of a search for variation in tolerance to low P and low K by CIAT scientists.

A commercial seed company in the United States corn belt offers a "variety nutrient consulting" service which matches soil test results with available corn hybrids to select a hybrid suited to the soil nutrients.

SUMMARY

Specific knowledge of the nature of soil mineral stresses and plant responses to these stresses have provided plant breeders with valuable new insights into crop adaptation to edaphic factors. The development of assay techniques to characterize genetic variation for tolerance to edaphic stresses provides breeders with powerful new tools in selection for crop improvement. Our growing understanding of the genetic control systems governing edaphic response provides needed guidance for the choice of the most efficient breeding systems for improving edaphic adaptation. The vigorous breeding efforts now underway at many locations hold promise for the development of valuable new crop cultivars intentionally bred for adaptation to recognized edaphic stress conditions.

REFERENCES

1 Abel, G. H. 1969. *Crop Sci.* 9:697–698.
2 Abel, G. H., and A. J. MacKenzie. 1964. *Crop Sci.* 4:157–161.
3 Andrus, C. F. 1955. *Tomato Genet. Coop. Rep.* 5:5.
4 Antonovics, J., A. D. Bradshaw, and R. G. Turner. 1971. *Adv. Ecol. Res.* 7:1–85.
5 Antonovics, J., J. Lovett, and A. D. Bradshaw. 1967. The evolution of adaptation to nutritional factors in populations of herbage plants. In *Proceedings of a Symposium on Isotopes in Plant Nutrition and Physiology*. Inter. Atomic Energy Agency, Vienna, pp. 596.
6 Armiger, W. H., C. D. Foy, A. L. Fleming, and B. E. Caldwell. 1968. *Agron. J.* 60:67–70.
7 Azcon- G. de A. C., R. Azcon, and J. M. Barea. 1979. *Nature* 279:325–327.
8 Barber, W. D. 1969. Ph.D. dissertation. The Pennsylvania State University, University Park.
9 Bell, W. D., L. Bogorad, and W. J. McIlrath. 1958. *Bot. Gaz.* 120:36–39.
10 Bernard, R. L., and R. W. Howell. 1964. *Crop Sci.* 4:298–299.
11 Bradshaw, A. D. 1970. *New Scientist* 497–500.
12 Brown, J. E., and J. E. Ambler. 1974. *Physiol. Plant* 31:221–224.
13 Brown, J. C., J. E. Ambler, R. L. Chaney, and C. D. Foy. 1972. Differential responses of plant genotypes to micronutrients. *In Micronutrients in Agriculture*, J. J. Mortvedt, P. M. Giordano, and W. L. Lindsay, Eds. Soil Sci. Soc. of Am., Madison, Wisconsin.
14 Brown, J. C., R. L. Chaney, and J. E. Ambler. 1971. *Physiol. Plant* 25:48–53.
15 Brown, J. C., R. B. Clark, and W. E. Jones. 1977. *Soil Sci. Soc. Am. J.* 41:747–750.
16 Brown, J. C., and T. E. Devine. 1980. *Agron. J.* 72:898–904.

17 Brown, J. C., and W. E. Jones. 1974. *Agron. J.* 65:624–626.

18 Brown, J. C., and W. E. Jones. 1976. *Agron. J.* 69:399–404.

19 Brown, J. C., and W. E. Jones. 1976. *Agron. J.* 69:405–409.

20 Brown, J. C., and W. E. Jones. 1977. *Agron. J.* 69:410–414.

21 Brown, J. C., and W. E. Jones. 1977. *Commun. Soil Sci. Plant Anal.* 8:1–15.

22 Brown, J. C., and M. E. McDaniel. 1978. *Crop Sci.* 18:551–556.

23 Bruetsch, T. F., and G. O. Estes. 1976. *Agron. J.* 68:521–523.

24 Butler, E. J. 1963. *J. Agric. Sci.* 60:329–340.

25 Cacco, G., G. Ferrari, and G. C. Lucci. 1976. *J. Agric. Sci. Camb.* 87:585–589.

26 Cacco, G., G. Ferrari, and M. Saccomani. 1978. *Crop Sci.* 18:503–505.

27 Campbell, L. G., and H. N. Lafever. 1976. *Ohio Rep. Res. Develop.* 61:91–93.

28 Carter, O. G., I. A. Rose, and P. F. Reading. 1975. *Crop Sci.* 15:730–732.

29 Centro Internacional de Agricultura Tropical, CIAT. 1977. Annual Report. Cali, Colombia.

30 Cook, L. C. A., C. Lefebvre, and T. McNeilly. 1972. *Evolution* 26:366–372.

31 Cox, R. M., and T. C. Hutchinson. 1979. *Nature* 279:231–233.

32 Croughan, T. P., S. J. Stavarek, and D. W. Rains. 1978. *Crop Sci.* 18:959–963.

33 Davies, M. S., and R. W. Snaydon. 1972. *J. Appl. Ecol.* 10:47–55.

34 Dessureaux, L. 1959. *Euphytica* 8:260–265.

35 Dessureaux, L. 1960. *Plant Soil* 13:114–122.

36 Devine, T. E. 1976. Genetic potentials for solving problems of soil mineral stress: Aluminum and manganese toxicities in legumes. In *Proceedings of Workshop on Plant Adaptation to Mineral Stress in Problem Soils*, M. J. Wright, Ed. Cornell University, Ithaca, New York.

37 Devine, T. E., C. D. Foy, A. L. Fleming, C. H. Hanson, T. A. Campbell, J. E. McMurtrey, III, and J. W. Schwartz. 1976. *Plant Soil* 44:73–79.

38 Devine, T. E., and B. H. Breithaupt. 1980. *Crop Sci.* 20:269–271.

39 Devine, T. E., C. D. Foy, D. L. Mason, and A. L. Fleming. 1979. *Soybean Gen. Newsletter* 6:24–27.

40 Dobereiner, J. 1966. *Plant Soil* 24:153–166.

41 Downton, W. J. S. 1978. *Aust. J. Agric. Res.* 29:523–534.

42 Eenink, A. H., and F. Garretsen. 1977. *Euphytica* 26:47–53.

43 Epstein, E. 1963. Selective ion transport in plants and its genetic control. Pp. 284–298 in Desalination Research Conference, National Academy of Sciences, Natl. Res. Coun. Publ. 942. Natl. Acad. Sci., Natl. Res. Coun., 2101 Constitution Ave., NW., Washington, D.C. 20418.

44 Epstein, E. 1972. Physiological genetics of plant nutrition. Pp. 325–344 in *Mineral Nutrition of Plants: Principles and Perspectives*. Wiley, New York.

45 Epstein, E. 1976. Genetic potentials for solving problems of soil mineral stress: adaptation of crops to salinity. In *Plant Adaptation to Mineral Stress in Problem Soils*, M. J. Wright, Ed. Cornell Univ. Agric. Exp. Sta., Ithaca, New York.

46 Epstein, E., and R. L. Jefferies. 1964. *Annu. Rev. Plant Physiol.* 15:169–184.

47 Foy, C. D. 1976. Tailoring plants for greater tolerance to mineral toxicities and deficiencies on hill land soils. In *Hill Lands*, J. Luchok, J. D. Cawthon, and M. J. Breslin, eds. West Virginia University Books, Morgantown.

48 Foy, C. D., R. L. Chaney, and M. C. White. 1978. *Annu. Rev. Plant Physiol.* 29:511–566.

49 Foy, C. D., A. L. Fleming, and J. W. Schwartz. 1973. *Agron. J.* 65:123–126.

50 Foy, C. D., and G. C. Gerloff. 1972. *J. Phycol.* 8:268–271.

51 Foy, C. D., H. N. Lafever, J. W. Schwartz, and A. L. Fleming. 1974. *Agron. J.* 66:751–758.

52 Foy, C. D., A. J. Oakes, and J. W. Schwartz. 1979. *Comm. Soil Sci. Plant Anal.* 10:953–968.

53 Foy, C. D., and N. C. Wheeler. 1979. *J. Am. Soc. Hort. Sci.* 104:762–767.

54 Frey, K. J. 1964. *Crop Sci.* 4:55–58.

55 Gallagher, P. H., and T. Walsh. 1943. *J. Agric. Sci.* 33:197–203.

56 Gerloff, G. C. 1963. *Annu. Rev. Plant Physiol.* 14:107–124.

57 Gerloff, G. C., D. G. Moore, and J. T. Curtis. 1966. *Plant Soil* 25:393–405.

58 Gorsline, G. W., W. I. Thomas, D. E. Baker, and J. L. Ragland. 1964. *Crop Sci.* 4:154–156.

59 Gorsline, G. W., W. I. Thomas, and D. E. Baker. 1964. *Crop Sci.* 4:207–210.

60 Gotoh, K. and S. Osanai. 1959. *Japan J. Breed* 9:173–178.

61 Greenway, H. 1973. *J. Aust. Inst. Agric. Sci.* 39:24–34.

62 Gregory, R. P. G., and A. D. Bradshaw. 1964. *New Phytologist* 64:131–143.

63 Haaland, R. L., C. B. Elkins, and C. S. Hoveland. 1978. *Crop Sci.* 18:339–340.

64 Halevy, J. 1976. *Agron. J.* 68:701–705.

65 Ham, G. E., L. R. Frederick, and I. C. Anderson. 1971. *Agron. J.* 63:69–72.

66 Hanson, W. D., and E. J. Kamprath. 1979. *Agron. J.* 71:581–586.

67 Heenan, D. P., and O. G. Carter. 1976. *Crop Sci.* 16:389–391.

68 Henning, S. J. 1975. Ph.D. dissertation. Oregon State University, University Microfilms, Ann Arbor, Michigan, USA 75-13,057.

69 Hickey, D. A., and T. McNeilly. 1975. *Evolution* 29:458–464.

70 Hill, R. R., Jr., and S. B. Guss. 1976. *Crop Sci.* 16:680–685.

71 Hill, J., A. D. Robson, and J. F. Loneragan. 1978. *Aust. J. Agric. Res.* 29:925–939.

72 Holding, A. J., and J. F. Lowe. 1971. *Plant Soil Spec. Vol.*:153–166.

73 Howeler, R. H., and L. F. Cadavid. 1976. *Agron. J.* 68:551–555.

74 Hovin, A. W., T. L. Tew, and R. E. Stucker. 1978. *Crop Sci.* 18:423–427.

75 Humphreys, M. O., and A. D. Bradshaw. 1976. In *Proceedings of Workshop on Plant Adaptation to Mineral Stress in Problem Soils*, M. J. Wright, ed. Cornell University, Ithaca, New York.

76 Hutton, E. M., W. T. Williams, and C. S. Andrew. 1978. *Aust. J. Agric. Res.* 29:67–79.

77 International Rice Research Institute. 1976. Annual Report. Los Banos, Philippines.

78 Johnson, G. R. 1967. *Crop Sci.* 7:257–259.

79 Johnson, G. R., and K. J. Frey. 1967. *Crop Sci.* 7:43–46.

80 Jung, G. A., Ed. 1978. *Crop Tolerance to Suboptimal Land Conditions*. Am. Soc. Agron., Madison, Wisconsin, 343 p.

81 Kadman, A., and A. Ben-Ya'Acova. 1976. *Acta. Hort.* 57:189–197.

82 Kemp, A. and M. L. t'Hart. 1957. *Neth. J. Agric. Sci.* 5:4–17.

83 Keser, M., B. F. Neubauer, F. E. Hutchinson, and D. B. Verrill. 1977. *Agron. J.* 69:347–350.

84 Keyser, H. H., and D. N. Munns. 1979. *Soil Sci. Soc. Am. J.* 43:500–503.

85 Keyser, H. H., and D. N. Munns. 1979. *Soil Sci. Soc. Am. J.* 43:519–523.

86 Kotze, W. A. G. 1976. *Agrochemophysica* 8:39–43.

87 Krull, C. F. I. Narvaez, N. E. Borlaug, J. Ortega, G. Vazques, R. Rodriquez, and C.

Meza. 1966. *Results of the Third Near East-American Spring Wheat Yield Nursery, 1963–1965.* Internat. Maize and Wheat Improvement Center Res. Bull. 5. Mexico 6, D.F.

88 Kruckeberg, A. R. 1959. Ecological and genetic aspects of metallic ion uptake by plants and their possible relation to wood preservation. Pp. 526–536 in *Marine Boring and Fouling Organisms*, D. L. Ray, Ed. Univ. Washington Press, Seattle.

89 Kudelova, A., E. Bergmannova, V. Kudela, and L. Taimr. 1978. *Acta Phytopath. Acad. Sci. Hungaricae* 13:121–132.

90 Lewontin, R. C. 1977. In *Proceedings of the International Conference on Quantitative Genetics*. E. Pollack, O. Kempthorne, and T. B. Bailey, Jr., Iowa State University Press, Ames.

91 Lafever, H. N. 1979. *Crop Sci.* 19:749.

92 Lafever, H. N., and L. G. Campbell. 1978. *Can. J. Genet. Cytol.* 20:355–364.

93 Lafever, H. N., L. G. Campbell, and C. D. Foy. 1977. *Agron. J.* 69:563–568.

94 Lauchli, A. 1976. Genotypic variation in transport. Pp. 372–393 in *Encyclopedia of Plant Physiology*, New Series, Vol. 2, Part B, V. Luttge and M. G. Pitman, Eds. Springer-Verlag, Berlin.

95 Lie, T. A. 1974. Environmental effects of nodulation and symbiotic nitrogen fixation. Pp. 555–582 in *The biology of nitrogen fixation*, A. Quispel, Ed. North-Holland Publ. Co., Amsterdam.

96 Loneragan, J. F. 1972. The soil chemical environment in relation to symbiotic nitrogen fixation. Pp. 17–54 in *Use of Isotopes for Study of Fertilizer Utilization by Legume Crops.* Tech. Rep. No. 149. Food and Agric. Org./Int. Atomic Energy Agency, Vienna.

97 Lopez-Benitez, A. 1977. Ph.D. dissertation. Oregon State University, Corvallis.

98 Masterson, C. L. 1968. *Int. Congr. Soil Sci., Trans. 9th* (Adelaide, Australia) 2:95–102.

99 Masterson, C. L., and M. T. Sherwood. 1974. *Irish J. Agric. Res.* 13:91–99.

100 McCormick, L. H., and K. C. Steiner. 1978. *Forest Sci.* 24:565–568.

101 McDaniel, M. E., and D. J. Dunphy. 1978. *Crop Sci.* 18:136–138.

102 McLean, F. T., and B. E. Gilbert. 1927. *Soil Sci.* 24:163–175.

103 Meredith, C. P. 1978. *Plant Sci. Letters* 12:17–24.

104 Meredith, C. P. 1978. *Plant Sci. Letters* 12:25–34.

105 Mesdag, J., and L. A. Slootmaker. 1969. *Euphytica* 18:36–42.

106 Mesdag, J., L. A. J. Slootmaker, and J. Post, Jr. 1970. *Euphytica* 19:163–174.

107 Mugwira, L. M., K. L. Patel, and P. V. Rao. 1976. *Acta Agron. Acad. Sci. Hungaricae* 25:365–380.

108 Munns, D. N., H. H. Keyser, V. W. Fogle, J. S. Hohenberg, T. L. Righetti, D. L. Lauter, M. G. Zaroug, K. L. Clarkin, and K. W. Whitacre. 1979c. *Agron. J.* 71:256–260.

109 Murray, J. J., and C. D. Foy. 1978. *Agron. J.* 70:769–774.

110 Myers, W. M. 1960. Genetic control of physiological processes: Consideration of differential ion uptake by plants. Pp. 201–226 in *Radioisotopes in the Biosphere*, R. S. Caldecott and L. A. Snyder, Eds. Univ. of Minnesota, Minneapolis.

111 Naismith, R. W., M. W. Johnson, and W. I. Thomas. 1974. *Crop Sci.* 14:845–849.

112 Neenan, M. 1960. *Plant Soil* 12:324–338.

113 Nielsen, N. E., and S. A. Barber. 1978. *Agron. J.* 70:695–698.

114 Novick, R. P., and C. Roth. 1968. *J. Bacteriol.* 95:1335–1342.

115 Nyborg, M. 1970. *Can. J. Plant Sci.* 50:198–200.

116 Ouellette, G. J., and H. Genereux. 1965. *Can. J Soil Sci.* 45:24–32.

117 Parsons, R. F. 1968. *Am. Nat.* 102:595–597.

118 Peedin, G. F., and C. B. McCants. 1977. *Agron. J.* 69:71–76.

119 Perdue, R. E. Jr. 1976. *Cancer Treat. Rep.* 60:987–998.

120 Polle, E., C. F. Knozak, and J. A. Kittrick. 1978. Rapid screening of wheat for tolerance to aluminum in breeding varieties better adapted to acid soils. *Agriculture Technology for Developing Countries*, Tech. Series Bull. No. 21. Agency for International Development, Washington, D.C.

121 Polle, E., C. F. Konzak, and J. A. Kittrick. 1978. Rapid screening of maize for tolerance to aluminum in breeding varieties better adapted to acid soils. *Agriculture Technology for Developing Countries*, Tech. Series Bull. No. 22. Agency for International Development, Washington, D.C.

122 Polle, E., C. F. Konzak, and J. A. Kittrick. 1978. *Crop Sci.* 18:823–827.

123 Prestes, A. M., C. F. Konzak, and J. W. Hendrix. 1975. *Agron. Abst.* 67:60.

124 Ratanadilok, N., V. Marcarian, and C. Schmalzel. 1978. *Agron. Abst.* 70:160.

125 Reid, D. A. 1970. Genetic control of reaction to aluminum in winter barley. Pp. 409–413 in *Barley Genetics II*, R. A. Nilan, Ed. Proc. 2nd Int. Barley Genet. Symp. Washington State Univ. Press, Pullman.

126 Rhue, R. D., C. O. Grogan, E. W. Stockmeyer, and H. L. Everett. 1978. *Crop Sci.* 18:1063–1067.

127 Safaya, N. M., and A. P. Gupta. 1979. *Agron. J.* 71:132–136.

128 Sartain, J. B., and E. J. Kamprath. 1978. *Agron. J.* 70:17–20.

129 Shannon, M. C. 1978. *Agron. J.* 70:719–722.

130 Sherwood, M. T. 1966. *Soils Div. Res. Rep.*, p. 64–65. An Foras Taluntais, Dublin.

131 Shukla, U. C., and H. Raj. 1976. *Agron. J.* 68:20–22.

132 Shuman, L. M., D. E. Baker, and W. I. Thomas. 1976. *Pennsylvania Agric. Exp. Sta. Bull.* 811.

133 Sleper, D. A., G. B. Garner, K. H. Asay, R. Boland, and E. E. Pickett. 1977. *Crop Sci.* 17:433–438.

134 Slootmaker, L. A. J. 1974. *Euphytica* 23:505–513.

135 Spain, J. M. 1975. *Am. Soc. Agron. Spec. Publ.* 24:1–8.

136 Tan, G., and J. A. Rajaratnam. 1978. *Crop Sci.* 18:548–550.

137 Theisen, A. A., E. G. Knox, and F. L. Mann. 1978. *Feasibility of Introducing Food Crops Better Adapted to Environmental Stress.* A report prepared for the National Science Foundation by Soil and Land Use Technology, Inc. U.S. Government Printing Office, Washington, D.C.

138 Vela-Cardenas, M., and K. J. Frey. 1972. *Iowa State J. Sci.* 46:381–394.

139 Vose, P. B. 1963. *Herbage Abst.* 33:1–13.

140 Vose, P. B., and P. J. Randall. 1962. *Nature* 196:85–86.

141 Wall, J. R., and C. F. Andrus. 1962. *Am. J. Bot.* 49:758–762.

142 Wann, E. V., and W. A. Hills. 1972. *Tomato Genet. Coop. Rep.* 22:28–29.

143 Wann, E. V., and W. A. Hills. 1973. *J. Hered.* 64:370–371.

144 Weiss, M. G. 1943. *Genetics* 28:253–268.

145 White, M. C., R. L. Chaney, and A. M. Decker. 1979. *Crop Sci.* 19:126–128.

146 White, M. C., A. M. Decker, and R. L. Chaney. 1979. *Agron. J.* 71:121–126.

147 Wild, H., and A. D. Bradshaw. 1977. *Evolution* 31:282–293.

148 Wright, M. J., Ed. 1977. *Plant Adaptation to Mineral Stress in Problem Soils.* Cornell University, Ithaca, New York, 420 pp.

Chapter 6

PLANT RESPONSES TO WATER STRESS

LAWRENCE R. PARSONS

Fruit Crops Department, University of Florida—AREC,
Lake Alfred, Florida 33850

Among the environmental variables affecting plant growth and development, water stress is one of the most important. All higher plants are exposed to desiccation at least once during their life cycle as the seed matures and dries (15). Commonly, plants are exposed to other periods of mild or severe drought during their vegetative or reproductive growth phases.

Numerous review articles (41) and books (59a, 59b, 65, 66, 66a) have been written on responses of plants to water stress. The topic is a broad one and new books and reviews arrive each year. The purpose of this chapter is not to cover the numerous aspects of crop responses to water stress but to update recent reviews with particular emphasis on topics that would be of potential interest to breeders or biologists who may not be stress physiology experts. Breeding will be covered in another chapter and some of the plant responses that may be useful in breeding programs will be briefly discussed here. Crop plants respond to water stress in a variety of ways. Mechanisms or responses that have evolved may be of adaptive value in terms of reducing further stress effects. However, distinguishing between mere responses and adaptations to water stress is difficult. Additional information can be obtained from reviews by others (15, 31, 6, 76b, 100). Selected topics that will be discussed include:

1 Morphological changes.
   a Leaf shedding.
   b Leaf angle changes.
   c Root factors.
2 Physiological changes.
   a Leaf cuticular wax.

    b   Osmotic adjustment.
    c   Leaf enlargement.
    d   Stomatal behavior.
    e   Photosynthesis.
    f   Translocation.
    g   Proline accumulation.

These areas represent the interests of the author; other equally important areas could be covered.

The old saying that the plant is caught between dying of thirst or dying from starvation is still valid when one considers responses to drought. Turner (99) discussed drought resistance mechanisms at high water potential and low water potential and pointed out that several of these mechanisms such as stomatal closure, leaf movements, leaf shedding, and pubescence will decrease photosynthesis and productive processes. However, Turner (99) suggested that other mechanisms such as increased cuticular resistance, root density, osmotic adjustment, increased elasticity, and decreased cell size may not reduce productive processes. Several of these areas will be discussed.

MORPHOLOGICAL CHANGES

Leaf shedding or the production of less leaf area is a common way of reducing water loss. Reduction in total plant leaf surface has been considered one of the most important factors in the survival of some desert plants (74). Examples are *Artemesia, Zygophyllum,* and *Haloxylon* (60). Loss of lower leaves is common. Developing plants with less total leaf area may be advantageous if they can also compete effectively with weeds and produce comparable yields. Reduced leaf area may, of course, reduce total plant photosynthate. With regard to drought deciduousness, Fischer and Turner (31) subdivide plants into aridopassive and aridoactive groups. The aridopassive plants include plants that have no active photosynthetic tissue during long dry periods and include annuals and drought-deciduous perennials. Aridoactive plants maintain photosynthetic tissue during long dry periods and include woody evergreen perennials and CAM plants. Turner and Begg (100) emphasized that in pasture plants, morphological responses such as the reduction in leaf area, tillering, and root growth were more sensitive to water stress than physiological processes such as photosynthesis, stomatal behavior, and translocation.

Leaf angle changes, either active or passive, are common responses to water stress. Good examples of active leaf movements under stress are often found in legumes. Active movements that lead to parallel orientation of the leaf to the incident radiation (parahelionasty) have been reported in stressed beans (29) and Townsville stylo (5a). When water stressed, the drought-resistant tepary bean, *Phaseolus acutifolius,* also shows leaflet rotation that results in leaflet orientation parallel to the incoming radiation. This rotation along with differences in leaf size leads to significantly cooler leaf temperatures under stress than is found in *P. vulgaris* (77). Passive flagging, rolling, or wilting also reduces the energy load on leaves (6).

Other morphological changes in response to water stress include a greater root/shoot ratio. This can be due to a decline in shoot growth or an increase in root growth or both. Mayaki et al. (69) reported that under water stress, shoot height of soybeans was reduced more than root depth. Hsiao and Acevedo (42) pointed out that in certain cases not only did water stress enhance root growth relative to shoot growth but in corn, root growth itself was enhanced. In this study, it was suggested that the stress was relatively mild so that shoot growth was reduced but photosynthesis was not. This would allow for more photosynthate which, because of the reduced sink strength in the shoot, would be available for osmotic adjustment and extra growth of the roots. Greater root proliferation would allow for a greater soil volume exploration and this would allow the plant to survive longer. They suggested that these absolute root increases could be useful in root crops such as sugar beets. However, the corn root is not a major storage organ, and storage roots may behave differently than absorbing roots.

Passioura (78, 79) argued that when plants must rely entirely on stored soil water, water conservation during the vegetative phase would be advantageous so some water would be available for use during the critical grain filling period. A high root resistance would aid in conserving water. Passioura (78) increased the yield of wheat growing on stored water by limiting water uptake to one seminal root which he indicated was analogous to increasing the total root resistance. By reducing water loss early in growth, final grain weight was increased. His arguments apply only to plants growing on stored water; he indicated that a high root resistance would not be advantageous for plants growing in sandy soil or soils with low water storage capacity.

PHYSIOLOGICAL CHANGES

Leaf Cuticular Wax

A thick or waxy cuticle would be advantageous in reducing water loss. Blum (7) pointed out that the Bm Bm genotype for epicuticular wax increased the drought resistance of sorghum. The improved drought resistance came from increased reflection of solar energy and decreased cuticular permeability and resulted in higher leaf water potentials. Chatterton et al. (19) reported that the bloom or waxy coating on sorghum was an advantage in semiarid regions and that bloomless lines yielded less than bloom lines. Net carbon exchange and transpiration were greater in bloomless lines and the water use efficiency was greater in lines having bloom.

Hull et al. (46) have reviewed the environmental factors that affect cuticle development. The presence of large amounts of leaf wax is apparently not a consistent xeromorphic feature since some plants can enhance wax production in response to certain environments (109). Conditions that favor wax production are low humidity and high radiant energy (3, 92).

Hull and Bleckmann (45) noted an unusual wax ultrastructure on *Prosopis* that could possibly enhance foliar uptake of dew or reduce cuticular transpiration. Hull et al. (47) found that large wax plates were unique to the leaf surfaces of drought-tolerant lines of *Eragnostis* but were always absent on drought-susceptible lines. They suggested that this characteristic could be used for selection of drought tolerance in this species. Clark and Levitt (21) showed that more lipids accumulated on the surface of water stressed soybean leaves than on plants supplied with adequate water. Weete et al. (109) showed that after a rehydration period, stressed cotton leaves produced more wax. They suggested that this wax production on rehydration could be explained on the basis of the accumulation of substrate, possibly palmitic acid. The increased wax synthesis caused by water stress would result in a higher cuticular resistance. Greater waxiness and increased cuticular resistance would only be advantageous if tight stomatal closure occurred in the species during stress. Theoretically, more wax could also increase stomatal resistance by partially plugging stomata.

Osmotic Adjustment

Gradients and changes in osmotic potential have been observed for a number of species and osmotic differences among desert plants have been well documented (104, 49, 105). Recent interest has focused on osmotic adjustment, turgor maintenance, and growth. These areas have been discussed by a number of authors (43, 39, 6, 100). Turgor is generally thought to be essential for cell enlargement and growth. Hence, maintaining turgor is essential for maintaining growth. Turgor can be maintained by increasing osmotic concentration, increasing elasticity or decreasing cell size. The relation between the different potentials is summarized in the familiar equation $\Psi w = \Psi s + \Psi p + \Psi m$ (reference 61) where Ψw = water potential, Ψs = osmotic or solute potential, Ψp = turgor or pressure potential, and Ψm = matric potential. Matric potential is usually considered to be small except in very dry tissue or dry seeds, and will not be discussed here.

Turner (99) uses the term "osmotic adaptation" and divides it into two subgroups: (*a*) the passive increase in solutes due to solute concentration or dehydration, and (*b*) the accumulation of solutes. He refers to the second as osmotic adjustment. Studies on osmotic adjustment have been accelerated by use of the pressure chamber method and analysis of pressure volume curves to measure water, osmotic, and turgor potentials (102, 20). Improvements and increased availability of thermocouple psychrometry have also aided in measurement of water and osmotic potentials.

It has been commonly reported that leaf osmotic potential decreases when osmoticum is added to the rooting medium or when plants are naturally subjected to saline water (50, 51, 6). However, the distinction must be made as to whether the decrease in Ψs was due to absorption of the osmoticum or if the solutes were generated internally. As Begg and Turner (6) point out, the solutes that accumulate during osmotic adjustment are generally unknown. Soluble sugars have been known to increase under water stress (49, 41) and it appears that soluble carbohydrates may be the main source of solute increase (6, 100). Other possibilities include proline involvement in osmotic adjustment of microorganisms and some halophytes (96). Other proposed osmotic adjusting agents that have been suggested include potassium, sugar alcohols, and organic acids (6).

In a field study, Cutler et al. (24) concluded that turgor in stressed cotton was maintained partly by the accumulation of sugars and malate and partly by high cell wall elasticity. In a related study, Cutler and Rains (22) conditioned cotton plants by several water stress cycles and found that prestressed plants maintained turgor to lower water potential values than did well watered plants. They found lower osmotic potentials in the preconditioned plants but noted that soluble sugars and malate accumulated to approximately the same levels on a dry weight basis in both prestressed and control plants. Leaves of prestressed plants had less water per unit dry weight than did control plants and this difference, rather than solute accumulation, largely accounted for the difference in osmotic potential of the prestressed and control plants.

Meyer and Boyer (72) demonstrated osmotic adjustment in soybean hypocotyls. Removal of the cotyledons indicated that the main cause of the osmotic change was the translocation of solutes from the cotyledons to the hypocotyl. Methods of determining osmotic adjustment have been debated, and there are conflicting reports on the presence or absence of osmotic adjustment in mature soybean plants (101, 110). Age and reproductive state of the plant may affect the degree of osmotic adjustment. In sorghum, little evidence for osmotic adjustment was seen before flowering, whereas genotypic differences in the amount of osmotic adjustment were seen after flowering (16). Osmotic adjustment is more easily observed when plants dry slowly under high radiation conditions. Hence, reports observing no osmotic adjustment may be changed later when plants at different development stages are studied under slowly drying test situations.

Turner (98) has suggested that osmotic adjustment has several major advantages and a few limitations. Included in the advantages were: (a) maintenance of cell turgor, (b) continued cell elongation, (c) maintenance of stomatal opening and photosynthesis, (d) survival of dehydration, and (e) greater soil exploration by roots. Limitations include the transience of osmotic adjustment; rewatered plants can lose most of their osmotic adjustment within 10 days. A second limitation is the finite limit of adjustment. Sorghum plants stressed at slow rates (1.5 or 7 bars per day) had equal solute accumulations of about 6 bars (54). Adjustment greater than 6 bars in sorghum was not observed in this study.

Osmotic adjustment can occur both on a seasonal and diurnal basis. Goode and Higgs (33) reported a seasonal osmotic adjustment in apples; unirrigated trees decreased leaf osmotic potential about 5 bars from July to September and turgor was maintained in spite of the decrease in water potential. Fereres et al. (30) demonstrated seasonal osmotic adjustment on sorghum growing in deep soils (Yolo clay loam) under high radiation conditions. They observed that midday leaf turgor (Ψp) of nonirrigated sorghum remained at levels similar to those of irrigated sorghum throughout the season because of the Ψs decrease. The net increase in solutes was the main reason for the turgor maintenance. They also observed osmotic adjustment in corn but Ψp was lower in the nonirrigated treatment during the final grain filling stages.

Cutler et al. (22) have argued that under water stress, smaller cell size allows for better turgor maintenance. With smaller cells or thicker cell walls, more of the water would be held in the wall and less would be available as a solvent in the protoplasm. Thus, the smaller the cell or the thicker the wall, the more negative would be the osmotic potential at a given relative water content and a given amount of solutes. Fereres et al. (30) agree with Cutler's argument but feel that the osmotic adjustment observed under their slow stress conditions was due mainly to solute accumulation. They felt a change in cell size or the amount of cell wall played a minor role in the adjustment they observed, but that further investigation on osmotic adjustment and cell size was warranted.

Another interesting aspect is the reported diurnal changes in osmotic adjustment. Growth and leaf enlargement are much more sensitive to water stress than is photosynthesis. At −4 bars, sunflower leaf enlargement was completely stopped and corn and soybean enlargement was greatly reduced (13). It has been suggested that because of this sensitivity to stress, enlargement would not occur during midday when leaves were stressed but at night when water potentials were nearer to zero. Boyer (12) demonstrated that sunflower leaf growth was greater at night than during the day and this correlated with the time of the highest leaf water potential. Earlier growth chamber work with corn showed that a leaf Ψw of −7 bars stopped corn leaf growth (1, 1a). However, growth in the field continued even though daily leaf Ψw values commonly went to −12 bars. Hsiao et al. (43, 44) explained this continued growth during midday water stress by the diurnal oscillation that was observed in leaf Ψs. This diurnal change allowed turgor maintenance and continued midday growth. They postulated that this daily osmotic adjustment allowed cold-sensitive crops to grow in areas that have cool nights. Low night temperatures will slow growth and if growth could not occur due to osmotic adjustment during the day, total plant growth in areas with cool nights would be greatly reduced.

Interest in osmotic adjustment has increased even more since Morgan (73) demonstrated that some species and cultivars of wheat showed osmotic adjustment and others did not. Adjustment to approximately −15 bars was demonstrated but no adjustment occurred below −15 bars in *Triticum durum* and certain other *Triticum* groups. Morgan (73) suggested that one could breed for osmotic adjustment to improve the drought tolerance of wheat. On the other hand, other studies have not yet demonstrated a clear relation between degree of osmotic adjustment and field observed drought tolerance. For example, Jones and Turner (55) studied two cultivars of sorghum ('RS 610' and 'Shallu') that differed in drought resistance but found no significant differences in the degree of osmotic adjustment or tissue water relations between the cultivars. Plants were stress preconditioned to predawn leaf water potentials of −4 or −16 bars. They found that zero turgor potential occurred at approximately the same relative water content value of 94% regardless of the previous stress history. They concluded that the superiority of 'RS 610' over 'Shallu' was due to factors beyond osmotic adjustment such as the extent of the root system, desiccation tolerance, or xylem water conductance. They also observed an approximate halving of tissue elasticity due to the stress preconditioning.

Differences in the amount of osmotic adjustment between cultivars have also been studied in rice. Cutler et al. (25, 23) showed that prestressing treatments resulted in a 3–5 bar osmotic adjustment but that turgor maintenance was similar for all the varieties tested. The osmotic adjustment was due to solute accumulation.

Thus, it appears that at present osmotic adjustment may be useful in breeding for drought tolerance in wheat but more work needs to be done on sorghum or rice. Investigations on different species or more diverse genetic material within a species may show greater differences in osmotic adjustment. Turner (personal communication) feels that osmotic adjustment does not "cost" the plant greatly; that is, osmotic adjustment does not slow photosynthesis as greatly as would stomatal closure. Hence, there will continue to be great interest in this phenomenon.

Leaf Enlargement

Leaf enlargement is particularly sensitive to mild water stress (41, 13). Boyer (13) found that corn leaf elongation rate was greatest when leaf water potentials were −1.5 to −2.5 bars but the enlargement rate dropped to 25% of the controls at −4 bars. Short desiccation periods can slow growth (1a) and normal growth rates can return when water potentials return to approximately −2 bars. However, prolonged drought can affect subsequent regrowth; leaves may not return to the original growth rate on rewatering (13). Ludlow and Ng (68) followed elongation rates of water stressed *Panicum maximum* leaves after rewatering. They found a short lived burst of elongation when plants were rewatered in the light. After rewatering, prestressed plant rate of elongation was greater than controls for up to 33 hours. Cell enlargement was generally considered to be more sensitive to water stress than was cell division and the greater expansion rate after rewatering was thought to be due to the expansion of cells that developed during stress. However, in spite of the greater elongation rates after rewatering, the final length of the stressed leaves was less than the controls. Barlow et al. (4) showed that elongation of wheat floral apices stopped around −12 bars and leaf elongation was strongly inhibited at the same level. Elongation rate is more sensitive to water stress than is photosynthesis (13). Osmotic adjustment is a suggested mechanism that would help maintain turgor, cell elongation, stomatal opening, and photosynthesis under lower water potentials. A simple test that has been suggested for breeders is to look for lines that maintain leaf expansion at reduced water potentials.

In addition to reducing leaf enlargement, water stress also can hasten leaf senescence. When stressed, *Phaseolus vulgaris* 'White Half Runner' showed more rapid leaf yellowing and senescence than did the drought-tolerant tepary bean *Phaseolus acutifolius* (77). Boyer and McPherson (15) showed that senescence was more rapid in desiccated corn plants than in controls. Also, plants that had been pretreated by low vapor pressure (VP) conditions maintained a greater viable leaf area longer than those pretreated with high VP conditions. As they point out, it is not yet clear to what extent this drought-induced senescense represents (*a*) a reduction in transpiring leaf area or (*b*) a movement of nitrogen compounds and carbohydrates out of the senescing leaves for continued grain formation.

Stomatal Behavior

Turner (97) has shown the relationship between leaf water potential and stomatal conductance for several species. A variety of factors will affect the water potential at which stomata close (6, 100) including age, growing conditions, leaf position, and stress history of the plant. Davis et al. (28) showed that leaf diffusion resistance increased with age under constant environment conditions. Davis and McCree (27) also showed there was a decline in CO_2 exchange rate with age and that there was a decline in internal leaf photosynthetic capacity. They suggested that the observed decline was due either to a decrease in the ability of the stomates to stay open or that the stomates were operating by a CO_2 feedback process that maintained substomatal CO_2 at a constant level. Wheat flag leaf stomata tend to close at lower Ψw values as they age (32). At tillering, wheat stomata closed around −13 bars but did not close until about −31 bars at grain filling. One result of this stomatal behavior would be conservation of water during the vegetative stage. Jordon et al. (56) indicated that there was not a single threshold leaf water potential value for stomatal closure in cotton. Leaf r_s was determined by individual leaf Ψw and modified by radiation and leaf age.

Removal of growing fruit can also cause partial stomatal closure. Kriedemann et al. (62) reported partial closure in grape and pepper after removal of fruit. Koller and Thorne (59) noted that in soybean ('Amsoy 71') the r_s of the upper leaf surface increased to approximately four times the normal r_s within 48 hours after pod removal. There was a very small increase in lower leaf surface stomatal resistance and the degree of closure on both leaf surfaces was roughly proportional to the number of pods removed. Depodding in the growth chamber also resulted in leaflet rotation so the leaflets were in a near vertical orientation. Interestingly enough, field grown plants of another cultivar ('Wells') did not demonstrate leaflet reorientation after pod removal.

Preconditioning has been recently emphasized as playing an important role in stomatal behavior. Brown et al. (16a) showed that repeated stress cycles on cotton resulted in the abaxial stomata closing at a water potential 14 bars lower than non-preconditioned plants. There was little effect of stress preconditioning on adaxial stomatal closing. They observed osmotic adjustment in the epidermis of the abaxial leaf surface but not in the adaxial leaf surface. Cutler and Rains (22) noted that preconditioned cotton plants had lower osmotic potentials at any given leaf water content and maintained turgor to lower leaf water potential values than did well watered plants. Klar et al. (58) also noted that the "threshold" water potential for stomatal closure in guinea grass was lower in prestressed plants than in well-watered controls.

Rate of drying also affects the water potential at which stomata close. Fereres et al. (30) showed that sorghum stomata did not close in nonirrigated field conditions when the leaf Ψw was as low as -20 bars. However, plants grown with a restricted root zone where the stress developed more rapidly showed closure at -14 to -16 bars. Lakso (63) also emphasized the effects of preconditioning and time of season on stomatal closure in apple. During the season, the leaf Ψw required to cause closure decreased by approximately 25 bars, and in late summer, net photosynthesis did not drop to zero until leaf Ψw went to -50 or -60 bars. Turner and Begg (100) present a table of different water potential values required to halve stomatal conductance and emphasized that there is not a unique water potential value for stomatal closure.

Jones and Rawson (54) compared different rates of water stress on osmotic adjustment, net photosynthesis, and water use efficiency in sorghum. They found that plants exposed to rapid stress rates of 12 bars per day demonstrated no solute accumulation whereas slower stress rates (7 or 1.5 bars per day) led to equal osmotic adjustments of about 6 bars. Leaf conductances at a given turgor potential were highest at the slower rates of stress induction. Recovery of osmotic potential took 6–11 days after rewatering but net photosynthesis and leaf conductance recovered in less than 3 days. Hsiao and Acevedo (42) pointed out that if stress is moderate or severe and lasts several days, it can take as long as 5 days after relief of the stress before stomates fully open again. Jones and Rawson (54) indicated that there was not a switchlike closure of stomata in sorghum at a threshold leaf water potential, and leaf conductances declined continuously over a range of potentials from -15 to -30 bars. At slower drying rates, stomatal closure was the most gradual.

In addition to the leaf water potential effect on stomata, humidity can control stomatal resistance directly and independently of the leaf water potential. This has been shown by several authors (84, 18, 76a). As the leaf-to-air vapor pressure deficit increases, stomates close and stomatal resistance increases (35, 2). Those species with a direct response to humidity averaged larger substomatal cavities than the other species tested (86). Lange (64) noted that this humidity response affected the diurnal stomatal resistance pattern and controlled the midday depression of transpiration and photosynthesis, especially under desert conditions. The response to humidity can also be affected by other environmental factors such as temperature, sun or shade, and whether plants had been exposed to prestressing cycles (57). The relation between stomatal response to humidity and guard cell potassium content was described as following a hysteresis curve (67). This humidity response has been described as an "early warning system" for the plant that allows for reduced transpiration before major leaf water loss occurs.

Begg and Turner (6) pointed out that there have been a number of comparisons of stomatal response between species but few comparisons within species. Henzell et al. (40) noted variation in stomatal sensitivity to soil water potential in 23 lines of sorghum. Blum (7) also showed large differences among sorghum genotypes in regard to stomatal action. Henzell et al. (40) found that stomatal conductance of 'Shallu' and 'Alpha' sorghum cultivars declined rapidly as leaf Ψw decreased, whereas conductance declined less rapidly in cultivars, 'M 35-1' and 'I.S. 1598C'. They reported that the F_1 hybrid stomatal behavior was similar to the more sensitive parent, but there was not a consistent relationship. Problems in preconditioning could lead to difficulty in interpreting differences between genotypes. Care should be taken when making comparisons, for stomata of plants grown in controlled environments behave differently than plants raised in the field (26).

Photosynthesis

As has been pointed out, the photosynthetic ability of a plant is mainly determined by the total leaf area and the photosynthetic activity of the various leaves (15). Stomatal closure will result in reduced water loss but also in reduced photosynthesis. The effect of water stress on photosynthesis via stomatal and nonstomatal factors has been studied a great deal. Begg and Turner (6) point out that the initial reduction in photosynthesis is due to stomatal closure and reduced CO_2 uptake. Transpiration often declines in parallel with the photosynthetic decline, suggesting that stomatal closure limits both processes.

However, other work suggests that nonstomatal factors that are affected by water stress also reduce photosynthesis. Studies on chloroplasts isolated from desiccated leaf tissue have shown that oxygen evolution, photophosphorylation and electron transport are reduced (13, 12a, 80). Boyer (14) concluded that sunflower photosynthesis rate was limited below -11 to -12 bars by reduced photochemical activity. O'Toole et al. (75, 76) stated that increased stomatal resistance was the primary reason for reduced photosynthesis but the nonstomatal factors of increased mesophyll resistance and decreased RuDP carboxylase activity also played a role. Bunce (17) also observed an increase in mesophyll resistance as leaves dried. However, Huffaker et al. (44a) found no effect of water stress on RuDP carboxylase activity. Begg and Turner (6) concluded that "photosynthesis declines initially as a result of stomatal closure, but prolonged and severe water stress can lead to depression of chloroplast and enzyme activity and to nonstomatal effects on photosynthesis."

Johns (52) studied four temperate herbage crops under water stress conditions and found that gross photosynthesis was reduced more because of reduced leaf area than because of stomatal closure. Considerable leaf death occurred when stomatal resistance was less than 10 sec/cm for perennial rye grass and tall fescue. Stomatal closure was incomplete when stressed, particularly in clover. The ability of fescue to yield under dry land conditions was related to its tight leaf rolling. Because leaf expansion is more sensitive than stomatal closure and photosynthesis, one cannot conclude that dry matter production is unaffected until plant water status reaches a level that reduces photosynthesis (42). Stress can reduce leaf area or leaf area index (LAI) and if stress occurs when LAI is low, dry matter production would be reduced because the development of the total leaf area would be delayed.

Rawson et al. (82) pointed out that during stress, the carbon fixed by soybean during early morning and late afternoon made up a greater proportion of the total daily photosynthesis. These periods corresponded with times of low evaporative demand. A calculated efficiency index showed that early morning was the time of most efficient assimilation.

Translocation and Distribution of Assimilates

Translocation is less sensitive to water stress than is photosynthesis (15, 71, 96a). Generally speaking, drought reduces the amount of photosynthate moved to the developing grain. Wardlaw (106–108) showed that drought reduced the translocation rate of ^{14}C. He suggested (107) that water stress reduced the rate of assimilate movement from the photosynthetic cells into the conducting system but the translocation pathway can operate efficiently under stress or is resistant to water loss. He concluded that the response to water stress was not due to a direct effect on photosynthate conduction, but due to the effects on photosynthesis and assimilate loading. Begg and Turner (6) supported Wardlaw's conclusion and felt that the reduced translocation caused by water stress was due to reduced source photosynthesis or sink growth, not due to effects on the conducting system. Sheikholeslam and Currier (85) noted a decrease in ^{14}C-assimilate translocation in the squirting cucumber and concluded that water stress caused a decrease in translocation by decreasing sieve tube turgor differences. They felt that this sieve tube turgor gradient was regulated by the availability of water and was the driving force necessary for translocation. Silvius et al. (87) found a complex pattern of ^{14}C distribution in soybean depending on when stress was applied. Before pod filling, more ^{14}C was found in the roots of water stressed plants than in the controls. Leaves of stressed flowering plants had less ^{14}C than the controls. They suggested that water stress reduced the carbon exchange rate and altered the assimilate distribution so that the percentage of ^{14}C in the leaves and reproductive structures increased at the expense of the roots and stems. Hsiao (41) suggested that stress effects on xylem and phloem gradients should be complex and thus stress effects would be hard to predict.

Boyer and McPherson (15) carried out experiments with corn to see how a stress pretreatment would affect photosynthesis and translocation during a later desiccation treatment. Plants were treated by growing them during the vegetative stage in high vapor pressure (VP) or low VP conditions. Pollination was allowed to occur in all plants during favorable high VP conditions. Plants were then desiccated by decreased irrigation. Desiccated plants that were pretreated at low VP produced 68% of the control yield which was substantially more than those pretreated at high VP. Boyer and McPherson suggested that (a) "plants can adapt to desiccation in some way that preserves grain production, and (b) plants can mobilize photosynthate produced before the grain-filling period and use it to fill the grain." The vegetative parts of the plant lost weight as photosynthate was moved to the grain during filling. Of the various factors measured, the single grain weight was decreased the most by the desiccation treatment. They concluded that the grain yield differences between the high and low VP pretreated plants were due to the total accumulated photosynthate, not to differences in sink strength or reserve mobilizing ability. In a similar comparison study, McPherson and Boyer (71) found that translocation and grain fill continued even though photosynthesis had stopped. They suggested that in corn, integrated photosynthesis over the growing season controlled yield and dry weight accumulation at low water potentials. McPherson and Boyer's (71) results differed somewhat from those of Brevedon and Hodges (16) who found a smaller effect of stress on translocation of recently fixed ^{14}C. The differences in results were explained by the fact that Brevedon and Hodges (16) were looking at recently fixed ^{14}C, whereas McPherson and Boyer (71) were measuring longer-term total translocation.

In another study, Rawson et al. (81) examined whether drought imposed on wheat and barley during the vegetative stage resulted in adaptations that allowed the plants to better deal with water stress during grain filling. Adaptations were not found to be persistent; previously stressed plants did not have greater water use efficiency during the final drying cycle. Barley plants stressed during grain growth did not show greater retranslocation of stored photosynthate than did well-watered plants. Rawson et al. (81) concluded that their results differed from those of Boyer and McPherson (15) because of the use of different species, different growth conditions, and the degree of leaf expansion when exposed to the hardening treatment. Rawson et al. (81) concluded that there was little or no evidence that drought improved water use efficiency.

Proline Accumulation

Free proline accumulates in water-stressed tissue (5, 83, 103) and net proline synthesis appears to come from carbohydrate via α-ketoglutarate and glutamate (95, 9). Oxidation of proline occurs rapidly in turgid tissue and Stewart et al. (94) suggested that proline oxidation could act as a control mechanism to maintain low levels of proline in turgid tissue. Under water stress conditions, proline oxidation was inhibited (94) and this was thought to maintain the high level of proline in the stressed tissue. A series of papers by several authors (88–91, 95, 94, 9, 11) describe the metabolism of proline.

Interest in proline was further stimulated when Singh et al. (88) indicated that the proline accumulating ability of 10 barley cultivars was correlated with their grain yield stability. These studies led to the suggestion that proline accumulation could be used as a screening tool for drought resistance breeding programs. However, later work has questioned the use of proline as a screening tool. Blum and Ebercon (8) did not find a simple relation between field drought resistance ratings and proline accumulation in sorghum. They did suggest that stress-induced free proline accumulation was related to the ability of the cultivar to recover after irrigation. Hanson et al. (38), using two of the cultivars examined by Singh et al. (88) found a gradient in leaf water potential along the barley leaf and observed that free proline followed the decline in water potential. Under water stress, leaf tips became irreversibly wilted and proline remained in this desiccated zone even after recovery from stress. Proline levels decreased in the viable zones after recovery from stress. 'Proctor', the barley cultivar that gave low yield under drought, desiccated more rapidly and accumulated proline faster than 'Excelsior', the cultivar that yielded better under drought. They suggested that selection for high proline could actually lead to development of varieties susceptible to stress. Because of the leaf gradients in water potential and proline levels and the environmental effects on development of stress, Hanson et al. (38, 38a) concluded that proline accumulation would not be of practical value in screening for drought resistance.

Whether proline has a beneficial role is unknown, but such roles as an osmoticum, a protectant, or a nitrogen reserve after recovery from stress have been suggested (38). However, Tully et al. (96b) indicate that proline is not of great importance as a nitrogen reserve after stress, nor is it a major form of translocated nitrogen during stress. Thus, they question if proline confers survival value during stress and suggest that proline accumulation would probably be a deleterious response to water stress rather than an adaptation with survival value.

Betaine is another compound that accumulates under water stress and has a time course of accumulation that is similar to proline accumulation. Hanson and Nelsen (37) suggested that *de novo* synthesis of betaine occurs from 1- and 2-carbon precursors during water stress. Betaine may be a metabolically inert end product. They suggested that selection *against* high betaine genotypes might be useful in a breeding program.

CONCLUDING REMARKS

Virtually all plant metabolic processes are affected by water stress if the stress is severe or of long duration. However, studying and duplicating water stress effects on metabolism is difficult. Simulating field drought conditions is a problem in growth chambers and greenhouses due to space and light limitations. Begg and Turner (6) have emphasized the importance of slow desiccation by growing plants in large containers. Rapid desiccation rates in small pots does not allow for overnight recovery or provide sufficient time for osmotic adjustment. As discussed earlier, stomata do not close rapidly at a threshold potential with slow drying but gradually close over a wide (10–15 bar) range of leaf water potential. It has been suggested that greater emphasis needs to be placed on approximating field stress conditions in growth chamber studies (6). Chamber plants will not duplicate field plants but the development of high irradiation chambers will improve the field simulation.

In addition to improving our knowledge of physiological factors, we will need to improve our knowledge of breeding for drought tolerance (48). It has been said that breeding for drought resistance can be done by breeding for characters that decrease yield (76b). Under well-watered conditions, drought resistant plants may produce less than higher yielding, more sensitive plants. Ideally, improved plants would produce higher yields than sensitive plants in dry years without sacrificing yield greatly during wet years. Several potentially useful screening tools such as leaf firing rate (15, 38a) and leaf elongation rate (23a) have been suggested. It can be argued (42) that breeders should not rely on a few screening tests but think in terms of the physiological processes related to tolerance. Also, physiologists should keep in mind the integrated behavior of the whole plant rather than isolated processes. For example, emphasis on stress effects on photosynthesis and stomatal closure should not overshadow the importance of stress induced leaf loss.

Comparative studies on different cultivars and species will help shed more light on the genetic transfer of drought tolerance and avoidance characteristics. This will require the coordinated efforts of physiologists and breeders along with the combination of growth chamber and field studies to increase our knowledge and to help develop more drought resistant plants.

REFERENCES

1   Acevedo, E., E. Fereres, T. C. Hsiao, and D. W. Henderson. 1979. *Plant Physiol.* 64:476–480.

1a Acevedo, E., T. C. Hsiao, and D. W. Henderson. 1971. *Plant Physiol.* 48:631–636.

1b Ackerson, R. C., D. R. Krieg, and F. J. M. Sung. 1980. *Crop Sci.* 20:10–14.

2   Aston, M. J. 1976. *Aust. J. Plant. Physiol.* 3:489–502.

3   Baker, E. A. 1974. *New Phytol.* 73:955–966.

4   Barlow, E. W. R., R. Munns, N. S. Scott, and A. H. Reisner. 1977. *J. Exp. Bot.* 28(105):909–916.

5   Barnett, N. M., and A. W. Naylor. 1966. *Plant Physiol.* 41:1222–1230.

5a Begg, J. E., and B. W. R. Torssell. 1974. In *Mechanisms of Regulation of Plant Growth*, R. L. Bieleski, A. R. Ferguson, M. M. Cresswell (Eds.) Royal Society of New Zealand Bull. 12:277–283.

6   Begg, J. E., and N. C. Turner. 1976. *Adv. Agron.* 28:161–217.

7   Blum, A. 1975. *Israel J. Bot.* 24:50.

8   Blum, A., and A. Ebercon. 1976. III. *Crop Sci.* 16:428–431.

9   Boggess, S. F., D. Aspinall, and L. G. Paleg. 1976. IX. *Aust. J. Plant Physiol.* 3:513–525.

10 Boggess, S. F., and C. R. Stewart. 1976. *Plant Physiol.* 58:796–797.

11 Boggess, S. F., C. R. Stewart, D. Aspinall, and L. G. Paleg. 1976. *Plant Physiol.* 58:398–401.

12 Boyer, J. S. 1968. *Plant Physiol.* 43:1056–1062.

12a Boyer, J. S., and B. L. Bowen. 1970. *Plant Physiol.* 45:612–615.

13 Boyer, J. S. 1970. *Plant Physiol.* 46:233–235.

14 Boyer, J. S. 1971. *Crop Sci.* 11:403–407.

15 Boyer, J. S., and H. G. McPherson. 1975. *Adv. Agron.* 27:1–23.

16 Brevedon, E. R., and H. F. Hodges. 1973. *Plant Physiol.* 52:436–439.

16a Brown, K. W., W. R. Jordan and J. C. Thomas. 1976. *Plant Physiol.* 37:1–5.

17 Bunce, J. A. 1977. *Plant Physiol.* 59:348–350.

18 Camacho-B, S. E., A. E. Hall, and M. R. Kaufmann. 1974. *Plant Physiol.* 54:169–172.

19 Chatterton, N. J., W. W. Hanna, J. B. Powell, and D. R. Lee. 1975. *Can. J. Plant. Sci.* 55:641–643.

20 Cheung, Y. N. S., M. T. Tyree, and J. Dainty. 1975. *Can. J. Bot.* 53:1342–1346.

21 Clark, J., and J. Levitt. 1956. *Physiol. Plant.* 9:598–606.

21a Cutler, J. M. and D. W. Rains. 1978. *Physiol. Plant.* 42(2):261–268.

22 Cutler, J. M., D. W. Rains, and R. S. Loomis. 1977a. *Physiol. Plant.* 40(4):255–260.

23 Cutler, J. M., K. W. Shahan and P. L. Steponkus. 1980. *Crop Sci.* 20:307–310.

23a Cutler, J. M., K. W. Shahan and P. L. Steponkus. 1980a. *Crop Sci.* 20:314–318.

24 Cutler, J. M., D. W. Rains and R. S. Loomis. 1977b. *Agron. J.* 69:773–779.

25 Cutler, J. M., K. W. Shahan, and P. L. Steponkus. 1978. *Agron. Abst.* 72.

26 Davies, W. J. 1977. *Crop Sci.* 17:735–740.

27 Davis, S. D., and K. J. McCree. 1978. *Crop Sci.* 18:280–282.

28 Davis, S. D., C. H. M. Van Bavel, and K. J. McCree. 1977. *Crop Sci.* 17:640–645.

29 Dubetz, S. 1969. *Can. J. Bot.* 47:1640–1641.

30. Fereres, E., E. Acevedo, D. W. Henderson, and T. C. Hsiao. 1978. *Physiol. Plant.* 44:261–267.

31 Fischer, R. A., and N. C. Turner. 1978. *Annu. Rev. Plant Physiol.* 29:277–317.

32 Frank, A. B., J. F. Power, and W. D. Willis. 1973. *Agron. J.* 65:777–780.

33 Goode, J. E., and K. H. Higgs. 1973. *J. Hort. Sci.* 48:203–215.

34 Hall, A. E., and G. J. Hoffman. 1976. *Agron. J.* 68:876–881.

35 Hall, A. E., and M. R. Kaufmann. 1975. *Plant Physiol.* 55:455–459.

37 Hanson, A. D., and C. E. Nelsen. 1978. *Plant Physiol.* 62(2):305–312.

38 Hanson, A. D., C. E. Nelsen, and E. H. Everson. 1977. *Crop Sci.* 17:720–726.

38a Hanson, A. D., C. E. Nelsen, A. R. Pedersen and E. H. Everson. 1979. *Crop Sci.* 19:489–493.

39 Hellebust, J. A. 1976. *Annu. Rev. Plant Physiol.* 27:485–505.

40 Henzell, R. G., K. J. McCree, C. H. M. Van Bavel, and K. F. Schertz. 1976. *Crop Sci.* 16:660–662.

41 Hsiao, T. C. 1973. *Annu. Rev. Plant Physiol.* 24:519–570.

42 Hsiao, T. C., and E. Acevedo. 1974. *Agric. Meteorol.* 14:59–84.

43 Hsiao, T. C., E. Acevedo, E. Fereres, and D. W. Henderson. 1976a. *Phil. Trans. R. Soc. Lond.* B 273:479–500.

44 Hsiao, T. C., E. Fereres, E. Acevedo, and D. W. Henderson. 1976b. Water stress and dynamics of growth and yield of crop plants. In *Water and Plant Life: Problems and Modern Approaches.* O. L. Lange et al. Eds. Springer-Verlag, Berlin.

44a Huffaker, R. C., T. Radin, G. E. Kleinkopf, and E. L. Cox. 1970. *Crop Sci.* 10:471–474.

45 Hull, H. M., and C. A. Bleckmann. 1977. *Am. J. Bot.* 64(9):1083–1091.

46 Hull, H. M., H. L. Morton, and J. R. Wharrie. 1975. *Bot. Rev.* 41:421–452.

47 Hull, H. M., L. N. Wright, and C. A. Bleckmann. 1978. *Crop Sci.* 18:699–704.

48 Hurd, E. A. 1976. Plant breeding for drought resistance. In *Water Deficits and Plant Growth*, T. T. Kozlowski, Ed. Vol. 4. Academic, New York.

49 Iljin, W. C. 1957. *Annu. Rev. Plant Physiol.* 8:257–274.

50 Janes, B. E. 1966. *Soil Sci.* 101:180–188.

51 Janes, B. E. 1968. *Physiol. Plant.* 21:334–345.

52 Johns, G. G. 1978. *Aust. J. Plant Physiol.* 5:113–125.

53 Jones, H. G. 1973. *New Phytol.* 72:1095–1105.

54 Jones, M. M., and H. M. Rawson. 1979. *Physiol. Plant.* 45:103–111.

55 Jones, M. M., and N. C. Turner. 1978. *Plant Physiol.* 61(1):122–126.

56 Jordan, W. R., K. W. Brown, and J. C. Thomas. 1975. *Plant Physiol.* 56:595–599.

57 Kaufmann, M. R. 1976. Water transport through plants: current perspectives. In *Transport and Transfer Processes in Plants*, I. F. Wardlaw and J. B. Passioura, Eds. Academic, New York.

58 Klar, A. E., J. A. Usberti, and D. W. Henderson. 1978. *Crop Sci.* 18:853–857.

59 Koller, H. R., and J. H. Thorne. 1978. *Crop Sci.* 18:305–307.

59a Kozlowski, T. T. (Ed.) 1968. *Water Deficits and Plant Growth*, Vol. I, Academic, New York.

59b Kozlowski, T. T. (Ed.) 1972. *Water Deficits and Plant Growth*, Vol. III, Academic, New York.

60 Kozlowski, T. T. 1976. Water relations and tree improvement. In *Tree Physiology and Yield Improvement*, M. G. R. Cannell and F. T. Last, Eds. Academic, New York.

61 Kramer, P. J. 1969. *Plant and Soil Water Relationships: A Modern Synthesis*. McGraw Hill, New York.

62 Kriedemann, P. E., B. R. Loveys, J. V. Possingham, and M. Satoh. 1976. Sink effects on stomatal physiology and photosynthesis. In *Transport and Transfer Processes in Plants*, I. F. Wardlaw and J. B. Passioura, Eds. Academic, New York.

63 Lakso, A. N. 1979. *J. Amer. Soc. Hort. Sci.* 104:58–60.

64 Lange, O. L. 1975. Plant water relations. In *Progress in Botany*, Vol. 37, H. Ellenberg et al., Eds. Springer-Verlag, Berlin.

65 Lange, O. L., L. Kappen, and E. D. Schulze (Eds.). 1976. *Water and Plant Life: Problems and Modern Approaches*. Springer-Verlag, Berlin.

66 Levitt, J. 1972. *Responses of Plants to Environmental Stresses*. Academic, New York.

66a Levitt, J. 1980. *Responses of Plants to Environmental Stresses* Water, Radiation, Salt, and Other Stresses. 2nd Ed. Vol. II, Academic, New York.

67 Loesch, R., and B. Schenk. 1978. *J. Exp. Bot.* 29(110):781–788.

68 Ludlow, M. M., and T. T. Ng. 1977. *Aust. J. Plant Physiol.* 42:263–272.

69 Mayaki, W. C., I. D. Teare, and L. R. Stone. 1976. *Crop Sci.* 16:92–97.

70 McMichael, M. P., and C. D. Elmore. 1977. *Crop Sci.* 17:905–908.

71 McPherson, H. G., and J. S. Boyer. 1977. *Agron. J.* 69:714–718.

72 Meyer, R. F., and J. S. Boyer. 1972. *Planta* 108:77–87.

73 Morgan, J. M. 1977. *Nature* 270.234 235.

74 Orshan, G. 1954. *J. Ecol.* 42:442–444.

75 O'Toole, J. C., R. K. Crookston, K. J. Treharne, and J. L. Ozbun. 1976. *Plant Physiol.* 57:465–468.

76 O'Toole, J. C., J. L. Ozbun, and D. H. Wallace. 1977. *Physiol. Plant.* 40.111–114.

76a Pallardy, S. G. and T. T. Kozlowski. 1979. *Plant Physiol.* 64:112–114.

76b Parsons, L. R. 1979. *Hort. Sci.* 14:590–593.

77 Parsons, L. R., and D. W. Davis. 1982. *Bot. Gaz.* In Press.

78 Passioura, J. B. 1972. *Aust. J. Agric. Res.* 23:745–752.

79 Passioura, J. B. 1976. *Aust. J. Plant Physiol.* 3:559–565.

80 Potter, J. R., and J. S. Boyer. 1973. *Plant Physiol.* 51:993–997.

81 Rawson, H. M., A. K. Bagga, and P. M. Bremner. 1977. *Aust. J. Plant Physiol.* 4(3):389–402.

82 Rawson, H. M., N. C. Turner, and J. E. Begg. 1978. IV. *Aust. J. Plant Physiol.* 5:195–209.

83 Routley, D. G. 1966. *Crop Sci.* 6:358–361.

84 Schultze, E. D., O. L. Lange, U. Buschbom, L. Kappen, and M. Evenari. 1972. *Planta* 108:259–270.

85 Sheikholeslam, S. N., and H. B. Currier. 1977. *Plant Physiol.* 59(3):381–383.

86 Sheriff, D. W. 1977. *J. Exp. Bot.* 28(107):1399–1407.

87 Silvius, J. E., R. R. Johnson, and D. B. Peters. 1977. *Crop Sci.* 17(5):713–716.

88   Singh, T. N., D. Aspinall, and S. F. Boggess. 1973. *Aust. J. Biol. Sci.* 26:57–63.

89   Singh, T. N., D. Aspinall, and L. G. Paleg. 1972. *Nature New Biol.* 236:188–190.

90   Singh, T. N., L. G. Paleg, and D. Aspinal. 1973. *Aust. J. Biol. Sci.* 26:45–56.

91   Singh, T. N., L. G. Paleg, and D. Aspinall. 1973. *Aust. J. Biol. Sci.* 26:65–76.

92   Skoss, J. D. 1955. *Bot. Gaz.* 117:55–72.

93   Stewart, C. R. 1972. *Plant Physiol.* 50:679–681.

94   Stewart, C. R., S. F. Boggess, D. Aspinall, and L. G. Paleg. 1977. *Plant Physiol.* 59:930–932.

95   Stewart, C. R., C. J. Morris, and J. F. Thompson. 1966. *Plant Physiol.* 42:1585–1590.

96   Stewart, C. R., and J. A. Lee. 1974. *Planta* 120:279–289.

96a Sung, F. J. M., and D. R. Krieg. 1979. *Plant Physiol.* 64:852–856.

96b Tully, R. E., A. D. Hanson, and C. E. Nelsen. 1979. *Plant Physiol.* 63:518–523.

97   Turner, N. C. 1974. *R. Soc. NZ Bull.* 12:423–432.

98   Turner, N. C. 1978. *Agron. Abst.* 87.

99   Turner, N. C. 1979. Drought resistance and adaptation to water deficits in crop plants. In *Stress Physiology in Crop Plants*, H. Mussell and R. C. Staples, Eds. Wiley-Interscience, New York, p. 343–372.

100 Turner, N. C., and J. E. Begg. 1978. Responses of pasture plants to water deficits. In *Plant Relations in Pastures*, J. R. Wilson, Ed. CSIRO, International Scholarly Book Service, Canberra, Australia, p. 50–66.

101 Turner, N. C., J. E. Begg, H. M. Rawson, S. D. English, and A. B. Hearn. 1978. *Aust. J. Plant Physiol.* 5:179–194. 343–377

102 Tyree, M. T., and H. T. Hammel. 1972. *J. Exp. Bot.* 23:267–282.

103 Waldren, R. P., I. D. Teare, and S. W. Ehler. 1974. *Crop Sci.* 14:447–450.

104 Walter, H. 1955. *Annu. Rev. Plant Physiol.* 6:239–252.

105 Walter, H. and E. Stadelmann. 1974. A New Approach to the Water Relations of Desert Plants. In *Desert Biology*, G. W. Brown (Ed.) Academic, New York.

106 Wardlaw, I. F. 1967. *Aust. J. Biol. Sci.* 20:25–39.

107 Wardlaw, I. F. 1969. *Aust. J. Biol. Sci.* 22:1–16.

108 Wardlaw, I. F. 1971. *Aust. J. Biol. Sci.* 24:1047–1055.

109 Weete, J. D., G. L. Leek, C. M. Peterson, H. E. Currice, and W. D. Branch. 1978. *Plant Physiol.* 62:675–677.

110 Wenkert, W., E. R. Lemon, and T. R. Sinclair. 1978. *Annals of Botany* 42:307–310.

Chapter 7

BREEDING FOR DROUGHT RESISTANCE AND PLANT WATER USE EFFICIENCY

J. E. QUIZENBERRY

U.S. Department of Agriculture, Science and Education Administration, Agricultural Research; Southern Plains Cotton Research Laboratory, Lubbock, Texas

Most places in the world are subject to drought, but the duration and intensity vary greatly from one climatic zone to the next. Losses incurred from an extended drought can amount to many hundreds of millions of dollars. Direct losses result from reduced crop yields. Estimates from indirect losses are more difficult to evaluate but would include losses from crops not planted, abandonment of land, and land use changes following the drought. Although the primary losses are borne by agricultural industries, the cost is ultimately spread over the whole nation when governments make relief grants to the agricultural sector and when consumer prices rise following the shortage of commodities.

Only precipitation or irrigation can completely alleviate the effects of drought on crop yields. During periods of drought there is presently no reliable way to increase the amount of precipitation. If irrigation water is not available, the only possible solutions are through cultural practices that increase the availability of stored soil moisture or by the development of crop varieties or hybrids that can more efficiently avoid or tolerate the drought period.

It is axiomatic that the productivity of a crop grown under moisture stress will be less than its productivity when it is grown with ample supplies of soil moisture. Therefore, biological immunity to the effects of drought is not a possibility. However, through plant breeding, it may be possible to develop some degree of tolerance to the effects of drought. In some ways, the situation with drought is analogous to breeding for disease or insect resistance in crop plants. For some diseases or insects, it is not usually possible to breed for immunity and in these cases, the breeder only attempts to find and develop high levels of resistance. Often, he must be satisfied with marginal levels of tolerance. Perhaps these same degrees of resistance to the effects of drought are possible within our major crop plants.

Extending the available soil moisture through more of the growing season by reducing the rates of transpiration or by developing more efficient use of transpirational water through photosynthesis should result in large increases in productivity. An additional centimeter of available soil moisture above that needed for plant maintenance can increase the yield of corn 18–44 kilograms per hectare, soybeans 11–20 kilograms per hectare (106), and lint yield of cotton 2–4 kilograms per hectare (83). Thus, the amount of improvement needed in water use efficiency does not have to be large to have a considerable influence on crop yields.

To the plant breeder, the term "drought resistance" is related to a moisture stress environment and means the ability of one genotype to be more productive with a given amount of soil moisture than another genotype. His primary concern is with the variability that exists within species rather than that which exists between species. Sometimes, species that are taxonomically closely related are useful, especially if they are not reproductively isolated. Some species of crop plants are better suited than others to culture in areas of low precipitation. At least 90% of the total crop production of semiarid regions, not counting the natural or improved perennial or annual grasses and herbaceous shrubs utilized by domesticated grazing animals, are derived from the four cereal crops of wheat, barley, sorghum, and millet (28). The bulk of the remaining acreage is made up of cotton, oilseeds, and leguminous pulses. This chapter will deal mainly with these crops. It is possible that the approaches presented may be extended to other crops.

DEFINITION OF THE ENVIRONMENT

A knowledge of the components of the environment in which the crop will be grown is required before an attempt can be made to develop better adapted crop varieties. The primary climatic components of interest are the amount of yearly precipitation, the variability in yearly precipitation, the expected seasonal precipitation distribution patterns, the relative humidities, the average temperatures, and type and depth of soils. Generally, average yearly precipitation is not a meaningful component because of the large amount of variation which occurs from year to year in most semiarid regions. Often, a more consistent measurement is the seasonal precipitation distribution patterns. Most of the precipitation a region receives often occurs during a definite period of the year. The amount of precipitation that occurs during this period generally determines a crop's productive potential. Some semiarid regions are associated with low relative humidities which cause rapid cooling after sunset. In other areas, relatively low average temperatures during the growing season interact with low soil moisture deficits to moderate the effect of drought through reduced growth and evapotranspiration. The response of plants to moisture stress depends to some extent on the type and depth of the soil. On coarse sands, most of the soil moisture is readily available and plants can make rapid growth until evapotranspiration depletes the available supply. Plants then suffer a sudden and severe moisture deficit and death may follow if the root zone is not resupplied. However, on heavy clay, soil moisture is held with progressively increasing tension as soil moisture content is reduced below field capacity. Transient water deficits reduce the rate of plant growth, but soil moisture continues to be released as the tension progresses towards the permanent wilting point. The reduced plant growth rate, the progressive drought-hardening, and the much greater amount of soil moisture available to the plant between periods of rewetting allow growth to continue for longer periods on clays than on sands. Adequate yields can often be obtained on clays especially if the crop has a deep and extensively branched root system and a high cuticular resistance (21).

A study of drought resistance requires an objective definition of drought, but to date no universally acceptable one has been developed. A report by the World Meterological Organization (21) on the definitions of drought included 14 based on precipitation, 13 on precipitation and mean temperatures, 11 on climatic indices and estimates of evapotranspiration, and 15 on soil-water and crop parameters. Some of the common criteria that they used were precipitation, air temperature, relative humidity, evaporation from a free water surface, transpiration from plants, wind, airflow, soil moisture, and plant conditions. In this chapter, drought will be defined as any period during which plant and/or soil water deficiencies affect the growth and development of crop plants. These deficiencies may result either from a small moisture supply or a large moisture demand. The duration of this drought period will determine the amount of damage done to the plant's growth and development.

The type of drought environment in which a crop will be grown should determine, to some degree, the type of resistance mechanisms that should be developed and the appropriate breeding methodology that should be used. In general, three types of environments can be associated with drought stress in plants; however, numerous combinations of these environments occur.

Stored Moisture Environment

In the first type of environment, the crop completes its entire life cycle on soil moisture that is stored in the soil during a prior wet season. In this evironment, distinct seasonal wet and dry periods occur. Generally, 70% or more of the total annual precipitation occurs in six months or less (106). The degree or intensity of the drought will be determined by the amount of water that is stored in the soil during this period. In this type of environment, any genetic, climatic, or cultural factor that lengthens the vegetative growth stage, increases the growth of the aerial plant parts, or enhances evapotranspiration, results in a smaller share of the stored soil moisture available and within reach of the roots for reproductive development and will reduce yields (89). Plant responses which affect the rate of evapotranspiration are rapid uptake of moisture, storing water in the tissue, retarding loss through the epidermis and stomata, increasing metabolic efficiency, and rapid growth. The potential plant mechanisms available to exploit this type of environment cover a broad spectrum of traits: phenological, morphological, functional, and metabolic. The potential for genetic improvement of drought resistance in this environment should have a high probability of success.

Variable Moisture Environment

The second climatic drought environment is one in which the crop is grown during the portion of the year when precipitation is expected to occur. Alternate dry and wet periods of varying lengths can be expected during crop growth. The soil moisture content throughout the profile will seldom be at field capacity. To improve the chance of establishing an adequate stand of seedlings, planting is generally attempted only after a rain has wet the soil. Plants, grown in this environment, must be able to take advantage of the periodic rainfall when it does occur. Plants are in constant competition with atmospheric evaporation for the available soil moisture. In this environment, it should be advantageous for plants to have high photosynthetic rates, sensitive stomatal responses to moisture deficits, dense but not necessarily deep roots, rapid osmotic adjustments, and indeterminate growth habits (102). The ability of a plant to survive periods of low cell turgor and to recover rapidly upon the relief of the moisture stress should be important (56). The variability in seasonal precipitation in this type of environment may hinder the development of adapted crop varieties and, thereby, reduce the probability of a successful breeding effort.

Optimal Moisture Environment

In the third type of moisture deficit environment, the crop is grown with adequate soil moisture during most of its life cycle but occasional periods of drought occur during the growing season. The imposition of a drought period in this environment can be particularly severe on crop productivity because of the inadequate amount of time available for the plants to become adjusted to the dry conditions. Droughts in these areas are highly unpredictable. In these production areas, high monetary inputs by producers are usually required and these producers expect relatively high productivity. This environment is considered to be optimal for crop production. The drought period can occur during a portion of one day when atmospheric evaporativity greatly exceeds root uptake; but ordinarily it is associated with a period when less than normal precipitation is received. Some mechanisms which may be of some value to a plant breeder would be extensive root development, rapid osmotic adjustment to maintain cell turgor, and indeterminate growth habits (i.e., increased tillering, progressive flowering, etc.) (102). Stomata that close quickly at a relatively high water content to maintain the plant water balance during drought should be useful (86). The absence of a predictable drought period to use in field screening and the possible negative relationships which may exist between drought-resistant traits and productivity under ample moisture conditions (18), should make breeding for drought resistance extremely difficult in this environment.

BREEDING FOR DROUGHT RESISTANCE

The breeder, whose prime responsibility is the development of drought-resistant genotypes, must take an interdisciplinary approach toward the evaluation of germplasm. He should establish working relationships with those disciplines that are concerned with plant and soil water relations, climatology, and plant growth and metabolism. An understanding of plant physiology is essential to the success of the breeding effort. A crop physiologist brings to the breeding effort a knowledge of plant mechanisms and experimental techniques that are necessary in the evaluation of drought resistance. Often these experimental techniques must be modified by the physiologist to fit the large number of measurements required in a successful breeding program. If the physiologist cannot modify the technique to give the necessary number of measurements, then perhaps only the parental material that enter into a crossing program can be tested. In this case, periodic evaluations should be made during the progress of the breeding program to be certain that the mechanism or trait is still in the population. Many times, it will be necessary for the crop physiologist to develop new techniques or even to identify new mechanisms to screen the available germplasm. The physiologist must maintain constant monitoring of soil and plant water throughout the growing season. Initially, the breeder should be prepared to search all the available sources of existing germplasm and to let yield take second place to the identification of superior sources of drought resistance. The breeder should recognize that if economic yield is anything other than total plant dry matter, then a second area of research becomes the partitioning of the products of photosynthesis between fruit and vegetative parts. Estimates of productivity, based strictly on the weight of reproductive units, can be misleading and may cause the discarding of potentially superior germplasm. Economically, only the usable portion of a crop plant is important; however, in a biological sense all plant dry matter is made through the processes of photosynthesis and, therefore, the total production of dry matter determines the response of a genotype to drought stress.

Traits Important in Drought Resistance

A number of recent papers and books have reviewed (a) the response of plants to water stress (28, 4, 44, 45, 66), (b) traits that adapt plants to a moisture deficit environment (86, 52, 78), (c) approaches to the evaluation of germplasm for drought resistance (102, 95, 96), and (d) methods of breeding for drought resistance (89, 3, 12, 13, 51, 53, 49). These reviews will be used extensively in this section without specific reference to the particular review. Emphasis will only be given to those traits which have been demonstrated to have some relationship to productivity under moisture deficit conditions and for which some within species variability has been identified.

Many traits have been identified that affect plant adaptation to climatic drought. Botanists have studied the survival of plants or plant parts during and after drought stress, while the main concern of agronomists has been maximum productivity of crop plants under moisture stress. Biologically, plant survival may not be associated with maximum productivity. Often, plant survival in a moisture deficit environment is achieved by partial or complete cryptobiosis, that is, reduction in leaf area, depression of metabolic activity, and so on. Usually, these types of survival mechanisms are associated with a reduction in yield potential, and therefore, are of limited practical value to the plant breeder.

Studies on the slender wild oat species, *Avena barbata,* have shown that the enzymatic constitution of plants can be genetically altered to adapt to a moisture deficit environment (1, 17, 35). These enzymatic changes are reflected in the phenotypes of the plants (36). Those genotypes that were adapted to moisture deficit (zeric) environmental conditions were shown to be genetically later in flowering and maturity, were taller in plant height, and had fewer tillers than were genotypes with enzymatic combinations associated with higher moisture (mesic) environments. Research on *Gossypium* and *Pinus* species suggest that population adaptation for increased productivity occurs when plants are grown under natural moisture deficit field conditions (84, 105). These results strongly suggest that sufficient genetic potential exists among plants to shift populations toward adaptation to moisture deficit environments.

Levitt (66) has classified drought-resistant traits into avoidance and tolerance categories. Most drought-resistant traits in crop plants can be placed in the avoidance category (i.e., they tend to aid the plant in avoiding the condition of low tissue water). Most of these avoidance traits are xeromorphic (i.e., they are developed by the plant as a result of moisture stress). Levitt (66) has stated that drought resistance is possible without the development of xeromorphic traits only if the resistance is due to tolerance. He has further concluded that among economically important crop plants, tolerance seldom seems to be the deciding factor in their drought resistance. Xeromorphic structure has been shown to be a quantitative character and is developed to different degrees in different plants (30). Thus, xeromorphy should be hereditarily fixed, but should vary in expression in response to the environment. The same genotype when grown under conditions of deficient moisture should develop xeromorphic traits to a greater degree than when it is grown under optimal supplies of moisture. An example of a xeromorphic trait would be epicuticular wax. A heavier layer of wax is deposited on a leaf grown in a stressful environment than on a leaf grown under optimal moisture conditions (31). Xermorphic traits should have high heritability in a moisture stress environment. Therefore, selection for these types of traits should be more successful in moisture deficit environments. Most of the traits discussed in the rest of this section are to some degree xerophilic (66).

Earliness of Maturity. In some crop plants, one of the advances made toward increased productivity in moisture stress environments has been made through increased earliness in crop maturity. In general, earliness results in an escape of the effects of drought and as such is not a true resistance mechanism. In wheat, a strong consistent negative correlation between grain yield and days to first ear emergence (late maturity) under simulated drought conditions has been observed (19). Between 40 and 90% of the variation in the yield of wheat under drought can be accounted for by earliness. Each day of earlier maturity in winter wheat has been shown to impart a yield advantage of from 54 to 120 kg/ha for varieties earlier than the Kharkof variety (88). In this study, it was noted that earliness afforded considerable escape from the degradation of rust, some insects, and the heat of summer. Early maturity and high yields in upland cotton are correlated when timely rainfall does not occur during the growing season, and the crop is forced to make its yield on stored soil moisture. This earliness-yield relationship in cotton reverses when adequate rain occurs during the growing season (83). Several studies have shown that yield may be positively correlated with maturity date when adequate water supplies are available (18, 32, 33). Care must be taken when earliness of maturity is used to select for increased productivity in moisture stress environments. If total crop production is expected to be made on stored soil moisture, then earliness should be advantageous; however, if rainfall occurs during the growing season or if above normal moisture is available, then productivity may be reduced. In those crops where breeding for earliness has been successful and the improvements associated with earliness have been exploited, additional advances for drought adaptation must come from other traits (89).

Root Growth. Root development and the amount of plant water absorption from the soil are closely related (51). A highly developed root system increases efficiency of absorption and relative resistance to drought (77). Ivanow (54) observed that a direct relationship did not exist between transpiration rate and drought resistance, but that the ability of a plant to resist the adverse effects of drought was directly proportional to the density and extent of root development. The relative productivity between sorghum and corn under soil moisture stress has been demonstrated to be proportional to the root development in relation to foliar areas (69). Generally, studies have shown that as the depth, width, and branching of root systems increases, plant water stress decreases (52). However, Passouri (81) suggested that reduced root development during the vegetative stage of growth increased wheat yields by making more water available during the grain filling stage.

Cultivar evaluations and breeding for root development have been carried out in several crop species. Lines of barley have been selected that exceeded the better parent in depth of rooting and amount of root branching (31). These selections were higher yielding than their parental lines. Hurd (47) has shown a strong genotype by moisture interaction for root development among several wheat cultivars. He concluded that rooting patterns partially explained the grain yield differences among the cultivars in drought stress environments. Hurd (46) further identified a wheat variety with a very extensive root system, especially at deeper soil depths. In a cross with a high quality durum wheat, he produced varieties that had extensive root patterns and superior quality. These new varieties had higher grain yield in moisture deficit environments than did either parent (50). Hurd (51) has concluded that selection for grain yield made over several seasons in a semiarid environment will produce the more extensive root systems of the better parental material. Comparisons between semidwarf and tall winter wheat varieties for root growth in contrasting soil types over several years have demonstrated little varietal differences in root growth, although there were some indications that at deeper depth the roots of the semidwarf types were more extensive (16, 68). Significant variation among 30 grain sorghum genotypes has been identified for modal root length and root dry weight (59). Blum et al. (8) have shown that greater root volume was associated with later maturity genotypes in sorghum. They further demonstrated heterosis in root length of the seminal roots, growth rate of adventitious roots, and total root volume (9).

Root systems in soybeans differ in their ability to absorb water from the soil and to provide the plant with water during the grain filling period (97). One soybean variety has been shown to have twice as large a root system as another variety (85). Taylor et al. (101) found that significant variation existed within and among 28 varieties of soybeans in the extension rate of the taproot and in the depth they could penetrate in 26 days. Eleven percent or less of this variation was accounted for by the relationship between taproot length and seed size and 42% or less by the relationship between top dry weight and taproot length at 26 days. In cotton, 'Acala' (*G. hirsutum* L.) extracted more water from the soil profile at the 180–210-cm depth than did 'Pima' (*G. barbadense* L.) (34). Roots of the 'Acala' plants extended 100 cm from the row at this depth whereas the roots of the 'Pima' extended only 33 cm.

The development of deep or extensive root systems, unlike mechanisms that reduce water use or increase metabolic efficiency, can only occur through the use of part of the photosynthate produced by the plant with a compensating reduction in other dry weight (28). The root-to-shoot ratio inevitably increases with water stress (64). This increase is usually a result of a reduced top growth relative to growth of the roots, but evidence of absolute increases in root growth have been reported (5). Several techniques have been described to enable the measurement of root growth (3, 47, 59).

Stomatal Control.   The stomata on the plant leaf are capable of influencing many aspects of plant metabolism. For example, photosynthesis is a process in which hydrogen is fixed in the form of a metastable carbon compound, carbon dioxide is taken up, and oxygen given off. Respiration reverses this process, and the energy built into the carbohydrate is released together with CO_2. The stomata are, therefore, of considerable importance to the vital functions of energy storage and utilization. The stomata further act as plant protective mechanisms by decreasing water loss through closure during periods of plant water deficits as well as influencing the rates of photosynthesis and respiration. These functions of the stomata cannot be readily dissociated. Structures that are associated with transpiration, respiration, and photosynthesis, all of which are important in plant growth, must have evolved under the influence of natural selection and be under genetic control (108). A corn plant that weighs 454 grams at harvest will have absorbed, transported, and transpired 204,228 grams, or 205 liters, of water during its life cycle (73). The total quantity of water required by this plant for actual dry weight production was relatively small, usually less than 5% of all water absorbed by its roots. Most of the water that entered the plant was lost through transpiration, directly contributing nothing to its growth (63). All of this water passed through the stomata except for that portion which escaped directly through the leaf cuticle.

A range of crop species has been shown to vary in the amount of leaf water stress necessary for stomatal closure. They ranged from a low stress of −8 bars in field beans to a high stress of −28 bars in cotton (102). The environment in which a plant is grown also has an effect upon the water content when stomatal closure occurs (58). The stomata of cotton grown in the field do not close at a leaf water potential of −27 bars; in contrast, greenhouse grown cotton plants exhibit marked closure at −16 bars. Furthermore, plants preconditioned with drying cycles can also change the threshold water potential where stomata close (90). If the ability of the stomata to regulate this water loss were under genetic control, then a mechanism should exist to either reduce the amount of water loss through transpiration or increase water use efficiency through relative stomatal closure during periods of high atmospheric stress (89). An inheritance study of stomatal behavior in upland cotton has shown that genetic separation of the parental lines could be detected only during the portion of the day with the highest leaf water stress (90). The environment consistently affected stomatal behavior over the entire day. Genetic analysis revealed that stomatal behavior was associated with both additive and dominance genetic variances and that although heritability was rather low, selection was possible. The genetic control was further shown to not be related to the source of the maternal cytoplasm. The tendency for the stomata to close during the period of highest leaf water stress was completely dominant to the capacity of the stomata to remain relatively open during this stress period. Varietal differences in stomata response to leaf water stress have been reported in sorghum (6, 40, 41). Sullivan (95) has outlined techniques to measure the response of the stomata to moisture stress.

Cuticular Resistance. The water-saving advantages of sensitive stomatal closure are lost if the plant does not have a high cuticular resistance to water loss. Schonherr (93) has demonstrated that water permeability of the leaf cuticle is determined completely by the amount of cuticular waxes. Chatterton et al. (15) have shown the significance of the leaf bloom (wax) trait in sorghum for reducing transpiration rates. Ross (91) has demonstrated the yield advantage of leaf bloom genotypes of sorghum in a moisture stress environment. Techniques to rapidly measure epicuticular wax content have been developed (22).

Although stomatal control and cuticle development may be effective in reducing water loss and thus reducing the effect of high tissue water deficits, when stomatal closure is complete the reduction in gas exchange also reduces photosynthesis (94). Thus, persistent stomatal closure should lead to a reduction in productivity. However, it seems very probable that under identical environmental conditions, plants with thick layers of cuticle and stomata that respond as soon as water deficits begin to develop are likely to resist the effects of drought better than those with less responsive stomata and less cuticle.

Stomatal Number. The number of stomata per unit leaf area has been shown to vary among genotypes within species (108, 74, 98) and to be under genetic control (39, 67, 99, 100, 109). In barley, stomatal frequency, stomatal resistance, and transpiration varied among a group of lines. The lines that had low stomatal frequencies transpired less water than did lines with more stomata; however, stomatal frequency did not affect the rate of photosynthesis (75). Low stomatal frequency was associated with high photosynthetic rates in beans (55) and maize (38), and with greater drought tolerance in blue panicgrass (20). Hesketh (42) and Freeland (29) did not find relationships between frequency of stomata and photosynthesis in several species. Muenscher (76) had earlier concluded that there was no relationship between frequency of stomata and transpiration in several genera. In cotton, stomata frequency of varieties presently grown on the Texas High Plains were compared with their ancestral varieties (87). The newer varieties had significantly more stomata than did their ancestral varieties. The relative difference was about the same for the top and bottom of the leaves and for greenhouse and field grown plants. The ratio of stomata numbers between the bottom and the top of the leaf was significantly higher in the ancestral varieties than in the new varieties. However, the number of epidermal cells per unit area was constant in the new and ancestral varieties; and therefore, the change in stomata frequency was not caused by a change in cell size but represented a change in the epidermal cell:stomata ratio. Measurement of leaf diffusive resistance showed that the stomata of the new varieties were generally more open throughout the day than the stomata of the ancestral varieties. It was postulated that in the selection of cotton varieties for adaptation to the Texas High Plains environment, which includes high winds, low humidity, and frequent moisture stress, leaves of the cotton plant have been modified to provide for increased gas exchange capacity.

Cell Turgor. The inhibitory effect of water stress on plant growth may be explained to some degree because cell growth depends on cell turgor pressure as its driving force. The degree of cell turgidity of a plant is based on the relative rates of root water absorption and of stomatal water loss. Turgor pressure can be affected by atmospheric, soil, and plant factors that modify the rates of absorption and transpiration. Turgor pressure is mathematically the result of the total leaf water potential less the osmotic potential and a small matrix component (44). Because the matrix component is very close to zero and does not become numerically significant until most of the tissue water is lost, it is for all practical reasons ignored. As the moisture deficit increases in a leaf, leaf water potential decreases with a subsequent decrease in turgor pressure. The decreases in leaf water potential and turgor pressure are not linear because the osmotic potential will increase as tissue moisture removal occurs. This increase in osmotic potential (osmotic adjustment) allows the plant to maintain cell turgidity while water potential is reduced. This relationship perhaps explains why attempts to relate seasonal changes in total leaf water potential to productivity have been unsuccessful (61, 62), although variability in the seasonal changes in water potential among wheat cultivars have been measured (60). Perhaps, estimates of turgor pressure will prove useful in measuring the ability of a plant to maintain water balance and, thereby, avoid the detrimental effects of moisture stress on crop productivity (56).

Chemical Traits. Various attempts have been made to define specific chemical or physiological traits which are a measurement of drought resistance. Among these are proline accumulation (7, 37, 70, 80, 111), increases in free abscissic acid (43, 65, 71, 110), and changes in nitrate reductase activity (79). At present, the results of these studies have not shown a consistent relationship with productivity, although the quantitative nature of the approach of many of the techniques have a certain appeal to the breeder.

Sullivan (96) has suggested three criteria that should be used to evaluate a plant's drought resistance. First, the plant should maintain high leaf water potential. He suggests that this indicates that a prolific root system is keeping the aerial portion of the plant well supplied with water. Secondly, stomata should be responsive for control of water loss. If the stomata are relatively closed and the leaf water potential is relatively high, the plant is effectively retarding water loss. And thirdly, a plant should have high heat and desiccation tolerance. He has developed a test for measuring heat and desiccation tolerance (3).

Relationship between Drought-Resistant Traits and Plant Productivity

After a trait that theoretically should be associated with adaptation to drought stress is determined and a suitable technique to measure this trait is developed, the available germplasm should be evaluated for variability. The initial evaluations should be conducted on agronomically acceptable crop varieties or on genotypes that are closely related to acceptable varieties. If significant variability can be found among this type of germplasm and if the trait can be shown to be related to productivity, then the time required for variety development can be reduced. If significant variability is not present among agronomic varieties, then wider sources of germplasm must be evaluated. These more exotic sources of germplasm will increase the breeding problems.

When significant variability in the expression of the selected trait is found, the relationship between this trait and productivity must be established. It is necessary to establish both theoretical and direct relationships. The assumption of a relationship between the trait and productivity should not be made if the trait is expected to be useful as a selection criterion. Differences in productivity between genotypes with extreme expressions of the trait should not be construed as the establishment of a relationship.

Composite Cross Analysis.   One method to establish a relationship between a trait and productivity is through the use of composite cross populations. Several genotypes that have extreme phenotypic expressions of the selected trait should be crossed in all possible combinations to develop the initial composite population. If some of the genotypes originate from exotic germplasm sources, then several cycles of recurrent crossing may be needed. However, in some cases, this will not be necessary since total plant dry matter production can be used as an estimate of productivity. The composite cross population should be inbred until most of the heterozygosity in the population is exhausted. This time period will be longer in crop species that have a high degree of outcrossing and will be shorter in crops with a high degree of self-pollination. The time can be reduced in outcrossing species by controlled self-pollination. A completely self-pollinated species will be homozygous at 87.5% of the gene loci by the F_4 generation. In many crop species, the time needed to develop composite cross populations will be reduced due to the large number of variable populations maintained by various crop breeders. These populations may also be useful to screen for different adaptive traits since many of them contain most of the available variability that exists within a crop.

To determine the relationship between the drought adaptive trait and productivity, seeds of random plants should be taken from the composite population. These seeds should then be planted in progeny rows in a moisture-stressed environment. Replications and locations or years should be included in the experimental design. Soil moisture and plant water status should be monitored on the test plots throughout the growing season. The selected trait and productivity should be measured on the test plots. Perhaps other growth and quality traits should also be measured. The use of regression (linear and nonlinear) along with covariance analyses should determine if the trait is related to productivity. If this test is repeated for 2 years, the seed produced on the first year test rows should be used the second year. The use of this seed will advance the population one generation, thus allowing regression estimates of heritability for both the selected trait and productivity. Analysis of variance techniques are also available to estimate heritability. The comparison of these heritability estimates will determine if the selected trait is more or less heritable than is productivity. If the heritability of the selected trait is less than the heritability of productivity, then the use of this trait in a breeding program may be extremely limited even if the trait is related to productivity. However, if the heritability of the selected trait is higher than the heritability of productivity and the trait is related to productivity, then the trait should be less influenced by the environment than productivity, and should be useful as a selection criteria. If quality traits are measured on these lines, then a regression between the selected trait and these quality parameters should determine if any detrimental relationships exist. When these studies are completed, selections from the evaluated lines may be useful directly in a breeding program. If the initial number of randomly chosen lines is large, it may be possible to select breeding strains that have good expression of the trait and relatively good productivity potential. However, as we have defined productivity (i.e., plant dry weight), some differences may exist between economic productivity and biological productivity.

Isoline Analysis.   Isolines may also be used to study the relationship between an adaptive trait and productivity (23, 24). Isolines are two genotypes that theoretically differ in only one trait. Usually this trait is controlled by a single gene pair (e.g., awned vs. awnless in barley, red vs. green plant color in cotton). In most crop plants, geneticists have developed a number of isolines, and these are available for isoline analyses (23). One approach to the use of isolines would be to grow plants of each allelic combination in a moisture-stressed environment, to monitor plant and soil moisture status throughout the growing season, and to measure productivity (either biological or economic yield or both). An important problem associated with the use of isoline analysis is that traits that are inherited through a polygenic system are difficult, if not impossible, to develop into isolines. Many of the traits that have been suggested as adapting a plant to a drought environment have a polygenic mode of inheritance (107). Eslick and Hockett (23) have suggested that in barley, the worth of a gene used in isoline analysis may be completely obscured by pleiotropic or compensating effects or both. As an example, light-colored foliage in barley is associated with delayed maturity, perhaps due to lower temperatures. A comparison of light-colored foliage with normal green foliage may be an evaluation of maturity date and not foliage color.

Breeding Approaches to Drought Resistance and Yield

Several studies (26, 2, 14, 30, 57, 104) have considered the effects of different growing environments on the response of selection for yield. Environmental and genetic variances have usually been higher when growing conditions were favorable (i.e., when yields were higher). Heritabilities for yield were higher in an optimal environment, and the rate of genetic advance through selection was usually greater (30, 57, 104). These studies indicate that selection for yield should be more productive in a favorable rather than a stress environment. The level of stress (not quantified in most studies) probably determined the magnitude of these selection differences (2).

In general, two breeding approaches are available to the plant breeder when he attempts to develop varieties that yield more in a moisture deficit environment. These approaches are to develop varieties that are highly adapted only to a moisture stress environment (49) or to develop varieties with adaptation to a wide range of environmental conditions (25).

Adaptation to a Specific Environment. The first available approach is to develop varieties that are specifically adapted to a moisture deficit environment. This approach should be most useful in an environment where plants must complete their life cycle on soil moisture stored during a previous season. Hurd (51, 53, 49) has suggested that varieties selected for high yield in optimal moisture conditions will not necessarily have high yield when they are grown in moisture stress conditions. He believes that a superior variety for a moisture stress environment must be selected and evaluated under this environment. Hurd (48) has demonstrated that the variety that has the greatest root development when grown in a nonstress environment does not have the most developed root system in a moisture stress environment. Reitz (89) has implied that to select for wide adaptability is to select for mediocrity or even low yield. Briggle and Vogel (10) have reported that the high yielding, widely adapted dwarf wheat varieties of the Pacific Northwest are, in general, not suited to drought conditions of the Great Plains. Cotton varieties that had the highest water use efficiency in an optimal moisture environment did not have the highest efficiency in a moisture stress environment (83).

The use of this approach may present several problems. Precipitation in most semiarid regions is quite variable from year to year (21). A variety developed through this approach may not be able to respond in years of above normal precipitation. This variability in precipitation also changes the breeders environmental selection index from year to year. The breeder must attempt to compensate for this moisture variability by increasing the number of test locations and by enlarging the population sizes (49). Extreme care must be exercised in field preparation and experimental designs. If below normal precipitation occurs, the yield of the selections may be smaller than is the precision of the available statistical analysis techniques to separate the selections. Heritabilities for yield and yield components will be extremely low, thereby greatly reducing the effectiveness of selection (57).

Adaptation to a Variable Environment. A second breeding approach is to develop crop varieties that are adapted to a broad range of environmental conditions. This approach should be most appropriate when plants receive precipitation during the growing season or in a more optimal growing climate where periodic droughts may occur. Finlay and Wilkinson (26) have described a technique for measuring adaptation in variable environments. They have used this technique to identify and describe the environmental responses of a range of different barley genotypes. Several genotypes were identified which were able to produce high mean yields over all the test environments. Finlay (27) has suggested that high yielding varieties can be selected in optimal environments, and that these varieties should retain their superior yield under less favorable environmental conditions. He has presented evidence from a combining ability study to show that the two components of adaptation, mean yield over all environments and yield stability, were largely independent of each other. He suggested that it should be possible in a breeding program to combine very high yield and wide adaptability. Barley genotypes that had wide adaptation tended to have traits that were intermediate in expression, whereas genotypes which were specifically adapted to a particular environment tended to have extreme trait expression (25). Roy and Murty (92) have suggested that the selection for yield and development characters in wheat should initially be performed under a range of optimal and moisture stressed conditions. Average yield over all these test environments should be used as the selection criteria to identify superior families. Within-family selection is only conducted under optimal moisture conditions. In soybeans, Mederski and Jeffers (72) have shown that optimal moisture conditions gave greater genotypic expression and, thereby, an increase in genotypic variance among varieties. They suggest that an optimal moisture environment appears to be the most useful for the selection of soybean yield attributes.

A Physiogenetic Approach. Another potentially useful approach toward breeding for drought resistance may be to combine the use of optimal and moisture stress environments. In this approach, the assumption is made that yield and drought resistance are different traits that are controlled by different genes or gene systems (i.e., they can be selected independently of one another). An attempt is made to select traits associated with drought resistance in a moisture stress environment and to select yield and quality traits in an optimal moisture environment.

The proposed approach can best be illustrated by using an arbitrarily chosen trait. The trait to be considered is cuticular transpiration as it is affected by the quantity of cuticular waxes. Cuticular waxes are developed on a leaf as the result of environmental stresses (78). The stomata on the leaf close at night; thereby, the rate of stomatal transpiration should be stopped. If the amount of wax on the leaf is high, the amount of water lost through the cuticle during this period should be greatly reduced. This trait has been shown to be related to productivity (91). Using this trait, a series of crosses are made between strains that have large quantities of cuticular wax and strains with superior yielding ability. Particular attention should be given to the choice of the parental strains which are chosen to complement the deficiencies associated with the strains that have high wax levels. A relatively few crosses should be made so that large segregating populations can be grown. In an advanced generation, numerous individual plants are chosen and seed saved. Attempts at individual plant selections are minimal, but obviously inferior plants are not selected.

In the next year, seed from each of the selections are planted in progeny rows in a moisture stress nursery. At least two locations should be used to insure moisture stress. From these progeny rows, a large number of selections are made, based entirely on the amount of cuticular wax. No effort is devoted to selection for agronomic performance; rather, all available effort is devoted to the measurement of leaf wax content.

The selections that are made, based on leaf wax content, are grown the next year in an optimal moisture nursery to allow maximum genetic expression of agronomic traits. Several replications of the material should be grown. In this environment, the strains are reselected for yield and quality traits and comparisons are made with the performance of the best available varieties. Several cycles of selection will probably be needed in the optimal moisture environment until strains with the high yield and acceptable quality are identified.

Further yield testing is conducted in a moisture stress environment similar to the one where the variety will be grown. Additional measurements for leaf wax should be made on the selected strains. The strains developed through this program can be released as varieties, recrossed and the procedure restarted, or used to incorporate additional drought adaptive traits.

In conclusion, the economic advantages associated with crop varieties or hybrids that have superior performance when subjected to moisture stress are tremendous. A small quantity of moisture either saved by the plant through reduced transpiration or more efficiently used by the plant in photosynthesis can mean a relatively large increase in economic yield. Many mechanisms which should aid the plant in adaptation to drought have been identified. Crop physiologists have developed some strong relationships between these mechanisms and adaptation to drought based on accepted approaches to plant growth and development. The plant breeder has, for the most part, failed to utilize these mechanisms because of (a) difficulty in measuring the mechanism, (b) lack of convincing evidence to support a relationship with increased productivity, (c) insufficient knowledge about inheritance, (d) poor communications with crop physiologists, and (e) a sense of helplessness associated with the phenomenon of drought. Water, like CO_2, light, and mineral nutrients is basic to the growth of plants. For many breeders, it is difficult to imagine that enough genetic variability for drought resistance exists within a species for plant breeding to be successful. They have accepted the inalienable occurrence of drought and have concluded that little effort should be expended in a lost cause.

An outbreak of corn blight causes a national alarm. Techniques of evaluation are available. Breeders immediately begin to search for germplasm resistance to the blight organism. Success is achieved. A serious drought occurs in the Great Plains of the United States. A national alarm is sounded, but techniques to evaluate germplasm for drought resistance are not available. Germplasm is not evaluated. We accept the consequences and hope for better weather the next year. We must develop better techniques to identify drought resistant traits and improved approaches towards the utilization of the identified traits in better adapted crop cultivars.

REFERENCES

1   Allard, R. W., G. R. Babbel, M. T. Clegg, and A. L. Kahler. 1972. *Proc. Natl. Acad. Sci.* 69:3043.

2   Allen, F. L., R. E. Comstock, and D. C. Rasmusson. 1978. *Crop Sci.* 18:747.

3   Atsmon, D. 1973. Breeding for drought resistance in field crops. Pp. 157–176 in *Agricultural Genetics Selected Topics,* 1st ed., Rom Moav, Ed. Wiley, New York.

4   Begg, J. E., and N. C. Turner. 1976. Crop water deficits. In *Advances in Agronomy,* Academic, New York-London, 28:161–217.

5   Bennett, O. L., and B. D. Doss. 1960. *Agron. J.* 52:204.

6   Blum, A. 1974. *Crop Sci.* 14:361.

7   Blum, A., and Adelina Ebercon. 1976. *Crop Sci.* 16:428.

8   Blum, A., G. F. Arkin, and W. R. Jordan. 1977. *Crop Sci.* 17:149.

9   Blum, A., W. R. Jordan, and G. F. Arkin. 1977. *Crop Sci.* 17:154.

10 Briggle, L. W., and O. A. Vogel. 1968. *Euphytica* Supplement No. 1:107.

11 Brown, K. W., W. R. Jordan, and J. C. Thomas. 1976. *Physiol. Plant* 37:1.

12 Burton, G. W. 1959. Crop management for improved water-use efficiency. In *Advances in Agronomy.* Academic, New York-London, 11:104–109.

13 Burton, G. W. 1964. The geneticist role in improving water use efficiency by crops. Pp. 95–103 in *Research on Water.* Am. Soc. Agron. Spec. Publ. Ser. No. 4, Madison, Wisconsin.

14 Byth, D. E., B. E. Caldwell, and C. R. Weber. 1969. *Crop Sci.* 9:702.

15 Chatterton, N. J., W. W. Hanna, J. B. Powell, and D. R. Lee. 1975. *Can. J. Plant Sci.* 55:641.

16 Cholick, F. A., J. R. Welsh, and C. V. Cole. 1977. *Crop Sci.* 17:637.

17 Clegg, M. T., and R. W. Allard. 1972. *Proc. Natl. Acad. Sci.* 69:1820.

18 Dalton, L. G. 1967. *Crop Sci.* 7:271.

19 Derera, N. F., D. R. Marshall, and L. M. Balaam. 1969. *Exp. Agr.* 5:327.

20 Dobrenz, A. K., L. Neal Wright, A. B. Humphrey, M. A. Massengale, and W. R. Kneebone. 1969. *Crop Sci.* 9:354.

21 Drought and Agriculture. (Report prepared by the C. Ag. M. working group on the Assessment of Drought.) 1975. by C. E. Hounam, J. J. Burgos, M. S. Kalik, W. C. Palmer, and J. Rodda. Tech. Note 138, World Meterological Organization, Geneva, pp. 51–58.

22 Ebercon, Adelina, A. Blum, and W. R. Jordan. 1977. *Crop Sci.* 17:179.

23 Eslick, R. F., and E. A. Hockett. 1975. Genetic engineering as a key to water-use efficiency. Pp. 13–25 in *Plant Modification for More Efficient Water Use,* 1st ed., J. F. Stone, Ed. Elsevier, New York.

24 Ferguson, Hayden. 1975. Use of variety isogenes in plant water use efficiency studies. Pp. 25–31 *Plant Modification for More Efficient Water Use,* 1st ed., J. F. Stone, Ed. Elsevier, New York.

25 Finlay, K. W. 1963. *J. Agric. Eng. Res.* 8:41.

26 Finlay, K. W., and G. N. Wilkinson. 1963. *Aust. J. Agric. Res.* 14:742.

27 Finlay, K. W. 1968. The significance of adaptation in wheat breeding. Pp. 403–409 in *Third International Wheat Genetics Symposium,* 1st ed., K. W. Shepherd, Ed. Butterworths, Sydney.

28 Fischer, R. A., and N. C. Turner. 1978. Plant productivity in the arid and semi-arid zones. In *Annual Review of Plant Physiology.* Annual Reviews, Inc., Palo Alto, California, 29:277–317.

29 Freeland, R. O. 1948. *Plant Physiol.* 23:595.

30 Frey, K. J. 1964. *Crop Sci.* 4:55.

31 Garkavy, P. F., P. V. Danilchik, and A. A. Linchevsky. 1970. *Vop., Genet. Selsksii, Semenovodstva.,* pp. 53–65.

32 Goldsworthy, P. R., and M. Colegrove. 1974. *J. Agric. Sci., Camb.* 83:213.

33 Gunn, R. B., and R. Christensen. 1965. *Crop Sci.* 5:299.

34 Gutstein, Y. 1969. *Plant Mater. Veg.* 17:347.

35 Hamrick, J. L., and R. W. Allard. 1972. *Proc. Natl. Acad. Sci.* 69:2100.

36 Hamrick, J. L., and R. W. Allard. 1975. *Evolution* 29:438.

37 Hanson, A. D., C. E. Nelsen, and E. H. Everson. 1977. *Crop Sci.* 17:720.

38 Heichel, G. H. 1971. *J. Exp. Bot.* 22:644.

39 Heichel, G. H. 1971. *Crop Sci.* 11:830.

40 Henzell, R. C., K. J. McCree, C. H. M. Van Bavel, and K. F. Schertz. 1975. *Crop Sci.* 15:516.

41 Henzell, R. G., K. J. McCree, C. H. M. Van Bavel, and K. F. Schertz. 1976. *Crop Sci.* 16:660.

42 Hesketh, J. D. 1963. *Crop Sci.* 3:493.

43 Hiron, W. P., and S. T. C. Wright. 1973. *J. Exp. Bot.* 24:769.

44 Hsiao, T. C. 1973. Plant responses to water stress. In *Annual Review of Plant Physiology.* Annual Reviews, Inc., Palo Alto, CA, 24:519–570.

45 Hsiao, T. C., E. Fereres, E. Acevedo, and D. W. Henderson. 1976. Water stress and dynamics of growth and yield of crop plants. Pp. 281–305 in *Water and Plant Life: Prob-*

lems and Modern Approaches, 1st ed., O. O. Lange, L. Kappen, and E. D. Schulze, Eds. Springer-Verlag, Berlin.

46 Hurd, E. A. 1964. *Can. J. Plant Sci.* 44:240.

47 Hurd, E. A. 1968. *Agron. J.* 60:201.

48 Hurd, E. A. 1969. *Euphytica* 18:217.

49 Hurd, E. A. 1971. Can we breed for drought resistance. Pp. 77–88 in *Drought Injury and Resistance in Crops,* K. L. Larson and J. D. Eastin, Eds. Crop Sci. Soc. Am. Spec. Publ. No. 2, Madison, Wisconsin.

50 Hurd, E. A., T. F. Townley-Smith, L. A. Patterson, and C. H. Owen. 1972. *Can. J. Plant Sci.* 52:689.

51 Hurd, E. A. 1975. Phenotype and drought tolerance in wheat. Pp. 39–57 in *Plant Modification for More Efficient Water Use,* 1st ed., J. F. Stone, Ed. Elsevier, New York.

52 Hurd, E. A., and E. D. Spratt. 1975. Root patterns in crops as related to water and nutrient uptake. Pp. 167–235 in *Physiological Aspects of Dryland Farming,* 1st ed., U. S. Gupta, Ed. Oxford and IBH Publishing Co., New Delhi.

53 Hurd, E. A. 1976. Plant breeding for drought resistance. In *Water Deficits and Plant Growth,* 1st ed., T. T. Kozlowski, Ed. Academic, New York, 4:317–353.

54 Ivanov, L. A. 1922. *Bull. Appl. Bot. and Plant Breeding* 13:31.

55 Izhars, S., and D. H. Wallace. 1967. *Crop Sci.* 7:457.

56 Johnson, D. A., and R. W. Brown. 1977. *Crop Sci.* 17:507.

57 Johnson, G. R., and K. J. Frey. 1967. *Crop Sci.* 7:43.

58 Jordon, W. R., and J. T. Ritchie. 1971. *Plant Physiol.* 48:783.

59 Jordon, W. R., F. R. Miller, and D. E. Morris. 1979. *Crop Sci.* 19:840.

60 Kaul, R. 1969. *Z. Pflanzenzuchtg.* 62:145.

61 Kaul, R., and W. L. Crowle. 1971. *Z. Pflanzenzuchtg.* 65:233.

62 Kaul, R., and W. L. Crowle. 1974. *Z. Pflanzenzuchtg.* 71:42.

63 Kramer, P. J. 1959. The role of water in the physiology of plants. In *Advances in Agronomy.* Academic, New York-London, 11:51–70.

64 Larcher, W. 1975. *Physiological Plant Ecology,* 1st ed. Springer-Verlag, Berlin Heidelberg-New York.

65 Larque-Sanvedea, A., and R. L. Wain. 1976. *Ann. Appl. Biol.* 83:291.

66 Levitt, J. 1972. *Responses of Plants to Environmental Stresses,* 1st ed. Academic, New York-London.

67 Liang, G. H., A. D. Dayton, C. C. Chu, and A. J. Casady. 1975. *Crop Sci.* 15:567.

68 Lupton, F. G. H., R. H. Oliver, F. B. Ellis, B. T. Barnes, K. R. Howse, R. J. Welbank, and P. J. Taylor. 1974. *Ann. Appl. Biol.* 77:129.

69 Martin, J. H. 1930. *J. Am. Soc. Agron.* 22:993.

70 McMichael, B. L., and C. D. Elmore. 1977. *Crop Sci.* 17:905.

71 McMichael, B. L., and B. W. Hanny. 1977. *Agron. J.* 69:979.

72 Mederski, H. G., and D. L. Jeffers. 1973. *Agron. J.* 65:410.

73 Miller, E. C. 1938. *Plant Physiology.* McGraw-Hill, New York.

74 Miskin, K. E., and D. C. Rasmusson. 1970. *Crop Sci.* 10:575.

75 Miskin, K. E., D. C. Rasmusson, and D. N. Moss. 1972. *Crop Sci.* 12:780.

76 Muenscher, W. L. C. 1915. *Am. J. Bot.* 2:487.

77 Newton, R. J., and W. M. Martin. 1930. *Can. J. Res.* 3:336.

78 Oppenheimer, H. R. 1961. Adaptation to drought. Pp. 105–138 in *Plant-Water Relationships in Arid and Semi-arid Conditions.* UNESCO, C. J. Bucher, Lucerne, Switzerland.

79   Pal, U. R., R. R. Johnson, and R. H. Hageman. 1976. *Crop Sci.* 16:775.

80   Palfi, C., and J. Juhasz. 1971. *Plant and Soil* 34:503.

81   Passioura, J. B. 1972. *J. Agric. Res.* 23:745.

82   Pyykko, M. 1966. *Ann. Bot. Fenn.* 3:453.

83   Quisenberry, J. E., and Bruce Roark. 1976. *Crop Sci.* 16:762.

84   Quisenberry, J. E., Bruce Roark, J. D. Bilbro, and L. L. Ray. 1978. *Crop Sci.* 18:799.

85   Raper, C. D., and S. A. Barber. 1970. *Agron. J.* 12:581.

86   Raschke, K. 1975. Stomatal action. In *Annual Review of Plant Physiology*. Annual Reviews, Inc., Palo Alto, California, 26:309–340.

87   Ray, L. L., C. W. Wendt, Bruce Roark, and J. E. Quisenberry. 1975. Genetic modification of cotton plants for more efficient water use. Pp. 31–38 in *Plant Modification for More Efficient Water Use*, 1st ed., J. F. Stone, Ed. Elsevier, New York.

88   Reitz, L. P., and S. C. Salmon. 1959. *U.S. Dept. Agric. Techn. Bull.* 1192:1.

89   Reitz, L. P. 1975. Breeding for more efficient water use—is it real or a mirage? Pp. 3–11 in *Plant Modification for More Efficient Water Use*, 1st ed., J. F. Stone, Ed. Elsevier, New York.

90   Roark, Bruce, and J. E. Quisenberry. 1977. *Plant Physiol.* 59:354.

91   Ross, W. M. 1972. *Sorghum Newsletter* 15:121.

92   Roy, N. N., and B. R. Murty. 1970. *Euphytica* 19:509.

93   Schonherr, J. 1976. *Planta* 131:159.

94   Slatyer, R. O. 1973. The effect of internal water status on plant growth development and yield. Pp. 177–191 in *Plant Response to Climatic Factors*, R. O. Slatyer, Ed. Proc. Uppsala Symp. UNESCO, Paris.

95   Sullivan, C. Y. 1971. Techniques for measuring plant drought stress. Pp. 1–18 in *Drought Injury and Resistance in Crops*, K. L. Larson and J. D. Eastin, Eds. Crop Sci. Soc. Am. Spec. Publ. No. 2, Madison, Wisconsin.

96   Sullivan, C. Y. 1972. Mechanisms of heat and drought resistance in grain sorghum and methods of measurement. Pp. 247–264 in *Sorghum in Seventies*, 1st ed. N. G. P. Rao and L. R. House, Eds. Oxford and India Book House, New Delhi.

97   Sullivan, T. P., and W. A. Brun. 1975. *Crop Sci.* 15:319.

98   Tan, Geok-Yong, and G. M. Dunn. 1975. *Crop Sci.* 15:283.

99   Tan, Geok-Yong, and G. M. Dunn. 1976. *Crop Sci.* 16:550.

100 Tan, Geok-Yong, Wai-Koom Tan, and P. D. Walton. 1976. *Crop Sci.* 16:722.

101 Taylor, H. M., E. Burnett, and G. D. Booth. 1978. *J. Agron. and Crop Sci.* 146:33.

102 Turner, N. C. 1979. Drought resistance and adaptation to water deficits in crop plants. In *Stress Physiology in Crop Plants*, 1st ed., H. Mussell and R. C. Staples, Eds. Wiley-Interscience, New York.

103 Van Volkenburgh, E., and W. J. Davies. 1977. *Crop Sci.* 17:353.

104 Vela-Cardenas, M., and K. J. Frey. 1972. *Iowa State J. Sci.* 46:381.

105 Venator, C. R. 1976. *Morelet, Turrialba* 26:381.

106 Viets, F. G. Jr. 1971. Effective drought control for successful dryland agriculture. Pp. 57–76 in *Drought Injury and Resistance in Crops*, 1st ed., K. L. Larson and J. D. Eastin, Eds. Crop Sci. Soc. Am. Spec. Publ. No. 2, Madison, Wisconsin.

107 Wallace, D. H., J. L. Ozbun, and H. M. Munger. 1972. Physiological genetics of crop yield. Pp. 97–146 in *Advances in Agronomy*. Academic, New York-London.

108 Walton, P. D. 1974. *Can. J. Plant Sci.* 54:749.

109 Wilson, D. 1971. *N. Z. Agric. Res.* 14:761.

110 Zabadai, T. J. 1974. *Plant Physiol.* 53:125.

111 Zingh, T. N., D. Aspinal, and L. C. Paleg. 1972. *Nature New Biol.* 236:188.

Chapter 8

PLANT RESPONSE TO LIGHT QUALITY AND QUANTITY

H. M. CATHEY AND LOWELL E. CAMPBELL

U.S. Department of Agriculture, Science and Education Administration, Agricultural Research, Beltsville, Maryland

INTRODUCTION

Development in higher plants is regulated by environmental and genetic factors. Under natural conditions one of the most important environmental factors is light. Light regulates the continuing organization, mobilization, utilization, and expansion of growth. All green plants have evolved or have been selected to be tolerant of light. Even though many aspects of the environment have changed or deteriorated in the past 50 years, we have made great progress with plants; we call it the "green revolution." We have accepted that particles and gases obscure the light that reaches plants. We have selected plants that utilize light in the presence of pollutants in the atmosphere, in the water, and in the soil. We have tended to ignore solar radiation except for visible light. We have selected plants that cope with the daily, seasonal, and yearly cycles and variations of light and dark intervals. Planting time and maturity time are varied to help fit these cultural needs according to seasons for different locations. The cultivars or clones of 50 years ago are seldom found in cultivation today. This disappearance is due in part to the accumulation of pests in the plant material from the rapid development of pests immune to chemical and cultural control. We also discovered the benefits of hybrids, homozygous breeding lines, and mulch-water-nutrient regulation. In altering the plant to cope with all of these factors, some alterations enhancing and others decreasing productivity, we have also altered the biochemistry of plants.

Paralleling the revolution in plant culture has been extensive research on the theoretical aspects of photosynthesis and photomorphogenesis. Most of the literature on photosynthesis has been concerned with unraveling the structure of chlorophyll and the complex biochemical pathways by which plants grow (17, 78, 79). Most of the research on photomorphogenesis has been concerned with the responses of etiolated (dark green) seedlings irradiated with relatively narrow wave band sources (69, 83, 91). Under natural sunlight conditions, out of doors, or in greenhouses covered with glass, fiberglass, and/or plastic, both photomorphogenesis and photosynthesis light-mediated systems must function together to regulate growth. We also anticipate the presence and ultimate discovery of many other light-regulated systems that may utilize or interfere with the known systems. Plants in agriculture are grown under a broad spectrum of light sources with varying irradiance, duration, and quality. The "real" environments are further complicated by varying levels of ultraviolet, visible, and infrared that have traditionally been modified or screened out of the light sources by glass coverings, water filters, and ventilating systems (1, 95). Also, plants themselves possess differing capabilities for reflecting, absorbing, and transmitting light and radiation (7). Thus, light can be radically changed when it encounters the canopy of plants. What strikes the plants initially may not be what creates the growth effects within the plants.

LIGHT—RADIATION AND MEASUREMENT

Of major concern to biologists are ultraviolet, visible, and infrared regions of the electromagnetic radiation spectrum (18).

Wavelength Classification of Radiation

The wavelength bands are arbitrarily separated into:

Classification	Wavelength (nm)
Ultraviolet (UV)	100–380
UV-C	100–280
UV-B	280–320
UV-A	320–380
Visible	380–780
Infrared (IR)	$780–10^5$
or	
Infrared (IR)	780–2500
Thermal	2500+

Traditionally the ultraviolet spectrum was divided into 100-nm bands from 100 to 400 nm by engineers and physicists. The classifications of UV into A, B, and C is from the International Commission on Illumination (CIE) for use by photobiologists. The spectral limits vary with authorities. The biological response is a gradual transition with overlap between regions without sharp delineation. The A, B, C system fits the peak emission of light sources better than 100-nm bands wherein the light sources may often peak at the wavelength division between two bands. UV and UV-A bands frequently use 400 nm as the upper limit (63). This is inconsistent with the established definition of visible light (380–780 nm). However, the biological response from UV-A may continue into the visible region above 400 nanometers (51) and into the UV-B range.

Almost all of these designated regions are referred to as light at times. The term "light" only refers to the visible (380–780 nm) portion of the electromagnetic spectrum, but all regions—ultraviolet, visible, and infrared—are electromagnetic radiation.

Sources of Radiation

The sources of radiation, both natural and artificial, are not limited to the discrete wavelength limits of the physical classification. The sun in outer space emits energy throughout all the wavelength bands although not equally at all wavelengths. Fluorescent lamps emit mainly in the visible region but have some energy in adjacent ultraviolet and infrared bands (18). Incandescent lamps emit relatively more infrared and a lower amount of visible than do sunlight or fluorescent lamps. Figure 1 shows the spectral emission of radiant energy from the sun.

Figures 2a–l show the spectral power emission of typical light sources used in horticulture. It should be noted that all lamps have infrared and thermal radiation not shown in the visible spectral power distributions.

Figures 3a,b show both the visible and infrared radiation from high pressure sodium lamps. Graphic displays of the power distribution of several lamps (cool white fluorescent, and the HID lamps—high pressure sodium, metal halide, mercury, and low pressure sodium) are shown in Figures 4a–e.

Measurement—Quantities and Units

We should first consider several concepts which may help in the comprehension of radiant power measurement. *Radiometry and radiometric terms are valid throughout the entire electromagnetic spectrum. "Light," photometry, and photometric terms are restricted to measurement in the region from 380 to 780 nanometers.* Light is a weighted response based on the relative stimulation of the human eye. Figure 5 shows the response for photopic vision (luminous efficiency, V) known as the "CIE (Commission Internationale de l'Eclairage) curve" or "eye sensitivity" curve.

Three systems of units for levels of radiation per unit area are:

Illumination E_v (photometric)
lumen per square meter = lux (lx)

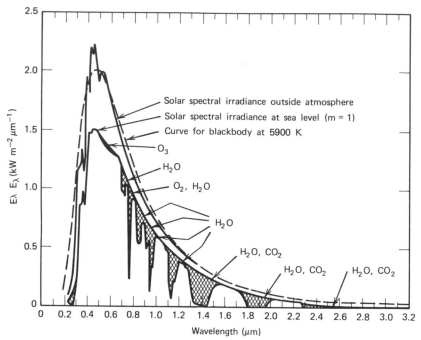

Figure 1 Spectral distribution curves related to the sun. The shaded areas indicate absorption at sea level due to the atmospheric constituents shown.

$$\text{Irradiance } E_e \quad \text{(radiometric)}$$
$$\text{watt per square meter (W/m}^2\text{) or (W} \cdot \text{m}^{-2}\text{)}$$
$$\text{Photon-flux density } E_p \quad \text{(photon radiometry)}$$
$$\text{quantum per second and square meter (q/s} \cdot \text{m}^2\text{)}$$

where subscript v refers to photometric quantities wherein luminous efficiency is included, subscript e refers to radiometric quantities, and subscript p refers to photon quantities.

Considerable confusion and disagreement exists about what units should be used to measure radiation in the plant sciences. Photometric units (lux, footcandle) are used by lighting engineers for describing irradiance of radiation sources and in specification or evaluation of lighting applications. This system is widely used by manufacturers and design engineers. It is generally agreed that plant response is different from that of humans and that these photometric units are unsuitable for directly describing photosynthetic and photomorphogenetic responses. Absolute, unweighted units correspond to plant response but direct measurement is complex.

Photon Units. Photon units are based on number of photons of quanta, which vary in energy with wavelength. For example, four photons at 400 nm

have the same energy as seven photons at 700 nm. Photon meters are cali-
brated to indicate the total number of photons over a wavelength interval,
for example, 400–700 nm.

In 1976 the *Crop Science Society of America* (80) defined the following:

Photosynthetically Active Radiation (PAR).   Radiation of the 400–700-nm
waveband (64, 63).

Photosynthetic Photon Flux Density (PPFD).   Photon flux density of PAR.
The number of photons (400–700 nm) incident per unit time on a unit surface. [Suggested units: nE (einsteins)/s·cm².]

Photosynthetic Irradiance (PI).   Radiant energy flux density of PAR. The
radiant energy (400–700 nm) incident per unit time on a unit surface (mW/
cm²).

These definitions permit PAR to be reported in either quantum or energy
units. SI units would be W/m² and E/s·m². In the past and in some current
literature PAR usage does not conform to these definitions.

Solar.   Outdoor solar radiation (in Langleys) was reported periodically by
the U.S. Weather Bureau (now NOAA) from 1950 until 1972. It was discon-
tinued when errors up to 100% in reported values were discovered. With
improved calibration this reporting was reinstated in 1978 in both W/m² and
Langleys. This radiation information is needed for all solar energy utilization. NOAA also reports percent sunshine, temperature, and other environ
mental parameters. (A large package describing available information is
available from the National Climatic Center, Federal Climatic Center, National Oceanic and Atmospheric Administration, Asheville, North Carolina
28801). Approximately 0.6 of the total solar radiation occurs in the 400–
850-nm region.

Radiometric.   Irradiance or radiometric units that indicate energy in watts
per square meter (W/m²) have equal response at all wavelengths. This is an
absolute unit but not easily measured simply and directly. Sensors with the
appropriate spectral response usually are not available. Radiometric units
are sometimes referred to as energy units. Nearly all standards and basic
calibrations are in radiometric units (watt). Spectral power distributions of
lamp manufacturers are also in radiometric units. We have adopted the use
absolute radiometric units (W/m²) in discussions of photoregulation of plants
in the subsequent portions of this chapter.

Photometric.   Lamp manufacturers rate lamps in lumens, the visually
weighted output. On request, most manufacturers will provide a spectral
power distribution which shows the emitted radiation in microwatts per
nanometer per lumen of emission. With this information, the absolute radia-

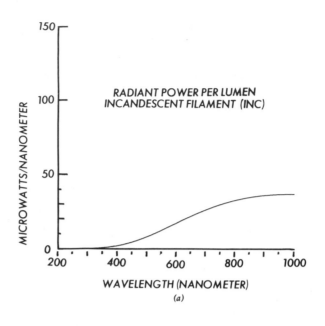

(a)

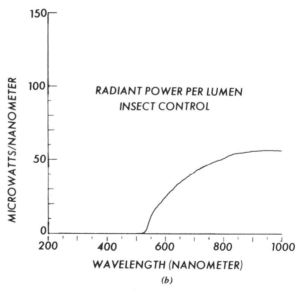

(b)

Figure 2 Spectral power emission of light sources used in agriculture. (*a*) Incandescent filament (INC). (*b*) Insect control. (*c*) Fluorescent cool white (FCW). (*d*) Fluorescent warm white (FWW). (*e*) Plant growth A (PGA). (*f*) Plant growth B (PGB). (*g*) Mercury (clear) (Hg). (*h*) Mercury deluxe white (HG/DX). (*i*) Metal halide A (MHA). (*j*) Metal halide B (MHB). (*k*) Low pressure sodium (LPS). (*l*) High pressure sodium (HPS).

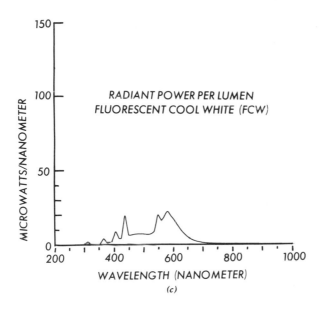

(c)

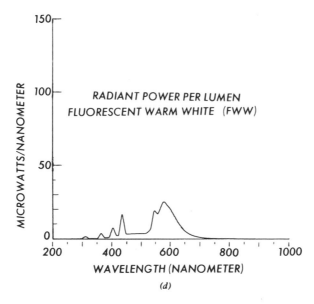

(d)

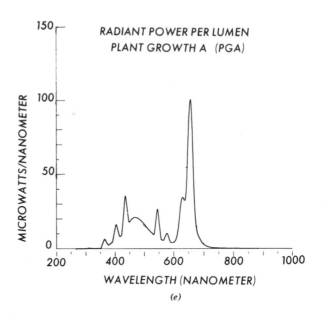

(e)

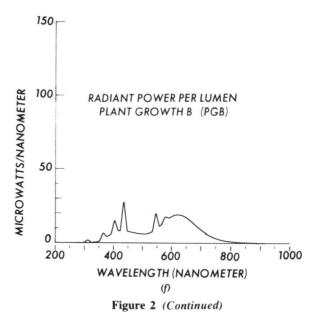

(f)

Figure 2 *(Continued)*

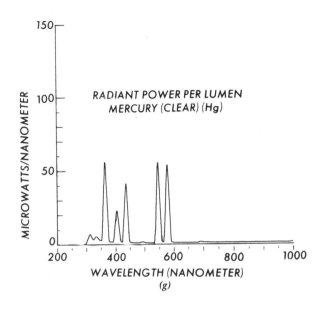

RADIANT POWER PER LUMEN
MERCURY (CLEAR) (Hg)

(g)

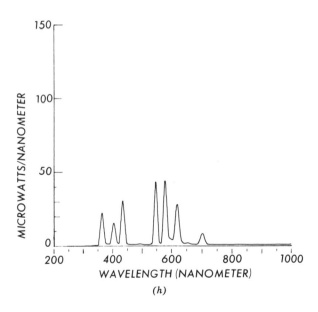

(h)

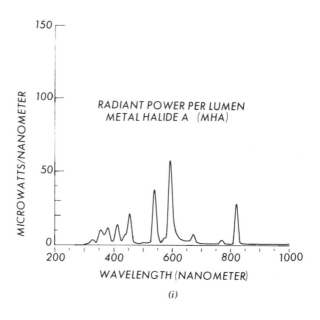

(i)

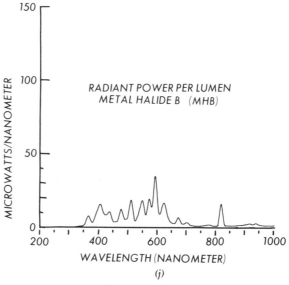

Figure 2 *(Continued)*

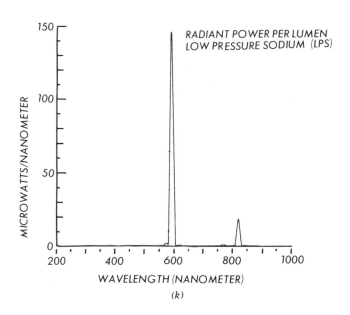

RADIANT POWER PER LUMEN
LOW PRESSURE SODIUM (LPS)

WAVELENGTH (NANOMETER)

(k)

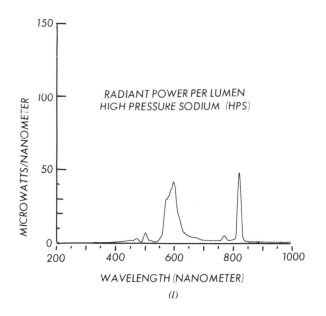

RADIANT POWER PER LUMEN
HIGH PRESSURE SODIUM (HPS)

WAVELENGTH (NANOMETER)

(l)

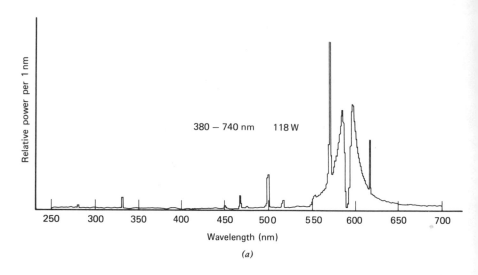

(a)

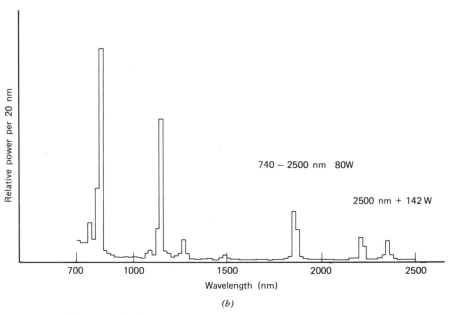

(b)

Figure 3 Visible and infrared radiation from high pressure sodium lamps: (a) 250–700 nm. (b) 700–2500 nm.

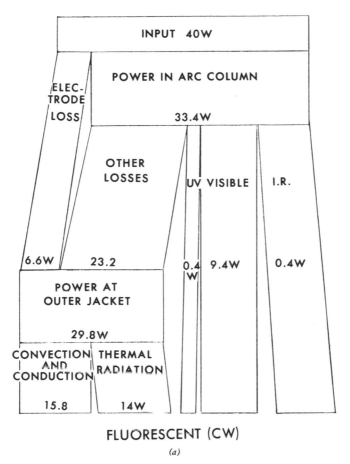

FLUORESCENT (CW)

(a)

Figure 4 Graphic display of the power distribution of (*a*) fluorescent (cool white) lamps (CWF), (*b*) high pressure sodium lamps (HPS), (*c*) metal halide lamps (MH), (*d*) mercury lamps (HG), and (*e*) low pressure sodium lamps (LPS).

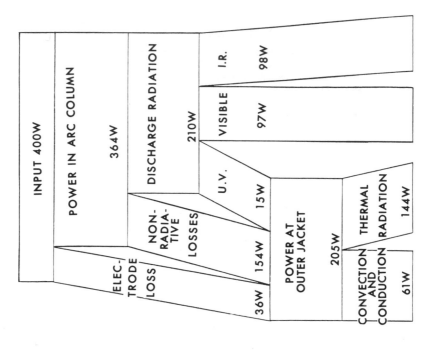

METAL HALIDE

(c)

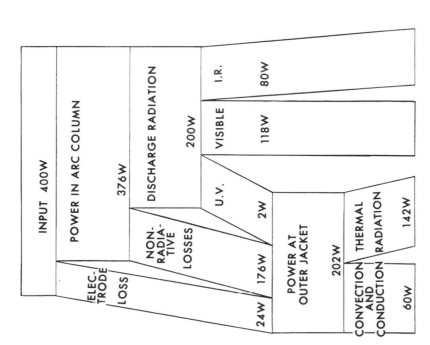

HIGH PRESSURE SODIUM

(b)

Figure 4. (Continued)

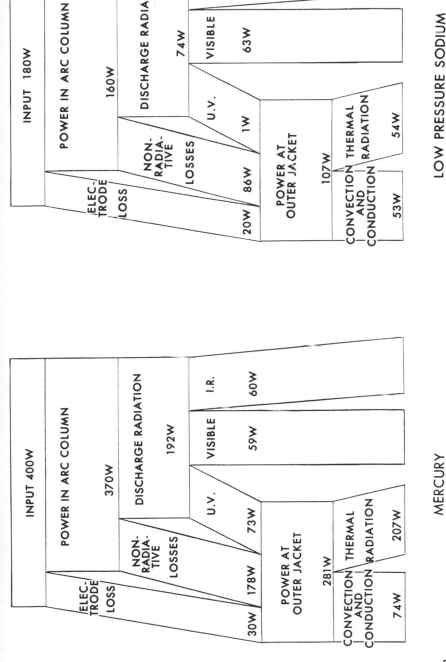

MERCURY
(d)

LOW PRESSURE SODIUM
(e)

227

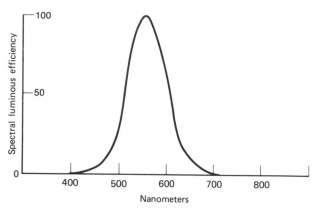

Figure 5 Standard Luminous Efficiency Curve (CIE).

tion in W/m² can be determined from lux measurements by summing the radiation over the interval desired.

LIGHT-SOURCE EMISSIONS

Graphic Data Description

The spectroradiometric system was used to determine the light-source emissions in graphic form (18). The graphs were then normalized to a per-lumen basis. The curves were compared with manufacturers data and with published information. Most comparisons were made within the wavelength region of 500–600 nanometer, since this region provides the most accurate measurement. Published information and manufacturers' data on wavelengths above 700 nanometers are limited because these wavelengths have little importance in vision lighting.

After the data in graphic displays were determined to agree with other known information at as many points as possible, data were compiled on irradiances in the wavelength intervals of concern in plant response. The data shown in the graphs and tables are essentially extensions of existing information, but they are presented in a manner that should be useful in horticultural lighting. Table 1 shows the electrical, photometric, and radiometric properties of a range of lamps used in horticultural lighting.

Table 2 shows the energy balance in the lamps. Input energy equals the sum of 400–850-nm radiation, other radiation, conduction, convection, and ballast loss. The power conversions are for lamps without luminaires or enclosures. Enclosures are expected to slightly decrease 400–850-nm radiation and increase other radiation, conduction, and convection.

Photometric Data

Columns 1 through 5 in Table 1 give the standard electrical and lumen ratings of the lamps. The lumen rating shown for the lamp FIR is an estimated value (30, 53, 54). Columns 6 through 20 give five repetitive wavelength intervals that describe the emission in three ways. Columns 6 through 10 are given in milliwatts per lumen, which can be used with illumination-meter measurements (footcandle or lux) to determine radiation in milliwatts per unit of area. Lux times milliwatts per lumen equals milliwatts per square meter. Footcandles times milliwatts per lumen equals milliwatts per square foot. When generic types of lamps are used in combination, measurements must be taken individually with only one type of lamp in operation at a time, and the measurements summed. Columns 11 through 15 give the total radiation output in the indicated wavelength interval for each lamp. The overall efficiency of the lamps—watts emitted per watt of electrical energy used, including ballasts—is shown in columns 16 through 20 for the wavelength interval indicated. These columns allow a comparison of the relative efficiency of the lamps for emissions at specific wavelength intervals.

Wavelength Intervals

The 400–700-nm wavelength interval is the traditional region of plant response in photosynthesis. The commonly accepted photosynthetic action spectra peaks in the blue and red regions. Reports indicate that all wavelengths between 400 and 850 nm, although not of equal energy, may be effective in photosynthesis (67) and the action spectra may be nearly flat for a plant canopy.

In photomorphogenesis, plants respond to emissions at wavelengths of 690 nm (red), 760 nm (far red), and 400–500 nm (blue). These wavelength intervals are important in photoperiod control and/or in the control of flowering and other light-mediated plant responses (23, 36, 47). The information shown in Table 1 on radiation emitted in the 800–850-nm wavelength interval permits comparisons of lamps with emissions above 800 nm with those with emissions below 800 nm. The importance of emissions above 800 nm is not known.

PHOTOMORPHOGENIC RESPONSE—ENERGY SOURCES

All light sources within the range of human vision exert influences on the growth of plants. Due to their spectral quality and quantity, they exert differing effects on how the plant appears to your eye, how they develop, and how long they persist. Few workers have grown plants simultaneously under a number of artificial light sources with similar cultural procedures and com-

Table 1 Electrical, photometric, and radiometric characteristics of selected lamps[a,b]

Column number		1	2	3	4	5	6	7	8	9	10
		Input Power		Output			Radiation per unit of luminous flux				
Lamp identification		Total (W)	Lamp (W)	Total (lm)	Total (lm/W)	Lamp (lm/W)	400–700 nm (mW/lm)	400–850 nm (mW/lm)	580–700 nm (Mw/lm)	700–850 nm (mW/lm)	800–850 nm (mW/lm)
Incandescent											
(INC)	100A	100	100	1,740	17	17	3.97	8.63	2.53	4.66	1.69
Fluorescent											
Cool white	FCW	46	40	3,200	70	80	2.93	2.99	1.02	0.06	0.009
Cool white	FCW	245	215	15,700	64	73	2.93	2.99	1.02	0.06	0.009
Warm white	FWW	46	40	3,250	71	81	2.81	2.86	1.23	0.05	0.006
Plant growth A	PGA	46	40	925	20	23	6.34	6.41	3.39	0.08	0.007
Plant growth B	PGB	46	40	1,700	37	42	3.96	4.37	1.95	0.41	0.03
Infrared	FIR	46	40	170	3.7	4.2	4.30	24.00	0.56	20.00	2.10
Discharge											
Clear mercury	HG	440	400	21,000	50	52	2.60	2.77	0.14	0.17	0.06
Mercury deluxe	HG/DX	440	400	22,000	50	55	2.62	2.81	0.73	0.19	0.05
Metal halide	MH	460	400	40,000	85	100	3.05	3.42	1.17	0.37	0.25
High pressure sodium	HPS	470	400	50,000	105	125	2.45	3.38	1.58	0.93	0.72
Low pressure sodium	LPS	230	180	33,000	143	183	1.92	2.18	1.89	0.26	0.25

[a]Columns 11–20 appear on the next page.
[b]Revised April, 1980 by R. W. Thimijan.

Table 1 Continued

Column number	11	12	13	14	15	16	17	18	19	20
	Radiation output					Efficiency				
Lamp identification	400–700 nm (W)	400–850 nm (W)	580–700 nm (W)	700–850 nm (W)	800–850 nm (W)	400–700 nm (mW/W)	400–850 nm (mW/W)	580–700 nm (mW/W)	700–850 nm (mW/W)	800–850 nm (mW/W)
Incandescent										
100A	6.90	15.00	4.41	8.11	2.94	69	150	44	66	29
Fluorescent										
FCW	9.38	9.56	3.27	0.18	0.03	204	208	71	4	0.6
FCW	46.00	47.00	16.00	0.88	0.14	188	192	65	4	0.6
FWW	9.13	9.28	4.00	0.15	0.02	199	202	87	3	0.4
PGA	5.86	5.93	3.13	0.07	0.01	127	129	68	2	0.1
PGB	6.73	7.42	3.32	0.69	0.05	146	161	72	15	1
FIR	0.73	4.13	0.10	3.40	0.35	16	90	2	74	8
Discharge										
HG	55	58	3	3.5	1.2	124	132	6	8	3
HG/DX	58	62	16	4.1	1.1	131	140	36	9	3
MH	122	137	47	14.8	10.0	265	297	102	32	22
HPS	123	169	79	46.5	36.0	261	360	168	99	77
LPS	63	72	62	8.6	8.3	276	313	271	37	36

Table 2 Input power conversion of light sources[a]

Lamp identification		Total input power (W)	Radiation (400–850 nm) (%)	Other radiation (%)	Conduction and convection (%)	Ballasts loss (%)
Incandescent						
(INC)	100A	100	15	75	10	00
Fluorescent						
Cool white	FCW	46	21	32	34	13
Cool white	FCW	225	19	34	35	12
Warm white	FWW	46	20	32	35	13
Plant growth A	PGA	46	13	35	39	13
Plant growth B	PGB	46	16	34	37	13
Infrared	FIR	46	09	39	39	13
Discharge						
Clear mercury	HG	440	13	61	17	09
Mercury deluxe	HG/DX	440	14	59	18	09
Metal halide	MH	460	30	42	15	13
High pressure sodium	HPS	470	36	36	13	15
Low pressure sodium	LPS	230	31	25	22	22

[a]Conversion efficiency is for lamps without luminaire. Values compiled from manufacturer data, published information, and unpublished test data by R. W. Thimijan.

pared them with plants grown under the sun (with comparable energy). A perspective based on our experience is presented in Table 3. All species do not respond exactly as we describe, but many species exhibit the basic growth characteristics described.

The statements on the sun require additional explanation. The sun, unlike all of the human vision sources, emits a continuous spectrum from the ultraviolet through the visible to the infrared. It is truly a broad spectrum source. Its inconsistencies due to daily and yearly changes in orientation and the varying filtering action(s) by the atmosphere and the plant canopy has promulgated many basic assumptions that only now are beginning to be challenged and/or discarded. The sun, as any human vision source, exerts both desirable and undesirable effects on the plants. Its primary benefits are:

1 Sun is the source under which most plants in agriculture are ultimately grown.

Table 3 Energy source and plant responses

Energy source	Plant characteristic	Plant response
Sun	Color	Green (light) foliage which tends to orient itself towards the source.
	Shape	Expands with cupping and deep ridges.
	Internode length	In response to intensity and duration, elongate and sturdy.
	Rate of leaf formation	In response to intensity and duration, rapid and continuous.
	Side shoot development	In response to intensity and duration, limited numbers and late in appearing.
	Rate of maturity and/or flowering	Prompt development towards flowering and flowering senescence.

Fluorescent
  Cool white (CWF)
  Warm white (WWF)
	Color	Foliage (greener than for SUN) which tends to orient itself parallel to the surface of the lamps.

	Shape	Expands somewhat smaller than SUN.
	Internode length	Less elongate and even more sturdy than SUN.
	Rate of leaf formation	Slower and not as frequent as SUN.
	Side shoot development	Increased numbers and earlier appearance than SUN.
	Rate of maturity and/or flowering	Development occurs over a period of time and rate of senescence is slower than SUN.

Fluorescent
  Gro Lux (GL)
  Plant Light (PL)
	Color	Foliage (greener than SUN, much greener than CWF) which tends to orient itself parallel to the surface of the lamps.

	Shape	Expands much less than SUN and less than CWF with flat surfaces.

Table 3   Continued

Energy source	Plant characteristic	Plant response
	Internode length	Much less elongate and even more sturdy (compact) than CWF in comparison with SUN.
	Rate of lead formation	Even slower as not as frequent than CWF in comparison with SUN.
	Side shoot development	Even more lateral shoots than CWF in comparison with SUN.
	Rate of maturity and/or flowering	Slowed development over a period of time, rate of senescence is slower than CWF in comparison with SUN.

Fluorescent
  Gro Lux WS
  Plant Light WS
    (GL-WS)
  Vita-Lite (VITA)
	Color	Foliage (not quite as light as SUN but not as dark as CWF) which tends to orient itself towards the source.

Agro-Lite (AGRO) and wide spectrum lamps	Shape	Expands more like foliage grown under SUN than under CWF or GL.
	Internode length	More elongate than CWF but less than SUN.
	Rate of leaf formation	Slower than SUN, but faster than CWF.
	Side shoot development	More than SUN but fewer than CWF.
	Rate of maturity and/or flowering	Development occurs less than SUN but more rapidly than CWF.

High intensity discharge
  Discharge mercury
    (HG)
  Metal halide (MH)
	Color	Foliage more green than SUN and similar to CWF and WWF, orients itself parallel to the surface of the lamp.

	Shape	Expands somewhat smaller than SUN and similar to CWF and WWF, with flat leaf surfaces.
	Internode length	Less elongate and even more sturdy than SUN, similar to CWF and WWF.

Table 3 Continued

Energy source	Plant characteristic	Plant response
	Rate of leaf formation	Slower and not as frequent as SUN, similar to CWF and WWF.
	Side shoot development	Increased numbers and earlier appearance than SUN, similar to CWF and WWF.
	Rate of maturity and/or flowering	Development occurs over a period of time, rate of senescence is slower than SUN and similar to CWF and WWF.
High intensity discharge High pressure sodium (HPS)	Color	Green (light) foliage which is similar to SUN, tends to orient itself towards source.
	Shape	Expands similar to SUN with cupping and deep ridges.
	Internode length	Elongates with a sturdy stem similar to SUN.
	Rate of leaf formation	Rapid and continuous development similar to SUN.
	Rate of maturity and/or flowering	Prompt development towards flowering and senescence.
High intensity discharge Low pressure sodium (LPS)	Color	Extra deep green foliage (or almost colorless according to species) which is similar to CWF, tends to orient itself parallel to the surface of the lamp.
	Shape	Expands larger and thicker than SUN and CWF.
	Internode length	Less elongate and even more sturdy than CWF.
	Rate of leaf formation	Rapid and continuous development similar to SUN.
	Side shoot development	Increased numbers and earlier appearance than SUN, similar to CWF.

Table 3   Continued

Energy source	Plant characteristic	Plant response
	Rate of maturity and/or flowering	Prompt development towards flowering and senescence.
Incandescent (INC) and incandescent mercury: (INC-HG) (Plant-light) (Tungsten-halogen)	Color	Paling of foliage which tends to orient itself towards the source.
	Shape	Expands with cupping and deep ridges similar to SUN.
	Internode length	Elongation may be excessive, eventually become spindly and breaking easily.
	Rate of leaf formation	Rapid and continuous, similar to SUN.
	Side shoot development	Suppressed in occurring and the number severely limited, plants expand primarily in height.
	Rate of maturity and/or flowering	Rapid development towards flowering and senescence, even more rapid than SUN.

2   Sun is a full spectrum source with energy throughout ultraviolet, visible, and infrared (Figure 1).

3   Energy from the sun is available to trigger plant responses even in winter months when the outside air temperatures are too low for growth.

4   Energy from the sun can be reflected, absorbed, transmitted, and still be reutilized.

Its primary limitations are:

1   Sun is a source of energy that varies in quality and quantity throughout every daily and yearly cycle.

2   Sun is a full spectrum source which can simultaneously activate and inhibit competing photoreactions in plants.

3   About 50% of energy from the sun is in the infrared and thus readily creating an environment with a rapidly rising temperature of the plant.

4   Facilities lighted with the sun require extensive and rapid ventilation and/or neutral filters to modify the supra-optimal quantities of light. Otherwise, plants cannot be grown successfully.

From the statements in Table 3, we conclude that none of the artificial light sources currently used simulate exactly the growth characteristics induced by the sun. Of the lamps currently available, the effects of INC and HPS on plants are much more like those of the sun than the other lamps. Both of these lamps have inherent problems resulting from high levels of infrared. INC, as described in Table 1, has traditionally been used with fluorescent light sources to provide additional red and infrared.

RESPONSE TO LIGHT—STAGES OF GROWTH

The quantity of light-energy required to regulate the growth of plants varies with the stage of growth. In fact many of the growth stages in nature occur in "overshoot" conditions. "Overshoot" means that the amount of incident light-energy greatly exceeds the quantity required to bring about the plant response(s). The extra light-energy may add undesirable side effects as the quantity exceeds the optimum level. An appraisal of light-energy requirements for each stage of growth follows.

Seed Germination

We classify seeds into three major types: (a) No specific germination response to light, (b) Specific light-energy requirement for seed germination, and (c) Specific light-energy inhibition of seed germination (9–11, 90).

Nonspecific Response. The majority of seed used to grow cultivated plants germinate equally well in the dark or light. The only requirements generally are an adequate period of after ripening and a warm, moist condition.

Light-Requiring Response. Many of the small-seeded species of plants (1000 seed or more per gram) require light for prompt seed germination. Light quality and quantity is extremely critical in promoting seed germination. Red (560–690 nm) with a peak at 660 nm is the region of maximum effectiveness. Its action can be reversed with far-red (690–760 nm), with a peak of 730 nm.
    There is about equal red (660 nm) and far-red (730 nm) in the radiation from the sun. Thus, there are types of seeds which respond to the sun by germinating promptly such as *Petunia* (petunia) and *Lactuca* (lettuce) (24). Others such as *Centaurea* (cornflower) and *Catharanthus* (vinca) are inhibited from germinating by sunlight. The sensitivity of the seeds to light is greatly affected by the moisture status of the seed and prior dark and/or irradiation treatments. In almost every batch of seed are several or many seeds that germinate under the unfavorable conditions and are companion seeds to the light-requiring ones. These dark-germinating seeds may result from premoistening under light induction conditions during the drying pro-

cess. They also may result from mechanical damage to the seed coats during the harvesting process. When seeds are saved from the dark-germinating types and grown for a second generation, the anticipated light requirement of the seeds can be demonstrated. Thus, harvesting and storage procedures can give slightly differing responses to light-energy in each seed lot, this situation can greatly confuse the identification and interpretation of the genetic characteristics of the plants. Hand harvesting seeds while avoiding damage to the seed coat, reducing their moisture content (to 8–10% moisture), and storage of the seeds at low temperatures (0–5°C) permits the evaluation of their light requirements.

Light-Inhibiting Response.   Few cultivated plant species are inhibited by light. The two most studied light-inhibited species are *Phacelia* (scorpion weed) and *Nemophila* (baby-blue eyes) (9–11). They have the same action spectrum curves as light-requiring species but the responses are reversed. Total inhibition of *Nemophila* seeds requires at least 20 hr of exposure to far-red (715 nm). Once the maximum inhibition has been reached, however, it is impossible to induce germination with red light or other inductive measures. The long exposure to far-red can be replaced by a simultaneous irradiation with 660 nm and 742 nm for 20 hr.

Photo-Inhibition Response.   Germination of three types of seed previously described are inhibited by long exposures to far-red (61, 62). This phenomenon would be only of academic interest if it were not for the differential screening which goes on in all vegetation. Chlorophyll selectively removes the red (660 nm) while transmitting most of the far-red (690–760 nm). Thus, the seeds on the surface of the soil under a canopy of green leaves are exposed to a low red/high far-red ratio. Inhibition with far-red, however, is extremely temperature dependent. There is usually a temperature range at which prompt germination occurs. Although the temperature may be favorable for germination as a given quality of light is filtered through the leaf canopy, seed germination is permanently inhibited. Other seeds, still viable but below the soil surface, may germinate weeks or years later when the correct combination of dark/light and temperature occurs.

Seedling Growth

Regardless of requirements for seed germination, very similar systems begin to function in germinating seedlings. In the absence of any light, the hypocotyl begins to elongate while development of the primary leave(s) and the epicotyl is depressed. All seedlings most effectively develop a compact growth form in response to red (660 nm) light. Further, with the suppression of hypocotyl elongation in response to red, the first leaves expand in size and the epicotyl begins to develop rapidly. When given immediately following red, far-red (730 nm) can reverse the inhibitory effects of red (36). Far-red reversal effects are progressively lost when the time between the red and far-red is extended beyond even a few minutes. These growth responses can be observed not only in the greening process of dark-grown seedlings but also with older plants grown in greenhouses or growth chambers. Seedlings grown in a room illuminated with light from cool white fluorescent lamps (CWF) develop into stocky young plants with deep green leaves. Red given at the beginning of the dark period has no observable effect on growth. Far-red given at the beginning of the dark period promotes internode elongation and suppresses the expansion and greening of the leaves. Red, given immediately, can reverse the promotive effects of far-red. Plants given multiple reversals over several hours can still demonstrate the regulatory action. When the last irradiation is red the plants will be compact. When it ends with far-red, the plants will be elongate. The addition of far-red or INC (equal red/far-red) to FLUOR does not significantly alter the internode length or the leaf color. Thus under the sun the mutual competition from shading and the selective screening of the light from the vegetation cannot alter dramatically the shape and color of the plants. Only during periods of dim, inadequate photosynthesis, intense light, excessive respiration and water loss in relation to photosynthetic activity do we observe visible changes in the growth characteristics of the plants. Over a relatively wide range of environments, plants of similar shape develop. The actual fresh and dry weights, and shoot/root ratios on close review may show trends, but the growth response is relatively fixed under a wide range of energy combinations. This response is based on the genetic potential of the cultivars and breeding lines and is a remarkable expression of the adaptability of plants to a wide range of environments. Yield of plants is a product of the ability of the plants to convert the products of photosynthesis into foliage and fruits. Although the plants may have very similar shapes from season to season, the yields may be different. Techniques for selecting breeding lines with high performance under episodes, or the continuous impact of stress, can thus be greatly modified by the photo-energy environment.

Flowering

Light-mediated systems rapidly begin to function in germinating seedlings and determine when the plants shift from a vegetative to a generative stage. Few plants can be induced to flower immediately. Most require some days or weeks of the proper combination of temperature, carbon dioxide, and light to induce flowering. Proposals for classifying plants into thermo- and photo-types have never been completed. Within every selected species genetic entities appear that are exceptions to the general rule for the species. Much of the classical physiological literature on the induction of flowering bears little relationship to the actual situation in fields or greenhouses when attempts are made to alter flowering patterns in plants. Recently introduced cultivars or breeding lines would not be classified as qualitative photoperiodic plants. The majority of wild or near wild species would be either quantitative or nonresponsive to photoperiod regulation. In fact, selection of plants for broad tolerance to environmental triggering of flowering has greatly extended the areas in which one specific cultivar or breeding line can be grown successfully and economically.

To understand how to isolate new environmentally tolerant cultivars and breeding lines, we must first understand the traditional broad classification of growth and flowering. The photosynthetic activity of plants is well documented. Plants are grouped into three basic types: C-3 sugar, high temperature and light sensitive; C-4 sugar, high temperature and light insensitive; and Crassulacean, dark fixation of CO_2. Plants can be grouped into four types by flowering response to daylength: (a) day neutral; (b) short day, increasing night length; (c) long day, decreasing night length; and (d) daylength intermediate, day-night limitations (47). All four groups can exhibit strong interactions with temperature, mineral nutrition, and carbon dioxide.

Most annual and perennial plants are quantitative in flowering response to the regulation exerted by the daily light-dark cycle (25). When an incandescent filament (0.9 W/m², 400–850 nm) was used as a light source, plants responded by flowering either during short or long day periods, but much earlier in the former than the latter. When exposed to at least 6 to 24 W/m² (400–850 nm) of supplemental light for 16 hr (0800–2400), both long day and short day plants flowered earlier and more abundantly than plants of a similar age and culture exposed daily to 8-hr supplemental light + 16-hr dark (short days) or to natural days (8.9–17 hr) + supplemental incandescent light for 4 hr (long days). These findings demonstrate that 24-hour radiant energy from artificial sources can override the classical photoperiod responses. Such conditions are close to those that induce flowering of plants in gardens and fields. The cited response (25) was relatively independent of the spectral composition of the light sources. Equal energies from 400 to 850 nm resulted in similar growth and flowering responses with cool white fluorescent (blue and red—CWF), high pressure sodium (yellow with shoulders into green and red—HPS), and low pressure sodium (monochromatic line at 589 nm LPS). The order of relative effectiveness of the various kinds of light sources is given in Table 4. A few species of lettuce and cultivars of soybean and cotton developed pale, almost colorless, foliage when grown entirely under LPS (14, 15). A normal color was partially or completely restored if low pressure sodium lamps were supplemented by low intensity incandescent or if the ambient temperature was raised to 28°C. Foliar sprays of minor elements, particularly ferrous (Fe^{2+}) iron, tended to correct some but not all of the foliar deficiencies observed. Pale foliage did not occur if plants were grown under LPS lamps in a greenhouse and were exposed to even the dimmest sun of winter months. There was sufficient broad spectrum radiant energy to mediate all stages of the growth of plants.

One notable group of plants, however, did not respond favorably to 24 hr of light. We continue to observe an overriding growth response of qualitative short day plants for flowering of such as *Euphorbia* (poinsettia), *Chrysanthemum* (chrysanthemum), and *Glycine* (soybean). Low intensity incandescent light at 0.9 W/m² (400–850 nm) delays or inhibits the flowering of these plants. In time, plants may initiate a potential inflorescence but they will not reach anthesis unless exposed to daily dark periods of 12 hr or greater.

"Overshoot" results from light for too many hours and/or too great an irradiance. It can occur with all species with all types of lamps. The appearance of plants showing overshoot is very different from plants suffering from deficiencies. Only a few species and cultivars exhibit abnormal growth responses to spectral deficiencies while the majority of plant species continue to produce deep green foliage and flower; when grown under unusual spectral compositions, "overshoot" paling of foliage occurs with most species. Although some cultivars and breeding lines may exhibit the responses earlier than others or in more varied environments, most plants grown in greenhouses supplemented with 24 W/m² (400–850 nm) for 24 hr daily develop chlorotic foliage and suppressed leaf and shoot development. When these chlorotic plants are shifted from overshoot conditions to a lighting regime with fewer hours (20 hr or less) and/or irradiance (less than 24 W/m² 400–850 nm), the top foliage immediately turns green and stem elongation rapidly resumes. A variety of fertilization schemes do not correct the chlorotic symptoms.

Dormancy

Light systems can also function in young seedlings to determine when they shift from vegetative or flowering to the dormancy stage. Few seedlings or rooted cuttings can be induced into dormancy. Most species require 3 weeks to several months of reduced daylength and/or temperature to induce such growth changes as formation of resting buds and cause abscission of the leaves.

Low intensity incandescent lighting is used to extend the end or beginning of the dark period or as an interruption during the night to delay the onset of dormancy (33, 73, 74, 76). The light need not be continuous but is equally effective when given 3, 6, and 12 seconds per minute for several hours. Light-dark cycles longer than 10 min are less effective than 1-min cycles in delaying the onset of dormancy.

When compared at a level of approximately 0.1 W/m² for 16 hr at a night temperature of 20°, human vision artificial light sources promote growth in the following order: Incandescent (INC) > high pressure sodium (HPS) >> metal halide (MH) = cool white fluorescent (CWF) >> clear mercury (Hg) (26). The intensity of light from HPS lamps must be from 4- to 8-fold that of INC to regulate vegetative growth of woody plants.

The favorable influence of extended daylength has been demonstrated with plants growing out-of-doors (27). Seedlings receiving extended daylength from late spring to late August in one year are equivalent in size to 2-

Table 4 Relative effectiveness of various kinds of light sources

Light sources	Radiant power 400–850 nm at plant level (W/m²)						
	0.3	0.9	3	9	18	24	50
Fluorescent—cool white							
40W single lamp 4 feet 3.2klm							
illumination, kilolux	0.10	0.30	1.0	3.0	NA	NA	NA
lamps per square meter	0.12	0.36	1.2	3.6			
distance from plants, meter	2.9	1.7	0.92	0.53			
40W 2-lamp fixtures (4 ft) 6.4klm							
illumination, kilolux	0.10	0.30	1.0	3.0	NA	NA	NA
fixtures per square meter	0.06	0.18	0.60	1.8			
distance from plants, meter	4.1	2.4	1.3	0.75			
215W 2 8-ft lamps 31.4klm							
illumination, kilolux	0.10	0.30	1.0	3.0	6.0	8.0	16.7
fixtures per square meter	0.01+	0.04	0.13	0.39	0.77	1.0	2.2
distance from plants, meter	8.8	5.1	2.8	1.6	1.1	1.0	0.7
High intensity discharge							
Mercury (1) 400W parabolic reflector							
illumination, kilolux	0.1	0.32	1.1	3.2	6.4	8.6	18.0
lamps per square meter	0.02	0.05	0.17	0.52	1.0	1.4	2.9
distance from plants, meter	7.6	4.4	2.4	1.4	1.0	0.8	0.6
Metal halide (1) 400W							
illumination, kilolux	0.09	0.26	0.88	2.6	5.3	7.0	15.0
lamps per square meter	0.01	0.02	0.08	0.24	0.47	0.63	1.3
distance from plants, meter	11.3	6.5	3.6	2.1	1.5	1.3	0.87

High pressure sodium 400W							
illumination, kilolux	0.089	0.27	0.89	2.7	5.3	7.1	15.0
lamps per square meter	0.005	0.015	0.05	0.15	0.30	0.39	0.82
distance from plants, meter	14.2	8.2	4.5	2.6	1.8	1.6	1.1
Low pressure sodium 180W							
illumination, kilolux	0.14	0.41	1.4	4.1	8.3	11.0	23.0
lamps per square meter	0.009	0.026	0.088	0.26	0.53	0.70	1.46
distance from plants, meter	10.7	6.2	3.4	2.0	1.4	1.2	0.83
Incandescent							
Incandescent 100W							
illumination, kilolux	0.033	0.10	0.33	1.0	2.0	2.7	5.6
lamps per square meter	0.056	0.17	0.56	1.7	3.4	4.5	9.4
distance from plants, meter	4.2	2.4	1.3	0.77	0.54	0.47	0.33
Incandescent 150W flood							
illumination, kilolux	0.033	0.098	0.33	1.0	2.0	2.6	5.5
lamps per square meter	0.035	0.10	0.35	1.0	2.1	2.8	5.8
distance from plants, meter	5.4	3.1	1.7	1.0	0.7	0.6	0.4
Incandescent-Hg 160W							
illumination, kilolux	0.050	0.15	0.50	1.5	3.0	4.0	8.3
lamps per square meter	0.07	0.22	0.74	2.2	4.5	6.0	12
distance from plants, meter	3.7	2.1	1.2	0.67	0.47	0.41	0.28
Sunlight							
illumination, kilolux	0.054	0.16	0.54	1.6	3.2	4.3	8.9

to 3-year-old seedlings given natural daylength. An aluminum reflective foil mulch and a systemic insecticide (disulfoton), as parts of an integrated system of growing plants, greatly increased the effectiveness of the extended light period. They reduced water- and heat-stress, reduced insect feeding, and increased radiant energy available to the plant had synergistic effects with the photoperiod manipulation and thereby resulting in significantly increasing the height, weight, and stem diameter of the plants.

RESPONSE TO LIGHT—QUANTITY OF ENERGY

We have arranged the energy requirements for displaying, handling, and growing plants into seven levels. All are well below what is recorded when plants are grown out-of-doors (64, 65). Much of the energy out-of-doors is "overshoot," supra-optimal levels of natural light (67). Due to frequent changes in irradiance as the result of clouds, rain, daylength, and the orientation of the sun, these levels are seldom sustained over extended periods of time (42). Further, reradiation of the shortwave energy from plants back into space effectively reduces the total energy that plants experience or tolerate.

We do not believe that indoor growth systems for plants should be designed to mimic out-of-door conditions in spectral energy distribution (6, 82). Practically it is nearly impossible to build indoor growth systems that simulate exactly outdoor conditions. As outdoor environment is constantly changing, one problem is what exactly to simulate. Even if certain standard conditions are agreed upon, it is doubtful they can be accomplished even with unlimited effort and expense. Supplemental or substitute lighting systems are at best a simulation of only part of what actually occurs in nature. We are fortunate, however, that lighting systems for plants can be designed which permit the culture of plants in controlled environments. Agronomic and horticultural research scientists must maintain the perspective, however, that they are creating only an approximation of the natural environment (42). One then should anticipate that some species, cultivars, or breeding lines will exhibit aberrant growth characteristics in controlled environments which may not be apparent when the same plants are grown in the natural environment. Even a trace or absence of a part of the spectrum (290–2500 nm) may limit the growth of a few cultivars or breeding lines. Other related types may develop normal looking plants in the same environment (14, 15).

Display—Exist: 0.3 W/m^2

Plants can exist in 0.3 W/m^2 (Table 4). Lamps of preference have changed with technological advances in energy efficiency and spectral distribution. The emphasis in lamp design, however, has always been directed towards color rendering and type of atmosphere created in the living spaces. Low wattage INC and FLUOR have been the lamps of preference. At irradiances delivered the plants can be displayed but one should expect little or no impact on the growth of the plants. Also, timing (light-dark durations) and temperature interaction would not be of concern.

Photoperiod: 0.9 W/m^2

The growth of plants can be regulated at an intensity of 0.9 W/m^2 (Table 4). By tradition, this irradiance has been tagged as the "low light intensity" system triggered by the photoreversible blue pigment phytochrome (31, 32, 35, 52, 93). The range of plant responses to low light that promotes or delays growth and flowering is extensive and is widely used by commercial growers (74, 76, 97–99). The relative order of activity in regulating photoperiod response is incandescent (INC) > high pressure sodium (HPS) > > metal halide (MH) = cool white fluorescent (CWF) > > clear mercury (Hg) (26). LPS is as effective as CWF in photoregulation of the daylength responses of plants (Cathey and Campbell, unreported). The photoregulatory effectiveness of any lighting system is increased by the use of reflective aluminum soil mulch (27).

Survival: 3 W/m^2

In general, plants survive at 3.0 W/m^2 (Table 4). This irradiance creates an environment in which many plants maintain their green color. Length of stems and reduction of leaf size and thickness, however, occur almost immediately following placement of plants under this irradiance. In time, the overall development of the plants falls behind that of other plants grown under higher irradiances. Photoperiod responses do not function well at this irradiance because all plants lengthen and seldom develop green foliage. There are, however, strong interactions between this irradiance and temperature, watering frequency, and nutrition. Temperatures less than 17°C tend to conserve the previously stored material while frequent watering and fertilizing increase the stem lengthening and aging of the older foliage.

Maintenance: 9 W/m^2

Plants can grow for many months when exposed to an intensity of 9 W/m^2 (Table 4). By tradition, this is the irradiance at which many indoor gardeners (12, 40, 41) (professional or hobbyist) start their plants from seeds, cuttings, or meristems. It has become a convenient base, particularly for those who use fluorescent lamps as a sole source for growing plants (8, 12). As one might anticipate, interactions of light source and intensity with the environment (temperature, air flow, relative humidity, pollutants) may vary greatly from installation to installation. When the light irradiance is within broad limits and the air exchange is adequate, simple facilities can be constructed to grow a wide range of plant species (85). The rate of development, particularly as the plants grow in size, can be slow in comparison with plants grown at higher intensities (28). During the development of seedlings and the rooting of cuttings, there appears to be little response to photoperiod. In fact, for most plants during the initial phases of development, continuous light (and heat) should be used to help compensate, in some part, for the limited light interception. Most plant species develop deep green foliage, large leaves, and may accelerate the transfer of nutrients and stored materials from the old to the young, rapidly developing leaves. The plants eventually begin to drop or lose an old leaf for every new leaf that develops. Adjustment of the light regime to a 12-hr light/12-hr dark cycle, coupled with reduced frequencies of watering and fertilizing creates an environment where growth is slowed, few new leaves are formed while most older leaves are retained. The growth of most container-grown foliage plants are now acclimatized for 4 to 16 weeks under an irradiance of 9 W/m^2 prior to their sale to consumers (28, 43). An acclimatized plant can be readily identified by its slowed growth, few if any new leaves, deep green foliage which is broad and flat, and leaves persistent to the soil line—all commercially desirable attributes.

Propagation: 18 W/m^2

Plants can be propagated rapidly when exposed to 18 W/m^2 for a minimum of 6–8 hr daily (Table 4). By tradition, this is the intensity many propagators attempt to achieve by their greenhouses in summer with one or several layers of neutral filters (films or coverings, plastic or other fabrics) to restrict the entry of light (and heat) into the propagation area. At least 50% of the incident sunlight reaching the plants is lost by reflection of the covering (glass, plastic, polyethylene) and by absorption or shading of the framing and supports in the greenhouse. Cuttings rooted at this intensity maintain a growth rate similar to that of the cutting when attached to the stock plant. Stem length, branching, and leaf color, however, can be regulated by manipulation of the temperature, moisture stress, and nutrients (55). Most plants grown for their flowers and fruits can be brought to maturity when exposed to 18 W/m^2 for 16–18 hr daily. However the growth rate under those conditions is relatively slow (56–58). For most rapid development (leaf number, number of branches, early initiation of flowers) the plants should be grown in a light intensity of 24–50 W/m^2.

Greenhouse: 24 W/m^2

Plants can be grown year round in a greenhouse in which the natural light is supplemented with 24 W/m^2 for 8–16 hr daily (Table 4). By tradition, this is the irradiance which best couples ambient sunlight with supplemental lighting. It simulates many of the growth responses and rates associated with studies in growth chambers (25, 39). The photomorphogenetic activity of sunlight, even under dim light conditions of midwinter, is essential to regulate growth responses. An irradiance of 24 W/m² from a wide range of artificial light sources combined with low winter daylight in greenhouses is sufficient to boost growth rates and create a growing environment for rapid development and early flowering (20–22, 94). The different phototypes (short-, long-, and neutral-day) and growth systems (regulation of flowering and dormancy) exhibit a wide array of responses. Because the most widely grown species and cultivars have been selected to be quantitative in their responses to daylength, supplemental lighting tends to lump the growth responses into one type—accelerated growth and early flowering (3). The plants grown in the greenhouse without supplemental lighting grow much slower and flower much later than the lighted ones. Duration (in hours) and placement (day-night) is extremely critical (37). Supplemental lighting for 8 hr, particularly during the day (0800–1600), is much less effective than lighting at night (2000–2400). Lighting of the short-day plants such as soybeans, chrysanthemum, and poinsettia is relatively ineffective because they can only be lighted during the 8–12-hr day and must be followed by an obligatory 12–16-hr daily dark period (2).

Deciduous trees lighted with INC (at 0.9 W/m²) maintain vegetative growth over many months. On the other hand, deciduous trees lighted with HID lamps, regardless of light spectral composition, develop abnormally colored leaves and become dormant. Species vary widely in their sensitivity to lighting with INC and HID lamps. Continuous lighting of some plants induces initially a paling of the foliage and then an abrupt loss of all visible pigments in the top-most leaves. The plants, however, do not die but survive many weeks with bleached leaves. The condition can be corrected in part by giving the plants at least 4 hr of dark each day, by increasing the temperature 2–4°C, and/or by spraying the foliage with minor element solutions.

Growth Chamber: 50 W/m²

Plants can be grown in growth chambers when the light intensity is a minimum of 50 W/m² (Table 4). This irradiance is approximately one-fourth normal ambient daylight and can be used to simulate many growth conditions (daylength, temperature range, relative humidity, air-flow, carbon dioxide concentrations). It has become the standard light level in growth chambers (56, 58, 84, 100). There are many sources used to light growth chambers (44, 87, 96). For convenience, readily available cool white fluorescent lamps have been widely used for more than 30 years (75, 89). More recently, HPS lamps have been substituted for fluorescent lamps (16, 80). For most consistent results, a glass or plastic barrier is used between the light sources and the plants and a ventilating system is necessary to remove heat generated by the lamps. Since water filters or air-flow cooling does not completely remove IR (infrared), chambers from different manufacturers are difficult to standardize (34). This lack of standardization often leads to confusing information on plant growth and flowering responses in relation to what is observed with plants grown in greenhouses and out-of-doors (88). When the total irradiance is 50–80 W/m² and 10–20% of the total watt input is provided with INC lamps, most kinds of plants can be grown successfully (58, 4). Plant morphology, flowering and fruiting responses are typical when the plants are subjected to daylength (8–24 hr), temperature (9–35°C), carbon dioxide (300–5000 PPM), relative humidity (20–100%), and free air exchange. Plant culture becomes progressively more difficult in chambers constructed to provide intensities greater than 50 W/m²; and the uncontrolled aspects become too complex to solve (66, 92).

RESPONSE TO LIGHT—QUALITY OF ENERGY

Incandescent (INC)

The standard light source for regulation of the photoperiod responses of plants is INC, providing equal red (660 nm) and far-red (730 nm) (60, 77) (Table 5). There are no differences in the plant responses to the INC when its basic 120 volt frosted covering is modified (38). Similar photoperiod effects are observed when the covering is changed from the traditionally frosted one to clear, ceramic coated (yellow, buff, orange, red) and colored glass (red, ruby, blue). These changes alter the lights visually in the yellow-green region but have only a slight effect on the red (600 nm) to far-red (730 nm) regions or ratio. The extended life lamps rated 750, 2500, to 8000 hours also are equally effective for dark interruption or for use in cyclic lighting. Most photoperiod responses can be regulated with 0.9 W/m² for 1, 2 or at most 4 hr—given continuously or cyclic (1–30 min) during the middle of a 12–16-hr dark period. Other light sources (fluorescent or HID) are never as effective or efficient as INC lamps (Table 1, columns 18, 19).

Fluorescent (FLUOR)

Fluorescent lamps, which emit more red, blue, infrared or ultraviolet radiation than the traditional cool white and warm white lamps, have been extensively tested. Although there are reports of exceptional performance of a specific plant under a special lamp (29), the prevailing conclusion is that total radiation output is a much better criterion for selecting fluorescent lamps for plant growth than any special spectral distribution in the visible range (Table 1, column 17; Table 2). Evaluating the effectiveness of a new fluorescent lamp can become very complex (28, 49). Fluorescent tubes are sensitive to interactions with the environment. The lamps are coated with phosphors which create the fluorescence. They vary in their spectral output and life. The glass used to make the lamp can alter the light emitted to the plants (59). Overall cool white and warm white flourescent lamps are anticipated to continue to be the standard fluorescent light source for plants (8, 28, 40, 49, 72, 86).

High Intensity Discharge (HID)

Mercury lamps lack the efficiency for high irradiance. Metal halide lamps are lower in efficiency than sodium lamps (Table 2). Metal halide lamps have the efficiency equal to cool white fluorescent lamps. The sodium lamps, high pressure (HPS) and low pressure (LPS), are the most efficient light sources available (16, 24). We have observed that many kinds of plants grow under HPS and LPS lamps as a sole source or as a supplement in greenhouses. When there is a special requirement for spectral composition for plant growth, HPS lamps are more satisfactory than LPS (71). The abnormal growth characteristics observed with some plants growing under LPS can be reduced by adding INC lamps and/or increasing the ambient temperature (14, 15). HPS lamps provide the required visible and infrared radiation to grow a wide range of plants. Even plants lighted with HPS benefit from the addition of INC. Again, the mixture of visible and IR more successfully simulates the action of sunlight than any single source of artificial light.

RESPONSE TO LIGHT—IDENTIFICATION OF PROBLEMS

Plants are the best indicators of cultural conditions. With proper conditions they will develop deep green foliage, compact stems with equal root and shoot growth, and will rapidly develop new, well-expanded leaves. Flowering and fruiting will occur uniformly throughout the population of plants. Plants under stress will not exhibit these growth responses. We suggest the following steps for identification of environmental problems. Each aspect is discussed under a separate section, accompanied with a set of drawings to illustrate the process(es). All aspects of the environment interact simultaneously to modify effects of individual parameters.

Temperature

The optimum temperature varies from location to location and among species. The ideal plant will have dark green foliage with equal root and shoot development (Figure 6). When it is too cool (day or night) growth is slowed and new leaves do not develop. When it is too warm, the foliage turns pale yellow and does not regreen regardless of other cultural procedures.

Daylength

The optimum daylength may be relatively elastic (Figure 7). Since there may be days of cloudy weather followed by intensely bright days, the sensitivity

Table 5 Lighting by stage of growth

Growth stage	Daylight	Artificial light[a]	Irradiance fc	W/m² (400–850 nm)	Duration (hr)
Seed germination	available	CWF, WWF	140	4.5	000–2400
Pretransplanted seedlings and cuttings	available	CWF, WWF	280		0800–2400
	available	MH	245	9.0	0800–2400
	available	HPS	245		0800–2400
	available	LPS	385		0800–2400
Transplanted seedlings and cuttings					
Daylength only	available	INC	20	1.9	2000–0400 (12 sec/min or 2 min/10 min)
Daylength and photosynthesis supplement (propagation)	available	CWF, WWF	560		2000–0400
	available	MH	490	18.0	2000–0400
	available	HPS	490		2000–0400
	available	LPS	770		2000–0400
Pre-interior preparation	none (and/or available)	CWF, WWF	280		0600–1800 (same or any 12-hr period)
	none	MH	245	9.0	
	none	HPS	245		
	none	LPS	385		

	Light	Lamp[a]			Period
Interior					
Survival	none (and/or available)	CWF, WWF	45		0800–1600 (same or any 8-hr period)
	none	MH	40	1.5	
	none	LPS	65		
	available daylight	INC	25		—
Maintenance	none (and/or available)	CWF, WWF	280		0600–1800 (same or any 12 hr period)
	none	MH	245		
	none	HPS	245	9	
	none	LPS	385		
	available	—	150		—
Propagation	none	CWF-WWF	560		0600–1800 (same or any 12 hr period)
	none	MH	490	18	
	none	HPS	490		
	none	LPS	770		
Growth chamber	none	CWF-WWF	1565		Any period required
	none	HPS	1390	50	
	none	LPS	2175		
Field					
Daylength	available	INC	10	0.9	2000–0400 (12 sec/min or 2 min/10 min)
Daylength and photosynthesis supplement	available	HPS	25	0.9	2000–0400
	available	INC	10	0.9	2000–0400 (12 sec/min or 2 min/10 min)
(Cover soil with reflecting aluminum-coated paper)	available	HPS	25	0.9	2000–0400

[a]Lamp code: CFW = Cool white fluorescent; WWF = Warm white fluorescent; MH = Metal halide; HPS = High pressure sodium; LPS = Low pressure sodium; INC = Incandescent.

INCREASING TEMPERATURE

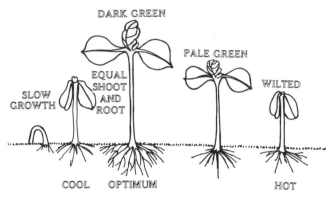

Figure 6　Growth responses of seedlings to increasing temperatures.

INCREASING DAYLENGTH

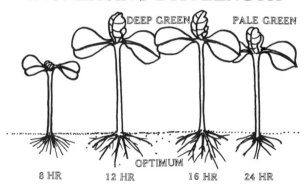

Figure 7　Growth responses of seedlings to increasing daylength.

to actual daylength may be greatly compounded. Photoperiod responses continue to be regulated even on the days with the dimmest natural light. As with temperatures, the optimum daylength is identified by the development of deep green plants with equal shoot-root ratios (Figure 7). Less than 12 hr of light or extremely dim light for some of the morning and evening hours slows growth and delays the development of new leaves. Lighting for more than 16 hr daily tends to cause paling of the foliage, the inhibition of stem elongation, and an early maturity/senescence of the plants.

Ratio of Red to Far-Red

Sunlight, which provides an equal red/far-red ratio, creates an environment that produces a plant like the one in the middle in Figure 8 (70). An environ-

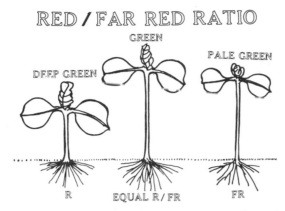

Figure 8   Growth responses of seedlings to red/far-red ratios.

ment that provides high red/low far-red, such as with a fluorescent lamp, produces plants which are compact and deep green (50). In contrast, an environment that provides low red/high far-red such as with a photographic safety lamp (incandescent lamps made with a covering of red or wine tinted glass) produces plants that are spindly, poorly branched, and pale green. New growth is slow. Crowding plants or shade from neighboring plants may have the same effect (45). Chlorophyll effectively screens out most of the red light permitting the far-red to pass through the plant canopy. Thus, the light entering the second and third leaves in the canopy are exposed to progressively less light but, of greater consequence, to low red/high far-red ratios. Adequate spacing to regulate the density of the canopy is essential to provide the optimum red/far-red ratio for growth and flowering.

Once you learn to read the signals from your plants, you can adjust the temperature and light (source intensity, duration) to grow them with a minimum of environmental stress.

SUMMARY: OVERVIEW

Light energy is absorbed by the many different pigments present in the leaves and the stems of the plants. In controlled facilities, we can accompany the proper lighting with all of the other environmental factors in near optimum combinations. We can do little to regulate the light available to plants grown under field conditions. Even with the time that has elapsed since we discovered the two major pigment systems of plants, we still can do little to regulate the activity and the products (results) of chlorophyll and phytochrome. Both are intimately bound together in the same tissues and must be present for the plants to function. Our theories on how light is utilized by intact plants grown in fields and greenhouses remain rudimentary but do serve to partially explain many light-evoked plant responses. We know the complex steps leading to the synthesis of simple sugars and how plants utilize them to grow. We also know in detail many of the growth systems triggered by phytochrome (red/far-red). The mystery remaining is how all of these synthetic and trigger systems function together in a green plant growing in a favorable or less than favorable environment. Agricultural sceintists have failed thus far to regulate light systems directly (5). We seek chemical substitutes under such names as florigen to regulate growth and flowering. We also seek growth regulators that increase the net productivity of plants. Our conclusions are that these approaches have a slim chance of success. There are just too many labile steps for its success in rainfall or protected agriculture.

Geneticists and breeders should be cautioned that any artificial environment used to select plants may be giving an unnecessary bias to the process. As we have discussed, none simulate precisely the radiant energy of the natural environment. The subtle differences in the ultraviolet and infrared regions undoubtedly have profound effects on the overall growth responses of the plants. Any final evaluation of breeding lines and selections must still be conducted under sunlight conditions. This chapter presents information as to how to relate and to interconvert from natural to artificial light sources and how to analyze the other aspects of the radiant sources.

We believe that there are many opportunities to use light sources to create energy efficient growth systems for plants. They can be used in warehouses to create growth facilities or they can supplement the light in solar-heated, film-covered frames to create simple greenhouses. A combination of natural and artificial light sources can give some of the crops used in agriculture a head start on and an extension of the growing season. Since most plants started at high temperatures and bright light conditions will continue to grow at low temperatures and with less light, we anticipate a reduction in the impact of the subsequent unfavorable environments. *Nicotiana* (tobacco) plants are started in a simple germination environment and transplanted to the field with automatic planting and cultivating equipment. This technique could be used for other field crops to improve plant establishment. Some species such as *Gossypium* (cotton) and *Phaseolus* (lima beans) are particularly sensitive to chilling and drought and often require replanting several times to ensure a good stand. The benefits of such a change in procedure include potential for multicropping, increased plant biomass production and extension of the range of arable land.

REFERENCES

1 Aldrich, R. A., and J. W. White. 1969. *Trans. ASAE* 12:1.
2 Anderson, G. A., and W. J. Carpenter. 1974. *HortScience* 9:58.
3 Austin, R. B., and J. A. Edrich. 1974. *J Agric. Eng. Res.* 19:339.
4 Bailey, W. A., H. H. Klueter, D. T. Krizek, and N. W. Stuart. 1970. *Trans. ASAE* 13:263.
5 Bailey, L. H. 1893. *Cornell Univ. Agric. Exp. Sta. Bull.* 55:127.

6 Balegh, S. E., and O. Biddulph. 1970. *Plant Physiol.* 46.

7 Bickford, E. D. 1972. *Lighting for Plant Growth.* Kent State University Press, Kent, Ohio.

8 Biran, I., and A. N. Kofranek. 1976. *J. Amer. Soc. Hort. Sci.* 101:625.

9 Black, M., and P. F. Wareing. 1959. *North Intl. Bot. Congr.* 2A:2.

10 Black, M., and P. F. Wareing. 1954. *Nature* 174.705.

11 Black, M. 1969. *Symp. Soc. Exp. Biol.* 23:193.

12 Boodley, J. W. 1970. *Plants and Gardens* 26:38–42.

13 Borthwick, H. A. *Proc. Intl. Seed Test. Assoc.* 30.

14 Brown, J. C., C. D. Foy, J. H. Bennett, and M. N. Christiansen. 1978. *Plant Physiol.* 63:692.

15 Brown, J. C., H. M. Cathey, J. H. Bennett, and R. W. Thimijan. 1979. *Agron. J.* 71:1015.

16 Buck, J. A. 1973. *Trans. ASAE* 16:121.

17 Calvin, M., and J. A. Bassham. 1962. *The Photosynthesis of Carbon Compounds.* W. A. Benjamin, Inc., New York.

18 Campbell, L. E., R. W. Thimijan, and H. M. Cathey. 1975. *Trans. ASAE* 18(5):952.

19 Canham, A. E. 1974. *Proc. XIX Intl. Hort. Congr.* 267.

20 Carpenter, W. J., and G. A. Anderson. 1972. *J. Am. Soc. Hort. Sci.* 97:331.

21 Carpenter, W. J., and G. R. Beck. 1973. *HortScience* 8:482.

22 Carpenter, W. J. Ramat. 1976. *J. Am. Soc. Hort. Sci.* 101:155.

23 Cathey, H. M., and L. E. Campbell. 1977. *Trans. ASAE* 30:360.

24 Cathey, H. M., and L. E. Campbell. 1974. *IES Lighting Design and Application* 4(10):41.

25 Cathey, H. M., and L. E. Campbell. 1979. *J. Am. Soc. Hort. Sci.* 104:812.

26 Cathey, H. M., and L. E. Campbell. 1975. *J. Am. Soc. Hort. Sci.* 100:65.

27 Cathey, H. M., G. G. Smith, L. E. Campbell, J. G. Hartsock, and J. Y. McGuire. 1975. *J. Am. Soc. Hort. Sci.* 100:234.

28 Cathey, H. M., L. E. Campbell, and R. W. Thimijan. 1978. *J. Am. Soc. Hort. Sci.* 103:781.

29 Corth, K., G. M. Jividen, and R. J. Downs. 1973. *J. Illum. Eng. Soc.* 9:139.

30 De Boer, J. B. 1974. *J. Illum. Eng. Soc.* 3(4):142.

31 Deutch, B., and B. I. Deutch. 1978. *Photochem. Photobiol.* 27:241.

32 Downs, R. J., and A. A. Piringer, Jr. 1958. *Forest Sci.* 4:185.

33 Downs, R. J., and H. A. Borthwick. 1956. *Bot. Gaz.* 117:310.

34 Downs, R. J., and V. P. Bonaminio. 1976. *Tech. Bull.* 244, North Carolina Agric. Exp. Sta., N. C. State Univ., Raleigh.

35 Downs, R. J., H. A. Borthwick, and A. A. Piringer, Jr. 1958. *Proc. Am. Soc. Hort. Sci.* 71.

36 Downs, R. J., S. B. Hendricks, and H. A. Borthwick. 1975. *Bot. Gaz.* 118:199.

37 Downs, R. J., W. T. Smith, and G. M. Jividen. 1973. *ASAE Paper* No. 73-4525, ASAE, St. Joseph, Michigan.

38 Downs, R. J. 1977. *HortScience* 12:330.

39 Duke, W. B. 1975. *Agron. J.* 67:49.

40 Dunn, S., and F. W. Went. 1959. *Lloydia* 22:302.

41 Dunn, S. 1975. *Florist Rev.* 156:41.

42 Evans, L. T. 1953. In *Environmental Control of Plant Growth*, L. T. Evans, Ed. Academic, New York and London, p. 421.

43 Fonteno, W. C., and E. L. McWilliams. 1978. *J. Am. Soc. Hort. Sci.* 103:52.

44 Frank, A. B., and R. E. Barker. 1976. *Agron. J.* 68:487.
45 Frankland, B., and R. J. Letendre. 1978. *Photochem. Photobiol.* 27:223.
46 Gaastra, P. 1959. *Meded v. d. LBHS to Wageningen* 59:11.
47 Garner, W. W., and H. A. Allard. 1920. US Department of Agriculture. *Yearbook of Agriculture*, p. 377–400.
48 Govingee, R. 1974. *Sci. Am.* 231:68.
49 Helson, V. A. 1963. *Can. J. Plant Sci.* 45:461.
50 Holmes, M. G., and H. Smith. 1975. *Nature* 254:512.
51 Incoll, L. D., S. P. Long, and M. R. Ashmore. 1977. *Current Advances in Plant Science*. Pergamon, New York and Oxford 9:331.
52 Jose, A. M., and D. Vince-Prue. 1978. *Photochem. Photobiol.* 27:209.
53 Kaufman, J. E., and J. F. Christiansen, Eds. 1972. *IES Lighting Handbook,* 5th ed. Illum. Eng. Soc. North America.
54 Kaufman, J. E. 1973. *Lighting Design and Application* 3(10):8.
55 Klueter, H. H., and D. T. Krizek. 1972. In *Landscape for Living*. Yearbook of Agriculture. U.S. Dept. of Agric., Washington, D.C. pp. 205–209.
56 Krizek, D. T., and R. H. Zimmerman. 1973. *J. Am. Soc. Hort. Sci.* 98:370.
57 Krizek, D. T., W. A. Bailey, and H. H. Klueter. 1972. *Proc. 3rd Natl. Bedding Plant Conf.,* p. 43–56.
58 Krizek, D. T., W. A. Bailey, H. H. Klueter, and H. M. Cathey. 1968. *Proc. Intl. Plant Prop. Soc.* 18:273.
59 LaCroix, L. J., D. T. Canvin, and J. Walker. 1966. *Am. Soc. Hort. Sci. Proc.* 89:714.
60 Lane, H. C., H. M. Cathey, and L. T. Evans. 1965. *Am. J. Bot.* 52:1006.
61 Mancinelli, A. L., H. A. Borthwick, and S. B. Hendricks. 1966. *Bot. Gaz.* 127:1.
62 Mancinelli, A. L., Z. Yaniv, and P. Smith. 1967. *Plant Physiol.* 42:333.
63 McCree, K. J. 1971. *Agr. Metrord.* 9:191.
64 McCree, K. J. 1972. *Agr. Metrord.* 10:443.
65 McCree, K. J. 1972. *Plant Physiol.* 49:704.
66 Measures, M., P. Weinberger, and H. Baer. 1973. *Can. J. Plant Sci.* 53:215.
67 Meijer, G. 1971. *Acta. Hort.* 22:103.
68 Menz, K. M., D. N. Moss, R. Q. Cannell, and W. A. Brun. 1969. *Crop Sci.* 9:692.
69 Mitrakos, K., and W. Shropshire, Jr., Ed. 1972. *Phytochrome*. Academic, London and New York.
70 Morgan, D. C., and H. Smith. 1976. *Nature* 262:210.
71 Morgan, S. F., and J. I. Cooke. 1971. *Rep. Ref. ECRC/R392*. Elec. Council, London, 19 pp.
72 Newton, D. 1973. *Acta. Hort.* 32:89.
73 Nitsch, J. P. 1957. *Proc. Am. Soc. Hort. Sci.* 70:526.
74 Nitsch, J. P. 1957. *Proc. Am. Soc. Hort. Sci.* 70:512.
75 Patterson, D. T., M. M. Peet, and J. A. Bunce. 1977. *Agron. J.* 69:631.
76 Perry, T. O. 1971. *Science* 171:29.
77 Piringer, A. A. 1962. In *Proc. Plant Sci. Symp.* Campbell Soup Company, Camden, New Jersey, pp. 173–185.
78 Rabinowitch, E. I., and Govindjee. 1969. *Photosynthesis*. Wiley, New York.
79 Rabinowitch, E. I. 1951. *Photosynthesis and Related Processes,* Vol. 1. Interscience, New York.
80 Raper, C. D., Jr., and J. F. Thomas. 1978. *Crop Sci.* 18:654.

REFERENCES

81 Shibles, R. 1976. *Crop Sci.* 16:437.

82 Singh, M., W. L. Ogren, and J. M. Widholm. 1974. *Crop Sci.* 14:563.

83 Smith, H. 1970. *Nature* 227:665.

84 Special Committee on Growth Chamber Environments, ASHS. 1977. *HortScience* 12:309.

85 Stoutmeyer, V. T., and A. W. Close. 1946. *Proc. Am. Soc. Hort. Sci.* 48:309.

86 Swain, G. S. 1964. *Proc. Am. Soc. Hort. Sci.* 85:568.

87 Thomas, J. F., and C. B. Raper, Jr. 1978. *Agron. J.* 70:893.

88 Tibbitts, T. W., J. C. McFarlane, D. T. Krizek, W. L. Berry, P. A. Hammer, R. H. Hodgson, and R. W. Langhans. 1977. *HortScience* 12:310.

89 Tibbitts, T. W., J. C. McFarlane, D. T. Krizek, W. L. Berry, R. W. Langhans, R. A. Larson, and D. P. Ormrod. 1976. *HortScience* 101:164.

90 Toole, V. K. 1973. *Seed Sci. Technol.* 1:339.

91 Vince-Prue, D. 1975. *Photoperiodism in Plants.* McGraw-Hill, London.

92 Wareing, P. F., M. M. Khalifer, and K. J. Treherne. 1968. *Naturwissenschaften* 220:453.

93 Whalley, D. N., and K. E. Cockshull. 1976. *Sci. Hort.* 5:127.

94 White, J. W. 1974. *Am. Soc. Agric. Eng.* 74–4043.

95 White, J. W. 1979. *Hort. Rev.* 1:141.

96 Wilson, D. R., C. J. Fernandez, and K. H. McCree. 1978. *Crop Sci.* 18:19.

97 Withrow, A. P. 1958. *Purdue Univ. Agr. Exp. Sta. Bull.* 533:27.

98 Withrow, R. B., and M. H. Richman. 1933. *Purdue Univ. Agr. Exp. Sta. Bull.* 380.

99 Withrow, R. B., and M. H. Richman. 1947. *Proc. Am. Soc. Hort. Sci.* 49:63.

100 Zimmerman, R. H., D. T. Krizek, W. A. Bailey, and H. H. Klueter. 1970. *J. Am. Soc. Hort. Sci.* 95:323.

Chapter 9

RESPONSE AND GENETIC MODIFICATION OF PLANTS FOR TOLERANCE TO AIR POLLUTANTS

R. A. REINERT

USDA-SEA-AR Plant Pathology Department,
North Carolina State University, Raleigh, North Carolina

H. E. HEGGESTAD

USDA-SEA-AR Plant Stress Laboratory, Plant
Physiology Institute, Beltsville, Maryland

W. W. HECK

USDA-SEA-AR Botany Department, North Carolina State
University, Raleigh, North Carolina

INTRODUCTION

Air pollutants cause plant disease and have been called plant pathogens (213). They injure plant foliage, significantly alter growth and yield, and are known to change quality of the marketable plant product. Plants vary in their susceptibility and resistance to air pollutants, just as they vary in resistance to pathogenic bacteria, fungi and viruses. Resistance to biotic pathogens is altered if more pathogenic strains develop, whereas sensitivity and resistance to air pollutants is altered when the concentration and duration of exposure exceed the plant's genetic capability to withstand the stress.

Regardless of genetic potential of the plant to withstand stress from air pollutants, effects on vegetaion can be devastating as revealed by barren areas near Copper Hill, Tennessee, and Sudbury, Ontario. At the borders of these areas, it is possible to recognize differences in sensitivity of plants to the pollutants present. Presently, sites near industries with no vegetation are rare, because current available technologies are more effective in controlling pollutants.

There is continuing concern about transport of pollutants long distances (50, 70, 212) and the subsequent injury to sensitive species from photochemical oxidants and sulfur oxides. In the San Bernadino Mountains of Southern California species such as Ponderosa pine are severely damaged by oxidants (143, 144, 154). Throughout the eastern United States there is a continuing decline of white pine and moderate damage to other tree species (12, 13, 70).

Numerous publications reveal the magnitude of differences in response between species to air pollutants (Table 1). The National Academy of Sciences has also published several excellent review documents concerning the effects of specific gaseous air pollutants on plants (149, 150, 151, 152).

Air Pollutants Considered

The most important air pollutants that damage plants are: (a) ozone (O_3) and other photochemical oxidants peroxyacetyl nitrate (PAN); (b) sulfur dioxide (SO_2); (c) oxides of nitrogen (NO_2 and NO); and (d) hydrogen fluoride (HF). Possible injury by ethylene, other hydrocarbons, particulates, and aerosols is of concern, but their interacting effects with the major gaseous pollutants may be more significant.

Ozone and PAN are strong oxidizing agents. They originate primarily from the action of ultraviolet radiation on nitrogen oxides (NO_x) in the presence of O_2 and reactive hydrocarbons (primarily from incomplete combustion—e.g., vehicular transportation). Sulfur dioxide originates from the combustion of fossil fuels and the smelting of ores. Oxides of nitrogen are produced from vehicular transportation or from combustion of fossil fuels. Acid precipitation is one manifestation of the SO_2 (sulfur oxides) and NO_x pollution (117). Fluoride is produced at point sources especially around aluminum manufacturing plants and fertilizer processing plants. Ethylene is released from incomplete combustion processes.

Ozone concentrations of 0.05–0.12 parts per million (ppm) for 2- to 4-hr exposure periods will injure leaves of the most sensitive cultivars of some

Table 1   References to crop and woody plant species sensitivity to air pollutants

Pollutant	References
Ozone (O_3)	40, 42, 57, 83, 92, 150, 153, 154, 158, 173, 201, 214
Peroxyacetyl nitrate (PAN)	83, 158, 173, 192
Sulfur dioxide (SO_2)	6, 42, 57, 61, 62, 63, 83, 152, 158, 169, 173, 194, 205, 218
Nitrogen dioxide (NO_2)	83, 151, 158, 173, 192, 193, 203
Hydrogen fluoride (HF)	63, 120, 149, 158, 173, 208, 218
Ethylene (C_2H_2)	77

crop species. During summer months areas throughout the United States frequently contain concentrations of O_3 ranging from 0.05–0.10 ppm. Peroxyacetyl nitrate is injurious to plants in the ppb range.

Sulfur dioxide will injure sensitive plant species at concentrations from 0.25 to 0.50 ppm for 6- to 8-hr exposures. Nitrogen dioxide is much less toxic, but a mixture of the two pollutants may injure some species and cultivars within species at concentrations that individually will not injure plants (172).

Fluoride can accumulate in plant tissue of many crop plants, but the magnitude of accumulation and subsequent damage is limited to areas around industries such as aluminum, superphosphate, and glass manufacture. Ethylene decreases the growth and productivity of some crops near urban industrial centers.

Research summarizing species sensitivity is available (Table 1). However, this information is difficult to evaluate for several reasons: (a) environmental conditions before and during exposure may differ; (b) highly sensitive or highly insensitive cultivars may have been used to test species sensitivity; (c) the plants may have been exposed at different stages of maturation; and (d) pollutant concentrations and exposure duration may not have been well defined.

Terminology Defined

A definition of certain terms as applied to plant response to air pollutants will help clarify research concepts developed in this chapter.

The terms "sensitivity" and "susceptibility" should not be used interchangeably. In this chapter the term "sensitivity" is defined as a physiological condition of plants or particular plant tissues, whereby they are prone to injury by air pollutants. The term "sensitivity" includes the concept that visible injury is produced. The magnitude of injury resulting from a given dose (concentration × duration of exposure, continuous or intermittent) further defines the range of sensitivity observed within or among a given group of cultivars.

The term "susceptibility" should be used in conjunction with the term "resistance." These terms (resistance and susceptibility) relate to the inherent and environmentally controlled capacity (resistance) or incapacity (susceptibility) of plants to prevent or lessen visible injury or other effects. These two terms help describe the genetics of plant response rather than the amount of injury present (sensitivity). In certain crops it may be useful to describe the genetic condition of the cultivar in terms of high and low susceptibility or high and low resistance. This does not require determination of the magnitude of visible injury over a wide range of doses when response of the cultivar or group of cultivars may be influenced by varying environments.

A closely related term, "immunity," is used in discussing biotic plant diseases. Immunity is the ability of the plant to prevent infection. In the case of air pollutants, it is the ability of the plant to absorb a pollutant without subsequent injury. Immunity to air pollutants does not occur; if the dose is high enough, all plants will be injured and eventually killed.

A set of related terms that should be defined are "tolerance" and "intolerance." In a pathological sense these terms describe the ability or inability of a plant to sustain (tolerate) the effects of a disease without dying or suffering serious injury or yield loss. Most plant species or cultivars within a species probably possess some degree of tolerance to air pollutants; that is, visible injury can be present without reducing yields of seeds or fruit. The degree of air pollution tolerance is difficult to define because of inadequate data bases.

In any plant breeding program concerning resistance to plant pathogens one needs to consider the plant (host) and its variability, the environment and its changing influence, and also the pathogen and its genetic variability. In the case of air pollutants, however, the susceptibility of the host, the environmental influences, and the dose of the toxicant are the primary concerns of the air pollution plant scientist.

Chapter Objectives

The primary objectives and goals of this chapter are: (a) to discuss genetic variability among cultivars within species including numbers of cultivars tested; (b) to discuss and evaluate sensitivity screens needed to assess tolerance; (c) to discuss plant injury factors, such as biomass, flowering, fruiting, and yield, in relation to tolerance; (d) to describe genetic control of resistance both in experimental studies and under native environments or ecosystems; and (e) to recommend research needed to develop a better understanding of breeding methods for the control of air pollution diseases.

GENETIC VARIABILITY

Intraspecific genetic variation in susceptibility of plants to air pollutants is known for many species. The variation among cultivars has been studied for a number of agronomic and horticultural crop species. Natural selection for air pollution tolerance occurs and has inadvertently appeared in some breeding programs.

Natural Evolution of Air Pollutant-Resistant Plants

Native Plants.   Spruce trees with more resistance to SO_2 and fluorides than those of the general population were identified in Europe by Rohmeder et al. (179). Resistant trees were found in forested areas near industries emitting pollutants. Differences in tolerance among spruce trees were more common than for pine species. Pelz (163) examined several phenotypic characteristics of spruce with varying sensitivity to SO_2. The needles of insensitive trees were longer, heavier and more numerous than those of the more sensitive trees. The insensitive trees seemed to have a smaller trunk diameter, lower height, and smaller crown dimensions. Insensitivity, or possibly tolerance, of spruce was not associated with more growth, because sensitive trees had wider growth rings in the early years of growth before industry operations.

In Britain, a native ryegrass from a high SO_2 area was more resistant to SO_2 than a commercial cultivar when exposed to SO_2 (see later section Changes in Resistance/Tolerance Levels). The effectiveness of natural selection in an annual species, *Geranium carolinianum*, was demonstrated recently in Georgia (189). Populations originating near a coal-fired power station were more tolerant to SO_2 than populations growing farther from the power plant. In a controlled 12-hr exposure to 0.76 ppm of SO_2, the SO_2-tolerant plants had about 10% less leaf injury than the plants growing away from the SO_2 source.

Cultivated Plants. Plant breeders working in locations with O_3 stress have inadvertently developed cultivars that are more tolerant to O_3 than those developed in locations without such stress. Plant breeding research at the U.S. Department of Agriculture Research Center, Beltsville, Maryland, has been carried out in elevated levels of O_3 (50, 207). 'Team' alfalfa and three other breeding lines routinely selected at Beltsville for the absence of foliar injury were more tolerant to O_3 than 10 cultivars selected elsewhere (102). Several potato cultivars developed at Beltsville were more tolerant to O_3 than cultivars developed in states with less O_3 air pollution (85). For example, 'Kennebec', 'Pungo', and 'Katahdin', developed at Beltsville, were tolerant enough to O_3, that culture in carbon-filtered air with reduced O_3 did not increase yield (85). Under similar circumstances, however, 'Norland' and 'Norchip' from North Dakota, 'Haig' from Nebraska, and 'La Chipper' from Louisiana produced significantly higher tuber yields in carbon-filtered air. A dry bean cultivar, 'California Small White 59', was more tolerant to O_3 than the cultivars 'Seafarer', 'Sanilac', and 'Gratiot', which are commonly grown in Michigan (86). Cotton cultivars that were developed in the San Joaquin Valley of California, such as 'Acala SJ-1', were more tolerant to O_3 than cultivars originating in other states, especially the cultivars 'Gregg 45' and 'Paymaster 202', which were developed in the high plains of Texas (87). Two sugar beet hybrids, 'USH6' and 'USH9B', originating in California, also had more tolerance to O_3 than other cultivars tested (139).

Plants may be stimulated by low levels of pollutants such as O_3 and SO_2. Bennett et al. (9) exposed barley and smartweed to low concentrations of O_3 and dry weight increased compared with plants grown in carbon-filtered air. A snap bean cultivar 'Gallatin 50' has produced significantly more beans (15%) in unfiltered open-top chambers than in filtered air (88). Plants in the filtered air were taller indicating a possible delay in maturity in the clean air.

Thor. H. Bertinusen, a cigar-wrapper tobacco breeder and tobacco specialist, stated "resistance to fleck is adequate to protect the crop" (Personal communication, H. E. Heggestad). Continuous selection pressure was used to maintain the resistance. A cultivar designated "L" has been grown since 1967 and in 1978 it was extensively used commercially. Seed was increased on a field known to be favorable for fleck development and only a few plants with little or no "weather fleck" were selected for seed harvest.

Controlled Studies of Species and Cultivar Variability

Studies of crop species have involved exposure to pollutants in ambient air and to different pollutant doses in controlled fumigations in greenhouse environments. Rankings of cultivar sensitivity based on leaf injury are especially important in those crops where foliar appearance is important to quality. To establish meaningful differences in susceptibility among cultivars to pollutants in ambient air or controlled fumigations, the cultivars must be evaluated in different stages of plant growth and changes in yield should be determined.

In controlled fumigations, cultivar response to two or more doses of the pollutant should be determined. Plant sensitivity is perhaps best examined when the various cultivars are in their log phase of growth and/or when the plants are in bloom. Some cultivars may not be visibly injured, but growth and development may be delayed or inhibited. Flowering may be delayed or flower and fruit size and number decreased. If little change in yield and marketable quality is identified, the cultivar should be classified tolerant even though it may display some visible injury. An evaluation of cultivar sensitivity to pollutants should also consider pollutant combinations. Guderian (63) states, "In heavily industrialized areas, a mixture of various pollutants always occurs, whereas in less industrialized areas, single pollutants may occur one after the other."

Various numbers of species and cultivars have been tested for each pollutant. Cultivars of O_3-sensitive species such as tobacco, soybean, alfalfa, tomato, snap bean, potato, grape, begonia, petunia, and others have been extensively studied. These studies have been reported from various parts of the country using different exposure systems and doses of artificially generated O_3. Controlled exposures using ambient O_3 have also been made. Foliar injury has been the primary response used to assess cultivar sensitivity. The number of cultivars tested for each pollutant are presented for agronomic crops, Table 2; vegetable and fruit crops, Table 3; and herbaceous and woody ornamentals, Table 4. Nearly all the studies of intraspecific sensitivity will provide information on the most sensitive and least sensitive cultivars. The range of foliar injury among cultivars within species may vary considerably (tomato, potato, tobacco, soybean, and others) or only by a small amount (radish).

Variations in the sensitivity of tree species to the pollutants O_3 and SO_2 have been identified: poplar (46, 108, 112, 215); spruce (112, 163, 179, 204);

Table 2   Number of agronomic cultivars tested for sensitivity to various air pollutants

Cultivar	Pollutant					Reference
	O_3	NO_2	SO_2	HF	Ambient air	
Alfalfa	23				2	35, 102, 196
Barley	3					183
Corn	11		4			74, 113
Cotton					8	87
Forage legumes	10			2		19, 118
Hops			Several			52
Rye	3					183
Safflower	12					104
Sorghum				15		93, 182
Soybeans	>40		21		4	88, 103, 142, 199
Sugar beets	5					139
Tobacco	>73		12		>48	5, 27, 43, 60, 80, 94, 129, 131, 133, 134, 135, 136, 137, 138, 187
Turfgrass	28		28			20, 147, 211
Wheat	3		7			76, 113, 183, 184

lodgepole pine (203); red maple (202); white pine (12, 13, 56, 57, 99); scotch pine (206).

Other Considerations

Plant Adaptation to Stress and Other Pollutants.   There is increasing evidence that plants can adapt to extreme environments, such as drought, salinity, heat, cold, and toxicity (15). The potential may be comparable to selection in corn for protein and oil content extending for 50 generations of segregating plant populations (217).

Plants of some species are known to adapt to elevated concentrations of copper, lead, and zinc in mine waste. The tolerant plants continue to root in the presence of elevated concentrations of these metals, whereas other populations die. With severe selection, a metal-tolerant population of *Agrostis* was produced from a nontolerant population in only one or two generations (15).

Characteristics of Cultivars Resistant to Fluorides.   With sensitive and tolerant cultivars of tulip Spierings (185) found that the tolerant cultivar 'Pre-

Table 3 Number of vegetable and fruit cultivars tested for sensitivity to various air pollutants

	Pollutant				
Cultivar	O_3	SO_2	HF	Ambient air	Reference
Bean	>80	33		>387	7, 23, 28, 41, 86, 91, 95, 96, 101, 125, 128, 166, 186
Blueberry		4			21
Celery	8			8	167
Citrus			7		25
Cucumber	4				24, 156
Grape	22			143	107, 110, 178
Lettuce	8				171
Onion	4				156
Pea	20				155, 157
Potato	>37			>75	16, 17, 53, 115, 85, 96, 97, 121, 156, 177
Radish	9				2, 171
Spinach	11				75, 91, 122
Strawberry	6	6			168
Sweet corn	2			50	31, 32, 53, 109, 195
Tomato	338[a]				38, 39, 55, 89, 160, 170

[a]A series of 1209 accessions was screened representing collections from 50 countries (38); however, only 24 accessions were reported.

Table 4 Number of flower and ornamental cultivars tested for sensitivity to air pollutants

	Pollutants						
Cultivar	O_3	NO_2	PAN	SO_2	HF	Ambient air	Reference
Azalea	10						59
Begonia	17			17			1, 114, 175, 176
Chrysanthemum	16		8	16			18, 22, 216
Coleus	2			2			1
Gladiolus					118		90, 127
Marigold	2			2			1
Petunia	69	15	8	17		15	1, 36, 45, 48, 51, 65, 67, 115
Poinsettia	15			6		2	37, 84, 115, 123, 124
Rose					14		26
Snapdragon	3			3			1
Tulip					2		185

ludium' had significantly more fluoride in the leaf tip than the sensitive variety 'Blue Parrot'. However, no relationship between the tissue accumulation of fluoride and susceptibility to leaf injury by fluorides was found in seven cultivars of sorghum (182). The entire plant was harvested and analyzed in the sorghum study but only 7.5 cm of leaf was used in the tulip study. That leaf was subdivided into 2.5-cm segments which were analyzed separately. Differences in fluoride content were found only in the tips of the leaves of the tulip cultivars.

SENSITIVITY SCREENS

Numerous screening procedures have been used to assess intraspecific variability in sensitivity. Much of the published literature represents results of exposure of plants for a few hours to acute concentrations of pollutant. The plant response most frequently measured has been foliar injury. It is usually estimated as a percent of each leaf surface injured. These studies help to characterize the germplasm sensitivity within a species and reveal potentially resistant and tolerant germplasm.

Many factors must be considered in setting up a screening program for resistance to air pollutants. These have been reviewed by Drummond and Pearson (47). The pollutant concentration and environmental condition (78) should be known and should not vary too much from the ambient conditions; duration of exposure is also important. Other factors such as morphological characters, plant age, maturity, stomate function, and soil nutrition may influence sensitivity. Thus, the credibility of a sensitivity screen depends on how carefully the researcher developed the experiments and evaluated the various factors affecting sensitivity.

Experimental vs. Ambient Exposures

An important consideration in any screening procedure is the establishment of credibility between experimental and ambient exposures. All evaluations of sensitivity to pollutants should demonstrate some relationship to an ambient screen.

Tree Species.   Young spruce trees have more resistance to SO_2 than old stands consequently, the sensitivity of conifer forests cannot be judged by the response of young specimens (210). Wentzel cautioned researchers about the limitations of chamber tests with forest species. For example, in chamber tests using high SO_2 concentrations, linden was more sensitive than spruce. However, in the field where concentrations of SO_2 were lower, linden survived long after the hardiest spruce. Thus, chamber tests were not able to assess the regenerative capacity of deciduous trees, such as linden. Wentzel further states that species which are the most sensitive to SO_2 in chamber tests (e.g., some types of clover, beans, and alfalfa) and to HF (e.g., gladiolas, tulips, and grapevines) can be cultivated profitably in areas where spruce and pine do not survive, except as young trees. Also, larch is very sensitive to high concentrations of SO_2 but it is one of the most resistant conifers to concentrations that produce chronic injury. Wentzel insists that leaf response in chamber tests has little value in predicting the ability of trees to remain useful and survive in industrial areas. Wentzel discusses four factors that influence plant response to pollutants: (a) the kind of pollutant and the variation in ambient air concentrations; (b) the stage of plant development both in relation to seasonal age and the conditions of health; (c) conditions affecting growth including climate, especially extremes of weather, soil, nutrition, and plant competition; and (d) the height of the tree in the stand. Severe frosts in the winter and late spring are known to increase the susceptibility of conifers to pollutants. Beech trees are more tolerant if grown in limed, fertile soils than in poor sandy soils. However, black spruce seems to be resistant to SO_2, HF, salt aerosols, and dust in sites that have poor or rich soil and also in sites with wet or dry soil.

In studies with poplar nine cultivars were evaluated for SO_2 sensitivity in a young plantation growing near a SO_2 source (112). The nine cultivars were also exposed to different doses of SO_2 in exposure chambers. Sensitivity ratings were obtained for each cultivar based on plant loss, leaf injury, height, and stem diameter. The overall sensitivity rating for each cultivar in each test is shown in Table 5. Rankings were the same in both tests for the least sensitive cultivar (#11) and the two most sensitive cultivars (#2 and #4).

Crop Species. Cigar-wrapper tobacco has been selected for tolerance to oxidants in fields for the past 20 years. Breeding lines are compared with

Table 5   Overall sensitivity ranking of nine poplar strains to SO_2[a]

	Rankings	
Strain	Ambient air exposure	Chamber exposure
11	1	1
1	2	7
13	3	6
16	4	4
15	5	3
18	6	2
5	7	5
2	8	8
4	9	9

[a]Data from reference 112.

known fleck-susceptible cultivars. Plants free of foliar injury are saved for seed with selection pressure continued until the seed is harvested. If, however, plant selection is based upon minimum leaf injury following exposure to a pollutant in a controlled environment, the pollutant dose should be similar to the maximum measured in the area where the crop is to be grown. Tobacco varieties showed the most difference in sensitivity when exposed to about 0.2 ppm O_3 for 2 hr and the least difference following exposure to 0.42 ppm O_3 for 3 hr (Table 6). The latter is a level which might occur in Southern California but not in the Connecticut Valley where the crop is grown.

Menser and Hodges (135) tested several flue-cured tobacco cultivars for susceptibility to O_3 in a controlled environmental chamber and under ambient air conditions. They concluded that the field exposures provided a somewhat better test for O_3 sensitivity than the experimental fumigations. They believed that only a limited possibility existed to improve O_3 resistance by progeny testing and selection. However, they used a small plant population compared with that used by the cigar-wrapper tobacco breeders, who made repeated observations during the season on thousands of plants in their efforts to maintain and improve the O_3 resistance in otherwise productive cultivars.

Meiners and Heggestad (128) observed a significant but low correlation (r = .3) for acute leaf injury to snap bean cultivars exposed to O_3 in either controlled environments or to ambient levels in field plantings. In the field, the maximum hourly concentrations were only about a third of those used in the chambers.

Chronic vs. Acute Exposures

Variation in sensitivity to gaseous air pollutants usually involves acute exposure (high concentrations of pollutant for short duration). An assumption is made that sensitivity to acute exposure will be a good indicator of sensitivity

Table 6 Response of tobacco cultivars to ozone[a]

| | % Foliar injury | | |
| | Ozone dose | | |
Cultivar	Low (0.09 ppm 4 hr)	Medium (0.20 ppm 2 hr)	High (0.42 ppm 3 hr)
Bel W_3	8	44	75
Bel C	1	13	43
Bel B	0	2	50

[a]Ozone was determined by the KI method; tobacco exposed to low and medium dose had six leaves and tobacco exposed to the high dose had nine leaves. Data from reference 82.

Table 7   Chronic and acute exposure of four tree species to SO_2[a]

Species	Chronic exposure[b] Number of days to chlorosis	Acute exposure[c] % Foliar injury
Chinese elm	7	90
Norway maple	12	30
Gingko	14	20
Pin oak	30	0

[a]Data from reference 194.
[b]Continuous exposure at 0.5 ppm SO_2.
[c]A single 6 hour exposure to 3 ppm SO_2.

of chronic exposure (low concentrations of pollutant are repeated exposure duration). In tests with four urban tree species, Temple (194) found excellent agreement in four urban tree species after acute and chronic exposure to SO_2 (Table 7). Chinese elm and Pin oak showed similar sensitivity to acute and chronic exposure, however the acute exposure gave a better separation of sensitivity among all four species. Heagle (72) examined the response of four soybean cultivars to acute (0, 15, 30, 45, 60, 75 pphm), and chronic (0, 3, 6, 9, 12 or 15 pphm) O_3 exposures. The cultivar ranking of injury from acute exposure differed markedly from the ranking obtained from chronic exposure. Rankings of cultivars to primary leaf injury also differed from rankings of cultivars to trifoliate leaf injury. The ranking also varied depending on the response measured; changes based on dry weight of shoots gave a different ranking of cultivar sensitivity than foliar injury (72).

Other Factors Influencing Screening

Cultivars sensitive to one pollutant may be sensitive or tolerant to another pollutant. The response of 15 petunia cultivars to O_3, SO_2, peroxyacetyl nitrate (PAN), NO_2, and irradiated auto exhaust were determined by Feder et al. (51). Some cultivars, such as 'Lilac Time', were sensitive to only one of the pollutants; other cultivars, such as 'White Cascade', were sensitive to several pollutants. However, in screening several species for sensitivity to SO_2 and HF there was evidence that species sensitive to one pollutant tended to be insensitive to the other pollutant (218).

Plant Age and Maturity.   The stage and rate of growth tend to regulate O_3 sensitivity, based on foliar injury (64). Dry beans are most sensitive after plants are in full bloom. At this time the production of new leaves decreases and there is a redistribution of carbohydrates from vegetative to reproductive organs. Markowski and Greziak (126) found that beans were most sensitive to both O_3 and SO_2 in the growth phase from bloom to fully developed pods. Plants showed variable recovery when injury occurred at earlier phases of growth.

Taylor (191) found a correlation of .78 comparing the reaction of very young tobacco cultivars and mature plant stages to ambient levels of pollutants. Reinert and Henderson (174) examined foliar injury differences in six tomato cultivars at three O_3 doses and three ages. Cultivar rankings of cotyledon injury at 2 weeks differed from rankings at 4 and 6 weeks of age but the rankings did not change with O_3 dose.

Morphological Characteristics. In tobacco, Dean (44) found the stomate density of 25–26/mm² on the lower leaf surface of two resistant varieties and a range of 35–41/mm² on the two susceptible varieties. The decrease in stomate density by one-third was associated with resistance, reducing the amount of O_3 entering the leaf. However, Menser (141) did not find differences in stomatal frequency using four different tobacco cultivars ranging in O_3 sensitivity from low to very high. Thorne and Hanson (197) hypothesized that tolerance to O_3 in petunia resulted from gas exchange potential more than from biochemical differences. If a genetic component in plants is influential in altering leaf morphology and thus stomate density, gas exchange and pollutant uptake rates would be altered. These areas of research need to be more intensively investigated.

Physiologic Age and Diurnal Sensitivity. The physiologic age determined the degree of injury to cotton. The leaves were most sensitive when 75% expanded and were progressively less sensitive with age (198). Tobacco leaves showed a diurnal response to O_3 (130); sensitive after 4 hr in the light and becoming increasingly less sensitive after about 6 hr in light.

General Environment. Environmental factors influence the overall sensitivity of plants to air pollutants but generally do not alter the relative sensitivity of cultivars. Soil moisture and nutrition are perhaps the most important soil factors affecting plant sensitivity to O_3 and SO_2. Investigators have examined factors which may influence the sensitivity of various screening methodologies (73, 130, 174) and how various methodologies may influence cultivar rankings.

Heagle (73) examined the sensitivity ranking of four soybean cultivars to O_3 as influenced by growth media, fertilizer rate and season of the year. Plants were more sensitive when grown in a sand Jiffy mix or peat-perlite-sandy loam soil medium than in a sandy-loam or peat-perlite growth medium. The sensitivity rankings of the four cultivars changed with the growth media, fertilizer rate, and season of exposure. Foliar injury was less from exposures in January and March than in June. The soybean cultivar 'Dare' became less sensitive and 'Scott' more sensitive from January to June.

INJURY, BIOMASS, YIELD, AND PLANT TOLERANCE

Sensitivity screens within species provide an assessment of germplasm response to pollutants. Cultivars that have genetic potential for high productivity and quality need to be tested for tolerance to air pollutant stress (by comparing performance in relatively clean air and in ambient air). It is impossible to identify a pollutant-free environment adjacent to a pollutant-stress environment without the use of some type of chamber. Open-top field chambers (71, 119, 148) and some greenhouse exposure systems (32, 85, 87, 159) permit the identification of tolerance to air pollutants among cultivars within a species based on a yield response, but there is little published literature on such studies that deal directly with the question of tolerance to air pollutants. Also related to this question is the fact that too much attention has been focused on annual averages of pollutants such as SO_2 or HF. Plant survival seems to be determined more by peak intermittent concentrations which may be several times higher than the monthly or annual mean concentrations.

Injury vs. Maturity—Yield

Early maturing cultivars of many plant species are usually the most susceptible to O_3 air pollution. However, plant breeders have developed early maturing cultivars of cigar-wrapper tobacco that have high tolerance to O_3 (81, 133, 137). The relation of O_3 injury to maturity and to yield was studied in 59 potato cultivars and breeding lines in replicated plots at two locations (146). A high correlation (4 = −.96) was found between maturity (1 = early to 5 = late) and O_3 injury (1 = no injury and 6 = severe injury). When average yields of all entries were compared with the O_3 injury ratings, the yields were highest for the cultivar grouping with lowest injury (late maturing cultivars) and lowest for cultivars with the most leaf injury (early maturing cultivars). The relative resistance (tolerance) was the same at both locations. Conclusions involving yield reductions were difficult, except for the highly sensitive entries, because yields could not be determined under O_3-free conditions. There were also a few cultivar exceptions to the maturity classification. 'Superior', a relatively early cultivar, was quite tolerant; 'Red Pontiac', a late maturing variety, developed considerable injury.

Dose vs. Yield

The relation of dose of pollutants to yield is illustrated in a field study with tomato cultivars (160). Five cultivars were planted both at Riverside and Irvine, California; the cumulative O_3 dose (concentration × exposure duration) at Riverside was 3798 pphm hours above 0.10 ppm, while at Irvine it was only 222 pphm hours. Fruit size and production were reduced for all cultivars at Riverside. At Irvine, all cultivars except 'Ace' produced the same yield; at Riverside 'H-11' and '6717 VH' yielded significantly more than the cultivars 'Ace', 'Earlypak 7' and 'Polepak F_4VF'. By contrast, considering foliar sensitivity, 'Earlypak 7' was the most resistant. 'H-11' tolerated considerable foliar injury with no affect on yield (160).

Highly sensitive cultivars of potato and snap bean produced lower yields in the presence of an O_3 stress than did tolerant cultivars. For example, in greenhouse studies (85) the susceptible potato cultivars 'Norland' and 'Haig' in unfiltered air produced only about 60% the yield of a resistant cultivar 'Kennebec' (Table 8). 'Bush Blue Lake 290', a susceptible cultivar of snap bean, produced 14% less than 'Astro' or 'Gallatin 50' in a 3-year study using open-top chambers supplied with either carbon-filtered or unfiltered ambient air (88).

There is evidence that some high-yielding potato cultivars do not reach their full yield potential in the presence of an O_3 stress. In greenhouse studies (85), the variety 'Norchip' produced the highest yield of 13 cultivars in both filtered and unfiltered air; it also produced significantly higher yield (35%) in the carbon-filtered air than in unfiltered air (Table 8). A snap bean variety, 'Bush Blue Lake 274', produced about the same yield as the resistant cultivars 'Astro' and 'Gallatin 50' in chambers supplied with unfiltered ambient air, but produced significantly more (28%) in the carbon-filtered air (88). In sweet corn studies (Heggestad, unpublished), 'Silver Queen' produced higher yields than three other sweet corn varieties, but it was the only variety that showed a significant yield increase in carbon-filtered air.

Clover and ryegrass were exposed in plastic greenhouse chambers to

Table 8 Yield of potato cultivars in carbon-filtered and unfiltered air at Beltsville, Md. 1971[a]

Cultivar	State where developed	Yield (g/plant)		
		Carbon-filtered air (CF)	Unfiltered air (NF)	% NF/CF × 100
Katahdin	Maryland	218	279	128
Penn 71	Pennsylvania	251	269	107
Pungo	Virginia	282	287	102
Norgold Russett	North Dakota	251	252	100
Superior	Wisconsin	318	304	96
Kennebec	Maine	369	339	92
Wausau	Maine	273	218	80
Norchip	North Dakota	466	346	74
La Chipper	Louisiana	380	276	73
Alamo	Texas	410	272	66
Haig	Nebraska	295	187	63
Norland	North Dakota	401	199	50

[a]LSD 0.05 = 71 g for cultivars within an environment and 81 g for cultivars between the two environments. Data from reference 85.

0.03 or 0.09 ppm O_3 for 8 hr/day for 6 weeks (11). Significant differences were observed in dry weight, leaf area, and root-shoot ratio for both species. Of the various parameters used, leaf area, plant dry weight and root-shoot ratio were most useful in showing the effects of O_3 on growth of these two species.

Sensitivity of Plant Compartments

Although total yields and productivity are important in understanding plant tolerance to air pollutants, recent data have been published that consider how various compartments of the plant react to pollutant stress. Several studies are important. Oshima et al. (161) exposed parsley 4 hr/day, 2 days/week for 8 weeks to 0.20 ppm O_3. Total plant dry weight and root dry weight were decreased 23% and 43% respectively with little effect on leaves. Ozone from the initial exposures had the greatest effect on top growth. The fumigated plants produced significantly more leaves by final harvest. The authors suggested that the effect on root growth was due to loss of lower leaves that are the main source of photosynthates for roots (161).

Carrots were exposed for 6 hr to either 0.19 or 0.25 ppm O_3 several times (1.5 times/week) during the growth period (10). Leaf dry weight was unaffected but root dry matter decreased 32–46%. Leaf chlorosis was increased by the O_3 treatments and root dry weight decreased significantly as the number of chlorotic leaves increased. Leaf length, root length, total number of leaves, and total leaf dry weight were not affected by the O_3 treatments.

Oshima et al. (162) have also grown cotton in a greenhouse having carbon-filtered air where plants were exposed to 0.25 ppm O_3 6 hr/day, biweekly, for 19 weeks. The largest reductions in dry weight of various plant parts occurred in roots and bolls. Ozone-treated cotton initially produced fewer leaves and less leaf area followed by a period of stimulation of axillary growth.

Influence of Pollutant Mixtures

The sensitivity of crops to pollutant mixtures has been reviewed by Reinert et al. (172). Some plant species show more than additive effects (synergism) and other species show a less than additive (antagonism) response. Synergistic effects of $SO_2 + O_3$ were first noted on tobacco by Menser and Heggestad (132) and the widespread occurrence of both SO_2 and O_3 in ambient air have stimulated interest in pollutant mixtures. Tingey et al. (200) examined the effects of O_3 and SO_2 singly and in combination on 11 species of plants. They found that lima bean and tomato were quite insensitive to the gas mixtures, and tobacco, alfalfa, and radish were sensitive.

The effects of 0.25 ppm O_3 and 0.5 ppm SO_2 singly and in combination were studied on five cultivars of Rieger begonia (176). Only one cultivar showed a significant interaction on flower weight; that is, the combined effect of the two pollutants was antagonistic. Some cultivars were more sensitive to SO_2 and others were more sensitive to O_3 based on changes in flower weight.

Several full-sib families of loblolly pine and half-sib families of American sycamore were screened for sensitivity to NO_2, SO_2, and/or O_3 (111). Significant height and growth suppression was noted in the susceptible family of both species if exposed to O_3, SO_2, $SO_2 + O_3$, and $NO_2 + SO_2 + O_3$. Ozone + SO_2 had a more significant effect than either pollutant alone and when NO_2 was added it further suppressed the height growth of the susceptible family of each species.

GENE CONTROL OF RESISTANCE

Changes in Resistance/Tolerance Levels—Natural

The effectiveness of natural selection for resistance and/or tolerance to air pollutants is not well documented. There are various studies suggesting that natural selection takes place in areas where severe stress of vegetation is occurring, or near stationary sources of the pollutant (e.g., results from *G. carolinianum;* earlier section Natural Evolution of Air Pollutant-Resistant Plants).

In Britain, a cultivar of ryegrass, 'Helmshore', grown for many years in an area exposed to SO_2 pollution, was more resistant to SO_2 than a commercial cultivar, 'S 23' (8). Infumigation experiments with SO_2 concentrations, similar to the ambient SO_2 stress, the yield of 'S 23' was reduced significantly without measurable effects on the indigenously derived 'Helmshore'. Although the resistance mechanism could not be identified, there was no reduction in the SO_2 penetration of leaves of the resistant plants and no difference in the growth rate of the two cultivars in the absence of stress from SO_2.

Various *Lycopersicon* spp. have been experimentally exposed to O_3. The least O_3-sensitive species was *L. esculentum*. Species with intermediate sensitivity were *L. peruvianum*, *L. hirsutum glabratum*, and *L. glandulosum*. The greatest sensitivity to O_3 was found in *L. pimpinellifolium* and in crosses of *L. esculentum* and *pimpinellifolium*. *L. pimpinellifolium* also showed the greatest sensitivity to O_3 in ambient air. Wild species of tomato, such as *L. pimpinellifolium,* are indigenous to the highlands of South America were O_3 concentrations would be similar or equal to background levels. Thus, the higher degree of O_3 sensitivity in *L. pimpinellifolium* may be due to a lack of selection pressure at low O_3 concentrations contrasted with the fact that cultivated species are guided through natural selection processes if bred in areas of high O_3 stress (55).

The processes of natural selection for air pollution resistance in forest ecosystems probably occurs over long periods of time. Ozone and other photochemical oxidants, SO_2, and fluoride have stressed forest species over large areas in many parts of the world (144).

It is widely accepted that woody vegetation growing under conditions of high humidity in the Eastern United States can be severely injured by O_3 at concentrations (0.15 ppm) below those (0.30 ppm) commonly experienced in the less humid inland sections of Southern California. Under these conditions of high humidity many species of woody plants may be under a continuing natural selection process. Various researchers have shown a strong genetic control over susceptibility and resistance of eastern white pine to O_3 and SO_2 (13). The genetic control of resistance and susceptibility of eastern forest tree species to air pollutants has been summarized by Houston (100).

Oxidant damage to ponderosa pine and associated species in the San Bernadino Mountains and perhaps other western forest ecosystems has been occurring at least since 1956 (143). Differing sensitivity among the species has been observed. It is too early to tell how natural selection processes may affect the resistance of western tree species to oxidants. Resistance of selected tree species to SO_2 and fluoride near pollution sources was greater in Europe (see earlier section Native Plants).

Changes in Resistance/Tolerance Levels—Experimental

Townsend and Dochinger (202) found significant variation in tolerance to O_3 among red maple seedling progenies from four seed sources. Seedlings from a southern location in the United States displayed the least foliar injury while seedlings from two northern locations displayed the most foliar injury. The seedlings showing the most injury also grew less in the presence of elevated concentrations of O_3. The authors concluded that genetic factors controlled O_3 susceptibility in red maple. The high correlation found between O_3 susceptibility of different-aged seedlings from the same seed source indicated selection would be effective at any growth stage (202).

Dochinger and Jensen (46) reported a correlation between foliar injury due to SO_2 and growth inhibition in hybrid poplar. Correlations between foliar injury and shoot growth were positive and significant. They concluded that enough genetic variation existed to permit selection of clones with significant tolerance to SO_2.

Genetic Studies

A few review articles summarize literature on the development of plants that are resistant or tolerant to air pollutants (14, 54, 58, 100, 145, 180). Gableman (54) points out that air pollutants have not been normal constituents of the environment in which plants have grown over the centuries. Thus, nature has not provided a continuous selection pressure to isolate genes for resistance. Genetic modification and recombination may provide a vehicle for permanent change that may provide morphological systems for excluding the pollutant or physiological systems that may tolerate excesses of the pollutant. Genetic studies of gene control over susceptibility and resistance in crops have been limited to alfalfa (35), tobacco (3, 105, 164, 165, 181, 188, 190), sweet corn (31, 32, 33, 34), bean (29, 30), onion (49), petunia (66, 69), and forest tree species (98, 100). Although this list of species is short, the amount of research is considerable.

Herbaceous Annual Plants.   Tobacco has been the most intensively studied due to its agricultural importance in O_3-stressed areas of eastern United States. Sand (181) studied weather fleck on shade-grown tobacco in the Connecticut Valley and found that the highest source of resistance came from a Sumatran and a Kentucky dark fire-cured cultivar. He made crosses of tobacco cultivars highly susceptible and highly resistant to weather fleck. The F_1 and F_2 generations were intermediate between the parents but the F_1 was significantly more resistant than the average for the F_2. A few selections in the F_4 generation exceeded the fleck resistance of the best parent. Since the genes controlling resistance in the parent could be reassembled in three generations, Sand concluded that only a few genes were involved. Transgressive segregation in the F_2 generation indicated the presence of resistance genes in the susceptible parent. After 5 years of study and examination of 22 inbred selections, Sand reached the following conclusions: (*a*) resistance to fleck was partially dominant, (*b*) the average resistance of the F_1 was significantly greater than that of the midparent, (*c*) the genes of the resistant parent could be reassembled after several generations of selection, and (*d*) the number of genes concerned with resistance may be as many as four.

Povilaitis (165) in Ontario, Canada studied progeny involving five flue-cured cultivars ranging from susceptible to highly O_3-resistant: 'White Gold', 'Hicks Broadleaf', 'Delcrest', 'Florida 22', and 'T. I. 694'. Six crosses among the five cultivars were made and six genetic populations were derived from each cross: (a) P_1 and P_2 = parental cultivars, (b) F_1 = single cross between the two parents, (c) F_2 = the F_1 selfed, (d) B_1 = $F_1 \times$ Parent 1 and (e) B_2 = $F_1 \times$ Parent 2. Povilaitis examined additive and dominant effects, interactions between additive effects, interactions between additive and dominance effects, and interactions between dominance effects. Since flue-cured cultivars were not as susceptible as cigar-wrapped cultivars, excessive irrigation was used to increase injury by ambient O_3. Povilaitis concluded that additive gene effects were the most important in the inheritance of resistance to weather fleck. Dominance was of little importance as were the epistatic or interacting gene effects. Mean values of weather fleck in the F_1 were higher than the midparent values. This differed from Sand's data which indicated the resistance of the F_1 was greater than the midparent values. However, two very different types of tobacco were involved in these studies.

More recently, Sung et al. (188) studied weather fleck in the progeny from F_1, F_2, and two back cross generations involving four tobacco cultivars of different tobacco types (flue-cured sensitive cultivars, 'Speight G3' and 'White Gold'; a sensitive burley, 'Burley 49'; and a very sensitive cigar-wrapper cultivar, 'Bel W_3') (Table 9). They concluded that the additive gene effects were highly significant in all three crosses, and that dominance was less important for genetic variation in tolerance to weather fleck. Epistatic gene effects were of minor importance. The F_1 performance for 'Speight G3' × 'Bel W_3' showed a small heterotic response for tolerance to weather fleck, but 'Speight G3' × 'White Gold' showed a heterotic response for susceptibility to weather fleck.

Aycock (3, 4), Aycock et al. (5), and Huang et al. (105, 106) have studied the performance of Maryland tobacco varieties and crosses with weather fleck indicator tobacco selections. Aycock (3) made a study of cultivars of Maryland tobacco and their hybrids and crosses with 'Bel W_3', 'Bel B', and 'Bel C' tobacco. Their data concerning crosses with the three 'Bel' cultivars are shown in Table 10. Weather fleck scores of the hybrids between four Maryland tobacco cultivars showed that no hybrid was more resistant to weather fleck than the most resistant parent. Similar conclusions were drawn from the results of the four Maryland tobacco cultivars crossed with the Bel cultivars. Aycock found that the crosses were stable considering both field sites and years (Table 11). There was no significant genotype × environment interaction. Thus, Aycock suggested that the utilization of the single most convenient location of breeding weather fleck resistance would be adequate since estimates of genotypic components would contain little bias from location (environment) effects. Menser and Hodges have also shown that considerable progress can be made in screening Maryland and

Table 9 Population means and standard error for ratings of weather fleck susceptibility in six genetic populations involving three parents and their crosses

Population[b]	Rating of weather fleck susceptibility[a]		
	Speight G3 × Bel W_3	Speight G3 × White gold	Speight G3 × Burley 49
P_1	1.97 ± 0.27	1.97 ± 0.27	1.97 ± 0.27
P_2	51.89 ± 1.43	14.90 ± 1.24	22.17 ± 1.20
F_1	18.81 ± 0.60	9.40 ± 0.78	11.76 ± 0.81
F_2	17.42 ± 1.01	11.22 ± 0.55	14.95 ± 1.04
B_1	7.25 ± 0.44	5.26 ± 0.38	6.56 ± 0.53
B_2	32.43 ± 1.00	13.34 ± 0.57	17.74 ± 0.75

[a]Leaf injury estimates were made with a scoring system of 0, 1, 2, 3, 4 where scores 1, 2, 3, 4 correspond to the upper limits of 25, 50, 75, and 100 percent of the leaf surface injured. Leaf scores were derived from an arcsin transformation.
[b]P_1 and P_2 = Parental cultivars, F_1 = P_1 × P_2, F_2 = F_1 selfed, B_1 = F_1 × P_1 and B_2 = F_1 × P_2. Data from reference 188.

Table 10 Mean performance of seven tobacco parents and 21 F₁ hybrids for weather fleck in tobacco grown at three locations for 2 years

Parent of hybrid	Weather fleck score[a]
Maryland 10	1.87
Maryland 59	2.36
Maryland 609	1.80
Maryland 64	1.53
Bel *W3* (high sensitivity)	4.53
Bel C (intermediate sensitivity)	2.79
Bel B (low sensitivity)	1.48
Md-10 × Md-59	2.09
× Md-609	1.84
× Md-64	1.93
× Bel *W3*	2.93
× Bel C	2.39
× Bel B	1.93
Md-59 × Md-609	1.89
× Md-64	2.04
× Bel *W3*	3.39
× Bel C	2.67
× Bel B	2.23
Md-609 × Md-64	1.83
× Bel *W3*	2.79
× Bel C	2.30
× Bel B	1.92
Md-64 × Bel *W3*	2.84
× Bel C	2.19
× Bel B	1.90
Bel *W3* × Bel C	3.82
× Bel B	2.62
Bel C × Bel B	2.18
LSD 0.05	0.27
Parent mean	2.34
Hybrid mean	2.37
Heterosis, %	1.28

[a]Scores were 1 for no damage and 5 for severe damage. Data from reference 3.

Connecticut shade-grown tobacco cultivars and breeding lines at a single site (139).

Because several investigators (3, 188) have used the tobacco indicators 'Bel *W3*', 'Bel B', and 'Bel C' in their research, a brief history of 'Bel *W3*', a very sensitive O_3 indicator (79), is presented. 'Bel *W3*' has been used extensively as an O_3 indicator and monitor and is a registered seed stock (140). This strain originated from two among 20 plants in a shade-grown, cigar-wrapper cultivar, 'CCC-*W3*', developed by the Consolidated Cigar Co.

Table 11   Location and year means for weather fleck injury in tobacco involving seven parents and 21 F_1 hybrids

| | Weather fleck score[a] | | | |
| | Location (Maryland) | | | |
Year	Davidsonville	Upper Marlboro	Dentsville	Mean
1970	2.55	2.37	1.64	2.19
1971	3.13	2.63	1.85	2.53
Mean	2.84	2.50	1.74	2.36

[a]Scores were 1 for no damage and 5 for severe damage. Data from reference 3.

'CCC-W_3' was too sensitive to be grown commercially. The 'Bel W_3' strain was more sensitive to O_3 than 'CCC-W_3' and O_3 lesions were larger, bifacial, and more frequent than lesions on the parent cultivar. Other selections 'Bel C' and 'Bel B' were less sensitive and developed more of the characteristic 'Fleck' symptom. In 1968, Taylor in Connecticut and Reinert in Ohio noted that 'Bel W_3' obtained from a Beltsville seed source developed flecks and white spots and was variable in response to ambient O_3. An inventory of seed at Beltsville disclosed seven containers of 'Bel W_3' seed. These seven seed sources were grown in the field and 20 plants of each source were evaluated. All seven seed sources were sensitive to ambient oxidant when compared with 'Bel B'. Only three, however, had percent injury means greater than 70% with low coefficients of variability. Because much of the seed distributed to plant scientists came from these three selections, it was concluded that the 'Bel W_3' tobacco used for air pollution monitoring studies was quite uniform in response (Personal communication, H. A. Menser and G. H. Hodges).

Campbell et al. (35) in Maryland examined the inheritance of resistance to O_3 in alfalfa. Test plants were exposed to ambient air in greenhouses. Two diallels were produced using clones previously selected for resistance and susceptibility. With alfalfa, general combining ability was highly significant as reported by Aycock (3) for tobacco. However, unlike tobacco, specific combining ability also was significant, thus progress in obtaining resistance in alfalfa might be made using hybrids.

Studies with vegetable crops provide some additional insight into the nature of inheritance of O_3 resistance. The most unique example of gene control of resistance in vegetable crops was found in onion. Engle and Gableman (49) found that resistance to tipburn in onion was controlled by a single dominant gene. Resistant onion inbreds 'W_4' and 'SW_{52}' and the susceptible inbreds 'TDYG' and 'SW34' were maintained for five generations without segregation for tipburn resistance and susceptibility and were assumed to be homozygous. Resistant by susceptible crosses of the inbreds yielded F_1

hybrids that were as resistant as the resistant parent. Crosses of susceptible × susceptible inbreds always resulted in susceptible F_1 hybrids. The resistant 'W4' and susceptible 'SW34' were crossed to test the hypothesis that a dominant gene pair was responsible for resistance to tipburn. Water loss and stomate closure were measured in both parents and in the F_1 hybrid. The authors concluded that the susceptible inbred line, 'SW34', was not able to close its stomates in the presence of O_3. They speculated that the probable mechanism for resistance involved the membrane of the guard cells. The guard cells of the resistant onion were sensitive to O_3 and lost their permeability allowing stomatal closure. The membranes of the guard cells in susceptible plants were not as sensitive to O_3, allowing O_3 to pass into the substomatal cavity (49).

Butler et al. (29) examined O_3 susceptibility in F_1 and F_2 populations from crosses between two O_3-sensitive snap bean cultivars ('Spurt' and 'Blue Lake Stringless') and two O_3-insensitive cultivars ('Black Turtle Soup' and 'French's Horticultural'). Their studies showed that F_1 plants were as susceptible as their more susceptible parent. Approximately 10% of the F_2 progeny (obtained from selfing F_1 plants) were as resistant as the resistant parent, whereas 90% of the remaining progeny could be equally grouped as susceptible or intermediate in relation to the parent plants. They concluded that O_3 resistance was recessive in snap bean and was probably regulated by more than two major genes. In other studies, Butler and Tibbitts (30) examined the number of stomates on the O_3-sensitive and insensitive cultivars mentioned previously. The insensitive cultivars had 25% fewer stomata per mm² leaf area than the sensitive cultivars. The stomata on the insensitive cultivars partially close following exposure to O_3 but stomates on the sensitive cultivars did not close. Leaf area and leaf expansion rate were not correlated with genetic resistance to O_3 in these bean cultivars.

In 1970 Cameron et al. (31) reported a marked differential sensitivity of a group of sweet corn hybrids to ambient O_3 at two locations in the Los Angeles air basin. They continued to study five inbred lines and three F_1 hybrids under ambient O_3 levels. Four inbreds were relatively resistant and the fifth was highly susceptible. The three hybrids showed some correlation with the resistance of their parents (32). They also studied the four inbreds and two F_1 hybrids following controlled exposures in the greenhouse. The O_3 sensitivity of the inbreds and F_1 hybrids exposed in the greenhouse was similar to the sensitivity in the field but the response was not as pronounced as under field conditions. They concluded that the field data suggested partially dominant susceptibility to O_3.

Nine sweet corn inbreds were studied by Cameron (34) for three summers under ambient O_3 conditions. Five inbred lines were resistant to O_3, two were intermediate, and two were susceptible. All retained the same sensitivity rating from season to season (Table 12). Three commercial F_1 hybrids ('Bonanza', resistant; 'Gold Winner', intermediate; and 'Monarch Advance', susceptible) were also grown for three summers. They retained their

Table 12   Scores of leaf injury by ambient O_3 to nine sweet corn inbreds in the field during three summers

Inbred	Ozone reaction[a]	Mean injury score			
		1971	1972	1973	3-year Avg.
NK6942	R	1.0	1.5	1.5	1.3
81-1	R	1.0	1.5	1.0	1.2
WQ9A	R	1.0	1.0	1.0	1.0
471-U6	R	1.5	1.0	1.0	1.2
CA-0333	R	1.5	1.5	1.0	1.3
NK7428	I	3.0	3.5	2.5	3.0
349/RS	I	3.0	3.0	—	3.0
NK6604	S	8.0	9.0	8.0	8.3
Od-3	S	6.0	6.5	5.5	6.0

[a]Scores of leaf injury were 1 for no visible injury to 9 for very severe tissue collapse (R = Resistance, I = Intermediate, and S = Susceptible and represent scores 1–2, 3–5, and 6–9, respectively). Data from reference 34.

sensitivity rating as established in an earlier study (33). One resistant inbred (NK 6942) and one susceptible inbred (NK 6604) were crossed with other inbreds from the nine studied (34). Cameron found the F_1 hybrids were susceptible when either parent was susceptible, suggesting additivity or incomplete dominance of genes. The F_2 populations segregated producing some resistant phenotypes. Segregation occurred even in susceptible × susceptible and intermediate × susceptible inbred crosses. In resistant × resistant inbred crosses there was very little F_2 segregation. Three generations of selection and inbreeding, beginning with two resistant populations, resulted in nearly fixed resistance but segregation continued in progenies from susceptible × resistant crosses.

Hanson and others have studied extensively the influence of air pollutants on petunia cultivars (66, 67, 68, 69). Hanson et al. proposed that differing sensitivities among petunia cultivars were associated with physiological age and ascorbic acid content. More recently they reported results in inheritance studies involving photochemical air pollution tolerance in petunia (69). Seven commercial inbred lines of pink-flowered multiflora petunia, differing widely in sensitivity to photochemical oxidants, were crossed in all possible combinations to yield a complete diallel cross. The 49 possible populations were separately subjected to controlled fumigations to O_3 and PAN, and exposed to ambient air. The petunia inbreds were more severely injured by PAN than by O_3 in ambient air. Of the four petunia hybrids most sensitive to PAN, three were also the most sensitive to O_3 in ambient air. The hybrids most PAN-tolerant were also most tolerant to ambient air pollutants. One inbred parent contributed sensitivity to O_3 and tolerance to PAN and ambient air while another inbred contributed tolerance to O_3 but sensitivity to PAN and ambient air. Hanson et al. concluded that genes which contribute photochemical oxidant tolerance operate in an additive manner. There was some indication of partial dominance for resistance. The expression of petunia susceptibility to PAN indicated that gene interaction was evident (69).

In summarizing the research with herbaceous annual plants several broad generalities are evident:

1 Povilaitis (164, 165) found evidence in flue-cured tobacco of additive genetic variance and suggested that susceptibility may be dominant over resistance.
2 Sung et al. (188) found that tobacco in Taiwan showed primarily additive genetic variance with genes for O_3 sensitivity showing relatively little dominance or interaction effect.
3 Aycock (3), working with Maryland-type tobacco, found the major part of the genetic variance to be additive.
4 Hanson et al. (66, 69) summarized O_3 resistance in petunia as additive and a partial dominance for resistance.
5 Cameron (33, 34) interpreted O_3 sensitivity in sweet corn to be either additive to incompletely dominant. Resistance appeared to be recessive and was nearly fixed in three generations.
6 Engle and Gableman (49) found that O_3-induced tipburn in onion was controlled by a single gene and that tolerance to O_3 was dominant to intolerance.

Thus, it is likely that tolerance to air pollutants is usually inherited in a polygenic manner and that the single-gene control found in certain onions (49) is a rare phenomenon.

Forest Species. Weir (209) investigated the inheritance of resistance and susceptibility to O_3 in loblolly pines originating as nursery-grown seedlings of open-pollinated half-sib families and as greenhouse-grown seedlings from controlled pollinations of full- and half-sib families. The open-pollinated study indicated that O_3 resistance, as measured by percent foliar injury, was under a moderate amount of additive genetic control. In the control-pollinated study there was little additive genetic variance for O_3 resistance. A significant amount of dominant variation was associated with ozone resistance. Foliar injury occurred without growth inhibition.

The genetics of air pollution tolerance in forest trees has recently been reviewed by Houston (100). He found considerable information in the literature dealing with both interspecific and intraspecific variability in forest tree species. Many studies involved controlled exposures in chambers; extrapolation of these results to field situations involves certain risks. Often, a tree species that is insensitive to one pollutant may be very sensitive to a second pollutant. He found much intraspecific variability in forest tree species to air pollutants (100). Research included variations among seedlings, saplings, clones, seedlots, and provenances.

Reports concerning pollutant impacts on forest ecosystems have suggested that long-term, low-level exposures to air pollutants may have profound effects on the structure and function of forest ecosystems that include effects on community composition and system stability (143, 144). Houston (100) also reviewed the impact of air pollutants on forest areas and ecosystems. He has discussed research reports dealing with abnormal tube elongation in pine pollen, reductions in cone dimensions, percent filled seed, number of seeds per cone seed weights, and seed and pollen germination (100). Several of these reported studies suggest that ambient air pollution is altering the genetic structure of forest-type populations over large areas. As previously described, such a shift in genetic composition of populations of an annual species in response to SO_2 was reported by Taylor and Murdy (189). Houston concluded that (a) efforts to improve forest tree species have concentrated on the identification of tolerant phenotypes and that (b) sufficient information exists to conclude that response to air pollutants in trees is under strong genetic control.

RESEARCH NEEDS AND SUMMARY

Much of the published literature uses injury as a means of assessing cultivar response to air pollutants. Cultivar sensitivity, and in some cases susceptibility and resistance, have been identified in at least some of the more widely grown field and vegetable crops. There is a need to determine cultivar response to air pollutants using plant variables other than leaf injury, such as fresh and dry weight changes of leaves, stems, roots and axillary growth; yield of fruit, seed or fleshy roots; and flower number and weight for ornamental crops. Determination of air pollution tolerance should become a part of plant breeding programs. In addition expanded efforts should: (a) describe characteristics of susceptible and resistant cultivars that may account for a variation in response to pollutants and (b) identify the mechanisms of resistance to air pollutants. Even though the characteristics and mechanisms of resistance may be identified, this understanding does not guarantee an understanding of tolerance. Thus, actual breeding progams may be needed to incorporate genes for resistance and tolerance to air pollutants into new cultivars.

Gableman (54) believes that "genetic change and recombination provides a vehicle for permanent change by which plants may either exclude the pollutant from entering the plant or, conversely, provide physiological systems which allow the plant to tolerate excesses of the 'toxic' pollutants." Our challenge, as air pollution plant scientists, is to work more closely with plant breeders in identifying plant responses to air pollutants. This will help the breeder detect and modify genotypes for resistance to air pollutants without endangering quality and yield characteristics that render existing and/or new cultivars important for crop production.

There is a continuing need to improve our understanding of environmental influences on plant response to pollutants and how the environment affects our ability to identify resistant and susceptible cultivars within species. One must continually keep the disease triangle (host, pollutant, and environment) in mind. Projecting this a step further we need to place all of our understanding in concert and consider air pollution effects as a plant disease and control problem that requires a pollution management system, including: (a) breeding and selection, (b) chemical control, and (c) cultivation practices. The costs of this management system need to be balanced against the costs of controlling air pollutants.

Lewis (116) has further expressed the ideas already presented. He suggested that the air pollution plant scientist and plant breeder can work together to improve plants for growth in air polluted environments or in stress environments using two major approaches. Lewis suggests: (a) indirect breeding or breeding for tolerance to stress environments that rely on a testing program involving exposure of plants to the stress. This would help identify known cultivars that perform well, giving satisfactory quality and yields, in the presence of the stress; and (b) direct breeding in stress environments. This second strategy can be broken into three areas: (a) deliberately choose areas with differing amounts of stress and evaluate plant performance. This may have limitations for air pollution studies, but it has been used and should be valid in certain areas; (b) establish screening experiments for air pollution resistance under precisely controlled laboratory conditions. A few such efforts have been made in air pollution studies. As existing knowledge increases, this approach will need to be developed with nearly all those plant species where air pollutant stress is a proven problem; and (c) breeding to incorporate factors that control stress resistance. This may become the most useful strategy, since it incorporates the fundamental cause of stress resistance. This approach will require an understanding of the biochemical and physiological relationships between the gene system and the phenotype. When specific plant processes can be correlated with the specific air pollutant stress, then the plant breeder can incorporate, eliminate, or modify the process to help the plant tolerate the air pollutant.

Regardless of the strategy or group of strategies used, we have enough evidence concerning plant response to air pollutants to conclude the following:

1 Germplasm within many crop species offers useful variation for breeding, improving, or recommending air pollution resistant and tolerant cultivars.
2 Considerable success in breeding for air pollution tolerance has already occurred through indirect methods.

3   Air pollution scientists are becoming more knowledgeable in the methodologies they use and more conscious of their responsibilities to improve and understand both the genotype and phenotype of the cultivars studied.

4   There is a continuing need to understand and evaluate existing literature, in terms of projected research plans.

5   Finally, more multidiscipline team research is needed. Such teams should include: a crop production specialist, a plant breeder, an air pollution scientist, a statistician, an engineer, and an economist. Scientists from these disciplines, working together, will be better able to assess the air pollution level our agricultural crop systems can tolerate in future decades and generations.

REFERENCES

1   Adedipe, N. O., R. E. Barrett, and D. P. Ormrod. 1972. *J. Am. Soc. Hort. Sci.* 97:341–345.

2   Adedipe, N. O., and D. P. Ormrod. 1974. *Z. Pflanzenphysiol.* 71:281–287.

3   Aycock, M. K., Jr. 1972. *Crop Sci.* 12:672–674.

4   Aycock, M. K., Jr. 1975. *Tob. Sci.* 19:102–103.

5   Aycock, M. K., H. A. Skoog, O. D. Morgan, C. G. McKee, J. H. Hoyert, and C. L. Mulchi. 1977. *Performance of Maryland Tobacco Varieties and Breeding Lines, 1975 and 1976.* Maryland Agric. Exp. Sta. Bull. MP915, 29 pp.

6   Barrett, T. W., and H. M. Benedict. 1970. Sulfur dioxide. In *Recognition of Air Pollution Injury to Vegetation: Pictorial Atlas,* J. S. Jacobson and A. C. Hill, Eds. Air Pollution Control Assoc., Pittsburgh, Pennsylvania.

7   Beckerson, D. W., G. Hofstra, and R. Wukash. 1979. *Plant Dis. Reptr.* 63:478–482.

8   Bell, J. N. B., and C. H. Mudd. 1976. Sulphur dioxide resistance in plants: a case study of *Lolium perenne.* Pp. 87–103 in *Effects of Air Pollutants on Plants,* T. A. Mansfield, Ed. Cambridge Univ. Press.

9   Bennett, J. P., H. M. Resh, and V. C. Runeckles. 1974. *Can. J. Bot.* 52:35–42.

10   Bennett, J. P., and R. J. Oshima. 1976. *J. Am. Soc. Hort. Sci.* 101:638–639.

11   Bennett, J. P., and V. C. Runeckles. 1977. *Crop Sci.* 17:443–445.

12   Berry, C. R. (1964). *South Lumberman* 209:164–166.

13   Berry, C. R. 1973. *Can. J. For. Res.* 3:543–547.

14   Bialobok, S. (1979). Identification of resistant or tolerant strains and artificial selection or production of such strains in order to protect vegetation from air pollution. In *Symposium on the Effects of Air-Borne Pollution on Vegetation.* Warsaw, Poland. Aug., 1979. 16 p.

15   Bradshaw, A. D. (1971). Plant evolution in extreme environments. Pp. 20–50 in *Ecological Genetics and Evolution,* R. Creed, Ed. Oxford, Blackwell Scientific Publications.

16   Brasher, E. P., D. J. Fieldhouse, and M. Sasser. 1973. *Plant Dis. Reptr.* 57:542–544.

17   Brennan, E., I. A. Leone, and R. H. Daines. 1964. *Plant Dis. Reptr.* 48:923–924.

18   Brennan, E., and I. A. Leone. 1969. *Plant Dis. Reptr.* 53:54–55.

19   Brennan, E., I. A. Leone, and P. M. Halisky. 1969. *Phytopathology* 59:1458–1459.

20   Brennan, E., and P. M. Halisky. 1970. *Phytopathology* 60:1544–1546.

21 Brennan, E., I. A. Leone, and R. H. Daines. 1970. *Plant Dis. Reptr.* 54:704–706.

22 Brennan, E., and I. A. Leone. 1972. *Plant Dis. Reptr.* 56:85–87.

23 Brennan, E., and A. Rhodes. 1976. *Plant Dis. Reptr.* 60:941–945.

24 Bressan, R. A., L. B. Wilson, and P. Filner. (1978). *Plant Physiol.* 61:761–767.

25 Brewer, R. F., R. K. Creveling, F. B. Guillemet, and F. H. Sutherland. 1960. *Proc. Am. Soc. Hort. Sci.* 75:236–243.

26 Brewer, R. F., F. B. Guillemet, and F. H. Sutherland. 1967. *Proc. Am. Soc. Hort. Sci.* 91:771–776.

27 Burk, L. G., and H. E. Heggestad. 1956. *Plant Dis. Reptr.* 40:424–427.

28 Butler, L. K., and T. W. Tibbitts. 1979. *J. Am. Soc. Hort. Sci.* 104:208–210.

29 Butler, L. K., T. W. Tibbitts, and F. A. Bliss. 1979. *J. Am. Soc. Hort. Sci.* 104:211–213.

30 Butler, L. K., and T. W. Tibbitts. 1979. *J. Am. Soc. Hort. Sci.* 104:213–216.

31 Cameron, J. W., H. Johnson, Jr., O. C. Taylor, and H. W. Otto. 1970. *HortScience* 5:217–219.

32 Cameron, J. W., and O. C. Taylor. 1973. *J. Environ. Qual.* 2:387–389.

33 Cameron, J. W. (1974). *HortScience* 9:279.

34 Cameron, J. W. 1975. *J. Am. Soc. Hort. Sci.* 100:577–579.

35 Campbell, T. A., T. E. Devine, and R. K. Howell. 1977. *Crop Sci.* 17:664–665.

36 Cathey, H. M., and H. E. Heggestad. 1972. *J. Am. Soc. Hort. Sci.* 97:695–700.

37 Cathey, H. M., and H. E. Heggestad. 1972. *J. Am. Soc. Hort. Sci.* 98:3–7.

38 Clayberg, C. D. 1971. *HortScience* 6:396–397.

39 Clayberg, C. D. 1973. *Evaluation of Tomato Varieties for Resistance to Ozone.* Connecticut Agric. Exp. Sta. Cir. 246.

40 Davis, D. D., and F. A. Wood. 1972. *Phytopathology* 62:14–19.

41 Davis, D. D., and L. Kress. 1974. *Plant Dis. Reptr.* 58:14–16.

42 Davis, D. D., and R. G. Wilhour. 1976. *Susceptibility of Woody Plants to Sulphur Dioxide and Photochemical Oxidants.* Environmental Protection Agency Ecol. Res. Ser. 3-76 102, 72 pp.

43 Dean, C. E. 1963. *Tobacco* 156:28–32.

44 Dean, C. E. 1972. *Crop Sci.* 12:547–548.

45 DeVos, N. E., R. R. Hill, Jr., R. W. Hepler, E. J. Pell, and R. Craig. 1980. *J. Am. Soc. Hort. Sci.* 105:157–160.

46 Dochinger, L. S., and K. F. Jensen. 1975. *Environ. Pollut.* 9:219–229.

47 Drummond, D. B., and R. G. Pearson. 1979. Screening of plant populations. In *Methodology for the Assessment of Air Pollution Effects on Vegetation,* W. W. Heck, S. V. Krupa, and S. N. Linzon, Eds. Air Pollution Control Assoc. Pittsburgh, Pennsylvania.

48 Elkiey, T., D. P. Ormrod, and P. L. Pelletier. 1979. *J. Am. Soc. Hort. Sci.* 104:510–514.

49 Engle, R. L., and W. H. Gableman. 1966. *Proc. Am. Soc. Hort. Sci.* 89:423–430.

50 Environmental Protection Agency. 1974. *Monitoring and air quality trends report, 1973.* EPA 450/1-74-007, 130 pp.

51 Feder, W. A., F. L. Fox, W. W. Heck, and R. J. Campbell. 1969. *Plant Dis. Reptr.* 53:506–509.

52 Fiala, V. 1967. *Hort. Abs.* 37:200.

53 Fieldhouse, D. J., and M. Sasser. 1970. *HortScience* 5:334.

54 Gableman, W. H. 1970. *HortScience* 5:250–252.

55 Gentile, A. G., W. A. Feder, R. E. Young, and Z. Santner. 1971. *J. Am. Soc. Hort. Sci.* 96:94–96.

56 Genys, J. B., D. Canavera, H. D. Gerhold, J. J. Jokela, B. R. Stephan, I. J. Thulin, R. Westfall, and J. W. Wright. 1977. *Intraspecific Variation of Eastern White Pine Studied in U.S.A., Germany, Australia, and New Zealand.* Maryland Agric. Exp. Sta. Tech. Bull. No. 189.

57 Genys, J. B., and H. E. Heggestad. 1978. *Plant Dis. Reptr.* 63:687–691.

58 Gerhold, H. D., and R. G. Wilhour. 1977. *Effect of Air Pollution on Pinus strobus L. Genetic Resistance. A Literature Review.* Corvallis Environ. Res. Lab. EPA 600/2-77-102, 44 pp.

59 Gesalman, C. M., and D. D. Davis. (1978). *J. Amer. Soc. Hort. Sci.* 103:489–491.

60 Grosso, J. T., H. A. Menser, G. H. Hodges, and H. H. McKinney. 1971. *Phytopathology* 61:945–950.

61 Guderian, R., and H. Stratmann. 1962. Field experiments to determine the effects of SO_2 on vegetation. I. Survey of method and evaluation of results. *Forschungsberichte Landes Nordrhein Westfalen* 1118. 102 Pp. (EPA Translation from German 4369).

62 Guderian, R., and H. Stratmann. 1968. Field experiments to determine the effects of SO_2 on vegetation. III. Threshold values of harmful SO_2 emissions for fruit and forest trees and for agricultural and garden plant species. *Forschungsber Landes Nordrhein Westfalen* 1920. 114 Pp. (EPA Translation from German 1130).

63 Guderian, R. 1977. *Air Pollution, Phytotoxicity of Acidic Gases and its Significance in Air Pollution Control.* Springer-Verlag, New York. 127 Pp.

64 Haas, J. H. 1970. *Phytopathology* 60:407–410.

65 Hanson, G. P., L. Thorne, and G. D. Jativa. 1971. Ozone tolerance of petunia leaves as related to their ascorbic acid concentration. Pp. 261–266 in *Proc. Second International Clean Air Cong.*, H. M. Englund and W. T. Berry, Eds. Academic, New York.

66 Hanson, G. P. 1973. *Genetics* 74:107.

67 Hanson, G. P., D. H. Addis, and L. Thorne. 1974. Smog tolerant petunias developed at LASCA. *LASCA Leaves* 24:117–119.

68 Hanson, G. P., L. Thorne, and D. H. Addis. 1975. *J. Am. Soc. Hort. Sci.* 100:188–190.

69 Hanson, G. P., D. H. Addis, and L. Thorne. 1976. *Can. J. Gen. Cytol.* 18:579–592.

70 Hayes, E. M., and J. M. Skelly. 1977. *Plant Dis. Reptr.* 61:778–782.

71 Heagle, A. S., D. E. Body, and W. W. Heck. 1973. *J. Environ. Qual.* 3:365–368.

72 Heagle, A. S. 1979. *Environ. Pollut.* 19:1–10.

73 Heagle, A. S. 1979. *Environ. Pollut.* 19:313–322.

74 Heagle, A. S., R. B. Philbeck, and W. M. Knott. 1979. *Phytopathology* 69:21–26.

75 Heagle, A. S., R. B. Philbeck, and M. B. Letchworth. 1979. *J. Environ. Qual.* 8:368–373.

76 Heagle, A. S., S. Spencer, and M. B. Letchworth. 1979. *Can. J. Bot.* 57:1999–2005.

77 Heck, W. W., and E. G. Pires. 1962. *Effect of Ethylene on Horticultural and Agronomic Plants.* Texas Agric. Exp. Sta. MP 613, 12 p.

78 Heck, W. W. 1968. *Annu. Rev. Phytopath.* 6:165–188.

79 Heggestad, H. E., and H. A. Menser. 1962. *Phytopathology* 52:735.

80 Heggestad, H. E., F. R. Burleson, J. T. Middleton, and E. F. Darley. 1964. *Int. J. Air Water Pollut.* 8:1–10.

81 Heggestad, H. E. 1966. *J. Air Pollut. Control Assoc.* 16:691–694.

82 Heggestad, H. E., and E. F. Darley. 1968. *Proceedings of First European Congress on the Influence of Air Pollution on Plants and Animals.* Wageningen, The Netherlands, pp. 329–335.

83 Heggestad, H. E., and W. W. Heck. 1971. *Adv. Agron.* 23:111–145.

84 Heggestad, H. E., K. L. Tuthill, and R. N. Stewart. 1973. *HortScience* 8:337–338.

85 Heggestad, H. E. (1973). *Am. Potato J.* 50:315–328.

86 Heggestad, H. E., W. L. Craig, J. P. Meiners, and A. W. Saettler. 1976. *Proc. Am. Phytopath. Soc.* 3:326.

87 Heggestad, H. E., M. N. Christiansen, W. L. Craig, and W. H. Heartley. 1977. Effects of oxidant air pollutants on cotton in greenhouses at Beltsville, Maryland. Pp. 101–127 in *Air Pollution and Its Impact on Agriculture*. Cottrell Centennial Symposium, California State College, Turlock.

88 Heggestad, H. E., R. K. Howell, and J. H. Bennett. 1977. The effects of oxidant air pollutants on soybeans, snap beans and potatoes. Ecological Research Series, EPA-600/3-77-128, 37 pp.

89 Henderson, W. R., and R. A. Reinert. 1979. *J. Am. Soc. Hort. Sci.* 104:754–759.

90 Hendrix, J. W., and H. R. Hall. 1958. *Proc. Am. Soc. Hort. Sci.* 72:503–510.

91 Hill, A. C., M. R. Pack, M. Treshow, and R. J. Downs. 1961. *Phytopathology* 51:356–363.

92 Hill, A. C., H. E. Heggestad, and S. N. Linzon. 1970. Ozone. In *Recognition of Air Pollution Injury to Vegetaton: A Pictorial Atlas*, J. S. Jacobson and A. C. Hill, Eds. Air Pollution Control Assoc., Pittsburgh, Pennsylvania.

93 Hitchcock, A. E., P. W. Zimmerman, and R. R. Coe. 1963. *Contrib. Boyce Thompson Inst.* 22:175–206.

94 Hodges, G. H., H. A. Menser, Jr., and W. B. Ogden. 1971. *Agron. J.* 63:107–111.

95 Hofstra, G., and D. P. Ormrod. 1977. *Can. J. Plant Sci.* 57:1193–1198.

96 Hooker, W. J., T. C. Yang, and H. S. Potter. 1972. *Air Pollution Effects on Potato and Bean in Southern Michigan*. Mich. State Agric. Exp. Sta. Rpt. 167.

97 Hooker, W. J., T. C. Young, and H. S. Potter. 1973. *Am. Potato J.* 50:151–161.

98 Houston, D. B., and G. R. Stairs. 1973. *Forest Sci.* 19:267–271.

99 Houston, D. B. 1974. *Can. J. For. Res.* 4:65–68.

100 Houston, D. B. 1976. The genetics of pollution tolerance in forest trees. pp. 56–72 in Proc. 4th North American Forest Ecology Workshop.

101 Howell, R. K. 1970. *HortScience* 5:344.

102 Howell, R. K., T. E. Devine, and C. H. Hanson. 1971. *Crop Science* 11:114–115.

103 Howell, R. K., and D. F. Kremer. 1972. *J. Environ. Qual.* 1:94–97.

104 Howell, R. K., and C. A. Thomas. 1972. *Plant Dis. Reptr.* 56:195–197.

105 Huang, T. R., M. K. Aycock, Jr., and C. L. Mulchi. 1975. *Crop Sci.* 15:785–789.

106 Huang, T. S. R., C. L. Mulchi, and M. K. Aycock, Jr. 1976. *J. Environ. Qual.* 5:352–356.

107 Ishikawa, H. 1976. Sensitivity of cultivated plants to air pollutants. In *Bio-environment Laboratory Report*. Central Research Institute, Electric Power Industry, 1646 Abiko, Chiba, Japan.

108 Jensen, K. F., and L. S. Dochinger. 1974. *Environ. Pollut.* 6:289–295.

109 Johnson, H., Jr., J. W. Cameron, and O. C. Taylor. 1971. *Calif. Agric.* 25:8–10.

110 Kender, W. J., and S. G. Carpenter. 1974. *Fruit Var. J.* 28:59–61.

111 Kress, L. W. 1978. Ph.D. dissertation. Virginia Polytechnic and State University, Blacksburg, Virginia.

112 Lampoduis, F., E. Pelz, and E. Pohl. 1970. *Biol. Zbl.* 89:301–326.

113 Laurence, J. A. 1979. *Plant Dis. Reptr.* 63:468–471.

114 Leone, I. A., and E. Brennan. 1969. *Hort. Res.* 9:112–116.

115 Leone, I. A., and D. Green. 1974. *Plant Dis. Reptr.* 58:683–687.

116 Lewis, C. F., and M. N. Christiansen. 1979. Breeding plants for stress environments. In *Plant Breeding Symposium*. II. Iowa State University.

117   Likens, G. E., and F. H. Borman. 1974. *Science* 184:1176–1179.

118   MacLean, D. C., R. E. Schneider, and L. H. Weinstein. 1969. *Contrib. Boyce Thompson Inst.* 24:165–166.

119   Mandl, R. H., L. H. Weinstein, D. C. McCune, and M. Keveny. 1973. *J. Environ. Qual.* 2:371–376.

120   Mandl, R. H., L. H. Weinstein, and M. Keveny. 1975. *Environ. Pollut.* 9:133–143.

121   Manning, W. J., W. A. Feder, I. Perkins, and M. Olickman. 1969. *Plant Dis. Reptr.* 65:691–693.

122   Manning, W. J., W. A. Feder, and I. Perkins. 1972. *Plant Dis. Reptr.* 56:832–833.

123   Manning, W. J., W. A. Feder, and I. Perkins. 1973. *Plant Dis. Reptr.* 56:814–816.

124   Manning, W. J., W. A. Feder, and I. Perkins. 1973. *Plant Dis. Reptr.* 57:774–775.

125   Manning, W. J., W. A. Feder, and P. M. Vardaro. 1974. *J. Environ. Qual.* 3:1–3.

126   Markowski, A., and S. Grzesiak. 1975. *Bull. De L'Acad. Polanaise Sci.* 22:875–887.

127   Marousky, R. J., and S. S. Woltz. 1971. *Florida State Hort. Sci. Proc.* 84:375–380.

128   Meiners, J. P., and H. E. Heggestad. 1979. *Plant Dis. Reptr.* 63:273–277.

129   Menser, H. A., H. E. Heggestad, and O. E. Street. 1963. *Phytopathology* 53:1304–1308.

130   Menser, H. A., H. E. Heggestad, O. E. Street, and R. N. Jeffrey. 1963. *Plant Physiol.* 38:605–609.

131   Menser, H. A., Jr. 1966. *Tobacco* 162:32–33.

132   Menser, H. A., and H. E. Heggestad. 1966. *Science* 153:424–425.

133   Menser, H. A., and G. H. Hodges. 1968. *Agron. J.* 60:349–352.

134   Menser, H. A., Jr. 1969. *Tobacco* 169:20–25.

135   Menser, H. A., and G. H. Hodges. 1969. *Tobacco Sci.* 13:176–179.

136   Menser, H. A., and G. H. Hodges. 1970. *Agron. J.* 62:265–269.

137   Menser, H. A., and G. H. Hodges. 1972. *Agron. J.* 64:189–192.

138   Menser, H. A., G. H. Hodges, and C. G. McKee. 1973. *J. Environ. Qual.* 2:253–258.

139   Menser, H. A. 1974. *J. Am. Soc. Sugar Beet Technol.* 19:81–86.

140   Menser, H. A., H. E. Heggestad, and J. J. Grosso. 1976. *Crop Sci.* 16:606.

141   Menser, H. A. 1977. *Tobacco Sci.* 21:39–42.

142   Miller, F. L., R. K. Howell, and B. E. Caldwell. 1974. *J. Environ. Qual.* 3:35–37.

143   Miller, P. R. 1973. Oxidant-induced community change in a mixed-conifer forest. pp. 101–117 in *Air Pollution Damage to Vegetation.* Adv. in Chem. Series No. 122.

144   Miller, P. R., and J. R. McBride. 1975. Effects of air pollutants on forests. pp. 195–235 in *Responses of Plants to Air Pollutants,* J. B. Mudd and T. T. Kozlowski, Eds. Academic, New York.

145   Moore, P. D. 1975. *Nature* 258:13–14.

146   Moseley, A. R., R. C. Rowe, and T. C. Weidensaul. 1978. *Am. Potato J.* 55:147–153.

147   Murray, J. J., R. K. Howell, and A. C. Wilton. 1975. *Plant Dis. Reptr* 59:852–854.

148   Musselman, R. C., W. J. Kender, and D. E. Crowe. 1978. *J. Am. Soc. Hort. Sci.* 103:645–648.

149   National Academy of Sciences. 1971. *Biologic Effects of Atmospheric Pollutants, Fluorides. Effects of Fluorides on Vegetation.* N. A. S., Washington, D.C.

150   National Academy of Sciences. 1976. Ozone and other photochemical oxidants. *Plants and Microorganisms; Ecosystems.* Chapters 11 and 12, Vol. 12. N. A. S., Washington, D.C.

151   National Academy of Sciences. 1977. *Nitrogen Oxides.* N. A. S., Washington, D.C.

152   National Academy of Sciences. 1978. *Sulfur Oxides. Effects of Atmospheric Sulfur Oxides and Related Compounds on Vegetation.* N. A. S., Washington, D.C.

153 O'Connor, J. A., P. G. Parberry, and W. Straus. 1975. *Environ. Pollut.* 9:181–192.

154 Ohmart, C. P., and C. B. Williams, Jr. 1979. *Plant Dis. Reptr.* 63:1038–1042.

155 Olsysk, D. M., and T. W. Tibbitts. 1974. *HortScience* 9:279.

156 Ormrod, D. P., N. O. Adedipe, and G. Hofstra. 1971. *Can. J. Plant Sci.* 51:283–288.

157 Ormrod, D. P. 1976. *Plant Dis. Reptr.* 60:423–426.

158 Ormrod, D. P. 1978. *Pollution in Horticulture.* Elsevier, New York.

159 Oshima, R. J., O. C. Taylor, P. K. Braegelmann, and D. W. Baldwin. 1975. *J. Environ. Qual.* 4:463–464.

160 Oshima, R. J., P. K. Braegelmann, D. W. Baldwin, V. VanWay, and O. C. Taylor. 1977. *J. Am. Soc. Hort. Sci.* 102:286–288.

161 Oshima, R. J., J. P. Bennett, and P. K. Braegelmann. 1978. *J. Am. Soc. Hort. Sci.* 103:348–350.

162 Oshima, R. J., P. K. Braegelmann, R. B. Flagler, and R. R. Teso. 1979. *J. Environ. Qual.* 8:474–479.

163 Pelz, E. 1962. *Wiss. Z. Tech., Univ. of Dresden* 11:595–600.

164 Povilaitis, B. 1964. *Can. J. Genet. Cytol.* 6:472–479.

165 Povilaitis, B. 1967. *Can. J. Genet. Cytol.* 9:327–334.

166 Prasad, K., J. L. Weigle, and C. H. Sherwood. 1970. *Plant Dis. Reptr.* 54:1026–1029.

167 Proctor, J. T. A., and D. P. Ormrod. 1977. *HortScience* 12:321–322.

168 Rajput, C. B. S., D. P. Ormrod, and W. D. Evans. 1977. *Plant Dis. Reptr.* 61:222–225.

169 Ranft, H., and H. G. Daessler. 1970. *Flora* 159:573–588.

170 Reinert, R. A., D. T. Tingey, and H. B. Carter. 1972. *J. Am. Soc. Hort. Sci.* 97:149–151.

171 Reinert, R. A., D. T. Tingey, and H. B. Carter. 1972. *J. Am. Soc. Hort. Sci.* 97:711–714.

172 Reinert, R. A., A. S. Heagle, and W. W. Heck. 1975. Plant responses to pollutant combinations. pp. 159–178 in *Responses of Plants to Air Pollutants,* J. B. Mudd and T. T. Kozlowski, Eds. Academic, New York.

173 Reinert, R. A. 1975. *HortScience* 10:495–500.

174 Reinert, R. A., and W. R. Henderson. 1979. *J. Am. Soc. Hort. Sci.* 105:322–324.

175 Reinert, R. A., and P. V. Nelson. 1979. *HortScience* 14:747–748.

176 Reinert, R. A., and P. V. Nelson. 1979. *J. Am. Soc. Hort. Sci.* 105:721–723.

177 Rich, S., and A. Hawkins. 1970. *Phytopathology* 60:1309.

178 Richards, B. L., J. T. Middleton, and W. B. Hewitt. 1959. *Calif. Agric.* 13:4, 11.

179 Rohmeder, E., W. Merz, and A. von Schönborn. 1962. Forstw. Cbl. 81:321–332.

180 Ryder, E. T. 1973. Selecting and breeding plants for increased resistance to air pollutants. pp. 75–84 in *Air Pollution Damage to Vegatation,* J. A. Naegle, Ed. Adv. in Chem. Series 122, American Chemical Society, Washington, D.C.

181 Sand, S. A. 1960. *Tobacco Sci.* 4:137–146.

182 Schneider, R. E., and D. C. MacLean. 1970. *Contrib. Boyce Thompson Inst.* 24:241–244.

183 Sechler, D., and D. R. Davis. 1964. *Plant Dis. Reptr.* 48:919–922.

184 Shannon, J. G., and C. L. Mulchi. 1974. *Crop. Sci.* 14:335–337.

185 Spierings, F. 1962. *Differences in Susceptibility to Damage by Hydrogen Fluoride between Tulip Varieties.* Inst. Voor Plantenzeikten-kundig, Onderzoek No. 387, 4 pp.

186 Starkey, T. E., D. D. Davis, and W. Merrill. 1976. *Plant Dis. Reptr.* 60:480–483.

187 Street, O. E., C. H. Sung, and H. Y. Wu. 1971. *Tobacco Abst.* 15:141.

188 Sung, C. H., H. H. Chen, O. E. Street, and Y. L. Yang. 1971. *Taiwan Asr. Quart.* 7:173–181.

189 Taylor, G. E., and W. H. Murdy. 1975. *Bot. Gaz.* 136:212–215.

190   Taylor, G. S. 1968. *Phytopathology* 58:1069.

191   Taylor, G. S. 1974. *Phytopathology* 64:1047–1048.

192   Taylor, O. C., and D. C. MacLean. 1970. Nitrogen oxides and the peroxyacyl nitrates. In *Recognition of Air Pollution Injury to Vegetation: Pictorial Atlas*, J. A. Jacobson and A. C. Hill, Eds. Air Pollution Control Assoc., Pittsburgh, Pennsylvania.

193   Taylor, O. C., C. R. Thompson, D. T. Tingey, and R. A. Reinert. 1975. Oxides of nitrogen. pp. 122–139 in *Responses of Plants to Air Pollutants,* J. B. Mudd and T. T. Kozlowski, Eds. Academic, New York.

194   Temple, P. J. 1972. *J. Air Pollut. Control Assoc.* 22:271–274.

195   Thompson, C. R., G. Kats, and J. W. Cameron. 1976. *J. Environ. Qual.* 5:410–412.

196   Thompson, C. R., G. Kats, E. L. Pippen, and W. H. Isom. 1976. *Environ. Sci. Tech.* 10:1237–1241.

197   Thorne, L., and G. P. Hanson. 1976. *J. Am. Soc. Hort. Sci.* 101:60–62.

198   Ting, I. P., and W. M. Dugger. 1968. *J. Air Pollut. Control Assoc.* 18:810–813.

199   Tingey, D. T., R. A. Reinert, and H. B. Carter. 1972. *Crop Sci.* 12:368–370.

200   Tingey, D. T., R. A. Reinert, J. A. Dunning, and W. W. Heck. 1973. *Atmos. Environ.* 7:201–208.

201   Townsend, A. M. 1974. *J. Am. Soc. Hort. Sci.* 99:206–208.

202   Townsend, A. M., and L. S. Dochinger. 1974. *Atmos. Environ.* 8:957–964.

203   Tzschacksch, O., M. Vogel, and R. Thummler. 1969. *Arch. Forstwes.* 18:979–982.

204   Tzschasksch, O., and M. Weiss. 1972. *Karst Beitr. Fostru.* 6:21–23.

205   Van Haut, H., and H. Stratmann. 1970. *Color Plate Atlas of the Effects of Sulfur Dioxide on Plants.* Essen, Verlag W. Gerardet. 206 pp.

206   Vogle, M. 1970. *Arch. Forstwes.* 18:3–12.

207   Wanta, R. C., and H. E. Heggestad. 1965. *Science* 130:103–104.

208   Weinstein, L. H. 1977. *J. Occupational Med.* 19:49–78.

209   Weir, F. J. 1977. Ph.D. dissertation. North Carolina State University, Raleigh.

210   Wentzel, K. F. 1968. *Forstarchiv.* 39:189–194.

211   Wilton, A. C., J. J. Murray, H. E. Heggestad, and V. F. Juska. 1972. *J. Environ. Qual.* 1:112–114.

212   Wolff, G. T., P. J. Lioy, G. D. Wight, R. E. Meyers, and T. Cederwall. 1977. *Atmos. Environ.* 11:797–802.

213   Wood, F. A. 1968. *Phytopathology* 58:1075–1084.

214   Wood, F. A., and J. B. Coppolino. 1972. *Mitt. Forslt. Bunderversuchanst. Wien* 97:233–253.

215   Wood, F. A., and J. B. Coppolino. 1972. *Phytopathology* 62:501–502.

216   Wood, F. A., and D. B. Drummond. 1974. *Phytopathology* 64:897–898.

217   Woodworth, C. M., E. R. Leng, and R. W. Jugenheimer. 1952. *Agron. J.* 44:60–65.

218   Zimmerman, P. W., and A. E. Hitchcock. 1956. *Contrib. Boyce Thompson Inst.* 18:263–279.

Chapter 10

PLANT RESPONSE TO ATMOSPHERIC STRESS CAUSED BY WATERLOGGING

DONALD T. KRIZEK

U.S. Department of Agriculture, Science and
Education Administration, Agricultural Research
Plant Stress Laboratory, Beltsville, Maryland

INTRODUCTION

The oxygen content of the soil atmosphere is of paramount importance for the successful growth and productivity of crop plants. Because of the need for oxygen in aerobic respiration and for various energy-requiring processes, no plant can survive under complete anoxia for prolonged periods (6, 40, 50, 84). Although few crop plants are subjected to complete anoxia, many are exposed to periods of oxygen deficiency caused by water-saturated or compacted soil.

Considerable attention has been given to studying the effects of water deficits on crop yield and to selecting plants for adaptation to drought (96, 125, 126, 145, 149, 191, 213, 249, 305, 319, 340, 354, 355). By comparison, studies on plant response to soil atmospheric stress caused by waterlogging have attracted much less interest, although there is still a vast amount of literature on the subject. During the past two decades a number of excellent reviews have been prepared. The reader is referred to these for further information (13, 14, 15, 25, 36, 40, 48, 51, 65, 68, 73, 85, 104, 111, 115, 140, 141, 183, 288, 297, 388).

The purpose of this chapter will be to summarize some of the properties of waterlogged soils, to present some of the latest findings on the morphological, anatomical, physiological, and biochemical responses and adaptations of higher plants to waterlogging and to review recent progress in breeding plants for flooding tolerance.

It should be pointed out that much of the literature is controversial and that there is no generally accepted basis for flooding tolerance in crop plants. Because of the complexity of the research in this field and the confounding effects often experienced under anaerobic conditions, unequivocal cause and effect relationships are difficult to establish. Hopefully this review will stimulate further interest in this important area of stress research and help with the design of experiments under carefully controlled conditions to answer many of the remaining questions.

NATURE OF ATMOSPHERIC STRESS IN THE SOIL

Definitions

Atmospheric stress in the soil has been referred to in the literature in a variety of ways. Terms used include oxygen deficiency (25), waterlogging (180, 181, 183, 184, 288, 291), and anaerobiosis (140). All involve a build-up of carbon dioxide, ethylene, and other potentially toxic gases, and oxygen depletion which leads to a reduction of aerobic respiration.

The response of plants to anaerobic soil conditions has also been described under various headings including flooding damage (158, 159, 160, 161, 162) and excess moisture injury (384). As stated by Vartapetian (361):

> Life under the conditions of oxygen deficiency, or even in an environment completely devoid of molecular oxygen is widespread on this planet. The phenomenon of anaerobiosis, permanent or temporary, complete or partial, can be observed at virtually every level of evolution of the biological system, from the primitive unicellular organisms up to man.

The Nature of Flooding Damage

The nature of flooding damage to plants depends upon many factors including species, cultivar, plant age, dormancy status, and type of soil (68). Oxygen deficiency is a feature common to all waterlogged soils (279). Anaerobiosis may develop within a few hours after flooding occurs because of the displacement of oxygen from the soil pore space (Figure 1) and the uptake of the remaining dissolved oxygen by microorganisms in the soil (308, 353). The diffusion coefficient of oxygen in water at 25° C (0.26×10^{-4} cm^2/s) is approximately 10,000 times smaller than that in air (0.23 cm^2/s) (13, 40) and the coefficient of oxygen in wet soil (1×10^{-5} cm^2/s) is nearly 20,000 times less than that for air (117). Consequently, the length of the pathway through the water film is of great importance. Crumb-structured soil horizons can become anaerobic without being completely saturated because water fills the crumb pore spaces (81, 117).

Magnitude of the Problem

Anaerobic conditions occur in vast regions of the world where fields are periodically flooded for irrigation (261). Temporary over-irrigation may cause localized flooding of the soil and thereby lower oxygen levels enough

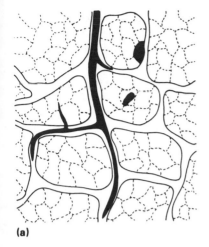

(a)

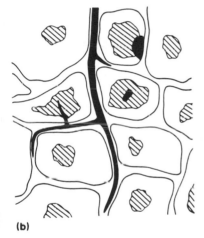

(b)

Figure 1 Schematic respresentation of onset of anaerobic soil conditions. (*a*) Soil well aerated: Pores between aggregates are air-filled and there are smaller air-filled pores (dotted) in aggregates: a growing root and two zones with abundant organic substrates are shown (shaded). (*b*) Increasing soil water has displaced air in the fine pores within aggregates: Anaerobic zones (lightly shaded) are developing within aggregates, especially where substrates are abundant. From R. S. Russell, 1977. *Plant Root Systems*. McGraw-Hill Book Co., New York.

to severely reduce plant growth (288). Leakage from irrigation ditches often causes soils to become waterlogged resulting in substantial reduction in crop yield (123). This problem is typical in western Canada where more than 280,000 hectares of land are irrigated. Of these nearly 24,000 hectares are permanently waterlogged due to water seepage from the irrigation canals (288).

In regions having a long and severe winter, hypoxia often affects the roots of perennial and winter crops when an ice crust seals the soil surface, thereby making it impervious to the diffusion of oxygen and carbon dioxide (361). Under these conditions, the roots of many wild and cultivated plants including trees and important crop species frequently experience problems with atmospheric oxygen supply in the soil. Because of the considerable acreage of excessively wet soils in countries throughout the world, the problem is global.

In Pakistan nearly 40,000 hectares of land are lost from production each year because of salinization and waterlogging of the soil (130). Efforts are being made to lower the depth of the water table by installing tube wells and providing drainage channels; however, these procedures are costly processes that may require several decades before the lost lands are returned to full productivity. In the meantime, in Pakistan, as in other countries, crops must be selected for the soils. This process will require a concerted effort by scientists representing various disciplines including plant physiology, biochemistry, soil science, genetics, hydrology, and agronomy.

Waterlogged soils are found in nearly any climatic zone from the tundra to the desert or humid tropics. Their ecological importance is emphasized by the fact that 72% of the earth's surface is estimated to be covered by submerged soils or sediments (279). As interest in nutrient flow techniques (NFT) and other commercial liquid culture practices increases, the need to avoid anaerobic stress also becomes an important consideration (157, 361).

PHYSICAL AND CHEMICAL PROPERTIES OF WATERLOGGED SOILS

Introduction

The physical and chemical characteristics of waterlogged and other submerged soils have been reviewed in great detail by Ponnamperuma (279), Grable (111), E. W. Russell (300), and others; only a brief summary will be presented here. Because heat transfer is more rapid in waterlogged soils they tend to show different temperature fluctuations than drier soils. Day temperatures tend to be lower and night temperatures tend to be higher in waterlogged soils than in drier soils (288). As a result there may be a pronounced interracting temperature effect on the growth and metabolism of roots and the soil microflora (2, 360).

As soils become waterlogged, changes occur in the gaseous composition, redox potential, pH, and mineral content. Significant changes in microbiological activity may also occur. These changes will be elaborated on subsequently in this section.

Measurement of Soil Aeration

When making measurements of soil aeration, the investigator must decide which parameter to measure, the gas phase or the water phase. The approaches used to characterize the atmospheric properties of the soil include measuring the redox potential (113), determining the rate of oxygen diffusion in the soil (210, 234, 235), and analyzing for substances produced under reducing conditions or as a result of anaerobiosis (40, 140). Measuring and characterizing the soil atmosphere in terms of plant behavior is not a simple task (111). In some cases plants grow equally well under nonaerated or aerated conditions (300). In some cases (9), artificial aeration may depress plant growth.

Localized Zones of Atmospheric Stress

Figure 1 illustrates schematically how localized anaerobic zones may develop during waterlogging. Russell (302) has shown that the average concentration of oxygen in the gas phase of soils partially saturated with water is a poor guide to the actual supply of oxygen available in the root system. It is also evident that aerobic and anaerobic processes can occur simultaneously in adjacent zones of soil. The implications of this are two-fold: (a) the growth of aerated roots may be influenced by diffusible products of anaerobiosis; and (b) substances that are formed in completely anaerobic conditions may be found in soils containing high amounts of oxygen (302).

Gas Exchange in Waterlogged Plants

The movement of respiratory gases within the plant and the directional exchange with the atmosphere are greatly influenced by the environmental conditions within the soil and in the aerial environment (15, 339). Oxygen may enter the plant in a number of ways. In nonaquatic species oxygen and carbon dioxide enter and leave the plant directly through the stomates and lenticels. In submerged aquatic plants the permeability of the leaf surface is high enough to permit gas exchange by liquid-phase movement across the epidermal layers.

In the yellow water lily, *Nuphar luteum*, the network of internal gas spaces or lacunae serves as a pressurized flow-through system which forces O_2 down the petioles of the young newly emerged leaves to the roots and rhizome buried in the anaerobic sediment. Simultaneously, CO_2 is forced from the rhizome up the petioles of the older leaves where it is fixed by photosynthesis or released to the external atmosphere. Appreciable quantities of methane from the lake sediment are also scrubbed by this unique ventilation system. By the physical processes of thermal transpiration and hydrometric pressure several liters of air per day enter the plant at rates up to 50 cm/min (82).

Plants in unsaturated soils are generally exposed to an oxygen-rich environment over most of their shoot and root surfaces. In these cases only limited longitudinal movement of gases may occur within the plant. Most of the gas exchange occurs by simple planar and radial movement, where the oxygen requirements of the root system are met largely by diffusive transfer from the soil atmosphere supplemented by transpirational flow (15). In most cases gaseous oxygen must enter the liquid phase at a point external to the root (114). Its passage across the cellular layer(s) of the root wall is thought to occur chiefly in the liquid phase. A large portion of the O_2 will in turn pass into the gas phase of the intercellular spaces at the outer surface of the root cortex (15).

In waterlogged soils the mode of gas exchange and transport is very different. Because little or no oxygen is available for radial entry to the root, longitudinal movement between shoot and root is the primary mode of transport. Longitudinal gas transport occurs in both the intercellular gas spaces and in the stele. The relative merits of each pathway are described in detail by Armstrong (15). Radial movements tend to be bidirectional—from root cortex to stele and from root to soil. The latter, radial oxygen loss (ROL) is believed to be very important for the survival of plants in waterlogged soils (15). Oxygen may also enter the plant in the combined state as water. In the xylem, water is transported from root to shoot where the O_2 is ultimately released within the chloroplasts during the photolysis stage of photosynthesis.

Waterlogging of the soil may or may not lead to anaerobiosis. If the hydraulic conductivity of the soil is sufficiently high and drainage is unimpeded, the movement of aerated surface water through the soil may provide sufficient oxygen to avoid adverse effects (302). This occurs in flooded meadows under cool conditions where relatively little oxygen is used in biological processes.

No consistent relationship exists between the air-filled pore space in a soil and the extent of anaerobiosis that develops (111). The transport of oxygen both in the gas and liquid phases to different zones within the soil are greatly influenced by the distribution and continuity of the pores (302).

Redox Potentials in Waterlogged Soils

Under oxygen deficiency soil microorganisms may use electron acceptors other than oxygen for their respiratory oxidations. This results in the conversion of a number of compounds in the soil to a reduced state and is reflected in a lowering of the redox potential (13, 279, 341).

The redox potential of a system is a measure of its tendency to accept or donate electrons and is governed by the nature and proportions of the oxidizing and reducing substances which it contains. A common redox couple in the soil is the reversible ferric system $Fe^{2+} = Fe^{3+} + e-$. In a pure solution containing equal concentrations of the ferrous and ferric ions, the system has a redox potential of -771 mV relative to the standard hydrogen electrode (13).

Redox potentials are commonly measured in the soil with a platinum electrode. However, because this electrode is not specific to a single redox couple, the measurements obtained generally represent a mixed potential that reflects the weighted average of the potentials contributed by each of the redox couples present in the system (101). Nevertheless, redox potential measurements in soils are widely used and are generally well accepted.

The use of these measurements depends greatly on the techniques employed and their valid interpretation. Despite theoretical limitations and several procedural difficulties, these measurements provide a rapid and convenient indicator of the intensity of reduction in waterlogged soils. In sediments and submerged soils, the redox potential ranges from about -400 mV (strongly reduced) to $+700$ mV (well oxidized). These values are in contrast to oxidized soils in which the redox potential ranges from about $+400$ to $+700$ mV (101).

The sequence of events that occurs in a waterlogged soil in relation to redox potential has been described by Patrick and Turner (260) and Russell (302) and summarized by Armstrong (13) (Table 1). The concentration of oxygen declines first, accompanied by nitrate reduction. This is followed by a reduction of insoluble but easily reducible manganese compounds and a corresponding increase in exchangeable manganese. After 4 days of waterlogging, iron reduction begins. At that time the redox potential has fallen to about $+150$ mV. The delay in Mn reduction appears to be related to the reserves of reducible Mn present in the soil. The iron concentration in solution then increases rapidly so that by the seventh day, the redox potential has fallen to about -150 mV.

Because soils generally contain more Fe than Mn, the iron hydroxides are the predominant redox system rather than the manganese system of oxides and carbonate. Just as Mn reduction does not occur until all free nitrate has disappeared, the presence of manganese dioxide and other Mn-containing compounds may delay or prevent Fe reduction to the ferrous form, and is commonly used to buffer rice soils against the development of extremely reducing conditions that are conducive to severe crop injury (13).

Toxicity effects from high levels of reduced Fe and Mn are frequently encountered on agricultural lands during waterlogging. This is frequently the case in rice culture where soils are not only flooded for long periods but also

Table 1   Oxidation-reduction (redox) potentials at which reactions occur in typical soil systems at 25°C and pH 5–7[a]

Product of reduction	Redox potential (mV)
H_2O (reduction of oxygen)	930 to 820
NO_2^-	530 to 420
Mn^{2+}	640 to 410
Fe^{2+}	170 to -180
H_2S	-70 to -220
CH_4	-120 to -240
H_2	-295 to -413

[a]Simplified from data of Ponnemperuma tabulated by E. W. Russell, 1973. From R. S. Russell. 1977. *Plant Root Systems*. McGraw-Hill, New York.

experience extremely low redox potentials because of the high levels of additional organic matter present (13). Differential sensitivity to iron toxicity has been shown to be responsible for the natural distribution of two heath plants, *Erica cinerea* and *E. tetralix*. Even brief periods of waterlogging were lethal to *E. cinerea* but *E. tetralix* was unaffected (18, 19, 20, 166–174).

Liquid culture studies demonstrated that cut shoots of the two species differed in sensitivity to high levels of Fe supplied in solution. The Fe-sensitive species showed the same symptoms as those sensitive to waterlogging while the Fe-tolerant species was also tolerant to waterlogging. The iron toxicity hypothesis was further confirmed by an experiment in which the sensitive species *E. cinera* was grown in two peat soils, one high in Fe, the other low in Fe. On waterlogging, those in the peat containing high iron developed the characteristic waterlogging syndrome which included leaf discoloration and excessive loss of turgor from the leaves, while those on the low iron peat survived (13).

Changes in Mineral Content in Waterlogged Soils

In arid areas, flooding caused by leakage of irrigation ditches or over-irrigation can frequently lead to a large accumulation of salts. Irrigation water can contribute up to 12 metric tons of salt per hectare on an annual basis (28). Waterlogging and salt accumulation may have been responsible for the collapse of the Babylonian and other ancient civilizations (288). Nearly 6.8 million hectares of arable land in India and 2 million hectares of land in Pakistan have been severely damaged by the twin problems of waterlogging and salt accumulation (288).

The deleterious effects of salt accumulation on plants are well known (132, 249, 278). If crusts form on the surface of the soil, soil aeration will be reduced. High osmotic potential resulting from salt accumulation is inhibitory to water uptake by seeds and roots. In sandy soils, however, salt accumulation may not be a problem; irrigation water with a salt concentration approaching that of sea water may be used because sand particles do not absorb salts as readily as clay particles (32).

Several investigators have shown that flooding can increase the availability of several elements to plants. These include Fe, S, Ca, Mo, Ni, Zn, Pb, and Co (136, 185). Manganese toxicity has been observed to be induced under flooding conditions (274).

The relationship of waterlogging damage to plant nutrient status has been described by a number of workers (87, 88). However, Reid (288) has suggested that effects of flooding on nutrient availability and status of soils cannot be generalized because different soils react in different ways.

Accumulation of Phytotoxic Substances in the Soil

Under anaerobic conditions, many phytotoxic substances are provided in the soil or released to the soil by stressed plants. These include organic acids; methane, ethylene, and other hydrocarbon gases; hydrogen sulphide, hydrogen, and carbon dioxide (302, 315). Decomposition of plant residues in the soil is often accompanied by formation of substances having phytotoxic properties (31, 118, 119, 231, 232, 233, 257, 258, 259, 315, 346). Studies have shown that the most severe phytotoxicity occurs in fields where decomposition of plant organic matter takes place in cold, wet soils (346).

Phytotoxic substances may also be produced by microorganisms present in the soil (229, 230, 231, 232). The production of phytotoxic substances in crop residues and in soil by soil microorganisms is of particular concern where stubble-mulch farming is practiced, because this system has been implicated in the reduced crop yield obtained in some years as compared with plowing (233).

Studies conducted at Lincoln, Nebraska, on the toxicity of residues left after the harvest of corn, wheat, oats, and sorghum (119) indicated a pronounced effect of weathering and decomposition on the extent of toxicity. Of the four crops tested, sorghum residues contained the most toxic material and wheat the least toxic. Nearly all of the poisonous materials disappeared from wheat and oat residues after 5 weeks of field exposure to the weather while corn residues remained quite toxic at this time. However, after 28 weeks in the field, even sorghum stalks had lost most of their phytotoxicity. McCalla and Nordstadt (233) and their co-workers demonstrated that nearly 40% of all soil microorganisms that were isolated from soils where stubble-mulching was practiced, produced phytotoxic substances that reduced growth.

Persidsky and Wilde (265) have shown that volatile substances released from humus in forest soils and from sawdust may affect the growth of roots but the importance of these compounds in soils is difficult to determine (52).

McNew (237) reported that anaerobic respiration in soils may result in the accumulation of salicyclic aldehyde to a concentration as great as 50 ppm. This concentration did not affect root growth of wheat or sugar cane but increased the susceptibility of the roots to attack by *Pythium arrhenomanes,* a root-rot fungus (52).

Organic Acids.   Under anaerobic conditions a large number of organic acids may arise (334). Most abundant of these are the volatile fatty acids, which include primarily acetic acid but also formic, propionic, buyric, and valeric acids. In the presence of ample substrates, the quantity of these acids produced per 100 g of waterlogged soil may exceed $2 \times 10^{-3} M$ (302).

Various aromatic acids have also been found in waterlogged soils. These include p-hydroxybenzoic, p-coumaric, and vanillic acids (368). Although many other acids have been found in waterlogged soils, they are of minor importance. Benzoic acid and phenylacetic acid were found to be the major components of ether-soluble phytotoxins present in samples of decomposing barley in the field and from cotton, cowpea, and soybean decomposing in soil in the laboratory (346). Under anaerobic conditions no consistent changes in pH of the soil occur despite the production of organic acids (302).

Sulfides. Hydrogen sulfide (H_2S) is formed in the mud of waterlogged soils by the reduction of sulfate and can accumulate in the form of iron sulfide (29, 333, 358). The gas may be released in large quantities if a period of warm weather is followed by a rapid cooling (as in early August), when the surface layer of the mud oxidizes and sulfuric acid is formed. Under certain conditions the H_2S can reduce the sulfuric acid and form sulfur dioxide, which is nearly fifty times more toxic than H_2S (359).

Hydrogen sulfide formation in the mud of rice fields is a serious problem in Hungary and in other countries. The damage to rice caused by H_2S is known in Hungary as browning disease. Since H_2S is a reduced compound formed under anaerobic conditions, steps have been taken to develop browning resistant varieties by selecting varieties that have roots with the greatest oxidizing ability (110, 359). Through large-scale selection, round-grained japonica rice varieties have been developed that are more resistant to H_2S than the long-grained indica varieties. The explanation for these differences is based in part on the differences in requirement for solar radiation.

A detailed analysis of weather data for 20 years revealed that the H_2S-induced browning disease destroyed the rice crops in years when little sunshine occurred during the critical stages of development, but had no effect in those years in which the summers were very sunny. Similar results were obtained in shading experiments. Vamos and Koves (359) suggest that H_2S may be oxidized by free OH radicals formed during the photolysis of water and that visible radiation may prevent the toxic effects of H_2S in the rice plant.

Hydrogen sulfide is generally not a problem, however, unless the redox potential of the soil is very low (Table 1) and ferrous iron is present. In the presence of ample quantities of soluble iron, toxic sulfides are unlikely to be produced.

Ethylene and Other Hydrocarbon Gases. Under anaerobic conditions, a number of unsaturated hydrocarbon gases including methane, ethylene, and a number of higher hydrocarbons may be formed (176, 177, 224, 321, 322, 325, 327).

Methane content of the soil air may be as high as 2.8% (364) in flooded soils under rice. Methane is familiar as the "marsh gas" of swamps (52). Its effects are poorly understood: in barley it has little effect, while in rice it may actually serve as a carbon source (301).

The formation of ethylene in waterlogged soils is of interest to plant scientists because this gas is an endogenous growth regulator that is normally produced by the plant in response to stress conditions, and can induce biological effects in extremely low concentrations (1, 79, 216, 224, 227, 326, 395). Ethylene concentrations that have been found in anaerobic soil are considerably in excess of those found to cause injury to roots of selected plants under laboratory conditions (327). Smith and Dowdell (324) measured greater than 10 ppm of ethylene in waterlogged field soils.

Numerous microorganisms have been shown to produce ethylene including *Escherchia coli* (281) and *Mucor hiemalis* (27, 223). Methionine and glucose which can result from decomposition of fresh organic material (108, 109, 216, 222, 239) are required substrates for ethylene formation in both of these organisms.

Russell (302) has shown that in order for ethylene to be produced, a restricted supply of oxygen in the soil (less than 0.01 bar) is normally required. This requirement appears to be related to the release of substrates rather than to the production of ethylene from them, because when ample substrates are added to well-aerated soil, abundant ethylene may be released (225). Ethylene gas has been detected not only in the fungal hyphae, but also in a diffusate collected from the hyphae (223). The amount of ethylene produced by the fungus is inversely related to its growth rate (226).

The concentration of ethylene in susceptible plants has been shown to increase markedly under anaerobic conditions. Comparative ethylene concentrations found in the shoots of tomato, pea, field bean, and dwarf bean after 4–5 days of waterlogging are shown in Table 2 (93, 158, 161, 162, 164, 176, 179, 183).

Although early work (164) showed that some ethylene in the plant may be derived directly from the soil, recent studies by Jackson et al. (164), Kawase (179, 180), and others have demonstrated that the shoot system of waterlogged plants may produce significant amounts of ethylene.

CO_2 and Other Gases in the Soil Atmosphere.　　Other gases commonly found under anaerobic conditions and in the presence of organic matter include hydrogen, nitrous oxide, and carbon dioxide (299). Flooded soils under rice may have a hydrogen content as high as 6.5% (364). In low concentrations, hydrogen is not very toxic to plant roots, although it depresses symbiotic nitrogen fixation (52).

Numerous studies have been conducted on the effects of CO_2 concentra-

Table 2　Effects of waterlogging for four to five days on the concentrations of ethylene in the shoots of four plant species[a]

	Ppm in the gas phase (v.v.)			
	Tomato (*Lycopersicon esculentum mill.*)	Pea (*Pisum sativum L.*)	Field bean (*Vicia faba L.*)	Dwarf bean (*Phaseolus vulgaris L.*)
Waterlogged	1.17	0.93	0.50	1.36
control	0.34	0.25	0.37	1.08

[a]From Jackson and Campbell, 1976. Personal communication. Cited by R. S. Russell. 1977. *Plant Root Systems*. McGraw-Hill, New York, p. 206.

tion on plant growth and development (197, 198, 199, 282, 378, 379) and various methods have been used to measure CO_2 levels in the soil (97, 208, 243, 323). Because CO_2 is approximately thirty times more soluble in water than oxygen (302), steep gradients from the soil surface to the center of water-filled zones do not occur.

In recently flooded soil, more than 50% of all gases dissolved in the soil water may be CO_2. Most plant roots are unable to tolerate such conditions (302). The threshold level of CO_2 at which root growth is impaired varies greatly with plant species. Cotton and soybeans were not injured by concentrations as high as 20% (195, 338) while other species showed reduced root growth at much lower levels (338).

In general it is thought that an excess of CO_2 in the soil is of minor importance in comparison with a deficiency of O_2 (111, 195, 302). In tobacco, for example, there was as much injury when plant roots were exposed to 20% CO_2 in the presence of 1% O_2 as when they were treated with 1% O_2 in the absence of CO_2 (379).

Enzymatic Activities of Flooded Soils.   Although extensive studies have been conducted on the chemistry (279) and microbiology (153, 318, 389) of flooded soils, our knowledge of enzyme activities is limited (43, 56). Recent experiments by Chendrayan et al. (60) demonstrated an increase in the dehydrogenase activity and a marked decrease in the invertase activity of soils following flooding. The addition of 0.5% rice straw greatly increased the dehydrogenase activity of flooded soils and invertase activity under both flooded and nonflooded conditions. This increase in dehydrogenase activity was felt to be related to the increase in population of anaerobic microorganisms.

Most dehydrogenases arise from anaerobic organisms (253). Because dehydrogenase has been implicated in the microbial reduction of ferric oxides (38) and iron reduction is one of the predominant of all redox reactions in a flooded soil (279), Chendrayen et al. (60) have suggested a possible relationship between iron reduction and dehydrogenase activity in flooded soils. Further studies are needed to determine the nature of this relationship.

OXYGEN REQUIREMENTS OF ROOTS AND SOILS

The oxygen consumption of a soil is greatly influenced by such factors as disturbance, compaction, and the addition of organic matter (41, 131). Plants vary widely in response to atmospheric stress in the soil. In general critical oxygen levels begin in the range of 5–10% O_2 by volume. Root growth in most plants is limited when less than 10% of the pore spaces are filled with air and the oxygen diffusion rates (ODR) are less than $0.2 \mu g/cm^2 \cdot min^{-1}$ (245, 277).

Data on the oxygen consumption of crop plants have been obtained by numerous investigators including Brown et al. (41), Hawkins (131), and others. A summary of these data is given in a review of Brouwer and Wiersum (40). Typical values for mean daily oxygen consumption rates for mature roots of crops grown at normal agricultural spacings are as follows: potatoes, 2.8 l/m^2; kale, 5–6 l/m^2; and tobacco, 3.0 l/m^2. On a per plant basis the oxygen consumption rates ranged as follows: potatoes, 0.108–0.608 l/day; kale, 0.169–0.569 l/day; and tobacco, 0.175–0.986 l/day (41).

Although measurements have been made on the oxygen consumption of root tissues adequate data are lacking and significant differences in value are frequently obtained for the same species or cultivar. The oxygen consumption of young roots, when expressed on a unit dry weight basis, tends to be high in comparison to older and thicker roots. Differences in temperatures at which the measurements are made influence the values obtained because of their effects on respiration rates (40, 135).

A 100% increase in respiration has been obtained for a 10°C rise in temperature. Thus within a range of 0–35°C in which many roots are viable, the oxygen consumption may vary eight-fold (40). As a result the adverse effects of poor aeration are less severe at low temperature than at high temperature (49). This is demonstrated by the seasonal decrease in sensitivity to flooding of many plants during the winter.

The growth of roots stops within a few minutes after withdrawal of oxygen. Provided that anaerobic conditions do not persist, growth of the roots resumes as soon as oxygen deficiency is overcome.

The mean daily consumption of oxygen by soils varies widely. Typical values reported by Brown et al. (41) for four soils examined at a depth of 91.5 cm ranged from 2.2 l/m^2 for an undisturbed sandy clay loam profile up to 10.8 l/m^2 for a peat top soil. The range of daily oxygen use by these soils were 1.8–5.3 l/m^2 in an undisturbed sandy loam profile and 7.7–11.0 l/m^2 for peat.

The oxygen consumption of a crop at normal spacing for a given area is remarkably consistent. However, the oxygen consumption of the soil varies greatly. In most cases the O_2 consumption of a normal crop is about one-third of that of the soil in which it is growing (131).

CROP RESPONSE TO WATERLOGGING

Morphological and Anatomical Effects of Waterlogging

The morphological and physiological effects of waterlogging have been described in detail by Reid (288), Brouwer and Wiersum (40), Kawase (183), and Bradford and Yang (36); therefore, only a brief summary will be given here. Typical effects of waterlogging on the growth of shoots include reduced elongation, chlorosis, senescence, abscission of the lower leaves, wilting, hypertrophy, epinasty, adventitious root formation on the lower portion of the stem, lenticel formation, aerenchyma formation, leaf curling and decline in relative growth rate (33, 133, 165, 175, 183, 184, 193, 194, 196, 283, 288, 350, 351, 352, 387).

Although investigators have used dry weight as the sole criterion of plant response to waterlogging, studies on wheat conducted by Trought and Drew (350) indicate that this factor may not be a reliable indicator of early waterlogging damage. During the first 4 days of waterlogging, dry weight of the shoots increased in comparison with the nonwaterlogged plants.

Root growth is also affected by waterlogging (133). Root elongation is slowed (250); root hair formation is inhibited (328, 390); and under prolonged flooding conditions, roots may blacken, die, and eventually rot (64, 130).

Under waterlogging conditions may plants form air spaces (aerenchyma) in the cortex of their roots (84, 183, 184, 316). After 2 days of waterlogging, aerenchyma formation occurs in the stems of sunflower and tomato plants (183). In soybean, flooding at or before the pod-filling stage results in the formation of aerenchyma and adventitious roots (121, 373).

In culture solution, aerenchyma tissue forms in the roots of maize and wheat whereas in well-aerated soil no aerenchyma are found (90). Investigators have reported differences in the growth of plants in aerated and nonaerated culture solution (42, 120, 250, 275, 350). The roots of barley plants which were grown for two months in nonaerated culture solution, formed aerenchyma in the root cortex, and were straighter, shorter, thicker, and more numerous than those from plants grown in aerated culture solution (42). There were no differences, however, in either shoot growth or in dry weight of the root system. The large diameter of barley roots grown in non-aerated solution has been attributed to an increase in cell width and not to an increase in cell number (275).

Plants grown in waterlogged soils and in nonaerated solution culture tend to develop root systems that are generally finer branched, have more lateral roots per unit of root length, and a greater density than those of roots grown in aerated solution culture (84, 103, 187).

The effect of O_2 deficiency in solution culture depends on the anatomical make-up of the plant, whether seeds are grown under conditions of little or no oxygen or whether plants are grown initially aerobically and then transferred to a nonaerated solution (84).

According to Brouwer [personal communication, as cited by de Wit (84)], corn roots maintain an equal growth rate in both aerated and nonaerated solution, whereas barley roots are unable to maintain their original growth rate in nonaerated solution. The difference is sensitivity to waterlogging in these two species appears to be related to their anatomy. The air spaces formed in the roots of maize grown in nonaerated solutions may occupy 30–45% of the cross sectional area, while in barley this value is less than 10% (84). A close correlation between flooding tolerance and aerenchyma development has been suggested for many agricultural crops (183, 184).

Several hypotheses involving indoleacetic acid, ethylene, and calcium have been proposed to account for aerenchyma formation (84, 111, 183). The lack of oxygen sets off a series of biochemical events that lead to the ultimate destruction and death of clusters of cells in the root cortex and subsequent formation of intercellular air spaces (84).

Flooding injury in most plants is manifested by a sequence of morphological and physiological changes. In tomato plants, wilting is observed within a few hours and reduction in growth rate occurs within 24 hr. By 36 hr, increased epinastic curvature takes place and lower leaves turn yellow. By 72 hr adventitious roots are formed (183, 288).

There is controversy in the literature as to whether adventitious root production is a symptom of flooding injury or a beneficial adaptation to waterlogging (104, 105, 183). Most investigators support adventitious rooting as an adaptive feature. For example, in three Eucalyptus species, Clemens et al. (63) found that 7–10 weeks of waterlogging caused more than 90% of the root system to be replaced with new adventitious roots; the greater the adventitious root formation, the greater the tolerance to waterlogging. Adventitious root development is important for waterlogging tolerance in sycamore, green ash, and water tupelo (138), and other woody plants (104, 143). Herbaceous species, such as tomato and cucumber, that produce adventitious roots near the soil surface, have also been found to be more tolerant to waterlogging than those species, such as melon and eggplant, that lack adventitious roots (144).

Physiological and Biochemical Effects of Waterlogging

The physiological and biochemical effects of waterlogging include changes in respiratory metabolism, root permeability, water and mineral uptake, nitrogen fixation, and endogenous hormones (Figure 2). A full discussion of these processes is included in articles and reviews by Brouwer and Wiersum (40), Reid (288), Bradford and Yang (36), Yang (385), Cannell and Jackson (48), and Jackson (157). Only the highlights will be described here.

Effects on Hormonal Metabolism.   Hormonal metabolism is altered greatly by waterlogging (Figure 2). All major groups of plant hormones have been investigated including the auxins, gibberellins (GA), cytokinins, ethylene, and abscisic acid (ABA). The biosynthesis and role of these hormones in the flooding syndrome, are discussed in detail by Kawase (183), Adams and Yang (4, 5), Bradford and Yang (34, 36), Reid (288), Lieberman (216), Elliott (94), Jackson (157), Cannell and Jackson (48), and Yang (385, 386).

One of the chief effects of waterlogging is to upset the hormonal balance between roots and shoots (36, 157, 183, 288, 372, 373, 394). This may result from: (a) a decreased supply of substances from the roots required for normal shoot growth; (b) an increased supply of substances from the roots that modify shoot growth; and (c) an accumulation in the shoot of substances normally exported to the roots (157). Plant hormones (or their precursors), water, and mineral nutrients all appear to be involved although their exact

Whole - plant Physiological Responses to Waterlogging

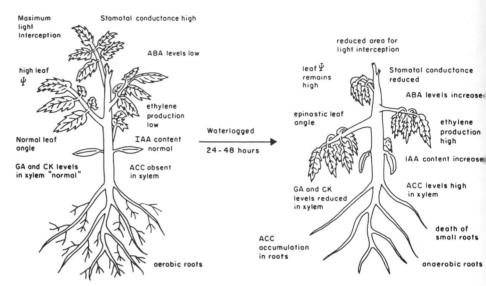

Figure 2 Typical physiological responses of intact plants to waterlogging. From Bradford and Yang, 1981. *HortScience*. Vol. 16, No. 1, pp. 28.

roles are still unclear (48). A number of workers have shown that cytokinin- and GA-like substances are produced in the roots and transported in the transpiration stream to the shoot (53, 54, 55, 80, 154, 200, 272, 289, 290, 291, 292, 293).

Conversely, auxins produced in the shoot are thought to move downward from the shoot and regulate root growth and development (94). Phillips (270) and Kramer (194) proposed that a lack of oxygen in the lower portions of the shoot might cause a decrease in auxin transport in the roots. Other workers (107, 376) showed that polar transport of auxin in oat coleoptiles was inhibited under anaerobic conditions, which could account for auxin accumulation in the shoots. In corn coleoptiles, however, indoleacetic acid (IAA) transport still occurred under anaerobic conditions; in this case auxin transport appeared to depend on active glycolysis (107, 376) (Pasteur effect).

Adventitious rooting and stem swelling in sunflower during flooding were attributed to auxin accumulation in the shoot (271, 367). Leaf epinasty induced by flooding in intact sunflower plants was overcome by decapitating the shoot. The addition of IAA in lanolin to the cut surface of decapitated plants restored the epinasty-inducing effect of flooding (270).

Several workers have suggested that reduced stem elongation of flooded plants is due to an inhibition of GA synthesis (270) or cytokinin production in the roots (84, 163).

Flooding greatly reduced GA levels in the xylem sap of tomato plants

(292); only one or two days were needed to cause a decline in GA content of shoots and roots (291). In sunflower plants flooding caused extensive death of the root apices and a sharp drop in the content of cytokinins transported in the xylem (44). A foliar application of benzyladenine (BA) delayed leaf senescence in flooded plants but had little effect on nonflooded plants (284). Treatment with BA also reduced petiole epinasty, completely prevented the formation of adventitious roots, and stimulated shoot growth in the flooded plants but had no observable effect on unflooded control plants.

Impairment of water uptake in bean leaves during waterlogging caused incipient wilting of the leaves and corresponding changes in ABA level (382, 383). In general, however, the role of ABA in waterlogging remains to be investigated (136, 365).

Auxin is known to induce ethylene production in a number of plant tissues. Many effects formerly attributed to auxin on plant growth are now attributed to auxin-induced ethylene production (1, 151, 169, 189, 391, 393).

Recent studies suggest that under anaerobic conditions a "signal" is produced in the roots and transported to the shoot where it stimulates ethylene synthesis (161). This signal has been recently identified by Bradford and Yang (35) as 1-aminocyclopropane-1-carboxylic acid (ACC), an intermediate in the conversion of methionine to ethylene (Figure 3). Evidence in support of this hypothesis was provided by determining the ACC content in xylem

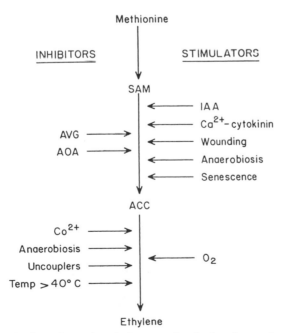

Figure 3 Biosynthetic pathway in ethylene synthesis showing various intermediates and inhibitors. From Bradford and Yang, 1981. *HortScience*. Vol. 16, No. 1, pp. 26.

sap from detopped tomato plants. A striking increase in ACC content which occurred in response to waterlogging and root anaerobiosis, preceded both the increase in ethylene production and epinastic curvature.

Plants flooded and then drained showed a rapid, simultaneous drop in ACC flux and rate of ethylene synthesis. When ACC was supplied through the cut stem of tomato shoots at concentrations comparable to those found in the xylem sap, typical epinasty and increased ethylene production was observed. These findings confirmed that ACC is synthesized in the anaerobic root and transported to the shoot where it is rapidly converted to ethylene (35).

Biosynthetic studies by Adams and Yang (4, 5) and others (81) indicated that S-adenosylmethionine (SAM) was an early intermediate in the biosynthetic pathway of ethylene (Figure 3). Independent studies in the United States and Germany (4, 219) suggested that ACC was derived from SAM.

Adams and Yang (4, 5) observed that aminoethoxy-vinylglycine (AVG), a potent inhibitor known to block the conversion of methionine to ethylene, was ineffective in blocking the conversion of methionine to SAM or the conversion of ACC to ethylene but was effective in blocking the conversion of methionine to ACC. They proposed that AVG inhibits ethylene synthesis by blocking the conversion of SAM to ACC (5).

A cell-free extract prepared from tomato fruit (30) has been found to contain ACC synthase, an enzyme capable of converting SAM to ACC. However, a natural cell-free system capable of converting ACC to ethylene has not been demonstrated (5). Although ACC synthase has not been isolated and characterized, Adams and Yang (5) believe this enzyme to be the rate-limiting step in most of the tissues examined. How diverse signals such as auxin treatment, flooding of roots, and wounding, can induce the same key enzyme, remains to be determined. The availability of methods for assay of ACC and ACC synthase should lead to exciting progress in stress physiology (45, 190, 392).

Effects on Respiratory Metabolism.   A shift in respiratory metabolism from aerobic to anaerobic pathways is one of the main effects of oxygen deficiency resulting from waterlogging (75, 288). This in turn inhibits electron transport and decreases ATP production and is followed by the accumulation of various end products of anaerobic respiration and rapid depletion of organic compounds (40, 78, 84, 124, 140, 298). An increase in ethanol and alcohol dehydrogenase (ADH) activity has been reported (72, 73, 74, 75, 100, 188). The effect of waterlogging on respiration rate depends on the plant species and cultivar and on its physiological age (245).

Rice seedlings grown under aerobic and anaerobic conditions revealed significant differences in lipid metabolism (362). On a dry weight basis, the total lipid content of coleoptiles grown under anaerobic conditions (darkness for 5 days at 27° C in distilled water bubbled with nitrogen) was nearly twice that of coleoptiles grown under aerobic conditions.

Although the same lipids appeared in both groups of seedlings, that is, phosphatidylcholine, phosphatidylethanolamine, and neutral lipids, the content of natural lipids and fatty acids was increased three-fold under anaerobic conditions. The distribution of acyl residues also differed significantly between the two treatments. Rice coleoptiles grown under nitrogen showed a decrease in the ratios of palmityl ($C_{16:0}$), palmitoleyl ($C_{16:1}$), and linolenoyl ($C_{18:3}$) residues and significant increases in oleyl ($C_{18:1}$) and linoleyl ($C_{18:2}$) residues. The same variations occurred in each class of lipids (362).

Effects on Water and Mineral Uptake. Low levels of oxygen in the soil during waterlogging influence the uptake of water and nutrients by: (a) changing nutrient availability in the soil; and (b) changing the metabolic status of the plant (39, 127, 152, 288, 320, 344, 370).

Decreased uptake of water in waterlogged soils has been reported (61, 192, 193, 194, 245, 256, 287, 303). Kramer (193) showed that transpiration of tomato plants was reduced when the soil air was replaced by CO_2 or nitrogen. Transpiration of 'Inia 66' wheat plants grown for 45 days at 0% O_2 was 50% that of plants grown at either 12 or 21% O_2 (3, 245). However, after 35 days the plants adapted to the low O_2 level and showed the same rate of transpiration for all O_2 treatments.

Two common signs of waterlogging damage, wilting and chlorosis, suggest a reduction in water and/or nutrient uptake (288). The need for energy from aerobic respiration for active uptake of water and minerals is well known (195, 207, 336).

A reduction in water uptake in anaerobic soils often occurs as a result of high salinity levels which can osmotically reduce the availability of water (37). In tomato, wilting was observed and water uptake was reduced by 50% within 6 hr of anoxia.

Drew and Sisworo (87) examined the effects of flooding on nitrogen status, leaf chlorosis and shoot growth in barley seedlings and found a rapid decrease in nitrogen uptake to the shoots within 2 days as oxygen concentration in the soil water dropped below 2%. At the same time the average concentration of nitrogen in the shoots decreased due to translocation from the older leaves to the younger expanding ones. These changes preceded the onset of chlorosis in the older leaves. Based on these results, these workers suggest that the inhibition of nitrogen uptake and the consequent redistribution of nitrogen within the shoot contribute to early leaf senescence and retard shoot growth in flooded plants.

The effects of oxygen supply to the root system on the accumulation of nutrients in fruit trees are well documented (142, 201, 202, 203, 209, 211, 320, 357). In peas, waterlogging sufficiently inhibited the uptake of nitrogen from the soil to reduce the concentration in vegetative and reproductive tissues. However, the uptake of K and P was less severely affected and the overall concentration of these elements in shoot tissues was slightly higher in waterlogged plants than in controls. The concentrations of Mn and Ca were not altered appreciably by waterlogging (156). In wheat, the concentrations of nitrate, Ca, and K in soil solutions fell during the waterlogging period while the concentrations of P, Mn, and Fe were variable and showed no clear trend. The decline in nitrate concentration was believed to result from denitrification. Changes in Ca and K concentration were directly propor-tional to that of nitrate (350).

Leyshon and Sheard (214) obtained a reduction in N, P, and K concen-tration of 51, 61, and 58%, respectively in barley plants exposed to short-term flooding conditions. In slash pine, Shoulders and Ralston (314) found that low O_2 levels reduced the uptake of P, K, Ca, and Mg but increased nitrate uptake. Nutrient interactions can occur chemically in the soil or physiologically within the plant. Toxic levels of Fe and Mn in the soil solu-tion are believed to be important factors causing reductions in growth in waterlogged soils (66, 244, 245).

Iron deficiency symptoms were relieved in tomato plants grown in non-aerated nutrient culture by adding Fe EDTA or leaf compost (which may have lowered pH) (120). However, some investigators (309) were unable to in-crease the growth of nonaerated tomato plants by foliar sprays or by addi-tions of various concentrations of nutrients to the culture media. Forced aeration of waterlogged soils has been reported to overcome potassium defi-ciency in corn (206).

The role of aeration in potassium absorption has been reviewed by Law-ton and Cook (207). Tomato plants grown in aerated solution culture were found to take up about 40% more potassium than those grown in nonaerated solution (17). Similar evidence for the importance of aeration in potassium uptake was reported by Chang and Loomis (58) for corn, and by Hopkins et al. (142) for tomato, soybean, and tobacco. However, studies by other workers (263) failed to show that potassium uptake was greatly dependent on O_2 level in liquid culture.

Because of large differences in response among flooding sensitive species, it is difficult to generalize as to the ability of plants to take up vari-ous nutrients in waterlogged soils (288, 306). Interpretation of experimental results is further complicated by the frequent difference in behavior of plants grown in water culture from that of plants grown in sand or soil (288).

Effects on Photosynthesis. The effects of waterlogging on photosynthesis are not well documented (61, 287, 317). In *Populus deltoides* Marsh. com-plete inundation of the root system for 28 days reduced photosynthesis by 50%. However, recovery occurred within one week after flooding ceased (287).

Effects on Nitrogen Fixation.   In legumes, excess moisture is not only detrimental to shoot and root growth (146, 302), it is also detrimental to the development and function of root nodules (332). This phenomenon has been attributed to reduced availability of oxygen to the nodule (331). Reduction in plant weight and nodule mass as a result of waterlogging has been reported for cowpea (137, 247), broadbean (163), and pea (246). Of all the legumes studied, soybean appears to be the least injured on wet soils (148, 262). Studies on nitrogen fixation in this species indicate that recovery from the effects of waterlogging is rapid.

Several workers have shown adverse effects of waterlogging on other symbiotic relationships, including mycorrhiza and Rhizobium root nodule associations, with subsequent reductions in nitrogen fixation and flow of nitrogen into higher plants (128, 337).

Although waterlogging reduces the metabolic activity of aerobic microorganisms, the presence of numerous anaerobic nitrogen-fixing organisms (337) enables nitrogen fixation to occur in waterlogged soils (294, 295). Because of rapid bacterial denitrification (111) and severe leaching of nitrates, waterlogging may result in large losses of available nitrogen from the soil (21, 288) (Figures 4, 5).

Figure 4   Comparative growth of corn plants grown in well-drained and waterlogged soil on the Eastern Shore of Maryland (Talbot County). Illustrates denitrification in waterlogged soil. Photograph taken in June 1979 by Dr. Robert Howell, Plant Stress Laboratory, U.S. Department of Agriculture, Beltsville, MD.

Figure 5 'Clark' soybean plants showing waterlogging damage in a poorly drained soil on the U.S. Department of Agriculture South Farm at Beltsville, Maryland. Plants were exposed to severe flooding intermittently during heavy rainfall in 1979. Photograph taken in early October, 1979 by Dr. Robert Howell, Plant Stress Laboratory, U.S. Department of Agriculture, Beltsville, MD.

Factors Influencing Waterlogging Damage

The extent of waterlogging damage depends on many factors including the duration of flooding, the stage of growth and the species and/or cultivar (212, 215). Environmental conditions such as temperature, carbon dioxide content, and other factors described earlier may also play a large role. Factors found to influence the flooding tolerance of woody species were summarized by Gill (104). These included species differences, interspecific and intraspecific differences, edaphic factors, timing and duration of flood, depth of flood, plant age, and size of stands.

With the exception of paddy rice, most agricultural crops experience only transient or short-term waterlogging (46, 47, 104, 245, 302, 307, 380). The duration of waterlogging may range from a few hours to several weeks (104, 302).

The effects of short-term flooding on plant growth are more difficult to evaluate than those of long-term flooding because the soil oxygen status is constantly changing (245). Some plants succumb to very short periods of low aeration while others such as rice can grow under conditions of 0% oxygen.

Even brief periods of waterlogging cause permanent plant injury. When root systems are flooded the lower leaves on the shoot begin to yellow and this yellowing generally proceeds up the stem (Figures 4, 5). In tobacco, epinasty of the leaves is observed after 24 hr, followed by wilting. In tomato, stem elongation slows down or ceases completely after 24 hr of flooding (84). In sunflower, roots emerge from the hypocotyl after 24 hr of flooding. In sunflower and tomato, the development of adventitious roots occurs after 2–4 days of flooding (183, 184).

Large decreases in yield have been reported in many legumes after only a 24-hr period of waterlogging. Species studied include pea (47, 95, 156), cowpea (248), bean (98), and soybean (375).

Short-term waterlogging of the soil has been shown to restrict the growth of a number of species including alfalfa, *Medicago sativa* (347); sugar beet, *Beta vulgaris* (311); corn, *Zea mays* (205); and wheat, *Triticum aestivum* L. (59).

Adverse effects of short-term flooding have also been reported for cotton, forage crops, strawberries, blueberry, hybrid tea roses, and conifers (349). Stone and pome fruit trees, grapes, cane fruits, and many ornamental woody species are also sensitive to short periods of waterlogging (349). A single week of flooding has been reported to kill apricot trees (349). Peach is nearly as sensitive (349). When the water table in orchards along the Nile rose to within 1 m (3 ft) of the ground surface, peach and other stone fruit trees developed gummosis, shoot and twig dieback, leaf shedding, and root rot and finally died (349). Under excessive irrigation citrus may also become chlorotic and develop dieback (349).

The extent of injury caused by short-term waterlogging depends on the stage of growth when it is experienced (8, 150, 245, 302). Germinating seeds are expecially sensitive to flooding injury (10, 47) because they must depend on the surrounding soil for oxygen. The severity of waterlogging damage to seedlings is often increased by the presence of plant residues (232, 233, 302, 346).

Although waterlogging depresses shoot growth at all stages, the most sensitive appears to be just prior to flowering or during early flower development (47, 95, 155, 156, 302). This is true for both cereals and legumes (47, 302). In general, grasses tolerate lower O_2 levels than other species (245).

The top growth of barley plants was most sensitive to short-term flooding at an early stage of vegetative growth (14 days) than at 21, 28, or 35 days; however, younger plants were able to recover more quickly than the older ones (214). Barley plants flooded 28 days after seeding showed a 55% reduction in grain yield but only a 35% reduction when flooded 35 days after seeding. Aceves-Navarro (3) obtained a 50% decrease in top growth of wheat when the root system was exposed to 0% oxygen (245).

The extent of waterlogging damage also depends on the amount of the root system affected. Corn and sunflower plants given treatments in which only half of the root system was flooded showed no decrease in dry matter production over those grown in well-drained containers (390).

Difficulty in Predicting Plant Response to Waterlogging

The response of plants, and particularly their root systems, to variations in water supply under natural conditions is perhaps the most difficult aspect of plant behavior to predict or to describe in quantitative terms (302). Uncertainty is caused by the rapid and generally unpredictable changes in water supply brought about by seasonal changes in the weather and by equally rapid responses of plants to these changing conditions.

Mathematical models have been developed to describe the growth of roots in uniform environments (122, 218). However, these models are of little help in understanding root function under field conditions where the plants are often subjected to transient and unpredictable stresses (306).

The difficulty of extrapolating the results of laboratory studies on anaerobiosis to field conditions was shown by Jackson (156). The injury to pea plants, which were grown in nutrient solution and supplied with oxygen and carbon dioxide at concentrations present in waterlogged soil, was less than that experienced by plants grown in waterlogged soil. Although premature quiescence of the apical bud of the main stem was induced by both treatments, the extent of dessication, chlorosis, and reduction in growth rate was much greater in plants grown in waterlogged soil than in anaerobic solution cultures.

These results on pea differ from those obtained with tomato (161), barley (89), and rape (161), where a decrease in the oxygen concentration in solution culture produced the same injury as that obtained under waterlogged conditions.

BREEDING FOR FLOODING TOLERANCE

Selection Criteria

Two criteria are generally used in selecting plants for tolerance to waterlogging: (a) how well does the plant survive and grow under actual waterlogged conditions, and (b) how well does it respond to flooding conditions under experimentally controlled conditions. Poststress growth and recovery time should also be a major consideration.

Differential flooding tolerance may be determined: by making measurements of the root's ability to penetrate flooded soil and to tolerate chemicals produced under waterlogged conditions; by measuring the plant's oxygen transport system; and by conducting biochemical tests on selected plant tissues (68).

To make progress the plant breeder must know which traits are important to preserve and enhance. Selection criteria based on generalized plant responses are often of little value. Those features that enhance oxygen transport to the root are likely to be the ones of greatest use to the breeder. Only after understanding the role of various adaptive mechanisms is it possible for the breeder to develop sound practical tests for flooding tolerance (68). Coutts and Armstrong (68) have concluded that

> If we are to progress beyond hit-or-miss selection procedures, the commonly employed techniques of suddenly flooding entire seedling root systems, and observing effects on the shoot, must give way to selection and experimentation which consider the entry points of oxygen into the plant, and the likely pathways of gas in the tissues.

A brief review of some of the known adaptive features of plants is given. The possession of porous tissues and the capacity to increase porosity under flooding conditions is closely correlated with tolerance to sudden and permanent waterlogging in herbaceous species (99, 183). A similar relationship exists in tree species such as cypress (111) but further tests on woody species are needed before generalizations can be made (68, 147). The genetic control of processes leading to increased porosity or other adaptive characteristics of flooding-tolerant plants is not understood. Basic studies in plant physiology, biochemistry, and ultrastructure are needed if significant advances are to be made in developing plants resistant to waterlogging.

Differential Sensitivity of Plants to Flooding Injury

Species and cultivars differ widely in their tolerance of anaerobic conditions in the soil (Tables 3 and 4) (134, 242, 264). Yu et al. (390) found in pot studies that corn and sunflower plants showed much less damage under prolonged flooding conditions that did tomato and barley plants. These investigators also found differences in sensitivity to flooding between 'Inia' and 'Pato', two dwarf high yielding cultivars of wheat developed in Mexico. The 'Pato' variety tolerated the flooded soil better than the 'Inia' variety. Higher root porosities were found in the tolerant plants of 'Pato' wheat, corn, and sunflower than in the sensitive plants, for example, 'Inia' wheat, tomato, and barley.

Table 3 Differences in flooding tolerance among herbaceous species and cultivars

Highly sensitive	Intermediate	Highly tolerant	Reference
Ladino clover		Strawberry clover	24
'Inia' wheat		'Pato' wheat	390
tomato		sunflower	
barley		corn	

Table 4   Differences in flooding tolerance among woody species and cultivars

Highly sensitive	Intermediate	Highly tolerant	Reference
Sitka spruce		Lodgepole pine	268, 269
sweet gum		water tupelo	138, 228
yellow poplar		green ash	
sycamore			
peach	plum	apple	298
apricot		pear	
almond		quince	
black walnut		Chinese walnut	57
English walnut			
		bald cypress	141
		black mangrove	

However, studies by Luxmoore et al. (221) involving 4-day flooding treatments of 'Inia' wheat plots near Sonora, Mexico at 15, 30, and 45 days after planting failed to show any differences between flooded and unflooded plots. Stolzy (338) suggests that the differences between these pot and field studies may be attributed to the amount of pore space filled with water. In pot studies, flooding fills most of the pore spaces, whereas in the field, flooding may saturate only a few centimeters of surface soil because of surface sealing and swelling phenomena.

Soybean cultivars also differ in their sensitivity to flooding (92). Over a 3-year period, yield of the cultivar 'Dorman' was reduced 8, 38, and 59% by flooding for 7, 14, and 21 days, respectively, at the time of flower initiation. In contrast, yield of the cultivar 'Lee' was unaffected by 7 days of flooding and reduced only 6 and 18% by 14 and 21 days of flooding, respectively (92).

Fruit tree species have a wide range of tolerance to long periods in water-saturated soils (25, 83, 298). The roots of apple, pear, and quince are relatively tolerant to flooding while those of peach, apricot, and almond are generally considered highly sensitive (298). 'Myrobalan' plum is intermediate in tolerance (83) (Table 4).

Tolerance of Anaerobic Conditions

Both morphological and physiological adaptive mechanisms occur in plants tolerant to anaerobic conditions (36, 69, 70, 139, 183, 342). One of the most common morphological adaptations is the formation of increased intercellular air spaces in the cortex that provide canals parallel to the axis of the root through which gases can diffuse longitudinally (184, 302). These air spaces, or aerenchyma, are well-developed in rice, ferns, and numerous aquatic plants (11, 12, 16, 67, 204, 369). In many species, aerenchyma may provide most if not all of the oxygen requirements of the roots as well as some of the requirements of the surrounding rhizosphere soil (26).

Another anatomical adaptation to waterlogged soils is the formation of adventitious roots close to the soil surface where the oxygen tension is usually higher or more quickly restored after transient waterlogging. This was shown to be common in tomato by Jackson (155) and in sunflower by Wample and Reid (366, 367).

Physiological characteristics associated with resistance to anaerobic conditions include the development of alternative respiratory pathways, which restrict the production of ethanol (302). The alternate production of organic acids, such as malic and shikimic, is a typical example of physiological adaptation (236).

There are numerous reports of specialized biochemical adaptations which contribute to flooding tolerance (7, 62, 71, 78, 102, 129, 141, 269, 181, 188, 217, 236, 241, 304, 362, 363, 374).

Evidence exists that flood-tolerant species from natural habitats differ from flood-sensitive species by being able to make more effective use of nitrate as an alternative electron acceptor to oxygen during periods of partial anaerobiosis. Flood-tolerant species show marked increases in nitrate reductase activity in the roots and leaves during waterlogging. Tolerant species also have a greater ability to synthesize amino acids under flooding conditions than do intolerant species, thereby facilitating the re-oxidation of $NADH_2$ under conditions of anoxia (102).

Philipson and Coutts (268) reported differences in the amount of oxygen that diffused out of the roots of waterlogging-tolerant Lodgepole pine seedlings and waterlogging-intolerant Sitka spruce seedlings; the greatest amount of O_2 came from the roots of the waterlogging-tolerant Lodgepole pine. Similar differences were obtained between waterlogging-tolerant rice and non-tolerant barley. Bartlett (21) reported that waterlogging-tolerant species of forage crops and rice had a greater ability to oxidize iron in the root rhizophere than intolerant species.

Anatomical Basis of Adaptation

Numerous studies have shown that roots of plants growing in a poorly aerated environment contain larger air spaces than those of plants grown in a well-aerated media (23, 40, 42, 90, 189, 238, 251). Aerenchyma development is felt to be an important adaptive feature because it provides an effective air transport system in the plant (115, 116, 184, 377). The development of aerenchyma in crop plants has been found to closely correlate with tolerance to waterlogging (384).

Although lysigenous aerenchyma have been demonstrated in the roots of several of the grain crops, for example, barley (42), wheat (40), and corn (23, 90, 238, 257), recent studies by Kawase and Whitmoyer (184) indicate that this phenomena is not confined to the grain crops. Waterlogged sunflower and tomato plants grown under greenhouse conditions were induced to form aerenchyma within two days of waterlogging treatment.

The formation of lysigenous aerenchyma in waterlogged corn roots has been attributed by McPherson (238) to the loss of turgidity and eventual collapse of cortical cells as a consequence of insufficient supply of energy through anaerobic respiration to maintain cellular integrity. An alternative explanation was proposed by Kawase (181, 182) who suggested that ethylene buildup in the waterlogged tissue was responsible for aerenchyma development by increasing cellulase activity.

Physiological Basis for Tolerance to Flooding

With the exception of citrus and avocado, most reports published through 1970 on tolerance to waterlogging were based on field observations. Careful studies conducted on flooding injury in the laboratory led to significant breakthroughs on an understanding of some of the physiological mechanisms involved in differential sensitivity of fruit trees to waterlogging (57, 298).

One of the key indicators of cellular damage and relative sensitivity to waterlogging in fruit tree species is cyanogenesis (298). Rowe and Catlin (298) suggested that, under anaerobic conditions, respiration and resultant energy transfer are inhibited initially. With insufficient energy from aerobic respiration to maintain membrane integrity, cellular disorganization occurs. As a consequence, the cyanogenic glycoside and hydrolytic enzymes "come in contact" and hydrolysis occurs. In turn, HCN is released which may cause additional inhibition and cellular damage in an autocatalytic manner. They suggest that the lower sensitivity of plum as compared to peach and apricot may be explained by differences in respiratory mechanisms, with secondary effects due to cyanogenesis.

Role of Hormones in Plant Adaptation to Waterlogging

Plant hormones have been shown to play an important role in adaptation to adverse environmental conditions, such as drought, salinity, flooding, and chilling injury (296). In general, increased resistance to these stresses is associated with a rise in endogenous abscisic acid (ABA) and a drop in giberellins (345).

The increased resistance to stress caused by ABA has been attributed chiefly to a better water balance brought about by induction of stomatal closure and by an increase in the root permeability to water through enhancement in hydraulic conductivity of the root (106, 276, 296, 343).

Jackson (157) has suggested that ethylene may aid root aeration indirectly by promoting the formation of aerenchyma within the roots (86). The role of ACC, an ethylene precursor, in plant adaptation to waterlogging has been described by Bradford and Yang (36).

Rice and Rice Weeds: Unique Examples of Flooding Tolerance

The rice plant is well known for its unique ability to germinate and grow for a limited time under anaerobic conditions (180). The rice coleoptile has been described as the only plant organ capable of growing in the absence of oxygen (280). However, recent studies by Kennedy et al. (180) indicate that barnyard grass (*Echinochloa crus-galli*), a common weed of rice fields, is equally well adapted to germinate and grow under these adverse conditions.

The seedlings of *E. crus-galli* are similar to those of rice. They are unpigmented, the primary leaves do not emerge from the coleoptile, and radicle emergence does not occur without oxygen. The ultrastructure of mitochondria from anaerobically grown *E. crus-galli* seedlings are of particular interest because they resemble the mitochondria of aerobically grown seedlings in size and shape and show cristae that are numerous and normal in appearance. This is in contrast to other species which show a loss of normal fine-structure when grown in the absence of oxygen (180).

A comparison of respiratory metabolism in rice and two rice weed varieties showed a sharp contrast in $^{14}CO_2$ evolution during ^{14}C-glucose feeding experiments. In all cases the rate of $^{14}CO_2$ evolution was lower in N_2 than in air especially for the first 24–48 hr. These findings were in contrast to the pattern observed in peas, a flooding-intolerant species. No difference in the rate of $^{14}CO_2$ evolution in N_2 occurred in peas.

Amelioration of Poor Soil Conditions

Conditions of poor soil aeration can be partially ameliorated by surface or subsurface drainage or by controlled water application in a portion of the root zone that may reduce the chances of water saturation (167, 266, 267, 348).

Additions of nitrogen fertilizer at certain stages of development may also partially ameliorate the effect of flooding and improve yields of certain crops growing in poorly drained soils (312, 313).

SUMMARY

It is now well established that for most agricultural crops and native plants not adapted to wetland conditions, waterlogging the soil reduces shoot and root growth, dry matter production, and total crop yields. As the soil becomes waterlogged, air space is displaced with water and the oxygen remaining in the soil, either dissolved in water or trapped in air spaces, is rapidly depleted by the respiration of roots and soil microorganisms.

Because of the slow diffusion of atmospheric oxygen into waterlogged soils, oxygen replenishment in the soil is inefficient. Therefore root systems are suddenly forced into anaerobic conditions. The soil does not have to be inundated for waterlogging damage to occur; damage may be seen on any soil having poor underground or surface drainage whenever sufficient air spaces of the soil become filled with water.

Oxygen deficiency in the root system disrupts root metabolism by forcing the plant to switch from aerobic to anaerobic respiration. This in turn results in a decrease in ATP production, accumulation of toxic end products of anaerobic respiration, and rapid depletion of organic compounds. The limited supply of available energy reduces the absorption and translocation of water and nutrients. A disruption of root metabolism caused by inadequate oxygen supply also adversely affects the hormonal balance of the shoot and suppresses the synthesis and translocation of hormones in the root such as gibberellins and cytokinins. Auxin concentration in the stem increases as a result of blocked transport to roots or by inhibited IAA-oxidase activity in the stems. One of the most striking hormonal changes found in waterlogged plants is a dramatic increase in ethylene concentration. Recent studies with tomato plants exposed to flooding conditions have demonstrated that anaerobiosis stimulates the synthesis of 1-aminocyclopropane-1-carboxylic acid (ACC), an immediate precursor of ethylene formation. At the same time ACC conversion to ethylene is blocked under anaerobic conditions. As a result it is hypothesized that ACC accumulates in the root, is exported to the shoot and is then converted to ethylene in tissues where the oxygen supply is adequate. Ethylene in turn promotes epinastic growth of the petioles and plays a role in other waterlogging responses.

In some species toxins may be produced by the damaged roots. Anaerobiosis has been reported to reduce the soil redox potential, change soil pH, and increase the potentially inhibitory concentrations of toxic elements (e.g., Mn, Fe). Organic acids (e.g., acetic acid, propionic, butyric), phenolic compounds (e.g., p-hydroxybenzoic, p-coumaric), and gases (carbon dioxide, ethylene, methane, hydrogen sulfide) may also accumulate and lead to plant damage. In addition, denitrification and leaching can result in a loss of nitrate from the soil solution so that plants may suffer from nitrogen deficiency.

However, few investigators have related the development of symptoms of waterlogging damage in plants to specific changes occurring in the waterlogged soil with the goal of identifying the causative factors. Progress is being made to develop criteria for selecting species and cultivars of plants for flooding tolerance. Although considerable progress has been made with woody plants, thus far relatively few breakthroughs have been made in developing cultivars of crop plants for flooding tolerance.

Survival of plants under prolonged flooding seems to be associated with the ability of plants to develop adventitious root systems and porous tissues. Both features serve to improve the ventilating system of the plant and appear to be the primary basis for the differential tolerance of species and cultivars to flooding.

I hope this review will provide the reader with an essential background on the physiological, biochemical, morphological, and anatomical responses and adaptations of higher plants to waterlogging needed to make further strides in breeding crop plants for tolerance to waterlogging.

ACKNOWLEDGMENT

I thank Dr. Douglas Adams, USDA, Post-Harvest Physiology Laboratory, Beltsville, Maryland; Dr. Robert K. Howell, Plant Stress Laboratory, USDA, Beltsville, Maryland; Mr. Kent J. Bradford, Department of Land, Air and Water Resources, University of California, Davis, California; Dr. Shang Fa Yang, Department of Vegetable Crops, University of California, Davis, California; Dr. R. S. Russell, Agricultural Research Council, Letcombe Laboratory; Dr. Howard M. Taylor, Agronomy Department, Iowa State University, Ames, Iowa; Dr. Makoto Kawase, Ohio Agricultural Research and Development Center, Wooster, Ohio, for kindly permitting me to examine unpublished manuscripts and/or to use figures or tables from their publications.

REFERENCES

1   Abeles, F. B. 1973. *Ethylene in Plant Biology.* Academic, New York. p. 302.
2   Abdel Rahman, A. A., P. J. C. Kuiper, and J. F. Bierhuizen. 1959. *Meded. Landbouwhogeschool, Wageningen* 59:1–12.
3   Aceves-Navarro, E. 1974. Ph.D. dissertation. University of California, Riverside.
4   Adams, D. O., and S. F. Yang. 1979. *Proc. Natl. Acad. Sci. USA* 761:170–174.
5   Adams, D. O., and S. F. Yang, 1981. *Trends in Biochem. Sci.* In press.
6   Adreeva, I. N., G. I. Kozlova, and B. B. Vartapetian. 1976. *Sov. Plant Physiol.* 23:89–98.
7   Ahmad, I., and S. J. Wainwright. 1977. *New Phytol.* 79:605–612.
8   Aleksandrova, A. Z., and F. D. Skazkin. 1964. *Dokl. Akad. Nauk.* 159:203–205.
9   Anderson, W. B., and W. D. Kemper. 1964. *Agron. J.* 56:453–456.
10 Andrews, F. M., and C. C. Beal. 1919. *Bull. Torrey Bot. Club* 46:91–100.
11 Arikado, H. 1955. *Bull. Fac. Agr. Mie. Univ., Japan.* No. 11.
12 Arikado, H. 1955. *Proc. Crop Sci. Soc. Japan* 24:53–58.
13 Armstrong, W. 1975. In *Environment and Plant Ecology*, J. R. Etherington, Ed. Wiley, New York, pp. 181–218.
14 Armstrong, W. 1978. In *Plant Life in Anaerobic Environments,* D. D. Hook and R. M. M. Crawford, Eds. Ann Arbor Science Publishers, Inc., Ann Arbor, Michigan, pp. 269–297.
15 Armstrong, W. 1979. In *Advances in Botanical Research*, H. W. Woolhouse, Ed. 7:225–332.
16 Armstrong, W., and D. J. Boatman. 1967. *J. Ecol.* 55:101–110.
17 Arnon, D. I., and D. R. Hoagland. 1940. *Soil Sci.* 50:463–486.
18 Bannister, P. 1964a. *J. Ecol.* 52:423–432.
19 Bannister, P. 1964b. *J. Ecol.* 52:481–497.

20 Bannister, P. 1964c. *J. Ecol.* 52:499–509.

21 Bartlett, R. J. 1961. *Soil Sci.* 92:372–379.

22 Barrow, N. J., and D. S. Jenkinson. 1962. *Plant Soil* 16:258–262.

23 Beals, C. C. 1917. *Proc. Indiana Acad. Sci.* 7:177–180.

24 Bendixen, L. E., and M. L. Peterson. 1962. *Crop Sci.* 2:223–228.

25 Bergman, H. F. 1959. *Bot. Rev.* 25:417–485.

26 Bidwell, O. W., D. A. Gier, and J. E. Cipra. 1968. *Trans. 9th Int. Congr. Soil Sci.* 4:683–692.

27 Bird, C. W., and J. M. Lynch. 1974. *Chem. Soc. Rev.* 3:309–328.

28 Black, C. A. 1968. *Soil-Plant Relationships.* 2nd ed. Wiley, New York, pp. 1–792.

29 Bloomfield, C. 1969. *J. Soil Sci.* 20:207–221.

30 Boller, T., R. C. Herner, and H. Kende. 1979. *Planta* 145:293–303.

31 Börner, H. 1960. *Bot. Rev.* 26:393–424.

32 Boyko, H., and E. Boyko. 1964. *N. Y. Acad. Sci. Trans.* Sect. II. 26, Suppl., 1087–1102.

33 Bradford, K. J., and D. R. Dilley. 1978. *Plant Physiol.* 61:506–509.

34 Bradford, K. J., and S. F. Yang. 1980a. *Plant Physiol.* 65:327–330.

35 Bradford, K. J., and S. F. Yang. 1980b. *Plant Physiol.* 65:322–326.

36 Bradford, K. J., and S. F. Yang. 1981. *HortScience.* 16:25–30.

37 Bresler, E. 1972. In *Optimizing the Soil Physical Environment Toward Greater Crop Yields*, D. Hillel, Ed. Academic, New York, pp. 101–132.

38 Bromfield, S. M. 1954. *J. Gen. Microbiol.* 11:1–4.

39 Brouwer, R. 1965. *Annu. Rev. Plant Physiol.* 16:241–266.

40 Brouwer, R., and L. K. Wiersum. 1978. In *Crop Physiology*, U.S. Gupta, Ed. Oxford & IBH Publishing Co., New Delhi, Bombay, Calcutta, pp. 157–201.

41 Brown, N. J., E. R. Fountaine, and M. R. Holden. 1965. *J. Agri. Sci.* 64:195–203.

42 Bryant, A. E. 1934. *Plant Physiol.* 9:389–391.

43 Burns, R. G. 1978. *Soil Enzymes.* Academic, London, pp. 1–380.

44 Burrows, W. J., and D. J. Carr. 1969. *Physiol. Plant.* 22:1105–1112.

45 Cameron, A. C., C. A. L. Fenton, Y. B. Yu, D. O. Adams, and S. F. Yang. 1979. *HortScience* 14:178–180.

46 Cannell, R. Q. 1977. *Applied Biol.* 2:1–86.

47 Cannell, R. Q., K. Gales, R. M. Snaydon, and B. A. Suhail. 1979. *Ann. Appl. Biol.* 93:327–335.

48 Cannell, R. Q., and M. B. Jackson. 1981. In *Modifying the Plan Root Environment*, G. F. Arkin and H. M. Taylor, Eds. Am. Soc. Agric. Engl. Madison, pp. 137–138.

49 Cannon, W. A. 1924. *Ecology* 5:319–321.

50 Cannon, W. A. 1925. *Carnegie Inst. Wash. Publ.* 368:1–168.

51 Cannon, W. A. 1940. *Science* 91:43–44.

52 Carr, D. J. 1961. In *Encyclopedia of Plant Physiology*, Vol. XVI, W. Ruhland, Ed. Springer-Verlag, Berlin, pp. 737–794.

53 Carr, D. J., and W. J. Burrows. 1966. *Life Sci.* 5:2061–2077.

54 Carr, D. J., and D. M. Reid. 1968. In *Biochemistry and Physiology of Plant Growth Substances*, F. Wightman and G. Setterfield, Eds. Runge Press, Ottawa. pp. 1169–1185.

55 Carr, D. J., D. M. Reid, and K. G. M. Skene. 1964. *Planta* 63:382–392.

56 Casida, L. E. Jr., D. A. Klein, and T. Santoro. 1964. *Soil Sci.* 98:371–376.

57 Catlin, P. B., G. C. Martin, and E. A. Olsson. 1977. *J. Am. Soc. Hort. Sci.* 102:101–104.

58 Chang, H. T., and W. E. Loomis. 1945. *Plant Physiol.* 20:221–232.

59 Chaudhary, T. M., V. K. Bhatnagar, and S. S. Praghar. 1974. *Agron. J.* 66:32–35.

60 Chendrayan, K., T. K. Adhya, and N. Sethunathan. 1980. *Soil Biol. Biochem.* 12:271–273.

61 Childers, N. F., and D. G. White. 1942. *Plant Physiol.* 17:603–618.

62 Chirkova, T. V. 1978. In *Plant Life in Anaerobic Environments*, D. D. Hook and R. M. M. Crawford, Eds. Ann Arbor Science Press, Ann Arbor, Michigan, pp. 137–154.

63 Clemens, J., A. M. Kirk, and P. D. Mills. 1978. *Oecologia* 34:125–131.

64 Clements, F. E. 1921. *Carnegie Inst. Publ.* 1. No. 315.

65 Conway, V. M. 1940. *Bot. Rev.* 6:149–163.

66 Cooper, L. N. H. 1937. *Proc. Roy. Soc.* B, 124:299–307.

67 Coult, D. A. 1964. *J. Exp. Bot.* 15:205–218.

68 Coutts, M. P., and W. Armstrong. 1976. In *Tree Physiology and Yield Improvement*, M. G. R. Cannell and F. T. Last, Eds. Academic, New York, pp. 361–385.

69 Coutts, M. P., and J. J. Philipson. 1978. *New Phytol.* 80:63–69.

70 Coutts, M. P., and J. J. Philipson. 1978. *New Phytol.* 80:71–77.

71 Crawford, R. M. M. 1966. *J. Ecol.* 54:403–413.

72 Crawford, R. M. M. 1967. *J. Exp. Bot.* 18:458–464.

73 Crawford, R. M. M. 1976. In *Tree Physiology and Yield Improvement*, M. G. R. Cannel and F. T. Last, Eds. Academic, New York, pp. 387–401.

74 Crawford, R. M. M. 1977. *New Phytol.* 79:511–517.

75 Crawford, R. M. M. 1978. In *Plant Life in Anaerobic Environments*, D. D. Hook and R. M. M. Crawford, Eds. Ann Arbor Science Press, Ann Arbor, Michigan, pp. 119–136.

76 Crawford, R. M. M., and Margaret A. Baines. 1977. *New Phytol.* 79:519–526.

77 Crawford, R. M. M., and M. McManmon. 1968. *J. Exp. Bot.* 19:435–441.

78 Crawford, R. M. M., and P. D. Tyler. 1969. *J. Ecol.* 57:235–244.

79 Crocker, W., P. W. Zimmerman, and A. E. Hitchcock. 1932. *Contrib. Boyce Thompson Inst. Plant Res.* 4:177–218.

80 Crozier, A., and D. M. Reid. 1971. *Can. J. Bot.* 49:967–975.

81 Currie, J. A. 1962. *J. Sci. Food Agric.* 13:380–385.

82 Daccy, J. W. H. 1980. *Science* 210:1017–1019.

83 Day, L. H. 1953. *Calif. Agric. Exp. Sta. Bull.* 736:76.

84 DeWit, M. C. J. 1978. In *Plant Life in Anaerobic Environments*, D. D. Hook and R. M. M. Crawford, Eds. Ann Arbor Science Publishers Inc., Ann Arbor, Michigan, pp. 333–350.

85 Drew, M. C. 1980. *Commentaries in Plant Science*, H. Smith, Ed. Pergamon, New York.

86 Drew, M. C., M. B. Jackson, and S. Giffard. 1979. *Planta* 147:83–88.

87 Drew, M. C., and E. J. Sisworo. 1977. *New Phytol.* 79:567–571.

88 Drew, M. C., and E. J. Sisworo. 1979. *New Phytol.* 82:301–314.

89 Drew, M. C., and M. C. T. Trought. 1977. *Rep. Agric. Coun. Letcombe Lab.* 1976, p. 75.

90 Dunn, G. A. 1921. *Am. J. Bot.* 8:207–211.

91 Efron, Y., M. Peleg, and A. Ashri. 1973. *Biochem. Genet.* 9:299–308.

92 Egli, D. B., and J. E. Leggett. 1973. *Crop Sci.* 13(2):220–222.

93 El-Beltagy, A. S., and M. A. Hall. 1974. *New Phytol.* 73:47–60.

94 Elliott, M. C. 1977. In *Plant Growth Regulation*, P. E. Pilet, Ed. Springer-Verlag, Berlin, pp. 100–108.

95   Erickson, A. E., and D. M. van Doren. 1960. *Trans. 7th Int. Congr. Soil Sci.* 3:428–434.

96   Etherington, J. R. 1975. *Environment and Plant Ecology.* Wiley, New York, pp. 347.

97   Fadeel, A. A. 1964. *Physiol. Plant.* 17:1–13.

98   Forsythe, W. M., A. Victor, and M. Gomez. 1979. In *Soil Physical Conditions and Crop Production in the Tropics*, R. Lal and D. J. Greenland, Eds. Wiley, New York.

99   Fukui, J. 1953. *Proc. Crop Sci. Soc. Japan.* 22:110–112.

100 Fulton, J. M., and A. E. Erickson. 1964. *Proc. Soc. Soil Sci. Am.* 28:610–614.

101 Gambrell, R. P., and W. H. Patrick, Jr. 1978. In *Plant Life in Anaerobic Environments*, D. D. Hook and R. M. M. Crawford, Eds. Ann Arbor Science Publishers, Inc., Ann Arbor, Michigan, pp. 375–423.

102 Garcia-Nova, F., and R. M. M. Crawford. 1973. *New Phytol.* 72:1031–1039.

103 Geisler, G. 1965. *Plant Physiol.* 16:241–266.

104 Gill, C. J. 1970. *For. Abst.* 31:671–688.

105 Gill, C. J. 1975. *Flora* 164:85–97.

106 Glinka, Z., and L. Reinhold. 1971. *Plant Physiol.* 48:103–105.

107 Goldsmith, M. H. M. 1968. In *Biochemistry and Physiology of Plant Growth Substances*, F. Wightman and G. Setterfield, Eds. Runge Press, Ottawa, pp. 1037–1050.

108 Goodlass, G., and K. A. Smith. 1978. *Soil Biol. Biochem.* 10:193–199.

109 Goodlass, G., and K. A. Smith. 1978. *Soil Biol. Biochem.* 10:201–205.

110 Goto, Y., and K. Tai. 1957. *Soil Pl. Fd., Tokyo.* 2:198–200.

111 Grable, A. R. 1966. In *Advances in Agronomy*, N. C. Brady, Ed., Vol. 18. Academic, New York, pp. 57–106.

112 Grable, A. R. 1967. *Effect of Tillage on Soil Aeration. Tillage for Greater Crop Production.* Conference proceedings, ASAE, St. Joseph, Michigan, pp. 44–55.

113 Grable, A. R., and E. G. Siemer. 1968. *Soil Sci. Soc. Am. Proc.* 32:180–186.

114 Greenwood, D. J. 1967. *New Phytol.* 66:337–347.

115 Greenwood, D. J. 1969. In *Root Growth*, W. J. Whittington, Ed. Butterworth's, London, pp. 202–223.

116 Greenwood, D. J. 1970. *Rep. Prog. Appl. Chem.* 55:423–431.

117 Greenwood, D. J., and D. Goodman. 1967. *J. Soil Sci.* 18:182–196.

118 Guenzi, W. D., and T. M. McCalla. 1966. *Soil Sci. Soc. Am. Proc.* 30:214–216.

110 Guenzi, W. D., T. M. McCalla, and F. A. Norstadt. 1967. *Agron. J.* 59:163–165.

120 Guminski, S., Z. Guminska, and J. Sulej. 1965. *J. Exp. Bot.* 16:151–162.

121 Gupta, P. C., and C. N. Hittle. 1970. *Agron. Abstr.* pp. 50.

122 Hackett, C., and D. A. Rose. 1972. *Aust. J. Biol. Sci.* 25:669–679.

123 Hagan, R. M., H. R. Haise, and T. W. Edminster. 1967. In *Irrigation of Agricultural Lands*. Am. Soc. Agron. Madison, Wisconsin.

124 Hageman, R. H., and D. Flesher. 1960. *Arch. Biochem. Biophys.* 87:203–209.

125 Hall, A. E. 1981. *HortScience.* 16:37–38.

126 Hall, A. E., K. W. Foster, and J. G. Waines. 1979. In *Agriculture in Semi-Arid Environments. Ecological Studies*, Vol. 34, A. E. Hall, G. H. Cannell, and H. W. Lawton, Eds. Springer-Verlag, Berlin.

127 Hammond, L. C., W. H. Allaway, and W. E. Loomis. 1955. *Plant Physiol.* 30:155–161.

128 Harley, J. L. 1969. *The Biology of Mycorrhiza.* 2nd ed. Leonard Hill, London, pp. 1–282.

129 Harris, D. G., and C. H. M. van Bavel. 1957. *Agron. J.* 15:89–92.

130 Hasnain, S., and K. H. Sheikh. 1976. *Biologia* 22:89–106.

131 Hawkins, J. C. 1962. *J. Sci. Food Agric.* 13:386–391.

132 Hayward, H. E. 1954. *UNESCO Arid Zone Res.* 4:37–71.

133 Heide, H., B. M. van der de Boer-Bolt, and M. H. van Raalte. 1963. *Acta Bot. Neerl.* 12:231–247.

134 Heinrichs, D. H. 1970. *Can. J. Sci.* 50:435–438.

135 Heinrichs, D. H. 1972. *Can. J. Plant Sci.* 52:985–990.

136 Hiron, R. W., and S. T. C. Wright. 1973. *J. Exp. Bot.* 24:769–781.

137 Hong, T. D., F. R. Minchin, and R. J. Summerfield. 1977. *Plant Soil* 48:661–672.

138 Hook, D. D., and C. L. Brown. 1973. *For. Sci.* 19:225–229.

139 Hook, D. D., C. L. Brown, and R. H. Wetmore, 1972. *Bot. Gaz.* 133:443–454.

140 Hook, D. D., and R. M. M. Crawford. 1978. *Plant Life in Anaerobic Environments.* Ann Arbor Science Publishers, Inc., Ann Arbor, Michigan.

141 Hook, D. D., and J. R. Scholtens. 1978. In *Plant Life in Anaerobic Environments*, D. D. Hook and R. M. M. Crawford, Eds. Ann Arbor Science Publishers, Inc., Ann Arbor, Michigan, pp., 299–331.

142 Hopkins, H. T., A. W. Specht, and S. B. Hendricks. 1950. *Plant Physiol.* 25:193–209.

143 Hosner, J. F., and S. G. Boyce. 1962. *For. Sci.* 8:180–186.

144 Hozumi, K. 1966. *Nogyo Gijyutsu* 26:352–357. Cited in reference 183.

145 Hsiao, T. C. 1973. *Annu. Rev. Plant Physiol.* 24:519–570.

146 Huck, M. G. 1970. *Agron. J.* 62:815–818.

147 Huikari, O. 1954. *Commun. Inst. Forest. Fenn.* 42:1–13.

148 Hunter, M. N., P. L. M. de Jabrun, and D. E. Byth. 1980. *Aust. J. Exp. Agric. Anim. Husb.* 20:339–345.

149 Hurd, E. A. 1971. In *Drought Injury and Resistance in Crops.* CSSA Spec. Publ. No. 2, 77088. Crop Sci. Soc. Am.

150 Ikeda, T., S. Higashi, and T. Kawaide. 1957. *Bull. Div. Plant Breed. Cultiv. Tokai-Kinki na. Agric. Exp. Sta.* 45:30–37. Cited in reference 302.

151 Imaseki, H., A. Watanabe, and S. Odawara. 1977. *Plant Cell Physiol.* 18:577–586.

152 Inden, T. 1956. *J. Hort. Assoc. Japan* 25:85–93. (Japanese with English summary).

153 Ioannou, N., R. W. Schneider, and R. G. Grogan. 1977. *Phytopathology* 67:651–656.

154 Itai, Ch., and Y. Vaadia. 1965. *Physiol. Plant* 18:941–944.

155 Jackson, M. B. 1979a. *Physiol. Plant* 46:347–351.

156 Jackson, M. B. 1979b. *J. Sci. Food Agric.* 30:143–152.

157 Jackson, M. B. 1980. *Acta Hort.* 98:61–78.

158 Jackson, M. B. and D. J. Campbell. 1974. *Agric. Res. Counc. Letcombe Lab. A. Rep.* 1973, pp. 23–26.

159 Jackson, M. B., and D. J. Campbell. 1975a. *Ann. Appl. Biol.* 81:102–105.

160 Jackson, M. B., and D. J. Campbell. 1975b. *New Phytol.* 74:397–406.

161 Jackson, M. B., and D. J. Campbell. 1976a. *New Phytol.* 76:21–29.

162 Jackson, M. B., and D. J. Campbell. 1976b. *Rep. Agric. Counc. Letcombe Lab.* 1975, pp. 42–43.

163 Jackson, M. B., and D. J. Campbell. 1979. *New Phytol.* 82:331–340.

164 Jackson, M. B., K. Gales, and D. J. Campbell. 1978. *J. Exp. Bot.* 29:183–193.

165 Jackson, W. T. 1955. *Am. J. Bot.* 42:816–819.

166 Jones, H. E. 1971a. *J. Ecol.* 59:167–178.

167 Jones, H. E. 1971b. *J. Ecol.* 59:583–591.

168   Jones, H. E., and J. R. Etherington. 1970. *J. Ecol.* 58:487–496.

169   Jones, J. F., and H. Kende. 1979. *Planta* 146:649–656.

170   Jones, L. H. 1963. *Hort. Res.* 3:13–26.

171   Jones, R. 1972a. *J. Ecol.* 60:131–140.

172   Jones, R. 1972b. *J. Ecol.* 60:141–146.

173   Jones, R. 1973. *J. Ecol.* 61:107–116.

174   Jones, R., and J. R. Etherington. 1971. *J. Ecol.* 59:793–801.

175   Kang, B. G. 1979. Physiology of movements. In *Encyclopedia of Plant Physiology*, New Series, Vol. 7, W. Haupt and M. E. Feinleib Eds. Springer-Verlag, pp. 647–667.

176   Kawase, M. 1972a. *J. Am. Soc. Hort. Sci.* 97:584–588.

177   Kawase, M. 1972b. *Proc. Int. Plant Prop. Soc.* 22:360–366.

178   Kawase, M. 1974. *Physiol. Plant* 31:29–38.

179   Kawase, M. 1976. *Physiol. Plant* 36:236–241.

180   Kawase, M. 1978. *Am. J. Bot.* 65:736–740.

181   Kawase, M. 1978. *Ohio Rep., Ohio Agric. Res. Dev. Center.* 63:14–15.

182   Kawase, M. 1979. *Am. J. Bot.* 66:183–190.

183   Kawase, M. 1981. *HortScience.* 16:30–34.

184   Kawase, M., and R. E. Whitmoyer. 1980. *Am. J. Bot.* 67:18–22.

185   Kee, N. S., and C. Bloomfield. 1962. *Plant Soil* 16:108–135.

186   Kennedy, R. A., S. C. H. Barrett, D. Vander Zee, and M. E. Rumpho. 1980. *Plant Cell Env.* 3:243–248.

187   Kleinendorst, A., and R. Brouwer. 1967. *Jaarb. I. B. S.*, pp. 29–39.

188   Kolloffel, C. 1968. *Acta Bot. Neerl.* 17:70–77.

189   Konings, H., and M. B. Jackson. 1979. *Z. Pflanzenphysiol.* 92:385–397.

190   Konze, J. R., and H. Kende. 1979. *Planta* 146:293–301.

191   Kozlowski, T. T. 1976. Soil water measurement, plant responses, and breeding for drought resistance. In *Water Deficits and Plant Growth*. Vol. IV. Academic, New York, pp. 1–55.

192   Kozlowski, T. T., and S. G. Pallardy. 1979. *Physiol. Plant* 46:155–158.

193   Kramcr, P. J. 1940. *Am. J. Bot.* 27:216–220.

194   Kramer, P. J. 1951. *Plant Physiol.* 26:722–736.

195   Kramer, P. J. 1969. *Plant and Soil Water Relationships. A Modern Synthesis.* McGraw-Hill, New York, pp. 1–390.

196   Kramer, P. J., and W. T. Jackson. 1954. *Plant Physiol.* 29:241–245.

197   Krizek, D. T. 1970. Proceedings: Controlled atmospheres for plant growth, ASAE Publ. 270. *Trans. ASAE* 13:237.

198   Krizek, D. T. 1979. Flowering, fruiting, and cut-out of cotton. Pp. 283–291 in *Cotton Physiology—A Treatise*, J. McD. Stewart, Ed. Proc. Beltwide Cotton Prod. Res. Conf., pp. 283–291.

199   Krizek, D. T. 1979. In *Controlled Environment Guidelines for Plant Research*, T. W. Tibbitts and T. T. Kozlowski, Eds. Academic, New York, pp. 241–269.

200   Kulaeva, O. N. 1961. *Fiziol. Rast.* 9:229–239.

201   Labanauskas, C. K., L. H. Stolzy, and M. F. Handy. 1972. *Soil Sci. Soc. Am. Proc.* 36:454–457.

202   Labanauskas, C. K., L. H. Stolzy, L. J. Klotz, and T. A. DeWolfe. 1965. *Soil Sci. Soc. Am. Proc.* 29:60–64.

203   Labanauskas, C. K., L. H. Stolzy, G. A. Zentmyer, and T. E. Szuszkiewicz. 1968. *Plant Soil* 29:391–406.

204   Laing, H. E. 1940. *Am. J. Bot.* 27:574–581.

205   Lal, R., and G. S. Taylor. 1969. *Proc. Soil Sci. Soc. Am.* 33:937–941.

206   Lawton, K. 1945. *Soil Sci. Soc. Am. Proc.* 10:263–268.

207   Lawton, K., and R. L. Cook. 1954. In *Advances in Agronomy*, A. G. Norman, Ed. Vol. VI. Academic, New York, pp. 253–302.

208   Lee, J. A., and H. W. Woolhouse. 1966. *New Phytol.* 65:325–330.

209   Leggett, J. E., and L. H. Stolzy. 1961. *Nature* 192:991–992.

210   Lemon, E. R., and A. E. Erickson. 1952. *Soil Sci. Soc. Am. Proc.* 16:160–163.

211   Letey, J., O. R. Lunt, L. H. Stolzy, and T. E. Szuszkiewicz. 1961. *Soil Sci. Soc. Am. Proc.* 25:183–186.

212   Letey, J., L. H. Stolzy, and G. B. Blank, 1962. *Agron. J.* 54:34–37.

213   Levitt, J. 1972. *Responses of Plants to Environmental Stresses*. Academic, New York.

214   Leyshon, A. J., and R. W. Sheard. 1974. *Can. J. Soil Sci.* 54:463–473.

215   Leyton, L., and L. Z. Rousseau. 1958. In *The Physiology of Forest Trees*, K. V. Thimann, Ed. Ronald Press, New York, pp. 467–475.

216   Lieberman, M. 1979. *Annu. Rev. Plant Physiol.* 30:533–591.

217   Linhart, Y. B., and Irene Baker. 1973. *Nature* 242:275–276.

218   Lungley, D. R. 1973. *Plant Soil* 38:145–159.

219   Lürssen, K., K. Naumann, and R. Schröder. 1979. *Naturwissenschaften* 66:264–265.

220   Luxmoore, R. J., R. A. Fisher, and L. H. Stolzy. 1973. *Agron. J.* 65:361–364.

221   Luxmoore, R. J., R. E. Sojka, and L. H. Stolzy. 1972. *Soil Sci.* 113:354–357. Cited in reference 338.

222   Lynch, J. M. 1972. *Nature* 240:45–46.

223   Lynch, J. M. 1974. *J. Gen. Microbiol.* 83:407–411.

224   Lynch, J. M. 1975. *Nature* 256:576–577.

225   Lynch, J. M., and S. H. T. Harper. 1974. *J. Gen. Microbiol.* 80:187–195.

226   Lynch, J. M., and S. H. T. Harper. 1974. *J. Gen. Microbiol.* 85:91–96.

227   Lynch, J. M., and S. H. T. Harper. 1980. *Soil Biol. Biochem.* 12:363–367.

228   McAlpine, R. G. 1961. *J. Forestry* 59:566–568.

229   McCalla, T. M. 1967. In *Tillage for Greater Crop Production*. Am. Soc. Agric. Eng., St. Joseph, Michigan, pp. 19–25.

230   McCalla, T. M. 1971. In *Biochemical Interactions among Plants*. Washington, D.C., Natl. Acad. Sci., pp. 39–43.

231   McCalla, T. M., and F. L. Dudley. 1950. *Soil Soc. Am. Proc.* 14:196–199.

232   McCalla, T. M., and F. A. Haskins. 1964. *Bacteriol. Rev.* 28:181–207.

233   McCalla, T. M., and F. A. Norstadt. 1974. *Agric. Environ.* 1:153–174.

234   McIntyre, D. S. 1966. *Aust. J. Soil Res.* 4:103–113.

235   McIntyre, D. S. 1970. In *Advance in Agronomy*, N. C. Brady, Ed., Vol. 22. Academic, New York, pp. 235–283.

236   McManmon, M., and R. M. M. Crawford. 1971. *New Phytol.* 70:299–306.

237   McNew, G. L. 1953. In *Plant Diseases, Yearbook of Agriculture*. U.S. Department of Agriculture, Washington, D.C., pp. 100–114.

238   McPherson, D. C. 1939. *New Phytol.* 38:190–192.

239   Mapson, L. W. 1969. *Biol. Rev.* 44:155–187.

240   Marshall, D. R., P. Broué, and A. J. Pryor. 1973. *Nature, New Biol.* 244:16–18.

241   Marshall, D. R., P. Broué, and R. N. Oram. 1974. *J. Heredity* 65:198–203.

242 Martin, M. H. 1968. *J. Ecol.* 56:777–793.

243 Martin, M. H., and C. D. Pigott. 1965. *J. Ecol.* 53:153–155.

244 Mason, A. C. 1950. *Rep. E. Malling Res. Sta.*, pp. 124–126.

245 Meek, B. D., and L. H. Stolzy. 1978. In *Plant Life in Anaerobic Environments*, D. D. Hook and R. M. M. Crawford, Eds. Ann Arbor Science Publishers Inc., Ann Arbor, Michigan, pp. 351–373.

246 Minchin, F. R., and J. S. Pate. 1975. *J. Exp. Bot.* 26:60–69.

247 Minchin, F. R., and R. J. Summerfield. 1976. *Plant Soil* 45:113–127.

248 Minchin, F. R., R. J. Summerfield, A. J. R. Eaglesham, and K. A. Stewart. *J. Agric. Sci. Cambridge* 90:355–366.

249 Mussell, H., and R. C. Staples. 1979. *Stress Physiology in Crop Plants*. Wiley, New York.

250 Nakayama, J. 1950. *Bull. Kyoto Univ. For. No.* 18:97–113. Cited in reference 288.

251 Norris, F. de la M. 1913. *Proc. Bristol Natl. Soc.* IV. 3:134–136.

252 Ohmura, T., and R. W. Howell. 1960. *Plant Physiol.* 35:184–188.

253 Orten, J. M., and O. O. Neuhaus. 1970. *Biochemistry*. 8th ed. Mosby, St. Louis, pp. 243–244.

254 Osborne, D. J. 1978. *Sci. Hort.* 30:1–13.

255 Pallaghy, C. K., and K. Raschke. 1972. *Plant Physiol.* 49:275–276.

256 Parker, J. 1950. *Plant Physiol.* 25:453–460.

257 Patrick, Z. A., and L. W. Koch. 1958. *Can. J. Bot.* 36:621–647.

258 Patrick, Z. A., T. A. Toussoun, and L. W. Koch. 1964. *Annu. Rev. Phytopath.* 2:267–292.

259 Patrick, Z. A., T. A. Toussoun, and W. C. Snyder. 1963. *Phytopathology* 53:152–161.

260 Patrick, W. H., and F. T. Turner. 1968. *9th Int. Congr. Soil Sci. Trans.* Vol. IV. Paper 6.

261 Pearsall, W. H. 1950. *Emp. J. Exp. Agric.* 18:289–298.

262 Pendleton, J. W. 1976. *A.S.P.A.C. Extension Bull.* No. 82, pp. 1–12.

263 Pepkowitz, L., and J. W. Shive. 1944. *Soil Sci.* 57:143–154.

264 Pereira, J. S., and T. T. Kozlowski. 1977. *Physiol. Plant* 41:184–192.

265 Persidsky, D. J., and S. A. Wilde. 1954. *Plant Physiol.* 29:484–486.

266 Phene, C. J., R. B. Campbell, and C. W. Doty. 1976. *Soil Sci.* 122:271–281.

267 Phene, C. J., G. J. Hoffman, and R. S. Austin. 1973. *Trans. ASAE* 16:773–776.

268 Philipson, J. J., and M. P. Coutts. 1978. *New Phytol.* 80:341–349.

269 Philipson, J. J., and M. P. Coutts. 1980. *New Phytol.* 85:489–494.

270 Phillips, I. D. J. 1964a. *Ann. Bot.* 28:17–35.

271 Phillips, I. D. J. 1964b. *Ann. Bot.* 28:37–45.

272 Phillips, I. D. J., and R. L. Jones. 1964. *Planta* 63:269–278.

273 Pierce, M., and K. Raschke. 1980. *Planta* 148:174–182.

274 Piper, C. S. 1931. *J. Agric. Sci.* 21:762–779.

275 Pitman, M. G. 1969. *Plant Physiol.* 44:1233–1240.

276 Pitman, M. G., and D. Wellfare. 1978. *J. Exp. Bot.* 29:1125–1138.

277 Poel, L. W. 1960. *J. Ecol.* 48:165–173.

278 Poljakoff-Mayber, A., and J. Gale. 1975. *Plants in Saline Environments*. Springer-Verlag, New York.

279 Ponnamperuma, F. N. 1972. *Adv. Agron.* 24:24–96.

280 Pradet, A., and J. L. Bomsel. 1978. In *Plant Life in Anaerobic Environments*, D. D. Hook and R. M. M. Crawford, Eds. Ann Arbor Science Publishers, Inc., Ann Arbor, Michigan, pp. 89–118.

281 Primrose, S. B. 1976. *J. Gen Microbiol.* 97:343–346.

282 Purvis, A. C., and R. E. Williamson. 1972. *Agron. J.* 64:674–678.

283 Rai, S. D., D. A. Miller, and C. N. Mittle. 1971. *Agron. J.* 63:331–332.

284 Railton, I. D., and D. M. Reid. 1973. *Planta* 111:261–266.

285 Raschke, K. 1979. Physiology of movements. pp. 383–441 in *Encyclopedia of Plant Physiology*, W. Haupt and M. E. Feinleib, Eds. New Series, Vol. 7, Springer-Verlag.

286 Rawson, H. M. 1979. *Aust. J. Plant Physiol.* 6:109–120.

287 Regehr, D. L., F. A. Bazzaz, and W. R. Boggess. 1975. *Photosynthetica* 9:52–61.

288 Reid, D. M. 1977. In *Physiological Aspects of Crop Nutrition and Resistance*. U. S. Gupta, Ed. Atma Ram & Sons, New Delhi, pp. 252–310.

289 Reid, D. M., and W. J. Burrows. 1968. *Experientia* 24:189–190.

290 Reid, D. M., and D. J. Carr. 1967. *Planta* 73:1–11.

291 Reid, D. M., and A. Crozier. 1971. *J. Exp. Bot.* 22:39–48.

292 Reid, D. M., A. Crozier, and B. M. R. Harvey. 1969. *Planta* 89:376–379.

293 Reid, D. M., and I. D. Railton. 1974. In *Mechanisms of Regulation of Plant Growth*, B. L. Bieleski, A. R. Ferguson, and M. M. Kresswell, Eds. Bulletin 12, The Royal Society of New Zealand, Wellington, pp. 789–792.

294 Rice, W. A., and E. A. Paul. 1971. *Can. J. Microbiol.* 17:1049–1056.

295 Rice, W. A., and E. A. Paul. 1972. *Can. J. Microbiol.* 18:715–723.

296 Rikin, A., D. Atsmon, and C. Gitler. 1979. *Plant Cell Physiol.* 20:1537–1546.

297 Rowe, R. N., and D. V. Beardsell. 1973. *Hort. Abstr.* 43:533–548.

298 Rowe, R. N., and J. B. Catlin. 1971. *J. Am. Soc. Hort. Sci.* 96:305–308.

299 Russell, E. J., and A. Appleyard. 1915. *J. Agric. Sci., Cambridge.* 7:1–48.

300 Russell, E. W. 1973. *Soil Conditions and Plant Growth.* 10th ed., Wiley, New York.

301 Russell, M. B. 1952. Soil physical conditions and plant growth. In *Agronomy*, Vol. 2, B. T. Shaw, Ed. New York, pp. 253–301.

302 Russell, R. S. 1977. *Plant Root Systems: Their Function and Interaction with the Soil.* McGraw-Hill, New York, pp. 90–112.

303 Russell, R. S., and V. M. Shorrocks. 1959. *J. Exp. Bot.* 10:301–316.

304 Sachs, M. M., and M. Freeling. 1978. *Molec. Gen. Genet.* 161:111–115.

305 Salter, P. J., and J. E. Goode. 1967. *Res. Rev. No. 2*, Commonwealth Agric. Bureaux, Farnham Royal, Bucks, England.

306 Samuels, G. 1972. *J. Agric. Univ. Puerto Rico.* 50:81–84.

307 Sanchez, P. A., G. E. Ramirez, and M. V. de Calderon. 1973. *Agron. J.* 65:523–529.

308 Scott, A. D., and D. D. Evans. 1955. *Proc. Soil Sci. Soc. Am.* 19:7–16.

309 Selman, I. W., and S. Sandanam. 1972. *Ann. Bot.* 36:837–848.

310 Shackel, K. A., and A. E. Hall. 1979. *Aust. J. Plant Physiol.* 6:265–276.

311 Shalhevet, J., H. Enoch, and S. Dasberg. 1969. *Israel J. Agric. Res.* 19:161–170.

312 Shalhevet, J., and P. J. Zwerman. 1958. *Soil Sci.* 85:255–260.

313 Shalhevet, J., and P. J. Zwerman. 1962. *Soil Sci.* 93:172–182.

314 Shoulders, E., and C. W. Ralston. 1975. *Forest Sci.* 21:401–410.

315 Shriener, O., and E. C. Shorey. 1909. *U.S. Dep. Agric. Bur. Soils Bull.* 53:53.

316 Sifton, H. B. 1945. *Bot. Rev.* 11:108–143.

317 Sij, J. W., and C. A. Swanson. 1973. *Plant Physiol.* 51:368–371.

318 Skinner, F. A. 1975. *In Soil Microbiology*, N. Walker, Ed. Butterworths, London, pp. 1–19.

319   Slavik, B. 1965. *Water Stress in Plants*. Dr. W. Junk Publishers, The Hague.

320   Slowik, K., C. K. Labanauskas, L. H. Stolzy, and G. A. Zentmyer. 1979. *J. Am. Soc. Hort. Sci.* 104:172–175.

321   Smith, A. M. 1976. *Annu. Rev. Phytopath.* 14:53–73.

322   Smith, A. M., and R. J. Cook. 1974. *Nature* 252:703–705.

323   Smith, K. A., and R. J. Dowdell. 1973. *J . Chromatog. Sci.* 11:655–658.

324   Smith, K. A., and R. J. Dowdell. 1974. *J. Soil Sci.* 25:217–230.

325   Smith, K. A., and S. W. F. Restall. 1971. *J. Soil Sci.* 22:430–443.

326   Smith, K. A., and P. D. Robertson. 1971. *Nature* 234:148–149.

327   Smith, K. A., and R. S. Russell. 1969. *Nature* 222:769–771.

328   Snow, L. M. 1905. *Bot. Gaz.* 40:12–48.

329   Sojka, R. E., and L. H. Stolzy. 1980. *Soil Sci.* 130:350–358.

330   Spooner, A. E. 1961. *Arkansas Agric. Exp. Sta. Bull.* 644:1–27.

331   Sprent, J. I. 1971. In *Plant and Soil*, Spec. Vol., T. A. Lie and E. G. Mulder, Eds. pp. 225–228.

332   Sprent, J. I. 1972. *New Phytol.* 71:603–611.

333   Starkey, L. R. 1965. *J. Soil Sci.* 101:297–306.

334   Stevenson, F. J. 1967. In *Soil Biochemistry,* A. D. McLaren and G. H. Peterson, Eds. Arnold, London, pp. 119–146.

335   Stevenson, K. R., and R. H. Shaw. 1971. *Agron. J.* 63:327–329.

336   Steward, F. C., W. E. Berry, and T. C. Broyer. 1936. *Ann. Bot.* 50:1–22.

337   Stewart, W. D. P. 1966. *Nitrogen Fixation in Plants*. Univ. Lond., Athlone Press, pp. 1–137.

338   Stolzy, L. H. 1971. In *The Plant Root and Its Environment,* E. W. Carson, Ed. Univ. of Virginia Press, Charlottesville, pp. 335–361.

339   Stolzy, L. H., J. Letey, T. E. Szuskiewicz, and O. R. Lunt. 1961. *Proc. Soil Sci. Soc. Am.* 25:463–467.

340   Stone, J. F. 1975. *Plant Modification for More Efficient Water Use*. Elsevier, New York. (Reprinted from *Agric. Meteor.* Vol. 14. 1974.)

341   Sturgis, M. B. 1936. *Bull. Louisiana Agric. Exp. Stn.* 271:1–37.

342   Swartz, G. L. 1966. *Queensland Agric. J.* 23:271–277.

343   Tal, M., D. Imber, and C. Itai. 1970. *Plant Physiol.* 46:367–372.

344   Taylor, S. A., and G. L. Ashcroft. 1972. *Physical Edaphology—The Physics of Irrigated and Non-Irrigated Soils*. Freeman, San Francisco, pp. 1–533.

345   Thorton, R. K., and R. L. Wample. 1980. *Plant Physiol.* 65:S–29. (Abstr.)

346   Tousson, T. A., A. R. Weinhold, R. G. Linderman, and Z. A. Patrick. 1968. *Phytopathology* 58:41–45.

347   Tovey, R. 1964. Trans. Am. Soc. Agric. Eng. 7:310–312.

348   Trafford, B. D. 1975. *Tech. Bull.* 29. Ministry of Agriculture, Fisheries and Food. HMSO, London, pp. 417–433.

349   Treshow, M. 1970. In *Environment and Plant Response*. McGraw-Hill, New York, pp. 153–174.

350   Trought, M. C. T., and M. C. Drew. 1980. *Plant Soil* 54:77–94.

351   Troughton, A. 1972. *Plant Soil* 36:93–108.

352   Turkova, N. S. 1944. *Comp. Rend. (Doklady) Acad. Sci. USSR* 42:87–90.

353   Turner, F. T., and W. H. Patrick. 1968. *Trans. Ninth Int. Congr. Soil.* 4:53–65.

354 Turner, N. C. 1979. In *Stress Physiology in Crop Plants*, H. Mussel and R. C. Staples, Eds. Wiley, New York, pp. 343–372.

355 Turner, N. C., and P. J. Kramer. 1980. *Adaptation of Plants to Water and High Temperature Stress*. Wiley, New York.

356 Tyler, P. D., and R. M. M. Crawford. 1970. *J. Exp. Bot.* 21:677–682.

357 Valoras, N., J. Letey, L. II. Stolzy, and F. F. Frolich. 1964. *Proc. Am. Soc. Hort. Sci.* 85:172–178.

358 Vámos, R. 1964. *J. Soil Sci.* 15:103–109.

359 Vámos, R., and E. Köves. 1972. *J. Appl. Ecol.* 9:519–525.

360 van Bavel, C. H. M. 1972. In *Optimizing the Soil Physical Environment Toward Greater Crop Yields*, D. Hillel, Ed. Academic, New York, pp. 23–33.

361 Vartapetian, B. B. 1978. In *Plant Life in Anaerobic Environments*, D. D. Hook and R. M. M. Crawford, Eds. Ann Arbor Science Publishers, Inc., Ann Arbor, Michigan, pp. 1–11.

362 Vartapetian, B. B., R. Bazie, and C. Costes. 1978. In *Plant Life in Anaerobic Environments*, D. D. Hook and R. M. M. Crawford, Eds. Ann Arbor Science Publishers, Ann Arbor, Michigan, pp. 539–548.

363 Vester, G. 1972. Ph.D. dissertation. University of Munich.

364 Waksman, S. A. 1932. *Principles of Soil Microbiology*. 2nd Ed. Williams and Wilkins. Cited in reference 288.

365 Walton, D. C. 1980. *Annu. Rev. Plant Physiol.* 31:453–489.

366 Wample, R. L., and D. M. Reid. 1978. *Physiol. Plant* 44:351–358.

367 Wample, R. L., and D. M. Reid. 1979. *Physiol. Plant* 45:219–226.

368 Wang, T. S. C., S-Y. Cheng, and H. Tung. 1967. *Soil Sci.* 104:138–144.

369 Watt, A. S. 1979. *New Phytol.* 82:769–776.

370 Weatherly, P. E. 1969. In *Ecological Aspects of the Mineral Nutrition of Plants*, I. H. Rorison, Ed. Blackwell Sci. Publ., Oxford, pp. 323–340.

371 Went, F. W. 1938. *Plant Physiol.* 13:55–80.

372 Went, F. W. 1943. *Plant Physiol.* 18:51–65.

373 Whigham, D. K., and H. C. Minor. 1978. In *Soybean Physiology, Agronomy and Utilization*, A. G. Norman, Ed. Academic, New York, pp. 77–118.

374 Whitlow, T. H., and R. W. Harris. 1979. *Environmental and Water Quality Operational Studies*. Tech. Rep. E-79-2, U. S. Army Corps of Engineers, Washington, D. C.

375 Wien, C., R. Lal, and E. Pulver. 1979. In *Soil Physical Properties and Crop Production in the Tropics*, R. Lal and D. J. Greenland, Eds. Wiley, New York, pp. 235–245.

376 Wilkins, M. B., and P. Whyte. 1968. In *Biochemistry and Physiology of Plant Growth Substances*, F. Wightman and G. Setterfield, Eds. Runge Press, Ottawa, pp. 1051–1062.

377 Williams, W. T., and D. A. Barber. 1961. *Symp. Soc. Exp. Biol.* 15:132–144.

378 Williamson, R. E. 1964. *Proc. Soil Sci. Soc. Am.* 28:86–90.

379 Williamson, R. E. 1970. *Agron. J.* 62:80–83.

380 Williamson, R. E., and G. J. Kriz. 1970. *Trans. Am. Soc. Agric. Eng.* 13:216–220.

381 Wong, S. C., I. R. Cowan, and G. D. Farquhar. 1979. *Nature* 282:424–426.

382 Wright, S. T. C. 1972. In *Crop Processes in Controlled Environments*, A. R. Rees, K. E. Cockshull, D. W. Hand, and R. G. Hurd, Eds. Academic, London, pp. 349–361.

383 Wright, S. T. C., and R. W.-P. Hiron. 1970. In *Proc. 7th Int. Conf. Plant Growth Subst.*, D. J. Carr, Ed. Springer-Verlag, Berlin, pp. 291–298.

384 Yamasaki, T. 1952. *Bull. Natl. Inst. Agric. Sci. (Japan)* B, 1:1–92. (Japanese with English summary)

385   Yang, S. F. 1980. *HortScience.* 15:238–243.

386   Yang, S. F., D. O. Adams, C. Lizada, Y. Yu, K. J. Bradford, A. C. Cameron, and N. E. Hoffman. 1980. In *Proceedings in Life Sciences,* F. Skoog, Ed. Univ. Wisconsin, Madison, pp. 219–229.

387   Yelonosky, G. 1964. *Proc. Intl. Shade Tree Conf.* 40:127–147.

388   Yoshida, S., and T. Tadano. 1978. In *Crop Tolerance to Suboptimal Land Conditions,* G. A. Jung, Ed. Am. Soc. Agron., Crop Sci. Soc. Am., Soil Sci. Soc. Am., Madison, Wisconsin, pp. 233–256.

389   Yoshida, T. 1975. In *Soil Biochemistry,* E. A. Paul and A. D. McLaren, Eds. Marcel Dekker, New York, Vol. 3, pp. 83–122.

390   Yu, P. T., L. H. Stolzy, and J. Letey. 1969. *Agron. J.* 61:844–847.

391   Yu, Y. B., D. O. Adams, and S. F. Yang. 1979a. *Plant Physiol.* 63:589–590.

392   Yu, Y. B., D. O. Adams, and S. F. Yang. 1979b. *Arch. Biochem. Biophys.* 198:280–286.

393   Yu, Y. B., and S. F. Yang. 1979. *Plant Physiol.* 64:1074–1077.

394   Zeroni, M., P. H. Jerie, and M. A. Hall. 1977. *Planta* 134:119–125.

395   Zimmerman, P. W., and A. E. Hitchcock. 1933. *Contrib. Boyce Thompson Inst.* 5:351–369.

PLANT PEST INTERACTION WITH ENVIRONMENTAL STRESS AND BREEDING FOR PEST RESISTANCE:

Plant Diseases

ALOIS A. BELL

U.S. Department of Agriculture, Science and Education Administration, Agricultural Research, National Cotton Pathology Research Laboratory, College Station, Texas

FACTORS AFFECTING DISEASE RESISTANCE

Disease resistance is a highly variable quality that depends on the specific host, pathogen, environment, and time interval in which these factors interact. Failure to recognize the interactions may frustrate efforts to develop stable, useful resistance, particularly for marginal environments. This chapter is concerned primarily with the effects of environmental variations on the resistance of plants to disease. To better appreciate these effects, however, we first must understand the various ways in which the host and pathogen, as well as environment, may alter resistance.

Effects of the Host

The resistance of a plant to disease depends on its genotype, age, and physiological condition (determined largely by environment). The resistance of a plant population to a disease epidemic further depends on the density and distribution plus the genetic and age variation of individual plants. Resistance refers to the ability of plants to prevent, restrict, or retard disease development. Plants may also be bred to escape disease. In this case, the morphology of the plant is altered to make the microenvironment less favorable for disease, or the developmental pattern of the plant is changed to shorten the time interval in which the pathogen and host may interact in an environment favorable for disease. Both resistance and escape can be enhanced by breeding for better adaptation to environmental extremes. For example, the potato cultivar 'Epicure' which can grow just above freezing temperatures (2°C) in Scotland is able to escape or resist infections by nematodes and *Phytophthora* for many weeks because the pathogens are largely inactive below 7 and 10°C, respectively (61).

Among cultivars within a species, there is frequently a wide array of resistance levels. Among related species or genera that can be hybridized, there usually is even greater variation (131, 108). The genes responsible for such differences in resistance are generally categorized into two types that I designate as specific and general. Specific resistance is effective against some but not all races of the pathogen, whereas general resistance is similarly, though not identically, effective against all races. Specific vs. general resistance also has been designated in the literature as vertical vs. horizontal, race-specific vs. race-nonspecific, monogenic vs. polygenic, nonfield vs. field, nondurable vs. durable, hypersensitive vs. nonhypersensitive, and major vs. minor. For this chapter, the reader needs to recognize that general resistance is usually more sensitive to environmental changes than specific resistance, but both types can be modified.

The resistance of a plant, or plant part, changes sequentially during growth and development (19, 119). For example, roots and hypocotyls are most sensitive to infection by facultative parasites when young but then become progressively more resistant with age. In contrast, leaves reach a peak in resistance when young, often before full expansion, and then progressively become more susceptible with age. The plant as a whole often is most resistant during middle age before extensive growth of reproductive structures (seeds, stolons, rhizomes, etc.) occurs. Environmental stresses may either increase or decrease overall resistance to disease by altering the pattern of sequential changes associated with ontogeny.

Failure to recognize ontogenic changes in resistance and their interaction with environment can lead to mistaken conclusions about levels of resistance in different cultivars. For example, determinant cultivars are usually considered more susceptible to foliar disease than indeterminate ones when they are compared on the same calendar date. If, instead, they are compared at the same stage of reproductive development under the same environment, the determinant cultivars may in fact have equal or greater resistance. The determinant cultivar may suffer the least from diseases favored by environmental conditions occurring late in the growing season. In this case, however, the difference is most likely due to disease escape. Thus, it is imperative that we recognize the individual contributions of genotype, ontogeny, and environment to disease severity in the field.

Effects of Pathogen

The pathogen population affects resistance to disease through its genetic potential for virulence, inoculum density, and physiological condition (determined largely by environment). Virulence is defined as the ability to overcome host resistance. Within pathogenic species there frequently is considerable variation in the specificity and degree of virulence shown by different isolates of the fungus to different host plants. Fungal isolates that differ in the specificity of their virulence to differential cultivars are called *physiological races*, whereas those that differ only in degree of virulence to all cultivars are called *races*. Isolates of a pathogenic species that vary in their specific virulence to different host species are usually designated as *formae speciales* (f. sp.), although the term "race" has been used sometimes when the plant species were closely related. Isolates should not be classified into any of the above subdivisions unless the differences in virulence can be clearly and reproducibly shown.

Physiological races should be clearly distinguished from races when breeding for disease resistance. Resistance to any given physiological race often is pronounced and not greatly influenced by environment, but such resistance often is of limited use in the field because of the build up of another virulent physiological race that was previously a minor part of the population or arose by mutation. Thus, the breeder must screen for resistance against all known physiological races in his area. In contrast, resistance to a race generally is more stable but under greater influence of environment. Thus, the breeder needs only to screen against the most virulent race to obtain stable resistance, but the resistance may vary in its effectiveness in different environments.

Resistance generally is inversely related to inoculum concentration (9). This relationship is evident particularly with soilborne pathogens, such as the wilt fungi, that may build to high population densities in some soils. The degree to which increases of inoculum density affect resistance depends on the specific host and pathogen and on the specific inoculum propagule. Low levels of resistance are overcome more easily than high levels of resistance. Thus, the polygenic intermediate resistance of potato to *Verticillium* wilt is overcome easily by increased inoculum, but the strong single gene resistance of tomato to *Fusarium* wilt is hardly affected by even the highest inoculum levels in natural soils. Increased numbers of chlamydospores or microsclerotia have a much greater effect on resistance than increased numbers of conidia. The interactions of the inoculum density of wilt fungi and resistance have been reviewed in detail by Bell and Mace (20). Rotem et al. (129) have reviewed the effects of host and environment on production of inoculum by various fungi.

Effect of Environment

The effects of environmental factors on disease development are extremely complex and, consequently, difficult to predict. A single environmental factor may alter the growth rate or metabolic capabilities of the pathogen or host, or it may alter other environmental factors that affect disease development. For example, drought invariably decreases bacterial antagonists around plant roots, retards pesticide breakdown, increases soil temperature, and usually is associated with greater light exposure. An environmental stress causes certain specific effects that aggravate the disease and others that alleviate it. Thus, the overall effect of an environmental stress is often the sum of numerous positive and negative specific effects.

Predicting the effects of environmental stress on disease is further complicated by the fact that a stress factor applied before inoculation may have an entirely different effect than the same factor applied after inoculation. For example, moisture stress before inoculation may increase resistance to *Verticillium* wilt, but it usually aggravates wilt when it occurs after inoculation. From an epidemiological perspective (23), an environmental stress may affect: (*a*) survival of inoculum, (*b*) germination of inoculum, (*c*) penetration and establishment of the pathogen, (*d*) rate and extent of infection, or (*e*) the rate and extent of sporulation (or reproduction). The term "predisposition" refers to changes in disease resistance that result from environmental factors varied before inoculation. Schoeneweiss (134) and Yarwood (159) have recently reviewed predisposition of plants to disease by environmental stress factors, while other reviews (42, 160) have described the general effects of environment on disease.

RESISTANCE ALTERATION BY SPECIFIC STRESSES

Temperature

Theoretical Considerations.   The resistance of plants to disease depends on active defense mechanisms that function after the pathogen or its metabolites make contact with the host (19, 17, 18). These active mechanisms include: release of constitutive antibiotics from vacuoles, hydrolysis of nontoxic glycosides to form antibiotic aglycones, synthesis of new antibiotics (phytoalexins), synthesis of enzyme-denaturing polyphenols, formation of morphological barriers, and detoxification of microbial-produced phytotoxins. Resistance is directly proportional to the speed of the active defense responses and inversely related to the speed of colonization of the pathogen in the host tissue. Thus, a simple formula for estimating resistance is as follows:

$$\text{resistance (R)} = \frac{\text{speed of active host resistance (HR)}}{\text{speed of pathogen colonization (PC)}}$$

In attempting to understand or predict how temperature may affect resistance, we need to know how temperature will affect HR and PC, and therefore the HR/PC ratio. In Figure 1, the theoretical effects of temperature

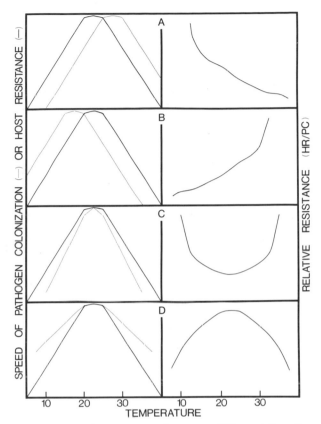

Figure 1 A model showing how temperature may differentially affect the speed of pathogen colonization (PC) and on the speed of active host resistance (HR), and thereby cause changes in relative resistance (HR/PC). Note that in each case, a temperature extreme (stress) causes a marked increase or decrease in resistance.

variations on the speed of active resistance in a single host (HR curve) has been compared with the speeds of colonization (PC curves) of four theoretical pathogens with different adaptations to temperature. According to the models (A–D), temperature extremes (stress) should cause a marked increase or decrease in resistance (R). The direction of the change will depend on the specific influences of the temperature changes on the pathogen relative to those on the host. Two observations are worthy of emphasis: (*a*) resistance (R) is rarely (only in Model D) maximum at the temperature optimal for active host resistance (HR), and presumably for plant growth; and (*b*) resistance (R) may change markedly with temperature extremes even though the plant and pathogen have the same optimal mean for active host resistance (HR) and colonization (PC), respectively (Model C and D).

Examples of diseases that follow the behavior predicted by models A, B, and C are given in the following three sections. Examples of model D are rarely encountered and therefore not discussed.

Resistance Decreases at High Temperatures (Model A). Crops adapted to cool temperate climates frequently show a decrease in disease resistance as temperature rises from 15 to 25°C. The resistance of peas to *Pythium* and *Fusarium solani* decreases at 18 and 24°C compared to 13°C (88), and resistance to *Fusarium oxysporum* decreases progressively from 21 to 24 to 27°C compared to 18°C (155). Similar progressive losses in resistance to *Fusarium* wilt with increasing temperature have been shown for cabbage, flax, safflower, and spinach. Cereals, such as oats (99) and wheat (79), show a progressive decrease in resistance to stem rust as temperature is increased from 20 to 25 to 30°C. Likewise, elms progressively lose resistance to Dutch elm disease (27) and carnations lose it to *Fusarium* stub dieback (147) as temperature is increased from 16 to 21 to 26° and 13 to 18 to 24°C, respectively. In most of the above cases the specific increase in temperature needed to overcome resistance is directly related to the level of genetic resistance in the host.

Resistance to certain pathogens almost always breaks down at high temperatures (30–45°C), regardless of the host, because of the pathogens unusual adaptation to high temperatures, which is usually coupled with an adaptation to moisture stress. *Macrophomina phaseoli,* the cause of charcoal rot of corn, cotton, pine, sorghum, soybeans, and other crops, is a classical example of this type of pathogen. Most plants under moisture stress progressively lose resistance to *Macrophomina* as temperature is increased from 30 to 35 to 40°C. At 40°C, the fungus grows optimally when the water potential is −13 to −40 bars (112); greater availability of water actually reduces growth of the fungus. Fluctuating temperatures may facilitate breakdown of resistance more rapidly than constant ones. When a mean temperature of 30°C was applied with 0, 4, and 8°C fluctuations to sugar pine, there was a progressive decrease in resistance to charcoal rot as the amplitude of the fluctuations was increased (143). Other fungi that cause little or no disease at 15–20°C but overcome resistance at 30°C or above, especially with low water potential, include *Rhizoctonia* (12), *Aspergillus* (5), *Botryodiplodia* (125, 97), and *Fusarium roseum* (44). *Rhizopus* species also overcome resistance at high temperatures but are favored by moist conditions.

Few studies have focused on the effects of temperature on resistance to bacteria. Resistance to *Pseudomonas solanacearum* in tomato progressively breaks down as temperatures are increased from 22 to 26 to 30 to 32°C (104). Three levels of resistance among tomato cultivars can be distinguished by the response to temperature; the temperature needed to break resistance increases as the level of cultivar resistance to the pathogen increases. Because of this difference, resistant tomato cultivars can be planted earlier in the fall in Florida than the more susceptible cultivars (144).

Warm temperatures before inoculation may also predispose plants to susceptibility at temperatures that normally allow resistance. For example, cotton seeds held in a moist atmosphere at 50°C for several days before planting become much more susceptible to seed rot and damping-off by many different fungi (31). Yarwood (159) has reviewed numerous examples of the breakdown of resistance to bacterial, fungal, and viral diseases caused by immersing stems and leaves in water at 40–55°C for intervals varying from a few minutes to several hours. Such high temperatures both slow defense reactions and increase exudation of amino acids by plant cells, thus favoring pathogen colonization. Loss of resistance caused by heat is often temporary, and normal resistance is reacquired after several days at low temperatures.

Temperatures highly favorable for plant growth may also predispose plants to disease more than cooler or widely fluctuating temperatures. For example, barley and wheat grown under ideal conditions for rapid growth in the greenhouse are more susceptible to mildew than comparable plants grown outdoors during cool weather. Likewise, clover grown at day/night temperatures of 30/25°C or 33/28°C before inoculation is more susceptible to *Kabtiella* (scorch) than plants grown at lower temperatures (67).

Resistance Decreases at Low Temperature (Model B). Crops adapted to warm tropical and subtropical climates, such as beans, canteloupe, corn, cotton, eggplant, pepper, poinsettia, loblolly pine, sugar cane, sorghum, soybeans, and yams, often are highly resistant to disease at 35–30°C but show progressive losses of resistance at lower temperatures. Cotton seedlings progressively lose resistance to seedling diseases caused by *Colletotrichum, Asochyta, Pythium,* and *Thielaviopsis* as temperatures are decreased from 30 to 27 to 24 to 21°C (4). Loss of resistance to the highly virulent *Pythium ultimum* becomes evident below 27°C (4), while plants remain resistant to the less virulent *Pythium irregulare* until temperatures are decreased below 23°C (127). Resistance to highly virulent forms of *Verticillium dahliae* in older cotton plants decreases markedly as temperatures are dropped by 2° intervals from 31° to 23°C (14, Figure 2). The specific temperature at which resistance begins to decrease depends on the level of cultivar resistance. Decreased resistance in susceptible, tolerant, and resistant cultivars is apparent first, with dropping temperatures, at 29, 27, and 25°C, respectively. Resistance in susceptible and tolerant cultivars to less virulent strains of *V. dahliae* also may be broken, but only by lower temperatures (10). Similar, though less striking, decreases of resistance with decreasing temperatures occur in soybean to *Phytophthora* root rot (87), loblolly pine to fusiform rust (91), clover to *Kabatiella* scorch (67), poinsettias to

Figure 2   Symptoms of *Verticillium* wilt in susceptible 'Stardel' (S), tolerant 'Acala 4-42-77' (T), and resistant 'Seabrook Sea Island 12B2' (R) cotton plants at mean temperatures of 25, 27, and 29°C. [Reproduced from *Phytopathology* 69:1141 (1969)].

Thielaviopsis root rot (12), beans to *Fusarium* wilt (53), and pepper (84) and eggplant (109) to *Verticillium* wilt.

   Resistance to certain host-specific toxins is lost progressively as temperatures decrease below 30 to 35°C. For example, sugarcane grown at 30 and 35°C, but not at lower temperatures, is resistant to the host-specific toxin helminthosporide, produced by *Helminthosporium sacchari* (36), and disease develops only at the lower temperatures. Sorghum grown at day/night temperatures of 37/30°C is insensitive to *Periconia* toxin, and resistant to the disease (33). Toxin-sensitive leaves of sorghum grown at 22°C become more resistant to the toxin after treatment at 35, 40 and 45°C for only 15, 4.5, and 0.8 min. Susceptibility to the toxin then returns after 3 days at 22°C. Oats also are more resistant to the *Helminthosporium victoriae* toxin at high than at low temperatures (33).

In some cases, increases of resistance to disease at 20°C and above are due mostly to the low temperature adaptation of the pathogen. Examples of such pathogens include *Pythium* species and *Puccinia striiformis*. Beans (116), cotton (77), and poinsettias (12) are most susceptible to *Pythium ultimum* at 15–18°C and become progressively more resistant at higher temperatures; all are highly resistant at 26–27°C. Shifting plants from 15 to 27°C quickly increases resistance and reduces disease loss. Resistance of wheat to stripe rust increases quickly as night temperatures are increased above 15°C (137). Germination of spores and infection are optimal at 2–7°C, and appressoria formation is best at 7–15°C. Increases in temperature during the day, up to 32°C, had little effect on resistance as long as night temperatures were below 15°C. Night temperatures above 15°C before inoculation also caused plants to have higher levels of resistance.

Susceptibility to diseases caused by viruses and related agents often is restricted to low temperatures. Wheat streak mosaic virus is readily transmissible at 10°C, but transmissibility is lost after plants are held 1 week at 20°C (141). Likewise, lettuce is highly susceptible to big vein disease at an air temperature of 14°C, but is resistant at 24°C; results are the same whether root temperatures are 14 or 24°C (156).

Cold applied before inoculation often causes plants to become more susceptible to disease. Rooted, but not unrooted, carnation cuttings are predisposed by cold storage to *Fusarium* root rot (147). Germinating cottonseed exposed to 10°C for several days show increased severity of seedling diseases (101) and exudation of sugars and amino acids (40, 65). Both characters increase progressively with degree of cold and length of exposure. 'Miragreen' pea seeds soaked at 10°C, compared with 20°, show enhanced seedling disease and greater magnitude and duration of nutrient exudation (88). 'Alaska' peas do not show the enhanced exudation or seedling disease.

Freezing injury also predisposes to susceptibility, especially if plants are not "cold hardened." For example, freezing temperatures of −10 to −30°C may cause white birch to become susceptible to *Botryosphaeria* cankers (65). Susceptibility increases progressively as predisposing temperatures are decreased from −10 to −20 to −30°C. Other examples of predisposition by cold are given in Schoeneweiss (134).

Resistance is Least at Moderate Temperatures (Model C). This pattern of resistance change is generally shown by plants adapted for growth over a wide range of temperatures, but favored by moderate temperatures. Optimum growth of the pathogen in each case also occurs at a moderated temperature. Tomato shows this pattern of resistance in response to bacterial canker (58) and *Fusarium* wilt (2). In each case plants are most susceptible at 23–28°C and resistance increases at 17–21°C and 29–32°C. The increase in resistance to *Fusarium* wilt at the high temperatures is shown by tolerant and resistant cultivars but not by highly susceptible ones. Predisposition to *Fusarium* wilt is also greatest at moderate temperatures and decreases as predisposing temperatures increase or decrease.

Other crops showing increased resistance as temperatures increase or decrease from the optimum for plant growth include corn, peanut, walnut, and barley. Resistance of corn to anthracnose is least at 30° and increases at 33°C or 28°C (95). Peanut has the least resistance to *Cylindrocladium* black rot at 25°C and is more resistant at 20 and 30°C; no disease develops at 35°C (115). Walnut resistance to *Gnomonia* is the least at 21°C, increases at 27°, and is complete at 32°C; symptoms are greatly delayed at 15 and 10°C (29). The susceptibility of barley to *Helminthosporium gramineum* is greatest at 14 or 22°C, depending on the cultivar, and resistance increases at 6 or 30°C (121). Resistance increases more when the plants are subjected to 6°C at night, compared to daytime, under diurnal fluctuations of temperature.

Moisture

Moisture Requirements of Pathogens. Fungal pathogens have critical moisture requirements for the production, survival, and germination of infective propagules, such as spores or sclerotia. Moisture levels are also critical for fungal and bacterial infection of host tissue. Inadequate moisture may reduce inoculum potential and percentages of infection because of reduced sporulation and increased death of microbial cells from desiccation. Excessive moisture may reduce infection by stimulating growth of bacterial antagonists that destroy the pathogens. In these ways moisture can indirectly affect resistance by changing inoculum potential or modifying the physiology of the pathogen.

Moisture requirements of sporulation vary considerably among pathogens. Obligate fungal parasites generally sporulate readily even at moderate or low relative humidities, whereas facultative pathogens that cause extensive tissue necrosis may require several days of free moisture for optimal sporulation. The optimal wet period then depends on temperature. For example, the optimal wet periods for sporulation of *Rhynchosporum secalis* in barley is 72, 72, 48, 24, 16, and 12 hr. at 5, 10, 15, 20, 25, and 28°C, respectively (128). When these wet periods are exceeded, lysis of spores increases markedly, probably because of bacterial antagonists. Soilborne fungal pathogens, such as *Fusarium, Verticillium, Macrophomina, Thielaviopsis,* and *Rhizoctonia,* generally form spores and sclerotia optimally at intermediate water potentials (−2 to −40 bars). High water potentials

(0 to −1 bars) are generally more deterimental to sporulation and particularly to survival than very low ones (−40 to −80 bars).

Moisture requirements for spore germination vary for different fungi. Spores of the powdery mildew fungus, *Leveillula taurica*, germinate and infect hosts at relative humidities only slightly above zero (110). Consequently, the fungus is a major pathogen in warm arid climates. Most pathogens, however, require at least several hours near 100% relative humidity or free moisture to germinate and infect their hosts.

The frequencies of infection often are directly related to frequency and duration of wet periods. For example, the incidence of *Guignardia* (black rot) on grape is directly correlated with numbers of rains, duration of rains, amount of rain, and hours of wetness following rain (57). High relative humidity or sprinkler irrigation, for example, with *Pyrenophora* in barley (103), or *Drechslera* in bluegrass (62), may also greatly increase frequencies of infection. *Fusarium* stub dieback of carnation is more severe with intermittent high humidity than with continuous humidity (147). Cultural practices that increase canopy density, and therefore humidity within the canopy, increase frequencies of infection by *Sclerotinia* in bean (30) or by *Cercospora* in celery (22). The length of wet or humid periods required for infection generally increase as temperature decreases.

Resistance Decreases with Wetness.   Most studies fail to distinguish whether moisture affects survival of the pathogen prior to infection or resistance of the host after infection. However, a few studies clearly show a decrease in resistance because of wetness.

The resistance of wheat to at least six fungal leaf spots declines progressively as the duration of the wet period is increased. For example, both the number of *Septoria* lesions and area of necrosis per lesion increase progressively with extension of the wet periods (56). Resistant cultivars showed less breakdown of resistance from wet periods than susceptible ones. Increasing the period of wetness from 48 to 121 hr also caused progressive breakdown of resistance to *Leptosphaeria* leaf spot in wheat, barley, and rye (72). The length of the wet period required to break resistance again was directly proportional to the level of cultivar resistance.

Some soilborne pathogens, such as *Pythium* and *Phytophthora*, are well suited for growth in free water and, in fact, form motile zoospores in their life cycle. Resistance to such pathogens often decreases markedly as water potentials in soil increase to near zero either as a result of rainfall or irrigation. Examples include *Pythium* root rot of poinsettia (13), pea (88, 138), and bean (116), and *Phytophthora* root rot of avocado (148), alfalfa (122), and soybean (87). Fluctuations in soil water potential of −12 to 0, −5 to 0, and −1 to 0 bars progressively decreased resistance of bean to *Pythium* as the range of fluctuation was narrowed. Accordingly, *Pythium* caused only slight damage to poinsettia roots at 30–40% moisture-holding capacity

(MHC); disease severity increased only slightly up to 70% MHC; then it increased quickly as moisture levels were increased above 70% MHC. Incidence of *Phytophthora* root rot in avocado increased from 4–8% at a matric potential of -0.25 bars to 50–100% at -0.10 bars.

High water potentials markedly increase sugar exudation by pea roots, which may stimulate *Pythium* colonization (85). Such exudation at similar water potentials is greater in loam than in sandy soils and in soils with high compared with low bulk densities.

Resistance to some diseases is greater at low than at moderate or high soil water contents. For example the percentage of mortality in chick-pea caused by *Fusarium* wilt was 83, 37, 27, and 13% at 25, 20, 15, and 10% soil moisture (39). Other diseases of this type include *Cylindrocladium* black rot of peanuts (115), *Fusarium* root rot of beans (106), *Thielaviopsis* root rot of poinsettia (13), *Verticillium* wilt of guayule (133) and powdery mildew of barley (6). Delaying irrigation until soil moisture was reduced to the wilting point helped maintain resistance to the diseases caused by *Verticillium* and *Cylindrocladium*. Excessive irrigation enhanced the loss of resistance in bean to the *Fusarium* root rot. The effect of excessive irrigation could be duplicated by depriving bean roots of oxygen for 24–72 hr at moderate moisture levels. Thus, the major effect of high moisture levels may be to reduce the availability of oxygen to roots.

Resistance Decreases with Dryness.   The effects of moisture stress on plant diseases have been reviewed by Cook (43), and predisposition, particularly of trees, by moisture stress is reviewed by Schoeneweiss (134).

Fungal pathogens that are favored by high temperatures generally also require moisture stress before they can readily attack the plant. Thus, *Macrophomina* can attack only drought-stressed sorghum plants (112), and *Botryodiploida* kills only drought-stressed sycamores (96). Resistance of white birch to *Botryosphaeria* canker is stable until the water potential of tissues falls below -12 bars; then resistance decreases rapidly (47). As the turgor of birch tissue is restored, however, the rate of canker expansion declines, and callus forms at the canker margins as resistance is restored. *Fusarium* root rot of wheat is negligible at -30 to -33 bars but increases greatly at -35 to -40 bars (113). Cultural practices, such as high nitrogen fertilization and high stand density may increase water stress in wheat and thereby root rot severity. Stalk rots of corn and sorghum caused by *Fusarium roseum* and *F. moniliforme* are also favored by high temperature and drought stress. At 35 and 40°C, all of the above fungi generally grow fastest at osmotic potentials of -20 to -60 bars. Drought stress may accentuate normal sequential losses of resistance that are associated with fruiting, since unpollinated male-sterile sorghum plants remain resistant to charcoal rot even under water stress (112).

Drought may predispose plants to certain pathogens. Sugar beet roots produced under moisture stress are more susceptible to storage rots caused by *Phoma, Botrytis,* and *Penicillium,* than beets grown with adequate moisture (35). Tomato plants grown with low soil moisture before inoculation are more susceptible to *Fusarium* wilt than those grown in very wet soil (59).

A few fungal pathogens are favored by intermediate to low water potentials in soil, and resistance increases as water potential is either increased or further decreased. Percentages of infection of barley seedlings by *Helminthosporium graminis,* for example, were 15–16, 64–84, and 27–36% at potentials of −1.0, −7.0, and −12.9 bars, respectively (121). Infection of poinsettia by *Rhizoctonia* also is maximum at 40% MHC and below, whereas the fungus causes little or no damage at 80% MHC or above (13).

Light

Resistance Decreases with Decreased Light. With few exceptions, decreases in either light intensity or photoperiod diminish plant resistance to disease. For example, short days or low light intensities predispose tomatoes to susceptibility to *Fusarium* wilt (59), and when applied after inoculation, decrease resistance to *Pseudomonas* wilt (90) and *Verticillium* wilt (80, 21). Low light intensity also decreases the resistance of wheat to stripe rust (149), potatoes to late blight (136), and corn to anthracnose (64). Other examples are given in reviews (20, 134, 159, 42).

Resistance in some cultivars may be "photostabile" while that in others is "photolabile." For example, tomato cultivars with the *Ve* gene for resistance to *Verticillium* remain resistant under a 4-hr photoperiod (80), whereas cultivars with lower levels of other types of resistance are overcome. General resistance to *Phytophthora* in the potato cultivars 'Katahdin' and 'Sebago' was progressively reduced by 47 and 80% shading, whereas the reaction of susceptible 'Russett Rural' was unchanged (136). Thus, high light intensity was necessary for maximum expression of the general resistance. Similar results were obtained for corn cultivars varying in resistance to anthracnose (64).

The effect of light intensity on resistance of wheat to *Puccinia striiformis* (stripe rust) varied with the time at which light intensity was reduced and with the cultivars (149). Reduction of light prior to inoculation had much less effect on resistance than that after inoculation; the reaction of only one wheat cultivar–*P. striiformis* race combination was affected by the predisposition treatment. Reduction of light intensity following inoculation generally decreased resistance in wheat, but two cultivars of wheat were "photostabile" showing no response to reduced light.

Resistance Decreases with Increased Light. Examples of this pattern of response are rare. Cabbage is less resistant to club root (*Plasmodiophora*) under high compared to low light intensity (42). The resistance of 'Topper' barley to stripe rust was much greater in the greenhouse when daylight intensity decreased in autumn and winter (149). However, in growth chambers, high light intensity gave the greatest resistance, indicating that light quality may have been more important in the greenhouse. High light intensities may predispose carnations to *Fusarium* stub dieback (147).

Soil Composition and Nutrients

The severity of plant diseases varies on soils of different porosity, bulk density, pH, and nutrient content. However, these soil factors affect other environmental variables, such as temperature, water potential, and microbial antagonists. Thus, it is almost impossible to determine the direct effects of soil composition and nutrients on disease resistance. This problem is further complicated by the fact that each plant and each pathogen tends to respond somewhat uniquely to various nutrients and pH.

A few generalizations can be made, recognizing that exceptions exist. Pathogens favored by wetness and cool temperatures generally are favored by clay or loam soils (87), whereas pathogens favored by dryness and warm temperatures are favored by sandy soils (38). A few pathogens, such as *Fusarium, Botrytis,* and *Plasmodiophora,* are favored by acid soils. Resistance to such pathogens increases considerably when soils are neutralized to pH 6.5–7.5 (42, 160). Other pathogens, such as *Streptomyces, Phymatotrichum,* and *Verticillium,* are most severe in neutral to alkaline soils and do little or no damage at pH 5.0–5.5 or below.

Fertilization with ammonium sulfate causes a drop in pH, whereas calcium, sodium, or potassium nitrate increase pH near roots (75). Ammonium, but not nitrate nitrogen, increases the severity of *Fusarium* wilt of banana, cotton, and tomato (158) and *Fusarium* root rot of bean; it increases resistance to *Verticillium* and *Gaeumannomyces* (160). Conversely, nitrate nitrogen increases severity of *Verticillium* wilt, *Phymatotrichum* root rot, and potato scab more than ammonium nitrogen. Thus, some of the effects of nitrogen fertilizers on disease may be due to localized pH changes at the root surface. Nitrapyrin [2-chloro-6-(trichloromethyl)-pyridine] inhibits the *Nitrosomonas* bacteria that are needed to convert ammonium to nitrate nitrogen. Accordingly, it also reduces severity of *Verticillium* wilt and potato scab but increases *Fusarium* root rot of bean when used with low levels of ammonium fertilizer (160, 75).

In pathogens favored by neutral pH, the effects of fertilizers depend on the age of plant attacked and the parasitic potential of the pathogen. High levels of balanced nutrition, and sometimes nitrogen alone, may increase resistance to weak facultative parasites that attack young seedlings or senescing plant parts. For example, resistance to *Diplodia* and *Colletotrichum* stalk rot of corn increases directly with increasing rates of nitrogen fertilization (157). Nitrapyrin gives an added increase in resistance at low rates of ammonium fertilizer. Diseases of this type are generally enhanced by multi-nutrient, or nitrogen, deficiencies. Fertilizers in these cases help seedlings grow through their natural susceptible stage more rapidly and slow the loss of resistance associated with senescence.

Resistance to obligate parasites often is decreased by high rates of nitrogen and phosphorus fertilization. Such an effect is shown in the resistance of pine to fusiform rust (130), and significant interactions occur between fertilizer treatments and cultivars. Thus, screening with and without fertilizer may be desirable. High nitrogen rates provide optimal conditions for detecting the "slow-mildewing" resistance of 'Knox' wheat (132). Disease susceptibility increases progressively with nitrogen rates in susceptible 'Vermillion' wheat but not in resistant 'Knox'. Thus, nitrogen fertilizers may be used as an aid in developing certain types of general resistance.

Potassium fertilizers show a marked effect on diseases caused by some facultative parasites but rarely affect those caused by obligate parasites. The severity of wilt diseases caused by vascular parasites often increases greatly when potassium is deficient (20). Conversely, excess potassium often increases resistance, particularly in cultivars which already have moderate levels of resistance. High nitrogen or nitrogen plus phosphate tend to decrease resistance to wilts, particularly when potassium levels are low. The resistance of tomato to bacterial soft rot (*Erwinia caratovora*) also is decreased when nitrogen fertilizer is doubled with constant potassium but not when both nutrients are doubled (11).

Phosphorus fertilizer shows strong interactions with cultivars, other nutrients, and soil conditions, making it difficult to predict its effect on disease. There are, however, exceptions. The resistance of pearl millet to downy mildew increases in direct proportion to superphosphate concentrations up to 50 kg/ha (54). Variable nitrogen and potassium treatments have no effect on the response to phosphorus. Likewise, treatments with triple superphosphate from 84 to 336 kg P_2O_5/ha progressively increase resistance to scab of potato (50); both lesion height and depth, as well as incidence, are decreased by phosphate. The effect may be due to an interaction with calcium. Phosphorus added to high nitrogen levels and low potassium levels generally decreases resistance to *Verticillium* wilts (20). Accordingly, adding the endotrophic mycorrhizal fungus *Glomus* to soils low in phosphorus increased the severity of *Verticillium* wilt on cotton (52). This and other mycorrhizal fungi facilitate uptake of phosphorus from soils low in this nutrient; thus, they may confound results obtained on the effect of phosphorus on disease.

The mechanisms by which nutrients change resistance are, for the most part, uncertain. The increased susceptibility of bluegrass to *Drechslera* as a consequence of an increased rate of nitrogen fertilization is associated with increased concentrations of free amino acids in leaves, which may enhance fungal colonization (126). In clover, there is an inverse relationship between nitrogen levels in nutrient solutions and the amount of the antibiotic medicarpin formed as an active defense response (48). Thus, both pathogen colonization and host defense appear to be affected by fertilizers.

Biotic Factors

Symbionts. Plant roots form symbiotic relationships with certain bacteria, such as *Rhizobium,* and with various endomycorrhizal and ectomycorrhizal fungi. *Rhizobium* generally affects diseases in the same way as moderate nitrogen fertilization, and endomycorrhizal fungi generally have the same effects as phosphorus fertilization. For example, *Glomus fasciculatus* increases severity of *Verticillium* wilt in cotton (52) and *Phytophthora* root rot of avocado (51) in soils with low, but not high, phosphorus content; however, it has no effect on *Phytophthora* root rot of citrus or alfalfa. *Glomus mosseae* does not change the incidence of *Thielaviopsis* root rot in cotton, but treated plants are more vigorous and recover more rapidly because of their improved nutrition (135). Ectomycorrhizae often greatly increase resistance to soilborne pathogens, such as *Pythium* and *Phytophthora,* because they both produce antibiotics and stimulate active defense systems in the host. These effects are reviewed by Marx (100).

Nonpathogenic Antagonists. The surface of plant roots and leaves are colonized by various bacteria and fungi that are neither symbiotic nor pathogenic under most environmental conditions. Many of these organisms may kill pathogens or reduce disease because of their competition for nutrients, production of antibiotics, or hyperparasitism. Cook (45) has reviewed the management of these "associated microbiota" for purposes of disease control.

Most of the promising biological antagonists are bacteria. For example, *Agrobacterium radiobacter* strain 84 greatly reduces crown gall caused by *Agrobacterium tumefaciens* (86, 107) in various fruit trees and ornamentals when used to inoculate transplants. *Bacillus subtilis* applied to potato seed-pieces reduces the frequency of charcoal rot caused by *Macrophomina* and *Botryodiplodia* (151). Likewise, *Pseudomonas fluorescens* strain PF-5 applied to cottonseed markedly reduces seedling diseases (73). In each case, the reduction in disease severity depends on production of an antifungal antibiotic by the antagonist.

The fungi *Gliocladium* and *Trichoderma* also reduce severity of soilborne diseases. These fungi reduce growth of pathogens both by hyperparasitism and antibiosis. However, they are generally somewhat less effective than bacteria, probably because of their slower growth rates.

Foliar diseases generally are not affected by antagonists, although antagonists do occur on leaves. For example, a green fluorescent *Pseudomonas* sp. on barley leaves apparently is responsible for lysis of spores of the pathogen *Rhynchosporium* (128). Activity of the bacteria depended on temperature and the duration of wet periods. Two of 13 isolates of bacteria from onion leaves slightly reduced numbers of lesions from *Botrytis,* but 10 others increased infection frequency (41). Similar inconsistent results have been obtained in other searches for antagonistic bacteria and fungi on leaves.

Disease Complexes.   Infection by one pathogen frequently increases or decreases resistance to subsequent infections by other pathogens. About half of the book *Plant Disease—An Advanced Treatise,* Vol. 5 (72) is devoted to these interactions.

Interactions between nematodes and fungi in disease complexes have been reviewed by Powell (120) and Bergeson (24). In general, nematodes decrease resistance to fungal wilt diseases and seedling diseases whether applied before or after the fungus. The degree to which resistance is decreased, however, depends on the specific nematode, the inoculum potential of the nematode and of the fungus, the resistance level of the cultivar, and environmental factors (20). Sometimes nematodes have failed to decrease resistance to pathogens in certain cultivars, even though resistance is broken in many other cultivars. For example, root-knot nematodes failed to break resistance to *Fusarium* wilt in 'Manapal', 'Florida MH-1', and 'Small Fry' tomatoes (81, 139). In 'Small Fry' this failure is due to the presence of two dominant genes of which one ($LMiR_2$) is effective against the nematode, while the other (FIR_1) is effective against the fungus. In progeny from 'Small Fry' that contain only the FIR_1 gene, resistance to *Fusarium* is readily overcome by nematodes. Thus, a clear definition of the genetics of resistance to each pathogen alone must be obtained before their interactions can be elucidated.

Virus diseases that cause extensive chlorosis frequently predispose plants to fungal pathogens. Seed from soybean plants infected with soybean mosaic virus have an increased incidence of *Diaporthe* infection (69). Either bean yellow mosaic virus or common pea mosaic virus reduces resistance of pea to root rots caused by *Fusarium* and *Aphanomyces* (25). Mustard infected with mustard mosaic virus is more susceptible to *Peronospora parasitica* (7).

Combined virus infections sometimes are synergistic in overcoming resistance. For example, cowpea stunt is caused by combined infections of cucumber mosaic virus (CMV) and blackeye cowpea mosaic virus (BICMV) (118). While stunt is a severe disease (86.4% yield loss), CMV or BICMV alone causes a very mild disease.

Fungal pathogens also may be synergistic. *Pythium ultimum, Fusarium solani,* and a combination of the two reduced total dry weight of bean seedlings by 52, 12, and 74%, respectively (117). *Rhizoctonia,* in contrast, increased resistance to *Pythium* so that combined infection caused less damage than *Pythium* alone.

There are hundreds of other examples of infection by one pathogen increasing resistance to another. This phenomenon is variously known as cross protection, induced immunity, or induced resistance. The highest

levels of induced resistance against virulent strains usually result from inoculations with avirulent or mild strains of the same pathogen. Such resistance has been induced in wheat against *Puccinia striiformis* (78), in tobacco
against *Alternaria alternata* (145) and *Phytophthora parasitica* (102), in
cotton against *Verticillium dahliae* (15), and in carnation against *Fusarium
roseum* (9).

Different pathogens that invade the same tissues also may give high
levels of induced resistance against each other. For example, infection by a
vascular wilt fungus generally causes resistance to other vascular pathogens
in many hosts (20, 140). *Pythium* infection likewise reduces *Fusarium* wilt of
peas (89); both pathogens initially invade plants through juvenile root tissues.

Localized infections, infections under environments that result in resistance, or infections that occur during a resistant stage of plant development
may cause the plant to be resistant to subsequent inoculations with the same
pathogen at ages and environments normally favorable for susceptibility.
For example, inoculation of a single cotyledon or the primary leaf of
cucumber, watermelon, or muskmelon with *Colletotrichum lagenarium*
causes systemic protection; both lesion number and size are reduced (37).
Infection of tolerant cotton with *Verticillium dahliae* for 7 days at 30°C,
which gives a resistant reaction, causes a high level of resistance to reinocu-
lation at 25°C, which otherwise would result in death (15). Inoculation of
Italian prune trees with *Cytospora cincta* in May gave strong, season-long
resistance, and inoculation in July gave unusually strong resistance to subsequent inoculations with *Cytospora* (123).

Cytospora and *Agrobacterium tumefaciens* induce reciprocal resistance
in prune (68). Expansion rates of both galls and cankers were significantly
depressed in the presence of the other, indicating that the bacterium induced
a nonspecific resistance to the fungus.

Viruses also may induce resistance to fungal diseases. Infection of the
first true leaf of cucumber with tobacco necrosis virus induced systemic resistance to *Colletotrichum* similar to that induced by localized *Colletotrichum* infection (76). Soybean mosaic virus induces resistance to *Cephalo-
sporium gregatum* (150). Because the virus is seed-transmitted it can cause
misleading impressions about levels of genetic resistance in infected breeding lines. Other examples of microbial interactions are given by Bowen and
Rovira (32).

Air Pollution

Around metropolitan areas, plants are sometimes injured by ozone. Ozone
may affect plant resistance to disease; diseases, in turn, may affect sensitivity to ozone.

Ozone generally increases resistance to obligate parasites that cause little
or no tissue necrosis, but decreases resistance to some facultative parasites

that cause necrosis. Lesions caused by *Helminthosporium maydis* on corn are larger on plants exposed to 6 hr of ozone either before or after inoculation than on control plants (66). Exposure to ozone 6 days before, but not after, inoculation enhances sporulation of the fungus. Ozone also increases infection of geranium by *Botrytis*, but only when the ozone causes visible necrotic tissues (98). These necrotic spots serve as infection courts for *Botrytis*. In contrast, ozone increases resistance to bacterial infections of soybean (92) and wild strawberry (93). Exposure of plants to 0.20 ppm ozone for only 3 hr either before or after inoculation inhibited the development of *Xanthomonas fragariae* in strawberry, whereas *Pseudomonas glycinea* was affected mostly by treatments before inoculation.

Ozone injury in tobacco is synergistically increased by prior infection with tobacco streak virus (124), but it is significantly reduced by prior infection with tobacco mosaic virus (TMV) (28). In TMV-infected plants the average percentage of leaf area injured by ozone was 5% compared to 11% in healthy plants. Four viruses, alfalfa mosaic (AMV), tobacco ring spot virus (TRSV), tomato ring-spot virus (tom-RSV), and TMV, reduced ozone injury in beans inoculated 5 days before ozone treatment (49). Percentages of tissues injured were: control, 44; AMV, 16; TRSV, 31; tom-RSV, 34; and TMV, 17. Reduction in ozone injury was directly related to virus concentration, and was induced in both primary leaves even when only one was inoculated. Possible differences in the response of cultivars to the interactions between ozone and pathogens have largely been ignored.

Cultural Practices

Agricultural Chemicals. Modern agriculture has become increasingly dependent on chemicals to control pests and to regulate plant growth. Herbicides, insecticides, and growth retardants have been shown to affect resistance to diseases. As with other environmental variables, resistance may be increased, decreased, or unchanged depending on the specific chemical, its concentration, the host, the pathogen, and other environmental variables.

Details of the effects of herbicides on plant diseases are contained in three recent reviews (3, 82, 83). Resistance to many seedling pathogens (*Pythium*, *Rhizoctonia*, *Fusarium solani*, and *Thielaviopsis*) is decreased by various herbicides. For example, three of eight herbicides increased the severity of *Thielaviopsis* root rot of soybean (94). In-depth studies with chloramben showed that it enhanced exudation of amino acids from roots and thereby stimulated germination of *Thielaviopsis* chlamydospores and colonization of roots by the fungus. Three different cultivars gave the same response to chloramben. In contrast, the dinitroaniline herbicides, with few exceptions, increase resistance to soilborne pathogens such as *Rhizoctonia*, *Fusarium oxysporum*, and *Verticillium dahliae* (63). In fact, resistance of eggplant and tomato to the latter two pathogens was greatly increased by trifluralin and nitralin, giving as much as a 97% reduction in disease. Such control occurs in spite of toxicity to the plant and lack of toxicity to the pathogens. Thus, the host defense system must be modified in some way. The chlorophenoxy herbicides may increase resistance to wilt diseases (20), but may also increase susceptibility to foliar diseases such as *Drechslera* leaf spot of bluegrass (70).

Growth retardants, like certain herbicides, increase resistance to wilt diseases. For example, three different growth retardants mitigate symptoms of *Verticillium* wilt in cotton (55). These chemicals markedly reduced the internal population of the fungus in the upper petioles of the plant. Apparently the host defense system is improved by these chemicals.

Few studies have been concerned with the effects of insecticides on diseases. Aldicarb significantly increases damping-off of sugar beet seedlings by *Rhizoctonia* at some, but not other, concentrations (153). In culture, aldicarb inhibits linear growth of the fungus which may explain why the highest concentrations do not increase disease. Similar inconsistencies have been found with other systemic insecticides and *Verticillium* wilt (16).

Failure to control severe insect attacks may cause a loss of disease resistance. Defoliation of elms by canker worms predisposes trees to infection by the Dutch elm disease fungus which is introduced through bark beetle injuries (20). Four weeks of defoliation of white birch predisposes it to attack by *Botryosphaeria* through wounds (47). Likewise, boll rot incidence in cotton is correlated with bollworm and bollweevil injury in many areas.

Cultivation and Harvest Practices.   Both cultivation and harvesting practices inflict numerous wounds on plants and may thus decrease their resistance to disease. In screening for resistance to *Fusarium oxysporum*, the fungus is usually introduced through severed roots because this practice increases uniformity and severity of infection (155). Severing of roots during cultivation also enhanced severity of *Verticillium* wilt in the field (160). The predisposition of carnations to *Fusarium* stub rot depends on the length of stubs left during harvest of flowers; short stubs are more susceptible and more apt to expand into severe lesions (147). Other examples of predisposition by wounds are given by Schoeneweiss (134) and Yarwood (159).

Disease severity often is most severe with monoculture, presumably because inoculum potential increases most on residues of the host crop. For example, *Thielaviopsis* infection of peanut is greater under monoculture than with sorghum rotation (74). Management of crop residues also affects diseases. Stubble-mulch of wheat straw residue causes poorer growth of wheat early in the fall in Washington probably because of increased seedling pathogens (46). *Cercosporella* foot rot, however, is less severe because its severity is directly proportional to plant size and vigor.

Organic amendments other than host residue often decrease both populations of soilborne pathogens and disease severity. For example, *Macrophomina* root rot of cotton is reduced appreciably by amending soil with alfalfa meal (60). Such amendments stimulate populations of actinomycetes and bacteria that are antagonistic to the pathogen.

BREEDING FOR DISEASE RESISTANCE UNDER ENVIRONMENTAL STRESS

Effects of Environment on Dominance and Segregation

Attempts to clearly define the genetic regulation of resistance to *Verticillium* wilt in cotton have repeatedly led to frustration. Different studies, reviewed by Barrow (10), have led to the conclusion that resistance is dominant, additive, or recessive, depending on the location and year of the study. Figure 2 illustrates how all of these conclusions may be correct even for the same resistance gene. When resistant 'Seabrook Sea Island 12B$_2$' is crossed with susceptible 'Stardel', the resulting heterozygous progeny show an intermediate level of resistance comparable to tolerant 'Acala 4-42-77' (108, 16). Consequently, the apparent degree of dominance of resistance is highly dependent on temperature. At 25, 27, and 29°C, respectively, the resistance in F_1 and F_2 progeny appears to be recessive, additive, and completely dominant. The actual additive nature of the resistance is seen only at 27°C, because of the strong temperature effect.

Resistance to plant pathogens, with the possible exception of obligate fungal parasites, probably is inherited as an incompletely dominant, largely additive, character much more than we realize. This happens because researchers frequently select a strain of the pathogen, inoculum concentration, plant age, and environmental conditions (temperature, moisture, light, and fertilization) that cause resistance to appear completely dominant or recessive, when it is actually additive. The few studies that have carefully considered most of these variables seem to verify this conclusion. For example, Barrow (10) found that the resistance of 'Acala 9519' cotton to *Verticillium* behaved as a single, completely dominant gene when temperature and light were rigidly controlled in an environment chamber, and a precise inoculum concentration of a mildly pathogenic strain was used. Greenhouse and field conditions, other environmental-chamber settings, use of a highly virulent strain, or other inoculum concentrations all gave F_1 and F_2 segregation patterns more consistent with quantitative inheritance. Alon et al. (2) studied factors that affect the penetrance (degree of dominance) of the *I* gene for *Fusarium* wilt resistance in tomato. Penetrance varies with plant age, the specific susceptible parent, inoculum concentration, and soil temperature. Thus, resistance appears completely dominant only under unique conditions. The resistance of cucumber to anthracnose (1) and safflower to *Phytophthora* (152) are other examples of resistance that appear due to a single recessive and dominant gene, respectively, when environment is rigidly controlled. Walker (154) has reviewed many additional examples of changes in F_2 segregation ratios as a consequence of temperature changes.

Disease complexes may cause resistance to appear polygenic when it is actually controlled by a single dominant gene. For example, the resistance of cotton to *Fusarium* wilt appears to be polygenic in fields infested with nematodes, but F_2 segregations typical of a single dominant gene are obtained after nematodes are killed by fumigation (142).

Sidhu and Webster (139, 140) studied the effects of disease complexes on F_2 segregation ratios in tomato. The cultivar 'Small Fry' contains two independent dominant genes for resistance, one ($LMiR_2$) effective against the nematode *Meloidogyne incognita,* and the other (FIR_1) effective against the fungus *Fusarium oxysporum.* The nematode can break down the FIR_1 resistance to *Fusarium,* when the $LMiR_2$ gene is absent, but not when both are present together. Consequently, when 'Small Fry' was crossed with a cultivar susceptible to both pathogens, the F_2 ratio observed for single infections (nematode or fungus) was 9 R R : 3 R S : 3 S R : 1 S S, where R R = resistant to both pathogens, R S = resistant to *Fusarium,* S R = resistant to the nematode, and S S = susceptible to both. However, when plants were inoculated sequentially with both pathogens a modified 9 R R : 3 S R : 4 S S ratio, characteristic of recessive epistasis, was obtained because of the nematodes' ability to break the resistance of the FIR_1 gene alone. If resistance to *Fusarium* alone was observed, a 3 : 1 ratio was obtained when plants were infected with only *Fusarium,* but a 9 : 7 ratio occurred when plants were infected with both nematodes and *Fusarium.* The latter ratio suggests complementary gene action for resistance to *Fusarium*; in fact, polygenic resistance is not present.

Crosses also were made between 'VFN-8' tomato, which has separate dominant genes for resistance to *Fusarium* and *Verticillium,* and a susceptible cultivar. *Fusarium* induces resistance to *Verticillium* in the presence of the *Fusarium*-resistance gene. Consequently, this cross gave F_2 ratios of 9 : 3 : 3 : 1 for single infections but 12 : 3 : 1, which is characteristic of dominant epistasis, for mixed infections. For *Verticillium* alone the corresponding ratios for resistance were 3 : 1 and 15 : 1. Again, the disease complex gave the impression of polygenic control for resistance to *Verticillium,* when resistance to this fungus, in fact, was controlled by a single dominant gene. Most reports of polygenic epistatic control of disease resistance have involved 9 : 7 or 15 : 1 ratios and were derived from field studies, where disease complexes were possible. Thus, single genes may be responsible for resistance in many cases where multiple genes are suspected.

The above studies show why the total pathogenic community, and especially its interactions, must be considered in breeding for disease resistance. This is best illustrated by a tomato cultivar which has homozygous recessive (susceptible) genes for nematode and *Verticillium* resistance but is homozygous dominant (resistant) for *Fusarium.* In a field containing only *Verticillium*, this cultivar will be susceptible; but when both *Verticillium* and

Fusarium are uniformly present, it will be resistant to both. When nematodes are then superimposed, the cultivar is no longer resistant to any of the three pathogens. In the latter situation we would be tempted to conclude that resistance had "broken down" or "a new, more virulent race of *Fusarium* had arisen," when, in fact, it was a consequence of a different disease complex. The example also emphasizes why it is important to develop multiple disease resistance in cultivars, giving special emphasis to those pathogens which may predispose plants to susceptibility towards other pathogens.

Use of Environment for Improving Resistance

Approaches to the management of host genes for disease control have been summarized in several reviews (129, 34, 114) and books (131, 108). These articles have treated the importance of environment in disease epidemics and the need to approach breeding for resistance with the purpose of preventing epidemics. A review by Walker (154) specifically emphasizes the use of environmental factors in screening for disease resistance. Environmental manipulations may be used to clarify the inheritance of resistance as discussed in the previous section, or they may be used to distinguish levels of resistance in different cultivars, strains, or hybrid progeny.

Temperature manipulations are particularly useful for distinguishing different levels of resistance, as illustrated in Figure 2. In this case intermediate levels of resistance to *Verticillium* wilt in cotton are only distinguished at 27°C; at 25°C or below the intermediate line is similar to the susceptible one, and at 29°C or above it is indistinguishable from the most resistant one. Similar manipulations may be used to distinguish the intermediate level of polygenic resistance of cabbage to *Fusarium* wilt from the higher level of monogenic resistance (154). At 24°C, susceptible and polygenic-resistant plants are killed, while homozygous or heterozygous monogenic-resistant ones survive. At 20–22°C, the polygenic-resistant plants also survive, while the susceptible ones are killed. Temperature manipulations have also been used to distinguish levels of resistance against *Fusarium* (2), bacterial wilt (104, 90), and bacterial canker (58) of tomato, Dutch elm disease (44), *Pythium* root rot of cotton (77), *Phytophthora* root rot of safflower (152), powdery mildew of soybean (105), *Fusarium* wilt of bean (53) and spinach (111), stripe rust of wheat (55), stem rust of oats (99), and many other diseases (154).

Changes in photoperiod or light intensity also may be used to distinguish resistance levels. Jones et al. (80) used 4-hr photoperiods to clearly distinguish high levels of monogenic resistance in tomato to *Verticillium* wilt; Ben-Yephet and Pilowsky (21) used low daylight intensity for the same purpose. Stubbs (149) used variable light intensity to distinguish "photostabile" and "photolabile" resistance to stripe rust in wheat cultivars. Long photoperiods and high light intensity were most desirable for distinguishing general resistance in potato to *Phytophthora* (136) and in corn to *anthracnose* (64).

Variable wet periods have been used to distinguish levels of resistance to leaf spots in wheat. Hosford (72) used wet periods of 48, 72, and 121 hr to distinguish three levels of resistance in wheat cultivars to *Leptosphaeria* leaf spot. Oats were not attacked even after 121 hr of wetness. Progressive increases in wet periods showed that winter wheats were more resistant to *Septoria* leaf spot than spring wheats, but differences also occurred among the winter wheat cultivars. The possible effects of variable water stress on resistance of cultivars to pathogens such as *Macrophomina* and *Fusarium* need to be investigated.

Cultural practices also may be used to facilitate screening for resistance. Deliberate pruning of roots prior to inoculation to increase severity of *Fusarium* wilt has been practiced for years in breeding for resistance (155). Rowan (130) found significant family × fertilization interactions in the resistance of pine to fusiform rust, and concluded that both fertilized and unfertilized seedlings should be screened to obtain maximum resistance. High nitrogen levels increase powdery mildew severity in susceptible wheat cultivars but not in cultivars with genes for general resistance. Consequently, the greatest differences in resistance of wheat cultivars occur with heavy nitrogen fertilization.

CONCLUSIONS

As more of our crops are grown in marginal environments, we can expect that levels of resistance needed for certain diseases will change. It is important to recognize that the changes in relative resistance due to environmental changes generally will not be linear (see Figure 1). For example, only a slight change in an environmental condition may cause a major decrease (or increase) in disease resistance as is shown in Figure 2 for *Verticillium* wilt of cotton.

By knowing the responses of both hosts and pathogens to the environmental extremes of a marginal environment, we can predict what changes in disease resistance are needed. For example, based on the effect of temperature on *Verticillium* wilt of cotton, we can predict that moving cotton into a cooler climate (e.g., 2°C cooler) would require a major increase in levels of cultivar resistance just to keep disease severity from increasing. Conversely, if we introduce cotton into a warmer climate, very low levels of resistance will suffice to completely control the disease.

Ideally we should develop cultivars that have broader adaptation to environmental extremes than do the pathogens that attack them as is shown for temperature in Model C of Figure 1. Plants with such broad environmental adaptation should show increased general resistance to nearly all pathogens, even though nothing has been changed in the specific resistance biochemistry of the plant. This is possible because the rate and speed at which host resistance develops, relative to that of secondary colonization by the pathogen, is generally the most important determinant of resistance to disease. We need to define the effects of environmental variables upon the speed of host defense reactions, and then we need to develop cultivars that have more rapid host defense reactions at environmental extremes.

One example of such an approach is the multi-adversity resistance (MAR) program for genetic improvement of cotton (26). Wild cotton species grow in warm arid areas of the tropics and subtropics. Consequently the crop is poorly adapted to cool and wet climates. In the MAR program special emphasis has been placed on obtaining resistance to seed and seedling diseases that occur in cool moist soils. Seeds must germinate slowly and remain free of mold at 14°C on water agar. The selected seedlings are then transplanted into soil naturally infested with *Rhizoctonia*, *Pythium*, and *Thielaviopsis,* and cotyledons are inoculated with races 1, 2, 7, and 18 of *Xanthomonas malvacearum.* Only seedlings with the highest levels of resistance to the fungal pathogens and immune to blight due to combinations of general and specific genes for resistance are advanced in the program. Subsequently, progeny from selected plants are planted in scattered nurseries where they are exposed to various insects, nematodes, and other diseases. Only lines showing some resistance to all adversities, and earliness for escape from late season pests, are advanced in the program. Cultivars developed in the MAR program have become predominant in many of the marginal areas for cotton production in Texas. Thus, development of improved resistance to disease complexes favored by environmental stress apparently is a key to crop production in marginal areas.

REFERENCES

1 Abul-Hayja, Z., P. H. Williams, and C. E. Peterson. 1978. *Plant Dis. Rep.* 62:43.

2 Alon, H., J. Katan, and N. Kedar. 1974. *Phytopathology* 64:455.

3 Altman, J., and C. L. Campbell. 1977. *Annu. Rev. Phytopathol.* 15:361.

4 Arndt, C. H. 1957. *Plant Dis. Rep. Suppl.* 246:63.

5 Ashworth, L. J. Jr., J. L. McMeans, and C. M. Brown. 1969. *Phytopathology* 59:669.

6 Ayres, P. G., and J. C. Zadoks. 1979. *Physiol. Plant Pathol.* 14:347.

7 Bains, S. S., and J. S. Jhooty. 1978. *Plant Dis. Rep.* 62:1043.

8 Baker, R. 1978. Inoculum potential. P. 137 in *Plant Disease—An Advanced Treatise*, Vol. 2, J. G. Horsfall and E. B. Cowling, Eds. Academic, New York.

9 Baker, R., P. Hanchey, and S. D. Dottarar. 1978. *Phytopathology* 68:1495.

10 Barrow, J. R. 1973. Genetics of *Verticillium* tolerance in cotton. P. 89 in *Verticillium Wilt of Cotton.* U.S. Dept. Agric. Publ. ARS-S-19.

11 Bartz, J. A., G. M. Geraldson, and J. P. Crill. 1979. *Phytopathology* 69:163.

12 Bateman, D. F., and A. W. Dimock. 1959. *Phytopathology* 49:641.

13 Bateman, D. F. 1961. *Phytopathology* 51:445.

14 Bell, A. A., and J. T. Presley. 1969. *Phytopathology* 59:1141.

15 Bell, A. A., and J. Presley. 1969. *Phytopathology* 59:1147.

16 Bell, A. A. 1973. Nature of resistance. Pp. 47 in *Verticillium Wilt of Cotton*, U.S. Dept. Agric. Publ. ARS-S-19.

17 Bell, A. A. 1974. Biochemical bases of resistance of plants to pathogens. P. 403 in *Biological Control of Plant Insects and Diseases*, F. G. Maxwell and F. A. Harris, Eds. The University Press of Mississippi, Jackson.

18 Bell, A. A., and R. D. Stipanovic. 1978. *Mycopathologia* 65:91.

19 Bell, A. A. 1980. Time sequence of defense. P. 53 in *Plant Disease—An Advanced Treatise*, Vol. 5, J. G. Horsfall and E. B. Cowling, Eds. Academic, New York.

20 Bell, A. A., and M. E. Mace. 1981. Biochemistry and physiology of resistance. P. 431 in *Fungal Wilt Diseases of Plants*, M. E. Mace, A. A. Bell, and C. Beckman, Eds. Academic, New York.

21 Ben-Yephet, Y., and M. Pilowsky. 1979. *Plant Dis. Rep.* 63:66.

22 Berger, R. D. 1975. *Phytopathology* 65:485.

23 Berger, R. D. 1977. *Annu. Rev. Phytopathol.* 15:165.

24 Bergeson, G. B. 1972. *Exp. Parasitol.* 32:301.

25 Beute, M. K., and J. L. Lockwood. 1967. *Phytopathology* 57:804.

26 Bird, L. S. 1975. *Beltwide Cotton Prod. Res. Confr.*, p. 150.

27 Birkholz-Lambrecht, A. F., D. T. Lester, and E. B. Smalley. 1977. *Plant Dis. Rep.* 61:238.

28 Bisessar, S., and P. J. Temple. 1977. *Plant Dis. Rep.* 61:961.

29 Black, W. M., and D. Neely. 1978. *Phytopathology* 68:1054.

30 Blad, B. L., J. R. Steadman, and A. Weiss. 1978. *Phytopathology* 68:1431.

31 Bollenbacher, K., and N. D. Fulton. 1959. *Plant Dis. Rep. Suppl.* 29:222.

32 Bowen, G. D., and A. D. Rovira. 1976. *Annu. Rev. Phytopathol.* 14:121.

33 Bronson, C. R., and R. P. Scheffer. 1977. *Phytopathology* 67:1232.

34 Browning, J. A., M. D. Simons, and E. Torres. 1977. Managing host genes: epidemiologic and genetic concepts. P. 171 in *Plant Disease—An Advanced Treatise*, Vol. 1, J. Horsfall and E. Cowling, Eds. Academic, New York.

35 Bugbee, W. M. 1979. *Phytopathology* 69:414.

36 Byther, R. S., and G. W. Steiner. 1975. *Plant Physiol.* 56:415.

37 Caruso, F. L., and J. Kuc. 1977. *Phytopathology* 67:1290.

38 Chakrabarti, D. K., and K. C. Basuchaudhary. 1978. *Plant Dis. Rep.* 62:776.

39 Chauhan, S. K. 1963. *Agra Univ. J. Res. Sci.* 12:271.

40 Christiansen, M. N. 1969. *Proc. Beltwide Cotton Prod. Res. Confr.*, p. 127.

41 Clark, C. A., and J. W. Lorbeer. 1977. *Phytcpathology* 67:96.

42 Colhoun, J. 1973. *Annu. Rev. Phytopathol.* 11:343.

43 Cook, R. J. 1973. *Phytopathology* 63:451.

44 Cook, R. J., and A. A. Christen. 1976. *Phytopathology* 66:193.

45 Cook, R. J. 1977. Management of the associated microbiota. Pp. 145 in *Plant Disease—An Advanced Treatise*, Vol. 1, J. Horsfall and E. Cowling, Eds. Academic, New York.

46 Cook, R. J., and J. T. Waldher. 1977. *Plant Dis. Rep.* 61:96.

47 Crist, C. R., and D. F. Schoeneweiss. 1975. *Phytopathology* 65:369.

48 Cruickshank, I. A. M., K. Spencer, and M. Mandryk. 1979. *Physiol. Plant Pathol.* 14:71.

49 Davis, D. D., and S. H. Smith. 1976. *Plant Dis. Rep.* 60:31.

50 Davis, J. R., R. E. McDole, and R. H. Callihan. 1976. *Phytopathology* 66:1236.

51 Davis, R. M., J. A. Menge, and G. A. Zentmyer. 1978. *Phytopathology* 68:1614.

52 Davis, R. M., J. A. Menge, and D. C. Erwin. 1979. *Phytopathology* 69:453.

53 de L. D. Ribeiro, R., and D. J. Hagedorn. 1979. *Phytopathology* 69:272.

54 Deshmukh, S. S., C. D. Mayee, and B. S. Kulkarni. 1978. *Phytopathology* 68:1350.

55 Erwin, D. C., S. D. Tsai, and R. A. Khan. 1979. *Phytopathology* 69:283.

56 Eyal, Z., J. F. Brown, J. M. Krupinsky, and A. L. Scharen. 1977. *Phytopathology* 67:874.

57 Ferrin, D. M., and D. C. Ramsdell. 1978. *Phytopathology* 68:892.

58 Forster, R. L., and E. Echandi. 1973. *Phytopathology* 63:773.

59 Foster, R. E., and J. C. Walker. 1947. *J. Agric. Res.* 74:165.

60 Ghaffar, A., G. A. Zentmyer, and D. C. Erwin. 1969. *Phytopathology* 59:1267.

61 Grainger, J. 1979. *Annu. Rev. Phytopathol.* 17:223.

62 Gray, P. M., and J. W. Guthrie. 1977. *Plant Dis. Rep.* 61:90.

63 Grinstein, A., J. Katan, and Y. Eshel. 1976. *Phytopathology* 66:517.

64 Hammerschmidt, R., and R. L. Nicholson. 1977. *Phytopathology* 67:247.

65 Hayman, D. S. 1969. *Can. J. Bot.* 47:1663.

66 Heagle, A. S. 1977. *Phytopathology* 67:616.

67 Helms, K. 1977. *Phytopathology* 67:230.

68 Helton, A. W. 1976. *Phytopathology* 66:212.

69 Hepperly, P. R., G. R. Bowers, Jr., J. B. Sinclair, and R. M. Goodman. 1979. *Phytopathology* 69:846.

70 Hodges, C. F. 1978. *Phytopathology* 68:1359.

71 Horsfall, J., and E. Cowling, Eds. 1980. P. 534 in *Plant Disease—An Advanced Treatise*, Vol. 5. Academic, New York.

72 Hosford, R. M. 1978. *Phytopathology* 68:591.

73 Howell, C. R., and R. D. Stipanovic. 1979. *Phytopathology* 69:480.

74 Hsi, C. H. 1978. *Phytopathology* 68:1442.

75 Huber, D. M., and R. D. Watson. 1974. *Annu. Rev. Phytopathol.* 12:139.

76 Jenns, A. E., and J. Kuc. 1979. *Phytopathology* 69:753.

77 Johnson, L. F. 1979. *Plant Dis. Rep.* 63:59.

78 Johnson, R., and D. J. Allen. 1975. *Ann. Appl. Biol.* 80:359.

79 Johnson, T., and M. Newton. 1941. *Can. J. Res.* 19:438.

80 Jones, J. P., P. Crill, and R. B. Volin. 1975. *Phytopathology* 65:647.

81 Jones, J. P., A. J. Overman, and P. Crill. 1976. *Phytopathology* 66:1339.

82 Katan, J., and Y. Eshel. 1973. *Residue Rev.* 45:145.

83 Kavanagh, T. 1974. *Sci. Proc. R. Dublin Soc.* Ser. B, 3:251.

84 Kendrick, J. B., Jr., and J. T. Middleton. 1959. *Phytopathology* 49:23.

85 Kerr, A. 1964. *Aust. J. Biol. Sci.* 17:676.

86 Kerr, A., and K. Htay. 1974. *Physiol. Plant Pathol.* 4:37.

87 Kittle, D. R., and L. E. Gray. 1979. *Plant Dis. Rep.* 63:231.

88 Kraft, J. M., and D. D. Roberts. 1969. *Phytopathology* 59:149.

89 Kraft, J. M. 1978. *Plant Dis. Rep.* 62:216.

90 Krausz, J. P., and H. D. Thurston. 1975. *Phytopathology* 65:1272.

91 Kuhlman, E. G. 1978. *Plant Dis. Rep.* 62:8.

92 Laurence, J. A., and F. A. Wood. 1978. *Phytopathology* 68:441.

93 Laurence, J. A., and F. A. Wood. 1978. *Phytopathology* 68:689.

94 Lee, M., and J. L. Lockwood. 1977. *Phytopathology* 67:1360.

95 Leonard, K. J., and D. L. Thompson. 1976. *Phytopathology* 66:635.

96 Lewis, R. L., Jr., and E. P. VanArsdel. 1978. *Plant Dis. Rep.* 62:62.

97 Lewis, R. L., Jr., and E. P. VanArsdel. 1978. *Plant Dis. Rep.* 62:125.

98 Manning, W. J., W. A. Feder, and I. Perkins. 1970. *Phytopathology* 60:669.

99 Martens, J. W., R. I. H. McKenzie, and G. J. Green. 1966. *Can. J. Bot.* 45:451.

100 Marx, D. H. 1972. *Annu. Rev. Phytopathol.* 10:429.

101 McCarter, S. M., and R. W. Roncadori. 1971. *Phytopathology* 61:1426.

102 McIntyre, J. L., and P. M. Miller. 1978. *Phytopathology* 68:235.

103 Metz, S. G., and A. L. Scharen. 1979. *Plant Dis. Rep.* 63:671.

104 Mew, T. W., and W. C. Ho. 1977. *Phytopathology* 67:909.

105 Mignucci, J. S, S. M. Lim, and P. R. Hepperly. 1977. *Plant Dis. Rep.* 61:122.

106 Miller, D. E., and D. W. Burke. 1975. *Phytopathology* 65:519.

107 Moore, L. W., and G. Warren. 1979. *Annu. Rev. Phytopathol.* 17:163.

108 Nelson, R. R., Ed. 1973. *Breeding Plants for Disease Resistance.* The Pennsylvania State University Press, University Park.

109 Nothmann, J., and Y. Ben-Yephet. 1979. *Plant Dis. Rep.* 63:70.

110 Nour, M. A. 1958. *Trans. Br. Mycol. Soc.* 41:17.

111 O'Brien, M. J., and H. F. Winters. 1978. *Plant Dis. Rep.* 62:427.

112 Odvody, G. N., and L. D. Dunkle. 1979. *Phytopathology* 69:250.

113 Papendick, R. I., and R. J. Cook. 1974. *Phytopathology* 64:358.

114 Parlevliet, J. E. 1979. *Annu. Rev. Phytopathol.* 17:203.

115 Phipps, P. M., and M. K. Beute. 1977. *Phytopathology* 67:1104.

116 Pieczarka, D. J., and G. S. Abawi. 1978. *Phytopathology* 68:403.

117 Pieczarka, D. J., and G. S. Abawi. 1978. *Phytopathology* 68:776.

118 Pio-Ribeiro, G., S. D. Wyatt, and C. W. Kuhn. 1978. *Phytopathology* 68:1260.

119 Populer, C. 1979. Changes in host susceptibility with time. Pp. 239 in *Plant Disease—An Advanced Treatise*, Vol. 3, J. G. Horsfall and E. B. Cowling, Eds. Academic, New York.

120 Powell, N. T. 1971. *Annu. Rev. Phytopathol.* 9:253.

121 Prasad, M. N., K. J. Leonard, and C. F. Murphy. 1976. *Phytopathology* 66:631.

122 Pratt, R. G., and J. E. Mitchell. 1976. *Phytopathology* 66:81.

123 Randall, H., and A. W. Helton. 1976. *Phytopathology* 66:206.

124 Reinert, R. A., and G. V. Gooding, Jr. 1978. *Phytopathology* 68:15.

125 Riffle, J. W. 1978. *Phytopathology* 68:1115.

126 Robinson, P. W., and C. F. Hodges. 1977. *Phytopathology* 67:1239.

127 Roncadori, R. W., and S. M. McCarter. 1972. *Phytopathology* 62:373.

128 Rotem, J., B. G. Clare, and M. V. Carter. 1976. *Physiol. Plant Pathol.* 8:297.

129 Rotem, J., Y. Cohen, and E. Bashi. 1978. *Annu. Rev. Phytopathol.* 16:83.

130 Rowan, S. J. 1977. *Phytopathology* 67:1280.

131 Russell, G. E. 1978. *Plant Breeding for Pest and Disease Resistance.* Butterworths, Boston.

132 Shaner, G., and R. E. Finney. 1977. *Phytopathology* 67:1051.

133 Schneider, H. 1948. *J. Agric. Res.* 76:129.

134 Schoeneweiss, D. F. 1975. *Annu. Rev. Phytopathol.* 13:193.

REFERENCES

135   Schonbeck, F., and H. W. Dehne. 1977. *Plant Dis. Rep.* 61:266.

136   Schumann, G., and H. D. Thurston. 1977. *Phytopathology* 67:1400.

137   Sharp, E. L. 1965. *Phytopathology* 55:198.

138   Short, G. E., and M. L. Lacy. 1976. *Phytopathology* 66:188.

139   Sidhu, G., and J. M. Webster. 1974. *J. Hered.* 65:153.

140   Sidhu, G., and J. M. Webster. 1979. *Physiol. Plant Pathol.* 15:93.

141   Slykhuis, J. T. 1975. *Phytopathology* 65:582.

142   Smith, A. L., and J. B. Dick. 1960. *Phytopathology* 50:44.

143   Smith, R. S., Jr. 1966. *Phytopathology* 56:61.

144   Sonoda, R. M. 1978. *Plant Dis. Rep.* 62:1059.

145   Spurr, H. W., Jr. 1977. *Phytopathology* 67:128.

146   Stack, R. W., R. K. Horst, P. E. Nelson, and R. W. Langhans. 1978. *Phytopathology* 68:423.

147   Stack, R. W., R. K. Horst, P. E. Nelson, and R. W. Langhans. 1978. *Phytopathology* 68:429.

148   Sterne, R. E., G. A. Zentmyer, and M. R. Kaufmann. 1977. *Phytopathology* 67:1495.

149   Stubbs, R. W. 1967. *Phytopathology* 57:615.

150   Tachibana, H., and L. C. Card. 1972. *Phytopathology* 62:1314.

151   Thirumalachar, M. J., and M. J. O'Brien. 1977. *Plant Dis. Rep.* 61:543.

152   Thomas, C. A., and R. F. Hill. 1977. *Phytopathology* 67:698.

153   Tisserat, N., J. Altman, C. L. Campbell. 1977. *Phytopathology* 67:791.

154   Walker, J. C. 1965. *Annu. Rev. Phytopathol.* 3:197.

155   Wells, D. G., W. W. Hare, and J. C. Walker. 1949. *Phytopathology* 39:771.

156   Westerlund, F. V., R. N. Campbell, and R. G. Grogan. 1978. *Phytopathology* 68:921.

157   White, D. G., R. G. Hoeft, and J. T. Touchton. 1978. *Phytopathology* 68:811.

158   Woltz, S. S., and A. W. Englehard. 1973. *Phytopathology* 63:155.

159   Yarwood, C. E. 1976. Modification of the host response—Predisposition. Pp. 73 in *Physiological Plant Pathology, Encyclopedia of Plant Physiology*, New Series, Vol. 4, R. Heitefuss and P. H. Williams, Eds. Springer-Verlag, New York.

160   Zentmyer, G. A., and J. G. Bald. 1977. Management of the environment. Pp. 121 in *Plant Disease—An Advanced Treatise*, Vol. 1, J. G. Horsfall and E. B. Cowling, Eds. Academic, New York.

PLANT PEST INTERACTION WITH ENVIRONMENTAL STRESS AND BREEDING FOR PEST RESISTANCE:

Insects

JOHNIE N. JENKINS

U.S. Department of Agriculture, Science and Education Administration, Agricultural Research, Crop Science and Engineering Research Laboratory and Mississippi State University, State College, Mississippi

RELATIONSHIPS OF RESISTANT AND SUSCEPTIBLE CULTIVARS WITH INSECTICIDE EFFICACY, BENEFICIAL INSECTS, AND PEST MANAGEMENT PROGRAMS

Environmental stresses, expressed as deficiencies or excesses, can affect the way resistant cultivars perform. There is a trait in cotton, *Gossypium hirsutum* L., controlled by a single recessive gene, called "frego bract." Frego bracts are flared and rolled away from the square (bud) exposing the square, in contrast to the enclosed square of normal bract cotton. Frego bract cotton is resistant to the boll weevil, *Anthonomus grandis* (Boheman). There is an interesting interaction between number of insects overwintering and the degree of resistance or control obtained with frego bract. Frego bract reduces boll weevil populations through a reduction in the number of eggs oviposited by female weevils. Depending upon the number of overwintering boll weevils entering a field, one can expect from one to three generations delay before the economic threshold is reached in frego bract cotton compared to normal bract cotton (17). Frego bract interacts in another way with the environment and other methods of control. Since the bud is exposed rather than enclosed, up to seven times more insecticide is deposited on the bud of frego bract cotton than on normal bract cotton (23). This, obviously, will affect control of insects when contact insecticides are being used or when pathogens are being sprayed onto the plants for control of the insects.

Boll weevil behavior is changed on frego bract cotton in such a way that seven times more movement by the boll weevils occurs on the frego bract plant than on normal bract cotton. This sets up an unfavorable environment for the boll weevil if insecticides are being used, because weevils would have more opportunity to contact the insecticide than they would in normal bract cotton. *Heliothis virescens* (Fab.) was controlled more efficiently on frego bract cotton than on normal bract cotton under high population levels (Jenkins et al., unpublished data, 1978). Thus, this interaction of frego bract with insecticides is a positive interaction for the grower, and one which gives him better control of *Heliothis* spp. and boll weevil.

Walker et al. (30) describe a cotton production system utilized in certain areas in Texas. This system involves the growing of cotton cultivars that will mature in about 130 days. The growing system is a complete management system involving cultivars which fruit fast and mature early, limited use of nitrogen fertilizer, limited use of irrigation water in those areas where irrigation is used, and use of insecticides for early-season boll weevil and fleahopper, *Pseudatomoscelis seriatus* (Reuter), control on a judicious basis. Reproduction diapause control for boll weevil allows the insecticide for boll weevil to be applied in the fall rather than in the spring, thus preserving natural enemies. This then allows about 30 days of blooming in the short-season varieties before boll weevil or *Heliothis* spp. reach damaging population levels. This system is contrasted with the old system of growing cotton in the area where plants were lush and green for 160–180 days, and insecticides were used early in the season for boll weevil and late in the season for *Heliothis* spp. This shortening of the growing season has placed severe stresses upon both *Heliothis* spp. and boll weevil; thus, their populations are smaller and control is more easily obtained. By using only one or two applications of insecticide with these fast fruiting cottons and reduced nitrogen and water, growers are also able to allow beneficial insects to build up in the fields and to aid in the control of *Heliothis* spp.

The greenbug, *Schizaphis graminum* (Rondani), is a serious pest in sorghum [*Sorghum bicolor* (L.) Moench.]. Recently, good resistance has been bred into commercial hybrids of sorghum (27). The type of resistance which these sorghum hybrids contain is primarily tolerance. This type of resistance allows large populations of greenbugs to develop in the field; however, the plants are not damaged economically by the toxin produced by the greenbugs. Studies by Teetes et al. (27) showed that beneficial insects are important in aiding the resistant sorghum hybrids in normal production. Thus, this situation in the environment of the insect and the plant is one in which large numbers of pests (greenbugs) are present in the field along with large numbers of beneficial insects. The dynamic balance achieved allows good sorghum production because the hybrids are tolerant to the attack by the greenbug. It is imperative that growers not use insecticides to reduce the number of greenbugs because this would also reduce the number of beneficial insects.

"Nectariless" is a trait in cotton controlled by two recessive genes. Nectariless cotton lines do not have the nectary on the leaf, nor at the base of the square and boll. This removes a favorite feeding site of several insects. Nectariless cottons produce an unfavorable environment for the reproduction of *Lygus lineolaris* (Palisot de Beauvais) (22) and *Lygus hesperus* (Hyer et al., personal communication). Since *Lygus* spp. feed in the nectary, which is absent on nectariless plants, this change in environment of the insects by growing the nectariless varieties places a great stress upon the insects and results in reduced populations of nymphs through reduced fecundity of females and reduced survival of nymphs (5).

The soybean, *Glycine max* (L.) Merr., accession ED73-371 has good resistance to the Mexican bean beetle, *Epilachna varivestis* Mulsant, as indicated by Van Duyn et al. (28). It was shown that the cultivar on which an insect feeds can have an interaction with the ability of insecticides to control these insects (18). Studies in which *Heliothis zea* (Boddie) and *Pseudoplusia includens* (Walker) were grown on the resistant genotype ED73-731 and the susceptible cultivar 'Bragg' indicated an interaction between the genotype on which the insects were grown and the susceptibility of the insects to two insecticides and *Bacillus thuringiensis* Berliner. *Heliothis zea* larvae grown on the resistant genotype were more susceptible to methomyl and *B. thuringiensis*; however, their LD_{50}'s to methyl parathion were similar to insects grown on 'Bragg'. In contrast, larvae of *Pseudoplusia includens* grown on ED 73-731 were more susceptible to methyl parathion, and their LD_{50}'s to methomyl and *B. thuringiensis* were similar on the two genotypes. In field studies *H. zea* was controlled better by methyl parathion and *B. thuringiensis* in plots of ED73-371 than in plots of 'Bragg'. The field control with methyl parathion does not agree with laboratory data; however there were about 50% fewer larvae on ED 73-731 than 'Bragg' which could have affected the degree of control with methyl parathion. Thus, this is an example of the genotype of the plants on which the insects are feeding stressing the insects in such a manner that on resistant genotypes it can affect the insect's susceptibility to other control agents.

EFFECTS OF LIGHT, TEMPERATURE, PHOTOPERIOD, AND RELATIVE HUMIDITY ON THE EXPRESSION OF RESISTANCE AND THE FECUNDITY AND GROWTH OF INSECTS AND NEMATODES

The spotted alfalfa aphid, *Therioaphis maculata* (Buckton), and the pea aphid, *Acyrthiosiphon pisum* (Harris), are major pests on alfalfa, *Medicago sativa* L. A number of studies relate light, temperature, photoperiod, relative humidity, and so on, to the reproduction of these two aphids and to the expression of resistance in various cultivars of alfalfa. When resistant and susceptible cultivars of alfalfa were compared at 10°C and 27°C, aphids generally exhibited good preference between resistant and susceptible plants at the higher temperature. However, at the lower temperature, some resistant cultivars behaved as though they were susceptible (25). However, in a study by Kindler and Staples (19), spotted alfalfa aphids at 10°C migrated from resistant to susceptible plants. Maximum production of aphids on susceptible plants occurred at 20–22°C, with survival and fecundity being significantly reduced at 24°C and 30°C. Fluctuating temperatures enhanced fecundity on susceptible, but not resistant, plants. In their study, soil moisture and photoperiod did not have an effect on susceptibility or resistance.

The stability of resistance to pea aphid and spotted alfalfa aphid in several alfalfa clones under various temperatures was studied by Isaak et al. (15). At 29°C, they could not separate resistant and susceptible clones; however, at 16°C, they could separate resistant and susceptible clones. All clones were resistant at 27°C, whereas, at 16°C, some suddenly became susceptible and others became susceptible more slowly. This interaction between resistant/susceptible clones and temperature was earlier studied by Hackerott and Harvey (12). In studies at 27°C and 16°C, they found that both resistant and susceptible lines were killed quicker at the higher temperature than at the low temperature. They also found that nymph development was slower at the low temperature. Plants which were resistant at 27°C would only support a limited population of aphids at 16°C. Isaak et al. (14) in a study of spotted alfalfa aphid and pea aphid found the highest resistance in a group of lines was expressed at 29°C; however, they did detect a significant clone × temperature interaction.

The effects of relative humidity and temperature were studied by Isaak et al. (14) on spotted alfalfa aphid and pea aphid resistant and susceptible clones. The clone × humidity interaction was not significant; however, there was a significant temperature × humidity × clone interaction. At 29°C, survival of adults and number of live nymphs produced by adults differed significantly with humidity, but the differences were insignificant at 13°C.

McMurtry (21) studied the environmental factors of temperature, light duration and intensity, soil moisture, plant mineral nutrition, and physiological age of the host leaves on resistant alfalfa cultivars to spotted alfalfa aphid. He found that the resistant cultivar, 'Lahontan,' which was composed of two clones, expressed its highest degree of resistance at high temperatures. However, one of the clones, 'C902,' in Lahontan was completely susceptible at 10–16°C but was resistant at high temperatures. The other clone, 'C84,' was intermediately resistant at 10°C and was almost immune at 20°C or higher. In comparing these with a susceptible strain, 'Caliverde,' aphid build-up on the susceptible strain was the highest at 29°C, whereas on the two resistant clones the optimum temperature for population increase was much lower. McMurtry believed that a physiological change, rather than a morphological change, in the resistant plants at the lower temperature caused them to be resistant. Photoperiod had no effect on aphid survival and reproduction. Mineral nutrition, however, did affect the resistance expression of the two clones C84 and C902. Both became more resistant when they were deficient in phosphorus and became more susceptible when they were deficient in potassium. The susceptible plants of Caliverde were not significantly altered by these deficiencies. Nitrogen had no effect upon the expression of resistance. Under field conditions, it appears that temperature will be the factor that will have the greatest effect on expression of resistance. It also seems that plants with intermediate levels of resistance are more affected by temperature changes than are the highly resistant plants. Kindler et al. (20) reported differences between the effects that the spotted alfalfa aphid and the pea aphid caused on the chemical composition of resistant and susceptible cultivars of alfalfa. Not only was yield reduced more in the susceptible cultivars, but also the chemical constituents such as protein, carotene, and digestible dry matter were affected more on the susceptible variety than on the resistant variety.

The alfalfa stem nematode, *Ditylenchus dipsaci* (Kuhn), and the root-knot nematode, *Meloidogyne hapla,* show an interaction with resistant and susceptible lines and temperature. Elgin et al. (9) evaluated a number of lines for resistance to the alfalfa stem nematode. He found greater cotyledonary node swelling at 16°C and 20°C than at 25°C. Alfalfa stem nematodes entered both resistant and susceptible plants; however, reproduction was less in the resistant plants, and less swelling of the cotyledonary node was reported in the study conducted by Wynn and Busbice (33). Relationship among temperature at which alfalfa plants were grown, age of plants, and their susceptibility to the root-knot nematode were reported by Griffin and Hunt (11). Resistant progeny were galled only at 32°C. Cuttings from their resistant line remained resistant to root-knot nematode regardless of temperature and age. However, when seedlings were grown of the resistant line, the older the plants were at time of inoculation, the greater the percentage of gall-free plants. Thus, in their study, they indicated that age of plant could have an effect upon how one classified the plants as to resistance and susceptibility. Jatala and Russell (16) found that temperature was an important factor governing the expression of sweet potato resistance to root-knot nematode. The number of females maturing was affected more at various temperatures in the resistant strain than in the susceptible strain. They also reported that the resistant strain, 'Nemagold,' produced a root exudate which was repellent to the larvae; thus reducing the amount of feeding and reproduction.

In a study of aphids and various factors affecting the expression of resistance, Cartier (7, 8) found that temperature, aphid biotype, field, and greenhouse environments affected the expression of antibiosis and had a direct bearing on the rating of pea varieties for resistance. When plants were 2.5–15.2 cm tall, pea aphids on peas, *Pisum sativum* L., were selective in their lighting response based on color of foliage. Yellowish-green foliage was the most attractive, whereas green and deep green were the least attractive. Plant height early in the season was also positively correlated with the interception of migrating insects; however, on full grown pea plants this seemed to change as the increased plant height had an exposure effect that reduced ensuing aphid populations because of taller stems, longer internodes, and less dense foliage.

In a study of the effects of temperature on the resistance of lima beans, *Phaseolus lunatus* L.; tomatoes, *Lycopersicon esculentum* (L.) Mill.; and chrysanthemums, *Chrysanthemum* spp.; to leaf miner, *Liriomyza munda* (Frick), Webb and Smith (31) reported that larvae developed more slowly at a given temperature on tomatoes and chrysanthemums than they did on lima beans. Lowering the temperature on resistant tomato and chrysanthemums increased mortality. In addition, with chrysanthemums they found a positive correlation between length of larval development and higher larval mortalities in resistant cultivars, but this was not true in tomato varieties.

The amount of light under which wheat varieties are grown affects the expression of resistance to the wheat stem sawfly, *Cephus cinctus* Norton. Resistance is reduced under shady conditions compared to more light conditions. For example, the resistant cultivar 'Rescue' would normally be susceptible in the greenhouse in the wintertime; however, when supplemented with 4000 foot candles (fc) of light it maintained its resistance. Fifteen hundred fc was not sufficient enough to maintain resistance. Even in the summer, light was not intense enough to cause 'Rescue' wheat to maintain its resistance when grown in the greenhouse (24). They concluded that high light intensities were required for the maximum expression of the solid-stem characteristic which confers resistance to wheat stem sawfly.

EFFECTS OF MINERAL NUTRITION ON THE EXPRESSION OF RESISTANCE AND THE FECUNDITY OF THE INSECTS

The rate and amount of nitrogen, phosphorus, and potassium have been shown to interact with the expression of resistance and with the fecundity of several insects. Bowling (3) found that nitrogen rates between 0 and 133 kg/ha did not affect adult rice water weevil, *Lissorhoptrus oryzophilus* Kuschel, feeding, and scars produced by adult feeding; however, larval populations of the rice water weevil increased with each rate of nitrogen.

The relationships between nitrogen and potassium with fecundity and reproduction rate of two aphids, *Myzus persicae* and *Brevicoryne brassicae*, were studied by Van Emden (29). He found that an increase in nitrogen or a decrease in potassium would increase the amount of soluble nitrogen in brussel sprout plant leaves. This was positively correlated with the fecundity and reproduction rate of *Myzus persicae*. The fecundity of the two-spotted spider mite, *Tetranychus urticae* Koch, on lima beans was positively correlated with nitrogen applied to the plant and nitrogen content of the leaves. Henneburg and Schiver (13) in a study on strawberries, found that nitrogen, potassium, and phosphorus levels had no effect on the fecundity of the obscure root weevil, *Sciopithes obscurus* Horn. However, when plants were deficient in nitrogen, there was a significant reduction in the fecundity of the black vine weevil, *Otiorhynchus sulcatus* (F.) (6).

MISCELLANEOUS FACTORS AFFECTING EXPRESSION OF RESISTANCE AND FECUNDITY OF INSECTS

Age affects the expression of resistance in sweet clover to the sweet clover weevil, *Sitona cylindricollis* Fahraeus. Resistance is influenced by a balance between stimulant A, deterrent A, and deterrent B (nitrate). The relative concentrations of the different factors vary with the stage of development of the leaves of the plant. Nitrate appears to be the predominant water-soluble factor in young leaves, while deterrent A and stimulant A assume increased importance in leaves of mature plants (1).

The level of infestation is important in rating plants for resistance and susceptibility. Blum (2) in a study of the sorghum shoot fly, *Atherigona varia soccata,* found that the level of infestation influenced the expression of resistance such that under low shoot fly populations, resistance seemed to be partially dominant, whereas under high populations susceptibility appeared to be dominant. He concluded that selection for nonpreferred genotypes should be established under extreme infestations. He preferred to see 13–15 eggs per plant on the susceptible lines.

The field environment involving the various types of cover crops and cultural practices, such as time of plowing under of cover crops, influences the population of the sugar beet wire worm, *Limonius californicus* Mann. Barley, *Hordeum vulgare* L.; mustard, *Brassicae juncea* (L.) Czern; and sweet clover, *Melilotus* spp., were responsible for buildup of wireworm populations. Early plowing tended to prevent increases in wireworm populations (Stone 26).

When one considers insecticides used for control of soilborne insects, there is an interaction between the soil type and the rate at which the insecticide must be applied. For example, to control cabbage maggots, *Hylemya brassicae* (Weidemann), Finlayson and Noble (10) found that the insecticide which was the most effective on sandy loam soil was not the one most effective on peat soil.

Resistant and susceptible plants can produce different types of plant structures or different nutrients in plants which can stress the insects. Brett et al. (4) found that D-glucose was associated in a positive way with resistance to the pickleworm, *Diaphania nitidalis* (Stoll). Concentrations above 1% increased resistance; concentrations below 1% decreased resistance. D-glucose content varied with age. They also found a lower concentration of D-glucose present in the nectar from flowers of susceptible plants than from resistant plants. D-glucose was twice as high in the rind of the fruit of resistant plants as in susceptible plants.

Heliothis zea feeds on corn, *Zea mays* L., and damages the kernels. It is a problem particularly in sweet corn production. The environment surrounding the ear, relative to husk length, is important in the resistance to corn earworm. Widstrom et al. (32) reported that corn lines with husk lengths in excess of that required to cover the ear tended to have more injury from *H. zea*; however, mechanical protection of tight husks reduced damage considerably. They also believed that feeding stimulation was involved in resistance and suggested that chemical factors involving feeding stimulation were also important in conjunction with the physical characteristics of husk length and tightness.

SUMMARY

In this chapter four topics have been discussed:

1   The relationship of resistant and susceptible cultivars with insecticide efficacy, with beneficial insects, and with pest management programs. All of these are important in the environment in which the plant and the insect are grown. Altering the environment for the insect by growing resistant cultivars can have positive interactions with insecticide efficacy, positive or negative interactions with beneficial insects, and positive interactions with pest management programs.

2   The environmental factors of light, temperature, photperiod, and relative humidity were discussed as they affected the expression of resistance in resistant and susceptible lines, and as they affected fecundity and growth rate of insects and nematodes.

3   Mineral nutrition, particularly nitrogen, phosphorus, and potassium, was discussed as it interacted with the expression of resistance in resistant and susceptible lines, and as it affected the fecundity of insects.

4   A group of miscellaneous topics was discussed which dealt with age of plants, number of insects in populations, cover crops, chemicals in plants, and physical factors, such as husk length, as they affected fecundity and survival of insects.

The reader should be able to obtain some idea of the types of environmental stresses that affect insects and can gain an appreciation for the interaction between the resistant or susceptible cultivars growing in different environments and affecting insects in different ways. Environmental stresses may affect the plant and indirectly affect the insect, or they may affect the insect directly, which in turn, can affect the damage done to the plants. It is a complex interrelationship among the environment, the host plant, the pest insect, and natural enemy insects. Changes in any one of the four components can have a major effect upon the growth of the plant or the growth of the insect.

REFERENCES

1 Akeson, W. R., G. L. Beland, F. A. Haskins, and H. J. Gorz. 1969. *Crop Sci.* 9(5):667–669.
2 Blum, A., 1969. *Crop Sci.* 9(6):695–696.
3 Bowling, C. C. 1963. *J. Econ. Entomol.* 56:826–827.
4 Brett, C. H., C. L. McCombo, W. R. Henderson, and J. D. Rudder. 1965. *J. Econ. Entomol.* 58:893–896.
5 Calderon, Mario C. 1977. Ph. D. dissertation. Mississippi State University, Missisisppi State.
6 Cram, W. T. 1965. *Can. J. Plant Sci.* 45(3):219–225.
7 Cartier, J. J. 1962. *J. Econ. Entomol.* 56:205–213.
8 Cartier, J. J. 1963. *Phytoprotection* 44(1):60.
9 Elgin, J. H., Jr., R. N. Peaden, L. R. Faulkner, and D. W. Evans. 1974. *Crop Sci.* 14(5):669–672.
10 Finlayson, D. G., and M. D. Noble. 1966. *Can. J. Plant Sci.* 46(5):459–467.
11 Griffin, G. D., and O. J. Hunt. 1972. *J. Nematol.* 4:87.
12 Hackerott, H. L., and T. L. Harvey. 1959. *J. Econ. Entomol.* 52:949–953.
13 Henneburg, T. J., and D. Schiver. 1964. *J. Econ. Entomol.* 57:377.
14 Isaak, A., E. D. Sorenson, and E. E. Ortman. 1963. *J. Econ. Entomol.* 56:53–57.
15 Isaak, A., E. D. Sorenson, and R. H. Painter. 1965. *J. Econ. Entomol.* 58:140–143.
16 Jatala, P., and C. C. Russell. 1972. *J. Nematol.* 4:1.
17 Jenkins, J. N., and W. L. Parrott. 1971. *Crop Sci.* 11(5):739–743.
18 Kea, W. C., S. G. Turnipseed, and R. G. Barner. 1978. *J. Econ. Entomol.* 71:58–60.
19 Kindler, S. D., and R. Staples. 1970. *J. Econ. Entomol.* 63:1198.
20 Kindler, S. D., W. R. Kehr, and R. L. Ogden. 1971. *J. Econ. Entomol.* 64:753.
21 McMurtry, James A. 1962. *Hilgardia* 32:501–539.
22 Meredith, W. R., Jr., and M. L. Laster. 1975. *Crop Sci.* 15(4):535–538
23 Parrott, W. L., J. N. Jenkins, and D. B. Smith. 1973. *J. Econ. Entomol.* 66:222–225.
24 Roberts, D. W. A., and C. Tyrrel. 1961. *Can. J. Plant Sci.* 41(3):457–465.
25 Schalk, J. M., S. D. Kindler, and G. R. Manglitz. 1969. *J. Econ. Entomol.* 62:1000.
26 Stone, M. W. 1951. *USDA Tech. Bull.* 1039:1–20.
27 Teetes, G. L., C. A. Schaefer, J. W. Johnson, and D. T. Rosenow. 1974. *Crop Sci.* 14(5):706–708.
28 Van Duyn, J. W., S. G. Turnipseed, and J. D. Maxwell. 1972. *Crop Sci.* 12:561–562.
29 Van Emden, H. F. 1966. *Entomol Exp. Appl.* 9:444–460.
30 Walker, J. K., R. E. Frisbie, and G. A. Niles. 1978. *ESA Bull.* 24(3):385–391.
31 Webb, R. E., and F. F. Smith. 1969. *J. Econ. Entomol.* 62:458–461.
32 Widstrom, N. W., W. W. McMillian, and B. R. Wiseman. 1970. *J. Econ. Entomol.* 63:803.
33 Wynn, J. C., and T. H. Busbice. 1968. *Crop Sci.* 8(2):179.

Chapter 11C

PLANT PEST INTERACTION WITH ENVIRONMENTAL STRESS AND BREEDING FOR PEST RESISTANCE:

Nematodes

J. N. SASSER

Department of Plant Pathology, North Carolina State University, Raleigh, North Carolina

The interaction between the host, the parasite, and the environment is complex and constantly changing. The relative influences of each component on the others are difficult to assess because of their interwoven nature and our lack of technology to monitor such influences accurately. Through experimentation, we have gained considerable evidence that the host affects parasite development, the parasite affects the host, and the environment affects both. The environment in which the parasite lives (host, soil, climate) may at times be favorable or unfavorable to the parasite. Conversely, the environment in which the host lives (parasite, soil, climate) may at times be favorable or unfavorable to the host. Rarely is the environment favorable to both for a prolonged period of time. Consequently, stresses are imposed on either the parasite or its host during most of the time the three components are interacting.

Effectiveness of control strategies for crop pests can be enhanced through a better understanding of these relationships, whether it be through the use of cultural practices, crop rotation, resistant cultivars, or through the use of pesticides. Emphasis in this section of Chapter 11 will be on plant-pathogenic nematodes and how they interact with their host and environmental stress. For purposes of presentation, consideration will be given to the following: (*a*) importance and nature of nematode diseases of crop plants; (*b*) progress and potential in the development of resistant crop cultivars; (*c*) interrelationships between nematodes, their host and the environment; (*d*) practical considerations and problems for the plant breeder in development of nematode resistant cultivars; (*e*) mechanisms of nematode resistance in plants and possible approaches for the plant breeder in the development of plants for less favorable environments; and (*f*) prospects for the future.

IMPORTANCE AND NATURE OF NEMATODE DISEASES OF CROP PLANTS

One of the major problems in the production of agricultural crops throughout the world is the damage caused by destructive plant-parasitic nematodes. For centuries, man has been plagued by these microscopic eelworms feeding on the roots, buds, stems, crowns, leaves, and even developing seed of crop plants essential to his survival and well being. The degree of damage to a particular crop is influenced by the crop itself, the nematode species, the level of infestation, and additional stresses imposed on the crop by unfavorable environmental conditions. Severe damage may occur if the nematode population density is high and the crop planted is susceptible to the nematode species present. Visible symptoms of nematode attack often include reduced growth of individual plants, varying degrees of chlorosis, and wilting of the foliage and sometimes death of plants. These deleterious effects on plant growth result in reduced yields and poor quality of such crops as potato, rice, coconut, pineapple, citrus, banana, coffee, vegetables, fruits, tobacco, cotton, soybean, and peanut. Decline or death of highly prized ornamentals and turf are often the result of nematode attack.

The above-ground, visible effects that nematodes have on plants are due to direct and indirect results of nematodes feeding on the fibrous root systems or other plant parts. For example, root lesions, root pruning, root galling, and cessation of root growth are common symptoms and effects of nematode feeding. Roots damaged in this manner are not efficient in the utilization of available moisture and nutrients in the soil. This inefficiency of the root system is evidenced by the fact that foliage of plants attacked by nematodes often shows symptoms of nutrient deficiency even when there is an adequate supply of nutrients in the root zone. Furthermore, roots so weakened and damaged are easy prey to many types of bacteria and fungi which invade the roots and accelerate root decay. Often, this type of damage may not be so spectacular as to draw immediate attention, and above ground the crop may appear to be healthy. Production is affected, however, and it is often estimated that crop losses resulting from this kind of damage, because of its wide occurrence, constitute a sizable portion of the overall damage incited by nematodes.

Frequently, plants developed for resistance to diseases caused by certain fungi and bacteria, succumb to attack by these organisms in fields where nematodes are not controlled. Also, many important soilborne virus diseases of crop plants are transmitted by plant-parasitic nematodes. The discovery of these interrelationships between nematodes and other soil-inhabiting organisms has been an important development of the last two decades.

Full recognition of the magnitude of crop damage resulting from nematode attack and development of effective control measures under field conditions have come about only in recent years. Control of nematodes in the soil is difficult and sometimes expensive. Nevertheless, control is necessary for profitable crop production throughout most of the temperate, subtropical, and tropical regions of the world (11, 20, 40). Control is usually a matter of reducing the nematode population, by one or more tactics, to a low level so that the damage is negligible or at an economically accepted level. Once nematodes are established in a field, eradication is impractical. Disease losses can be minimized through crop rotation in which a susceptible host is grown only once every 2–8 years. In some instances, resistant varieties may be available which will decrease the reproductive and survival capacity of certain nematode species. In many cases, however, it may be both more expedient and economical to use a nematicide. Choice of control method will depend on the nematode-crop combination, the availability of land free of damaging numbers of nematodes, and the value and nature of the crop to be grown.

PROGRESS AND POTENTIAL IN THE DEVELOPMENT OF RESISTANT CROP CULTIVARS

Breeding for nematode resistance is the most promising long-range method of nematode control. Much success has already been realized in the development and use of important crop plants with resistance to major nematode species. Some of the important crops which have been developed with resistance to one or more species of nematodes are shown in Table 1. Some of these crops are widely planted and because of their resistance to certain species of nematodes have been of tremendous economic importance. In fact, one would have to conclude that far too little research effort is being devoted to the development of resistant cultivars considering the potential economic gain which could result from the widescale use of resistant cultivars of such crops as cowpeas, beans, peanut, soybean, cassava, yams, and many others. Hopefully, this treatment of nematodes will stimulate more effort in the search for resistant germplasm and the development of highly resistant cultivars for farmer use.

INTERRELATIONSHIPS BETWEEN NEMATODES, THEIR HOST AND THE ENVIRONMENT

In breeding for nematode resistance, consideration must be given to the interaction between the host, the parasite and the environment in which each is competing. Plant-pathogenic nematodes are microscopic and live their entire lives either in the soil or in plant parts such as roots, tubers, buds, or seed. Generally, they inhabit the root zone of plants and greatest populations

Table 1 Examples of crops for which resistant or tolerant varieties have been developed

Crop	Nematode
Alfalfa	*Ditylenchus dipsaci*
	Meloidogyne spp.
Almond rootstock	*Meloidogyne spp.*
Barley	*Heterodera avenae*
Bean	*Meloidogyne spp.*
Bean (lima)	*Meloidogyne spp.*
Bermudagrass	*Meloidogyne spp.*
Cassava	*Meloidogyne spp.*
Citrus	*Radopholus similis*
	Tylenchulus semipenetrans
Clover	*Ditylenchus dipsaci*
	Meloidogyne spp.
Corn	*Meloidogyne spp.*
Cowpea	*Meloidogyne spp.*
Cotton	*Meloidogyne spp.*
Crown vetch	*Meloidogyne spp.*
Eggplant	*Meloidogyne spp.*
Grape rootstock	*Meloidogyne spp.*
Lespedeza	*Meloidogyne spp.*
Muskmelon	*Meloidogyne spp.*
Okra	*Meloidogyne spp.*
Pea	*Meloidogyne spp.*
Peach rootstock	*Meloidogyne spp.*
Pepper	*Meloidogyne spp.*
Plum rootstock	*Meloidogyne spp.*
Potato	*Globodera rostochiensis*
	Globodera pallida
Soybean	*Heterodera glycines*
	Meloidogyne spp.
	Rotylenchulus reniformis
Strawberry	*Meloidogyne spp.*
Sweet potato	*Meloidogyne spp.*
Tobacco	*Meloidogyne incognita*
Tomato	*Meloidogyne spp.*
Watermelon	*Meloidogyne spp.*

are found in close proximity to roots. This greater population level of nematodes in the root zone is true because plant-parasitic nematodes are obligate parasites and can reproduce only when living plant tissues are available as a source of food. In general, if nematodes are present, the more abundant the root system of the host plant, the greater the population; and conversely, when the root system of a host plant is destroyed or in poor condition, the population of nematodes decreases. This response is simply a matter of food supply. In the absence of a suitable host, the population of most plant parasites will decrease by as much as 90% over a period of 4–6 months. Cyst nematodes, and a few others because of protective mechanisms, have the capacity to remain viable in the soil for longer periods of time without feeding. However, when a suitable host is again available, the population builds up rapidly because of abundant food and little competition. Thus, the cycle of high and low populations is repeated over and over.

The population increase and damage caused by nematodes depends on many factors. The host is perhaps the most important. Some plants are highly susceptible to infection and development of nematodes and are severely affected when the population reaches a certain level. Other plants may be quite tolerant, and even in the presence of high populations, show few or no symptoms of decline. Conversely, some plants are very sensitive to nematode attack and may be severely stunted by only a few nematodes.

In addition to the host, there are many factors of the environment which affect the severity of nematode attack. Soil texture, soil moisture and temperature, host nutrition, and presence or absence of other soil organisms which may interact with nematodes could affect the development of the parasite and the amount of damage caused.

Although certain environmental factors influence the severity of nematode attack, nematodes, except for certain above-ground foliar parasites, are not sensitive to minor changes in the microclimate as are some fungal and bacterial pathogens. Most nematodes are well buffered by the soil and/or the host tissues. Small changes in salt concentrations of the soil solution, soil pH, moisture level, and temperature, have little or no effect on the nematode. Extremes of salt concentration, prolonged high or low temperatures, and prolonged dry or wet soil can have serious effects on both the nematode and the host.

In susceptible plants, the general rule is that any condition (nutrition, moisture, temperature) favorable to the host, is also favorable to the nematode pest. Similarly, those factors of the environment (poor nutrition, inadequate moisture or unfavorable temperature) which are detrimental to good plant growth, will also impede development and reproduction of the nematode. Damage is most severe when an unfavorable change in the environment such as drought, nutrition, or temperature follows several months of favorable conditions for the host and the parasite resulting in a rapid buildup of the nematode. The high level of infestation (infection), coupled with the additional stress imposed on the plant because of lack of moisture, low fertility, or unfavorable temperature, results in serious damage to the crop. For example, cold injury and death of small peach trees can be attributed to the general unthriftiness of the trees brought about by infection and damage to the roots by various parasitic nematodes. Small trees in the same orchard, not infected with nematodes, survive the cold weather. Of course, all the trees may be killed if the cold period is severe and prolonged.

Various ornamentals are hosts for several species of parasitic nematodes.

Severe infections result in chlorosis of the foliage and general stunting of plant growth. Dieback or death of valuable plants can result if the plants are subjected to additional stresses such as inadequate fertilizer, inadequate moisture, or severe cold. Life of the plants can be prolonged if these conditions of stress are corrected.

Brown spot of tobacco, a leaf disease caused by *Alternaria alternata*, a weak pathogen, is often more severe when the plants have been weakened (as evidenced by chlorotic leaves) by nematode infection of roots.

There are some environmental conditions in which the nematode parasite or pathogen will be more aggressive and deleterious to the host. The root-knot nematode (*Meloidogyne spp.*) is more damaging in soil containing 50% or more of sand. The exact reasons for this are not known but probably are associated with ease of migration of the nematode through the larger pore spaces of the soil and better aeration resulting in a greater reproductive rate of the nematode and lower populations of other microorganisms which feed on nematodes.

PRACTICAL CONSIDERATIONS AND PROBLEMS FOR THE PLANT BREEDER IN DEVELOPMENT OF NEMATODE RESISTANT CULTIVARS

Evaluation or Measurement of Resistance and Tolerance

Rohde (37) defined resistance to nematodes as a set of characteristics of the host plant which act to the detriment of the parasite. In plant nematological literature, resistance is referred to in two ways: (*a*) resistance to the nematode and its development (reproduction) and (*b*) resistance to the disease (damage) caused by the nematode. For example, with root-knot and cyst nematodes (endoparasitic, sedentary forms), resistance may manifest itself in the following ways: (*a*) resistance to infection or invasion of the roots by the second stage larvae (none or few nematodes enter the root); (*b*) resistance to development of the parasite within the root (takes longer for the larvae to develop to maturity; there are fewer of them, and ultimately less reproduction); or (*c*) resistance to damage (nematodes enter the roots freely, feed and develop to maturity and reproduce normally but little damage occurs as evidenced by good root and shoot growth; quality and yield of crop are not affected. The latter is generally referred to as tolerance. In practice, resistance is often based on the degree of reproduction of the nematode. In most situations, this is adequate because of the high positive correlation between reproduction of the nematode and crop damage. Furthermore, reproduction is easily measured by root examination or through soil assays, whereas disease development directly attributable to the nematode is more difficult to evaluate. In cases where high or normal reproduction of the nematode does not cause crop damage, the plant is said to be tolerant.

Obtaining Sufficient Quantities of Viable Inoculum

In screening and testing progenies of crop plants for resistance to nematodes, it is essential to have sufficient quantities of viable inoculum. For certain nematode genera, this is not too difficult and techniques developed for culturing of nematodes on callus tissue (19) have made it easier to obtain sufficient quantities of single species for test purposes. There still are many problems, however, associated with obtaining and using inoculum correctly. Ectoparasites, for example, are more sensitive to culturing conditions and reproduce more slowly.

Pathogen Variability and Development of New Pathotypes

Discovery of new nematode pathotypes which will attack resistant cultivars is the most serious problem for the plant breeder as it decreases the longevity and usefulness of the resistant variety. Crops with a high degree of resistance to specific nematodes have been developed and have been widely planted without serious difficulty from the development of resistance breaking pathotypes. These include potato varieties highly resistant to *Globodera rostochiensis,* soybeans highly resistant to *H. glycines,* and tomato and tobacco varieties highly resistant to *Meloidogyne incognita.* Resistance-breaking pathotypes, however, have been discovered within each of these three important genera. The reasons for this are not fully understood but probably are associated with the limited numbers of populations of the species of nematode used in developing the resistant variety, resulting in a narrow base for resistance in the germplasm released. When such varieties are planted in widely separated geographical regions, naturally occurring populations of the species may attack them. For example, populations of *Meloidogyne incognita* from Peru and other parts of the world readily attack the tobacco variety 'N.C. 95' which is practically immune to *M. incognita* populations tested from various parts of the United States (39). This is an immediate response and does not appear to be a result of selection pressure. Of equal importance, however, is the inherent capacity for physiological changes in the nematode when subjected to selection pressures exerted from the continuous planting of resistant varieties (14) or changes brought about by conditions of the environment. Breeding programs aimed at specific species rarely, if ever, result in complete resistance, even to a single local population. New pathogenic races develop, presumably from the increase of the few individuals that can mature and reproduce on the resistant variety. The resistant variety becomes widely adopted, and once the new race occurs, it has little competition and a large plant population of the new variety in which it can develop. The time required for new races to develop will depend upon how widespread the new variety is planted and the frequency of planting in an infested field. Table 2 shows how a new pathogenic race may be detected. The hypothetical case illustrated would be

Table 2   The interaction of five cultivars of host (H) and five pathogenic races of pathogen (P)[a]

	H_0	H_1	H_2	H_3	H_4
P_0	+	−	−	−	−
P_1	+	+	−	−	−
P_2	+	+	+	−	−
P_3	+	+	+	+	−
P_4	+	+	+	+	+

[a]+ = Susceptible reaction, − = resistant reaction.

true if each new pathogenic race of the pathogen maintained its ability to attack the old cultivars as well as the new. This may not be true, however, since the ability of the pathogen to reproduce on each cultivar may be independently controlled and races that can reproduce on one cultivar may or may not be able to reproduce on another.

Based on past experience, we cannot rule out the possibility that a race will develop that can attack host 4 as illustrated in Table 3. Thus resistance in a single cultivar to all possible races may never be obtained. Resistance, however, to the predominant races can be of great economic importance.

Within certain genera there appear to be naturally occurring host races, that is, there is no evidence that they have recently developed through selection pressure brought about by the continuous planting of resistant cultivars. For example, *Radopholus similis* is primarily a parasite of citrus in the United States, and populations from citrus do not attack banana. In Central and South America, this same species is a serious parasite of banana but does not attack citrus. A third race attacks both crops. Similar host races occur in other nematode species: *Belonalaimus longicaudatus* (35), *Ditylenchus dipsaci* (43), and others.

A most striking example of naturally occurring host races is found in the root-knot nematodes, *Meloidogyne* species. Studies conducted at North Carolina State University, as part of the International *Meloidogyne* project, revealed the presence of host races in at least two of the important species.

Table 3   Host races of *Meloidogyne incognita*

| Race | Differential hosts[a] | | Percent of 298 |
designation	Cotton	Tobacco	populations studied
1	−	−	67
2	−	+	18
3	+	−	11
4	+	+	4

[a]Cotton, 'Deltapine 16'; tobacco, 'NC 95'.

When standard differential host plants were inoculated to detect pathogenic variation, a pattern of host response to cotton ('Deltapine 16') and tobacco ('NC 95') indicated at least four host races in *M. incognita*. Two host races occur in the species *M. arenaria*, one that attacks peanut, 'Florrunner,' and one that does not. As we study these nematodes more thoroughly, we may find additional races among other species. The presence of this phenomena must be considered in any control program, including crop rotation and especially in the development of resistant cultivars.

Association of Nematodes with Other Plant Diseases

In many instances, the control of nematodes through plant resistance results in the added effect of controlling other diseases caused by soilborne bacteria, fungi, and viruses. Plants, if attacked by nematodes, often become more susceptible to invasion and development of other disease causing organisms. This increase in susceptibility to other diseases may result from mechanical injury or openings made by the nematode which facilitate entry and growth of other pathogens (21), but more often it may be due to changes in host physiology (32, 33).

　　In contrast to the above, there are reports of soilborne pathogens such as *Verticillium* (23) and *Fusarium* (6) that increase the populations of plant-parasitic nematodes around host roots.

Maintaining Other Desirable Characteristics in Resistant Varieties

Quite often in the development of a variety resistant to a specific nematode, resistance to other pathogens and insects may be lost. Also, undesirable characters can be introduced. For example, Henderson and Troutman (16) observed that a strain of the potato Y virus, which in its most severe form causes vascular necrosis, infected 100% of the 'N.C. 95' and 'Fla. 22' tobacco varieties, which carry a gene for root-knot resistance. The least severe symptoms (mottling and vein-banding) were observed on older commercial varieties and selected breeding stock. Oftentimes a new variety may not be adapted to the climatic conditions of the area—such as temperature, day length, and maturity dates—and thus may have limited use. In crops like tobacco, a newly developed resistant variety may be more sensitive to fertilization, harvesting operations, and curing, and may yield less than susceptible varieties. Thus, the task of the breeder is to incorporate or select for resistance but not lose desirable qualities. This sometimes is not an easy task, and certain compromises must be made.

Multiple Infestations

Agricultural soils, with few exceptions, are infested with several plant pathogenic nematode genera and many species. The kinds and population density of the various forms vary with soil type, other environmental factors, and with cropping history. The importance or potential damage of each species of nematode present is dependent upon the resistance or susceptibility to them of the crops to be grown in that area. With most crops, several species are capable of attacking the roots and causing damage. Usually one species appears to be the most damaging and it becomes the *target* nematode in breeding for resistance. Selection of the *target* nematode is based on previous experience in the area or elsewhere. The possibility that the new variety will be susceptible to the other nematodes present is recognized by the breeder. Justification for the new variety, however, is based primarily on its value in controlling the *target* nematode.

Hypersensitive Reaction in Resistant Varieties

Resistance in plants usually does not include resistance to invasion of the roots by infective larvae of the nematode. This invasion by large numbers of larvae, although they do not develop because of a hypersensitive reaction, frequently causes considerable stunting of the host plant and has to be reckoned with in the growing of resistant crops in heavily infested soil. For example, it may be profitable to fumigate the soil the first year prior to growing a resistant variety of tobacco to avoid such damage.

Difficulty of International Mobility of the Investigator and the Germplasm Used

One of the limitations of most resistant varieties to nematodes is that they do not possess a broad base for resistance even to populations within the species for which resistance was developed. As a result, these varieties are useful primarily in the small geographic area in which they were developed. Obviously, it is not possible to expose new selections to all possible races of the pathogen. It does seem feasible, however, with present-day transportation and available facilities for studying nematode pests, that the breeder could obtain a broader base for resistance by evaluating his genetic material simultaneously in many regions of the world. Or, if the investigator has a worldwide collection of the pathogen, his program may be more productive and permanent, if there were more freedom in the movement of seed and propagules to him for study. Reasonable precautions, however, must be taken not to inadvertently introduce into new areas destructive pathogens on plant materials.

Need for Team Approach

The plant breeder, in the development of a resistant crop variety, must consider two distinct biological systems—the plant and the pathogen, both trying to survive and remain healthy, while interacting in a complex soil system. Although a specialist in his field, the plant breeder may have limited knowledge of the pathogen with which he is working. Similarly, the nematologist, a specialist in his field, may have limited knowledge about plant breeding. Because of this, the team approach, involving at least the breeder and the nematologist, would be more productive.

MECHANISMS OF NEMATODE RESISTANCE IN PLANTS AND POSSIBLE APPROACHES FOR THE PLANT BREEDER IN THE DEVELOPMENT OF PLANTS FOR LESS FAVORABLE ENVIRONMENTS

Resistance in plants varies with the host and the species of nematode. The nematode must locate its host, penetrate (in the case of endoparasites), feed, and be able to mature and reproduce itself. If the nematode can accomplish these things, the plant is said to be susceptible. We know that in resistant plants, these processes are impeded to varying degrees. Furthermore, it is known that factors governing resistance and susceptibility are inherent within individual cells in the roots or other plant parts, and are not translocated across union grafts (10, 17, 34, 4).

General Mechanisms of Resistance

The following general mechanisms have been proposed by Peacock (29) to explain the nature of resistance. The plant may: (a) lack or inhibit the nematode hatching factor, (b) lack an attractive root diffusate, or may give off a diffusate which repels nematodes, (c) inhibit nematode penetration, (d) kill the nematode upon entry or inhibit its development, (e) inhibit the effectiveness of enzymes or other constituents of the nematode's excretions which cause giant cells (nurse cells or feeding sites), (f) upset the sex ratio of the nematode.

Rohde (37) indicated similar reasons but added that genetic reactions of the plant may be modified by the environment. Such effects include temperature, mineral nutrition, soil texture, and age and vigor of plant. There probably are other mechanisms, and as pointed out by Rohde (37), a resistant plant is resistant usually for several different reasons.

Some Specific Factors Associated with Resistance or Susceptibility

1 Chemical.
   a Cholinesterase inhibitors in plants (36, 3).
   b Hydrocyanic acid resulting from hydrolysis of amygdalin by nematode enzymes or injury to the host (24).
   c Phenolic compounds, including chlorogenic acid (46, 30, 12).
   d Cytokinins (8).

    e   B-glycosidase in *Globodera rostochiensis* larvae (47).
    f   Potassium fertilizer (1, 28, 5).
    g   A-terthienyl (27, 38).
    h   Enzyme neutralizers (25).
    i   IAA-oxidase (13).
    j   Glyceollin (18).
    k   Root necrosis (2, 22).

2 Morphological.
    a   Endodermis (24, 31).
    b   Wound periderm (45).

3 Environmental influences.
    a   Soil texture (26).
    b   Temperature (15, 7, 42).
    c   Alteration of sex ratios (44).

Although many physical and chemical factors are reported in the literature to be associated with resistance, we know very little about the nature of resistance. The most effective kind of resistance is that in which the nematode does not feed on plant tissues or if feeding occurs, it is minimal and does not stunt the growth of the plant or predispose it to infection by various pathogenic fungi and bacteria.

In this book, there are chapters devoted to breeding for tolerance to heat and cold, drought and plant water efficiency, and various soil conditions. All of these, if accomplished, can play a role in nematode control. It is known that certain parasitic nematodes are inactive during the winter months because low temperatures inhibit their activities. For example, in California, sugar beet production in fields infested with sugar beet nematodes is much higher if the beets are planted in January or February than if planted in March or April. In the British Isles, the golden nematode may be controlled effectively by early planting of early-maturing varieties. Apparently, potato roots develop and grow at a lower temperature than that which is favorable for development of the nematode. In North Carolina, the root-knot nematode seldom damages the spring potato crop which is planted in February/March and harvested in June/July. On the other hand, potatoes planted in the late spring (April) are harvested in the autumn and make most of their growth during the hot summer months. Under these conditions, root-knot nematode can be a serious problem unless controlled. If vegetable crops such as tomato, beans, and squash could be developed that would grow and mature at lower temperatures (10–15°C), they would escape most nematode damage.

One of the effects of root damage caused by nematodes is that under drought conditions wilting occurs because of the inability of the root system to absorb adequate moisture. Symptoms of drought do not generally appear on noninfected plants, unless the drought is severe and prolonged. Crop plants tolerant to drought conditions would therefore be less affected when the roots are parasitized by nematodes than crop plants not tolerant to drought.

Another possible innovation the breeder might consider is to develop plants which, when infected by nematodes, are triggered to produce additional roots. Such roots would compensate for possible damage caused by infection of the original root system.

PROSPECTS FOR THE FUTURE

Success in breeding crop plants resistant to parasitic nematodes is somewhat proportional to the amount of effort devoted to this task. Certainly a number of successes should encourage additional effort. The potential for increasing food and fiber production through development of resistant varieties of certain crops is immeasurable. For example, if crop plants could be developed which had resistance to two species of root-knot nematodes, *Meloidogyne incognita* and *M. javanica,* this would provide resistance to 84% of the root-knot nematode populations encountered in the tropical and subtropical regions of the world. A large number of crop plants with resistance to one or more species of *Meloidogyne* have been developed (41) but additional work is urgently needed. For many important nematode species there are no resistant cultivars. In most cases, this lack of resistance cultivars is not because resistant germplasm does not exist, but rather that time and effort have not been expended to find and incorporate resistance into commercial cultivars. It may be necessary to travel to countries or regions which are indigenous for the crop species in question and this is costly and time consuming. Such regions however are the most likely sources for resistance, especially if the pest for which resistance is sought is also widespread in the area. Of course, this is not always necessary as there are other sources of resistance, and mutations, whether natural or artificially induced, are always possibilities for increased resistance in plants.

Research concerning the biochemical nature of resistance in plants is progressing and holds considerable promise. Unfortunately, only a few scientists are actively pursuing this line of investigation. Not only is it important to know what the plant possesses that makes it susceptible or resistant, it is equally important to know what the nematode possesses which enables it to be successful in establishing a pathogenic relationship. It is obvious that resistance in plants is not simple and that it may vary for each nematode-host combination. Also, factors for resistance are not always directly but rather indirectly involved in the host-pathogen-environment complex. In spite of this, an understanding of the physiology and biochemistry of the host and parasite should provide some valuable guidelines for the plant breeder in developing a plant capable of withstanding the potential genetic and biochemical changes of the nematode.

In summary, let me emphasize that many weapons are in the arsenal of the nematologist, and they all must be considered in designing nematode disease control systems. One method by itself seldom gives lasting control, and a multiple integrated approach is necessary. We hear a great deal about overuse of chemicals, but their use is only one of the weapons to be used along with established cultural and biological methods. Plant resistance has in the past and will in the future constitute an important segment of nematode control. Its effectiveness will be dependent upon how well we learn to use it in combination with other control measures.

REFERENCES

1   Bessey, E. A. 1911. *U.S. Dept. Agric. Bur. Plant Ind. Bull.* 217: 89.
2   Brodie, B. B., L. A. Brinkerhoff, and F. B. Struble. 1960. *Phytopathology* 50:673–677.
3   Casida, J. E. 1964. *Science* 146:1011–1017.
4   Chambers, A. Y., and J. M. Epps. 1967. *Plant Dis. Rep.* 51:771–774.
5   Crittenden, H. W. 1954. *Phytopathology* 14:388. (Abstract)
6   Davis, R. A., and W. R. Jenkins. 1963. *Phytopathology* 53:745. (Abstract)
7   Dropkin, V. H. 1963. *Phytopathology* 53:663–666.
8   Dropkin, V. H., J. P. Helgesen, and C. D. Upper. 1969. *J. Nematol.* 1:55–61.
9   Flor, H. H. 1971. *Annu. Rev. Phytopathol.* 9:275–296.
10 Forster, A. R. 1956. *Nematologica* 1:283–289.
11 Franklin, Mary T. 1979. In *Root-knot Nematodes* (Meloidogyne *species) Systematics, Biology, and Control,* Lamberti and Taylor, Eds. Academic, New York, pp. 331–341.
12 Giebel, J. 1970. *Nematologica* 16:22–32.
13 Giebel, J., and A. Wilski. 1970. *Proc. IX Int. Nematol. Symp.,* Warsaw, 1967:239–245.
14 Graham, T. W. 1969. *Tobacco Sci.* 12:43–44.
15 Grundbacher, F. J., and E. H. Stanford. 1962. *Phytopathology* 52:791–794.
16 Henderson, R. G., and Troutman, J. L. 1963. *Plant Dis. Rep.* 47:187–189.
17 Huijsman, C. A. 1956. *Nematologica* 1:94–99.
18 Kaplan, D. T., N. T. Keen, and I. J. Thomason. 1978. *J. Nematol.* 10:291–292. (Abstract)
19 Krusberg, L. R. 1961. *Nematologica* 6:181–200.
20 Lamberti, F. 1979. In *Root-knot Nematodes* (Meloidogyne *species) Systematics, Biology, and Control,* Lamberti and Taylor, Eds. Academic, New York, pp. 341–359.
21 Lucas, G. B., J. N. Sasser, and A. Kelman. 1955. *Phytopathology* 45:537–540.
22 Minton, N. A. 1962. *Phytopathology* 52:272–278.
23 Mountain, W. B., and C. D. McKeen. 1962. *Nematologica* 7:261–266.
24 Mountain, W. B., and Z. A. Patrick. 1959. *Can. J. Bot.* 37:459–470.
25 Myuge, S. G. 1964. *Plant parasitic nematodes. Feeding of phytohelminths and their relationships with plants.* Izd. Kolos, Moskva, pp. 47.
26 O'Bannon, J. H., and H. W. Reynolds. 1961. *Soil Sci.* 92:384–386.
27 Oostenbrink, M., K. Kuiper, and J. J. s'Jacob. 1957. *Nematologica Suppl.* 2:424–433 S.

REFERENCES

28 Oteifa, B. A. 1953. *Phytopathology* 43:171–174.

29 Peacock, F. C. 1959. *Nematologica* 4:43–55.

30 Pi, C. L., and R. A. Rohde. 1967. *Phytopathology* 57:344. (Abstract)

31 Pitcher, R. S., Z. A. Patrick, and W. B. Mountain. 1960. *Nematologica* 5:309–314.

32 Powell, N. T., and C. J. Nusbaum. 1960. *Phytopathology* 50:899–906.

33 Powell, N. T. 1971. In *Plant Parasitic Nematodes*, Zuckerman, Mai, and Rohde, Eds. 2:119–136.

34 Riggs, R. D., and N. N. Winstead. 1958. *Phytopathology* 48:344 (Abstract)

35 Robbins, R. T., and Hedwig Hirschman. 1974. *J. Nematol.* 6:87–94.

36 Rohde, R. A. 1960. *Helminthol. Soc. Proc. Wash.* 27:121–123.

37 Rohde, R. A. 1965. *Phytopathology* 55:1159–1162.

38 Rohde, R. A., and W. R. Jenkins. 1958. *Maryland Agric. Exp. Sta. Bull.* A-97, 19 p.

39 Sasser, J. N. 1972. *OEPP/EPPO, Bull.* 6:41–48.

40 Sasser, J. N. 1979. In *Root-knot Nematodes* (Meloidogyne *species) Systematics, Biology, and Control,* Lamberti and Taylor, Eds. Academic, New York, pp. 359–375.

41 Sasser, J. N., and M. F. Kirby. 1979. *Crop Cultivars Resistant to Root-Knot Nematodes* (Meloidogyne *species) with Information on Seed Sources.* Department of Plant Pathology, North Carolina State University, 24 pp.

42 Slana, L. J., and J. R. Stavely. 1978. *Phytopathol. News* 12:133. (Abstract)

43 Sturhan, Dieter. 1971. In *Plant Parasitic Nematodes,* Zuckerman, Mai, and Rhode, Eds. Academic, New York, pp. 51–71.

44 Trudgill, D. L. 1967. *Nematologica* 13:263–272.

45 Van Gundy, S. D., and J. D. Kirkpatrick. 1964. *Phytopathology* 54:419–427.

46 Wallace, H. R. 1961. *Nematologica* 6:49–58.

47 Wilski, A., and J. Giebel. 1966. *Nematologica* 12:219–224.

Chapter 12

PLANT GERMPLASM RESOURCES FOR BREEDING OF CROPS ADAPTED TO MARGINAL ENVIRONMENTS

JAMES A. DUKE

U.S. Department of Agriculture, Science and Education Administration, Agricultural Research, Economic Botany Laboratory, Beltsville, Maryland

Modern agriculture depends on the rather narrow but extremely important genetic bases of the crop plants of the world. Although I deal superficially with over 75 plant genera, most agricultural production is based on closer to a dozen genera which by and large feed the world.

Recent top level studies indicate the need to broaden our genetic baselines, that is, to widen our germplasm resources by salvaging some of the biotic and abiotic tolerances that can be found in some of the persisting but often endangered land races. The International Board of Plant Genetic Resources and the National Plant Genetic Resources Board are two organizations whose function is to further the collection, conservation, documentation, evaluation, and use of plant germplasm. The Economic Botany Laboratory (EBL) of the United States Department of Agriculture (USDA) maintains computerized files of geographic and ecological information on economic plants of the world. These data, available to the Germplasm Resources Information Project of the National Plant Germplasm System, are useful to determine where and when to collect germplasm for marginal environments.

EBL does not maintain germplasm collections but does maintain computerized files which identify correspondents who have reported stress-tolerant germplasm. Such data are usually provided to new correspondents who are willing, in exchange, to fill out EBL's Phytomass Questionnaire, available from this author.

Researchers without direct contacts for germplasm may contact the Chief Plant Introduction Officer, SEA, USDA, Beltsville, Maryland 20705. The Introduction Officer manages vast holdings of the U.S. Plant Germplasm System and has good contacts with other national and international collections cooperating with the United States. Many of these collections also identify germplasm tolerant to various environmental stresses.

Only a fraction of my ecosystematic data, that is, those relating to cultivars of more common food plants possessing tolerance to abiotic factors in marginal environments, are reported here. An expanded version is being prepared for presentation elsewhere. Limited data on abiotic tolerances follow, listed under the more important food-producing genera. Important fiber and forage species have been omitted in this chapter but are of great concern to modern agriculture, especially as our petrochemical feedstocks become more expensive. If American agriculture is destined to produce fuel via botanochemical farming in addition to fiber, food, and forage, new germplasm tolerant to intercropping scenarios for marginal environments may be imperative. Currently 90% or more of our legumes and cereals are consumed by livestock.

Abelmoschus (okra). Originating in the Old World, domesticated in Ethiopia or West Africa, okra may be polyphyletic, arising also in India. Drought-tolerant but intolerant of cold, okra is being bred for short-season cropping and for disease and pest resistance.

Allium (chive, onion, garlic).   A genus of more than 600 species, *Allium*'s New World species are mostly $x = 7$, the Old World predominantly $x = 8$. Except for chives, most of the cultivars come from Asia or the Near East. The primary gene center for onion (*A. cepa* L.) ($2n = 16$) is near Afghanistan, with several diploid relatives (*A. galanthum* Kar. & Kir., *A. oschanini* O. Fedtsch., *A. pskemense* B. Fedtsch., *A. vavilovii* Popov & Vved.) in Central Asia (41). The Near East is regarded as a secondary center. Today's onion is a self-compatible outbreeder that suffers inbreeding depression. Experimental hybrids with its relatives are highly sterile. Hence, selection within species for tolerance has generally been more productive than crossbreeding. Garlic, *A. sativum* L., is also diploid ($2n = 16$) like onion, and known only in cultivation, possibly derived from *A. longicuspis*. Garlic's primary center of origin is considered to be central Asia, with a secondary center in the Mediterranean. Like its presumed ancestor, garlic is viviparous and seedless, and perhaps more drought tolerant than onion. Chive (*Allium schoenoprasum* L.) has diploid, triploid, and tetraploid races ranging to the Arctic in North America and Eurasia. Said to have originated in the Mediterranean, perhaps more than once from wild populations, chives embrace both outcrossers and inbreeders.

In *Allium cepa* L., tolerance to cold has been reported in 'Winter-over', tolerance to dry muck in 'Michigan sweet spanish', tolerance to intermediate photoperiod in 'F.M. Harvest Pak-A', tolerance to long day in 'Burpee sweet spanish', 'F.M. Hybrid Span Pak-A', 'Hypak', 'Mohawk', 'Ontario', 'Paydirt', 'Spartan bounty', 'Topaz'; tolerance to long photoperiod in 'Apache', 'Cherokee', 'Sioux', tolerance to muck soils in 'Benny's red',

'Burpee yellow globe hybrid', 'Cooperskin', 'Elba globe', 'Mucker', 'Prospector', 'Simcoe', 'Spartan banner', 'Spartan era', tolerance to ozone in Gabelman, 'Spartan germ', tolerance to peat in 'Iowa 4_4', tolerance to short day in 'Ferry's early white', 'Red creole hybrid', 'Red granex', and 'Tropex'.

Amaranthus (amaranth). Sauer (55), after years of research on *Amaranthus*, divides the cultivated species into three geographically distinct "species": *A. caudatus* L. in the Andes, *A. cruentus* L. (including *A. paniculatus* L., etc.) in Central America and southern Mexico, and *A. hypochondriacus* L. (including *A. frumentaceus* Buch.-Ham., *A. leucocàrpus* S. Wats., etc.) in central and northwestern Mexico. Colchicine-induced autoploids and amphiploids have much larger seed. C-4 cultivars may hold promise for rapid production of cereals (and greens and biomass) under conditions of relative heat and drought, but all seem to be frost sensitive. Heat-tolerant cultivars assigned to *A. gangeticus* L., the so-called "Tampala" include 'Fordhook tampala', 'Red tampala', and 'Tampala'. Many cultivars are photoperiod sensitive.

Anacardium (cashew). The cashew tree, *Anacardium occidentale* L., ($2n = 42$) is noted for its tolerance to acidic soils, hardpans, drought, and all the attendant stress found in tropical American savanna, its apparent center of diversity. Intolerant of frost, it apparently cannot be grown at higher latitudes.

Ananas (pineapple). Native to the New World, drought-tolerant pineapple ($2n = 2x = 50$) is now widely cultivated in the tropics and subtropics. Most improvement results from selection of somatic mutants within clonal cultivars and by intercultivar hybridization. Germplasm collections of wild material at the dry and cold extremes of the wild pineapple's distribution could be valuable in extending the ecological range of the cultivar. Pineapple is already valuable in its tolerance, similar to that of cashew, to the vicissitudes of the tropical savanna habitat. It is also a useful crop for highly acid, tropical, peat soils, and savanna. *A. ananassoides* (Bker) L. contributes disease resistance, like *A. bracteatus* Schult, but also tolerance to cool moist climate.

Annona (Soursop). *Annona* is largely a tropical American genus with about 125 species. *Annona muricata* L., ($2n = 2x = 14$) the guanabana or soursop, is extremely sensitive to cold, suffering leaf-fall and die-back near 4°C. *Annona glabra* L. ($2n = 4x = 28$) and *A. montana* Macf., however, endure the normal winters in southeastern Florida. The latter has been crossed with *Annona muricata*, offering hope for increased cold tolerance in the soursop. *Annona glabra* can contribute tolerance to dry soils. 'Hardy' is a relatively hardy cultivar of *Annona cherimola* Mill. ($2n = 2x = 14$), native to the uplands of Ecuador and Peru. If the cold tolerance of the pawpaw (*Asimina*) could be transferred to *Annona,* it would offer great potential for moving *Annona* to higher latitudes (14).

Arachis (peanut). With 15 to 70 species, the peanut genus has its center of diversity in the Matto Grosso of Brazil. Most species are diploid ($2x = 2n = 20$). Annual species are more characteristic and tolerant of semiarid areas, perennials of humid high-rainfall areas. Peanuts are quite tolerant of acid soils, and aluminum, requiring a minimum of lime for acceptable yields.

Armoracia (horseradish). Horseradish (*Armoracia rusticana* Gaertn., Mey. & Scherb.) ($2n = 4x = 32$), largely root propagated, has many seed-sterile strains. Native to temperate Eastern Europe and Turkey, it has been cultivated for about 2000 years. In the United States, most horseradish production has a narrow genetic base, mainly on three cultivars: 'Big Top Western', 'Common', and 'Swiss'. Tolerant of poor soil, and drought, horseradish is so weedlike, once established, and the market so small, that no search for abiotic tolerance seems indicated.

Asparagus (asparagus). From an Old World genus of some 300 species, mostly of rather dry habitat, only *Asparagus officinalis* L. ($2n = 2x = 26$) provides a commercial vegetable. A dioecious perennial, of necessity outpollinated, asparagus produces highly variable seed crops. Hence, useful variations are propagated vegetatively. Suspected to have originated in European salt steppes, asparagus already possess some salt and cold tolerance. With alfalfa, beets, and palms, asparagus is one of the most tolerant crops to boron toxicity; cereals, cotton, and truck crops are intermediate, while citrus, deciduous fruits and nuts are most sensitive to B toxicity (7). Cultivars within asparagus and others vary in B tolerance.

Avena (oats). Fourth among temperate cereals, oats have played a big role in both moist maritime and Mediterranean climates. There are diploids, for example, *A. strigosa* Schreb. and *A. brevis* Roth, minor forage oats, tetraploids, *A. abyssinica* Hochst, ex A. Rich., and the more important hexaploids, for example, *A. sativa* L. Recent European land races were very probably heterogeneous mixtures of homozygotes capable of adaptive responses to the condition of their habitat. Spring- and autumn-sown oats probably existed in these land races before selection for heat and cold tolerance began. *A. sterilis* L., with many weedy wild ecotypes, is a likely source of new variation (29). It is generally accepted in the United States that all the important hexaploids have been derived from wild *A. sterilis* L. Winterhardiness derives from resistance mostly to low temperature but also to alternating temperatures, unfavorable moisture conditions, smothering, diseases, and desiccation resulting from high wind.

Marshall (39) describes his search for freezing resistance and winterhardiness in oats. Of 58 populations, 34 increased significantly in freezing resistance and winter tolerance. Of the F_3 lines derived from crosses between certain winter and spring cultivars, 17% exceeded the winter parents in hardiness. Further, wild hexaploids are suggested as an untapped reservoir of divergent genes to combine with those of cultivars to produce transgressive types.

High levels of available P in soils lowered transpiration rates, so cultivars able to make more P available to themselves might be preadapted to the drought habitat. As a rule, oats are more tolerant to low pH and high Al than many barley cultivars. 'Park' is an Mn-efficient cultivar that tolerates low Mn soil levels. Selection for heat and drought tolerance frequently involves germplasm from *A. byzantina* K. Koch., for example, 'Montezuma'. Winter-hardy cultivars include 'Dade', 'Hickory', 'Nora', 'Norline', 'Pennlan', 'Pennwin', and 'Windsor'. Cultivars differ in susceptibility to Fe chlorosis.

Beta vulgaris (beet, chard). *Beta vulgaris* L. ($2n = 2x = 18$) is the only economically important species of the largely European genus, with subspecies *vulgaris* embracing sugar beet, beet, fodder beet, and mangold, subspecies *cicla* embracing the chards, and subspecies *maritima,* thought to be the progenitor. Useful tolerances are scattered in the genus, and may be incorporated into *Beta vulgaris* L., depending on the proximity of genetic relationships.

Tolerant cultivars reported include 'Crosby Green Top' for adverse weather, '67-9166', 'Mongerm 2205', 'SP66406 0 mmPf', 'Sp68503 0 mmPF', and 'US401' for aluminum; 'Select Dark Red B & C', and 'Seneca Detroit' for B deficiency; 'Early Wonder Staygreen' for cool weather; 'King Red' for muck and peat soils. Differential tolerance to Al exists among sugar beet cultivars and among individuals of the same cultivar. Genetic materials exist for expanding sugar beet into areas of acid soils by developing new cultivars tolerant to Al toxicity (32).

Beets are highly tolerant to salt during most of their life cycle, but peculiarly more sensitive during germination. Most crops, on the other hand, seem equally or more tolerant during germination than at other stages (45). Germination of sugar beet seed is strongly depressed by salinities of 3–8 mmhos/cm at 25°C and above but not at 10–15°C.

Brassica (cabbage, cauliflower, mustard, etc.). A complex genus, *Brassica* has at least six interconnected groups of economic species: three diploids, the *nigra* group ($2n = 16$), the *oleracea* group ($2n = 18$), and the *campestris* group ($2n = 20$); the allotetraploids, the *carinata* group ($2n = 34$), the *juncea* group ($2n = 36$), and the amphiploid *napus* group ($2n = 38$). Among the so-called mustards (see also *Sinapis*), *B. nigra* (L.) Koch, the black mustard, is the only species with $x = 8$. *B. carinata* A. Br., the Ethiopian mustard, is of little importance, whereas *B. juncea* (L.) Czern. & Coss., the brown mustard, is an allotetraploid involving an $x = 8$ species and an $x = 10$ species. The black-mustard probably originated in the Persian area, but has been widespread as a spice (and weed), with the development of many land races.

B. carinata A. Br. is a local population of little importance in the Ethiopian center of diversity. Machine-harvestable *B. juncea* (L.) Czern. & Coss., which has all but replaced *B. nigra* (L.) Koch, is thought to have migrated to secondary diversity centers in the Caucasas, China, and India, from a primary center in Central Asia, perhaps near the Himalayas. Breeding work among the mustards has been all but restricted to pure lines of *B. juncea* (L.) Czern. & Coss., a self-pollinating species. The long-lived seeds and wide distribution of land races serve as good germplasm reservoirs, with many short cold-season frost-tolerant races.

The *oleracea* group contains a diverse group of vegetables with flowers, buds, stems, or leaves, being eaten in one cultivar or another. Wild cabbage for Atlantic and Mediterranean coastlines is an important type in the origin of the vegetables, probably commingling germplasm from such interfertile species as *Brassica cretica* Guss., *B. insularis* Moris, *B. macrocarpa* Guss., and *B. rupestris* Rafin. Cultivars were probably developed independently around the Mediterranean, but they are all interfertile. Though derived in warmer climes, many cultivars have been selected for cold tolerance. In broccoli, cold tolerance is reported for 'Early C-T' and 'Zenith'. In cabbage, cold tolerance is reported in 'Allyear', 'Huguenot', and 'Madison'; drought tolerance in 'Jersey Queen' and 'Racine Markey'; heat tolerance in 'September 6' and 'Zavolzhskaya 3', and "muck" tolerance in 'Badger Belle', sand tolerance in 'Badger Blueboy'. In cauliflower, cold tolerance is reported in 'Armado Clio', 'Primo', and 'Svale', heat tolerance in 'Pua Kea'. In India's rays [*Brassica juncea* (L.) Czern. & Coss.], 'RL 198' and 'T51' are regarded as tolerant to Zn deficiency.

The polymorphic *campestris* group ($2n = 20$) contains the turnip and its relatives, a number of subspecies differing probably by a few genes. Pakchoi or Chinese mustard is a Chinese leaf vegetable, as is pe-tsai or Chinese cabbage with inflated petioles. Other leafy vegetables include sspp. *laciniata, narisosa,* and *nipposinica.* Indian oilseeds include ssp. *dichotoma* ("toria" or "Indian rape") and ssp. *trilocularis,* the yellow-seeded sarson, which are interfertile, and *Brassica tournefortii* Gouan ("rai" or "wild turnip") ($2n = 20$), not interfertile. Ssp. *oleifera,* turnip-rape seems closest to the wild types. Within this group there are hardy cultivars like turnips and tender cultivars. Within turnip 'Broccoli Raab' and 'Purple Top White Gold' are cold-tolerant cultivars.

The *napus* group is an amphiploid derivative of the *campestris* and *oleracea* group. Hence it probably originated in an area of overlap of these two groups. Indeed amphiploidy could have been repeated between differing populations of the parental group. Unlike its parents, it lacks the sporophytic incompatibility system. Oriental *"B. napella"* is completely interfertile with *B. napus* L. Some artificial oil-seed rapes are promising sources of cold tolerance.

Cajanus (pigeon pea). With $x = 11$, most cultivars are diploid, but tetraploids and hexaploids occur. *Cajanus* is closely related to and interfertile with several species in the genus *Atylosia.* Both genera have their center of diversity in India. Self-compatible and self-pollinated, *Cajanus* is often crossed by insects. Tolerating very poor conditions (except shade and waterlogging), *Cajanus* also offers promise for late maturation, and perennial habit with tolerance for drought (54).

Camellia (tea). *Camellia sinensis* (L.) Ktze. ($2n = 30$) is economically the most important species in this genus of some 100 species. Wild tea, if it still exists, is masked by introgression with cultivated types. The centers of origin and diversity are the subtropical highlands of southeast Asia, perhaps near the headwaters of the Irrawady or the lower Tibetan mountains, perhaps even central Asia. *Camellia japonica* L. is a source of cold tolerance. Selected clones yield 50–100% more than seedling teas; similar variation might exist in tolerances. Seedling populations should be sought near ecological extremes for tolerance, before clonal reproduction largely supplants the seedling reproduction and its inherent variability. Tea is tolerant of aluminum.

Capsicum (chili, peppers). About 25 species, five of them rather important economically, constitute the American genus *Capsicum*. The cultivated species are self-compatible, tolerant of inbreeding and usually bred as pure lines. Sweet peppers, as well as many chili and paprika peppers, belong to *C. annuum* L. Rarely cultivated outside South America, *C. baccatum* L. is readily distinguished from the preceding by the spots on the corolla. *C. frutescens* L. contains the cultivar 'Tabasco' and others with blue anthers and fascicled peduncles frequent. *C. chinense* Jacq. is the most commonly cultivated species in the Amazon. The rocoto, *C. pubescens* Ruiz & Pavon, is an upland Andean species, also cultivated in Central America and Mexico.

'Burpee's Fordhook' is said to be tolerant to adverse conditions, 'Merrimack Wonder' to cold weather during fruiting, 'Early Bountiful' to drought and heat, 'Morgold' to prairie conditions, 'Early Bountiful' and 'Pacemaker' to short season, 'Idabelle' to intense sun, and 'Puerto Rico Perfection' and 'Puerto Rico Wonder' to tropical conditions.

Carica (papaya). *C. papaya* L. ($2n = 2x = 18$) is the main economic species but edible fruits are reported also in *C. chilensis* (Planch.) Salms, *C. goudotiana* Solms-Laub., *C. monoica* Desf., and *C. pubescens* Lenne & K. Koch. Only three species in the family are not strictly dioecious: *C. monoica, C. papaya,* and *C. pubescens.* The center of origin of the papaya is considered to be eastern Central America. Disease and fruit flies are serious biotic enemies; frost is an abiotic enemy. Increased cold tolerance may be donated by montane populations of *C. papaya* L.

Carthamus (safflower). Cultivated safflower ($2n = 2x = 24$) probably stems from the Near East where *C. flavescens* Willd., *C. oxyacanthus* Bieb., and *C. palaestinus* Eig occur. It hybridizes readily with these weedy species. Safflower tolerates drought and heavy soils. Breeding has aimed at disease resistance to *Alternaria, Botrytis, Fusarium, Phytophthora, Puccinia,* and *Verticillium,* which will allow the movement of safflower into wetter areas.

Desert species like *C. palaestinus* might contribute more drought tolerance, lowering water requirements. In Iran and the United States tolerance to $-15°C$ has been developed, pointing toward a potential winter safflower. *Carthamus flavescens* and *C. lanatus* L. may be donors for cold tolerance. 'N-S' and 'WO-14' are relative cold-tolerant cultivars. Knowles (34) mentions winter-hardy introductions from Iran, and the transfer of winter hardiness from wild species to cultivated safflower.

Carya (pecan).   With 25 species from temperate Asia and North America, the genus *Carya* is important for pecans, hickory nuts, and timber. The pecan, *Carya illinoensis* (Wangenh.) K. Koch, is most important. Among tolerances reported in pecan: 'John Garner' is said to tolerate aridity; 'Apache', drought; 'Elliot', 'Hasting', 'Kibler', 'Maramec', [6]'Mount', 'Select', 'Shoshoni', 'Stark Hardy Paper Shell', 'Stark Surecop', 'Starking Hardy Giant', 'Steuck', 'Stuart', 'Wilson', and 'Witte', cold; 'Curtis', 'Desirable', 'Humble', and 'Moreland', low chill situations; 'Select', variable climate; and 'Owens', 'Shoshoni', and 'Tejas', due to strong crotch development, may be better adapted to tolerate wind.

Castanea (chestnut).   With some 10–15 species, *Castanea* is a hardy north temperate genus, producing some nuts of economic importance. Hardier cultivars include: 'Essate-Jap', 'Kelsey', 'Manoka', 'Penoka', 'Sleeping Giant'.

Chenopodium (lambsquarter, quinoa).   From American centers of diversity, one Andean diploid, *C. pallidicaule* Aellen ($2x = 18$) and two tetraploids, Andean *C. quinoa* Willd. and Mexican *C. nuttalliae* Safford ($4x = 36$) are cultivated as cereals. All had been developed from unknown progenitors before the arrival of Columbus. Growing high in the Andes, some cultivars possibly have greater cold tolerance than the widespread small-seeded weed, *Chenopodium album* L.

Cicer (chick pea).   With some 20 to 40 species, the Asian genus *Cicer* consists largely of diploid $2n = 2x = 16$ species. The center of diversity may have been the Caucasus or Asia Minor. India and Ethiopia seem to represent secondary centers of diversity (50). Cicer has largely been selected at the land race level with low inputs. Little or no effort has been devoted to improvement of abiotic tolerances, but there is genetic potential for drought, salt, and poor-soil tolerance. In India 'BR 78', 'N 59', 'P 6628', and 'S 26' are said to tolerate Zn deficiency (51).

Cichorium (chicory, endive).   Chicory and endive belong to the genus *Cichorium*, a small genus with about 10 species. Chicory (*Cichorium intybus* L.) ($2n = 2x = 18$) stems from the EuroSiberian center of diversity. Widely cultivated in Europe as a perennial salad, chicory has become a weed in the United States. Autotetraploid chicories vary more than the diploid parent. Endive or escarolle (*Cichorium endivia* L.) ($2n = 2x = 18$) has its primary center of diversity in the Mediterranean with secondary centers in the

EuroSiberian Center. An annual or biennial, it may have derived from a cross between *C. intybus* L. and *C. pumilum* Jacq.

Citrullus (watermelon). The drought-tolerant watermelon is native to the sandy dry areas of the Kalahari Desert. Watermelon tolerance to cool weather is reported in 'Merrimack Sweetheart'; to drought in 'Calhoun Sweet', 'Chris-Cross', and 'Sugar Baby'; to sandy soil in 'Alabama Giant' and 'All Heart'; and to sunburn in 'Charleston Gray', 'Graybelle', 'Purdue Hawkesbury' and 'Spalding'.

Citrus (lemon, lime, orange, etc.). *Citrus* is a complex genus, of 20 to 200 species depending on your taxonomic point of view. Most of the common cultivars and scions are diploids ($2n = 18$). Some *Citrus* selections tolerate higher maximum temperatures than others and certain hybrids sunburn or fail to set fruit in hot desert areas. Florida breeding programs sought cold-hardy rootstocks in *Citrus* × *Poncirus* crosses like those in Japan which produced 'Natsudaidai' and 'Yuzu'. Crosses for cold-hardy scions have involved *Fortunella*, 'Meyer' lemon, 'Natsudaidai' and 'Satsuma' mandarin.

Seedling trees of *Poncirus trifoliata* (L.) Raf. can survive in the northeastern United States. *Citrus, Eremocitrus*, and *Fortunella* are more cold tolerant than the commercially important citrus types, including 'Satsuma' mandarin. Each is a source for cold tolerance in rootstock breeding and has been hybridized. Requirements for breeding cold-hardy citrus hybrids include (*a*) suitable cold-tolerant parents, (*b*) heritable cold hardiness, and (*c*) ability to produce hybrid seedlings. Numerous trifoliolate orange (*Poncirus trifoliata*) hybrids were more hardy than the least hardy parent. Several hardy but ill-flavored citranges were developed. Cold-hardy kumquats (*Fortunella* spp.) were crossed with lemons and limes. The 'Lakeland' and 'Eustis' limequats, with lime-like fruits were more cold-hardy than their lime parent, demonstrating *a, b,* and *c* above, but mostly illustrating the heritability of cold hardiness. Some rather cold-hardy tangerines and their hybrids were bred into oranges, and so forth, to produce cold-hardy tasty fruits such as 'Robinson', 'Page', 'Lee', 'Nova', and 'Osceola' in Florida and 'Fairchild', 'Fortune', and 'Fremont' in California. (Practically no difference in cold hardiness between cultivars exists when the plants are in an unhardened condition. Young seedlings are comparable to the adults in cold hardiness, and can be screened easier for cold hardiness.)

Large differences in cold hardiness among scion cultivars and small to moderate differences among rootstocks with a common scion have been reported in most citrus-growing areas following exposure to both natural and artificial freezes (67). Russian studies (CAB abstract 10759) suggest that, in many cases, citrus polyploids are more tolerant to frost (and disease) than their diploid analogues.

Tolerance of flooding by rough lemon may be associated with tolerance to sulfide injury by H_2S. 'Poorman' orange (New Zealand grapefruit) rootstock appears adapted to heavier soil types.

For deep sandy soils, rough lemon, 'Milam', 'Carrizo', and 'Rangpur' lime are the most adaptable rootstocks. Sour orange, 'Cleopatra mandarin' and *Poncirus trifoliata* are best rootstocks for the poorly drained organic flatwoods soils. *Eremocitrus glauca* (Lindl.) Swingle has shown exceptional drought tolerance. *Microcitrus australis* (Planch.) Swingle and *M. australasica* (F. Muell.) Swingle are adapted to high rainfall, low fertility areas, and hence are candidates for breeding rootstock adapted to these stresses (27).

'Rangpur' lime, 'Cleopatra', 'Shekwasha', and 'Sunki' mandarins (*C. reticulata* Blanco), *Eremocitrus glauca* and *Severinia buxifolia* Tenore are salt tolerant, and all except the last have been tried for breeding salt-tolerant rootstocks. Salt tolerance is inherited quantitatively in citrus.

Tolerances reported for citrus include: drought tolerance in 'Yuma' citrange; cold tolerance in 'Glen' citrangedin, 'Thomasville' citrangequat and 'Swingle' "citrumelo"; high pH in 'Sacaton' citrumelo; salt in 'Swingle'; frost in 'Newell' line; salt in 'Rangpur' lime; cold in 'Eustis' and 'Lakeland' limequats; cold in 'Fairchild', 'Fremont', 'Kimbrough', 'Satsuma', and 'Silverhill' mandarins; heat in 'Fairchild', 'Fortune', and 'Fremont' mandarins; salt in 'Cleopatra', 'Shekwasha', and 'Sunki' mandarins; cold in 'Fukuhara' and 'Parson Brown' orange; desert in 'Rufert' orange; frost in 'Diller' orange; sand in 'Choate Navel' orange; cold in 'Nippon' orangequat and 'Gill' tangelo; and cold in 'Changsha', 'Clementine', 'Kara', 'King', 'Kinno', 'Orlando', and 'Wilkins' tangerines.

Cocos (coconut). The diploid coconut ($2n = 2x = 32$) belongs to the monotypic genus *Cocos,* of undetermined cytogenetic relationships with South American relatives (66). Common (or tall) palms are referred to, incorrectly, as *typica,* the dwarfs as *nana.* The talls are mainly outbreeding, dwarfs usually inbreeding. Southeast Asia, usually regarded as the main center of diversity, is where variations and colloquial names are most numerous. Breeding has made little impact on coconut evolution. Although local ecotypes have evolved to fit local ecosystems, the genetic bases are narrow. I would recommend crossing of strains from different ecosystems, hoping for transgressive segregation and heterosis in the offspring, the former for wider tolerance, the latter for higher yields.

Coffea (coffee). *Coffea arabica* L. and *C. canephora* Pierre ex Froehner are the chief sources of our main nonalcoholic stimulating beverage. Of species studied *C. arabica* is the only polyploid, being a tetraploid $2n = 4x = 44$. Others appear to be diploid and self-incompatible. Ethiopia is the primary center of diversity. The widespread cultivars are based on a small fraction of this diversity. Breeding programs recently and in the future will probably emphasize disease tolerance, but frost tolerance would be quite valuable. *C. arabica* is more tolerant of cooler climates than many of the other species. 'Mundo Novo' tolerates heat better than several cultivars.

Cola (cola). The genus *Cola,* with about 50 species, is centered in the humid tropics of Africa. The two most important species *Cola nitida* (Vent.)

Schott & Endl. and *Cola acuminata* (Beauv.) Schott & Endl. are tetraploid ($2n = 4x = 40$). The cultivated species tolerate poor soil, low pH, and some shade as seedlings and saplings.

Corylus (filbert, hazels).   Most cultivated species of *Corylus* have $2n = 2x = 28$, but some also have $2n = 22$ (*C. avellana* L., *C. maxima* Mill., and *C. tibetica* Batal.) *Corylus chinensis* Franch. tolerates heat and drought better than *Corylus avellana* L. Frost-hardy cultivars include 'Bixby', 'Filazel', 'Graham', 'Pearson Early Red', 'Pontica', 'Potomac', 'Rush', and 'Skinner'. Most of the hazel species tolerate frost. Numerous selections from native stand are considered to be more frost resistant, cold hardy and disease resistant, and to have larger nuts than some of the named cultivars. *Corylus colurna* L. tolerates shade, drought, and poor soil. *Corylus heterophylla* Fisch. ex Besser seems to tolerate acid soils better than some other species. *Corylus cornuta* Marsh. seems to tolerate smog and air pollution, especially SO_2 and temperatures down to $-50°C$ (37).

Cucumis (melon, cucumber).   Melons are native to India (though never found in the wild state). Selection has been directed toward disease resistance (downy mildew, powdery mildew, mosaic, scab). Cucurbitaceins repel some insects and attract others. A bitter species, *C. hardwickii* Royle (also $2n = 14$) grows wild in the Himalayan foothills, and hybridizes readily with the cucumber. Tolerance to cold wet conditions is reported in 'Hycrop Pickling', to drought in 'Morden Early', and to tropical conditions in 'PR-10', 'PR-27', and 'PR-39'.

India may be regarded as a secondary center of diversity for the melon, *Cucumis melo* L., but the $x = 12$ species are centered in southern Africa. Many embrace drought-tolerant germplasm. All muskmelon cultivars investigated have $2n = 2x = 24$. Tolerance to cold weather is reported in 'Pioneer' and 'Topset'; to drought in 'Texas Resistant #1'; to heat in 'Short'n'Sweet'; to salt in 'PMR 45'; and to sulfur burn in 'Gold Cup', 'Sierra Gold', 'SR91', 'SR1463', 'Top Mark', and 'V-1'.

Cucurbita (pumpkin, squash).   Of the more than 25 species of *Cucurbita*, five are domesticated, genetically isolated coenospecies whereas the many wild ecospecies, though separated geographically, are cross-compatible. With a generic center of diversity near southern Mexico, all have $2n = 2x = 40$. *Cucurbita ficifolia* Bouche is restricted to the highland subtropics of Latin America, and is tolerant of 12-hour days. *Cucurbita maxima* Duch., *C. mixta* Pang., *C. moschata* (Dutch.) Duch. ex Poir, and *C. pepo* L. contain types of pumpkins, squashes, and zucchinis. *Cucurbita pepo* has some tolerance to colder weather. Considerable interest accrues to the domestication of wild perennial species with extreme drought tolerance. There are genetic barriers to hybridization between species of the xerophytic species groups, DIGITATA group (*C. californica* Torre. ex S. Wats., *C. cordata* Wats., *C. cylindrica* L.H. Bailey, *C. digitata* A. Gray, *C. palmata* S. Wats.) and FOETIDISSIMA group (*C. foetidissima* H.B.K., *C. galeottini* Cogn., and *C. scabridifolia* L. H. Bailey). However it is possible to construct a genetic bridge using *C. moschata* as the bridging species (65). The buffalo gourd, *Cucurbita foetidissima,* has been singled out as a promising crop for arid lands.

With a tendency to shade out weeds, *Cucurbita* was intelligently used as intercrops by American Indians. Some shade tolerance must have been incorporated in the land races. 'Fortuna' is adapted to the tropics, 'Peraoro' to the humid tropics. 'Small Sugar Pumpkin' is eight times more tolerant of SO_2 than inbred cucumber lines [CAB 11055 (048)].

Cydonia (quince). *Cydonia* is a monotypic Near Eastern genus containing the drought-tolerant quince, *Cydonia oblonga* Mill ($2n = 2x = 34$). Quince is relatively tolerant to salt compared to most stone and pome fruits. Clonal *Cydonia* selections have been the most widely used rootstocks for pears in Europe. One of the most hardy rootstocks is 'Melitopolskaya'.

Daucus (carrot). A genus of some 60 species, *Daucus* has only one economically important species, the carrot, *Daucus carota* L. ($2n = 2x = 18$). Afghanistan is the center of diversity for anthocyanin-pigmented carrots and may be carrot's center of origin. Tolerance to long day probably was developed by prolonged and intensive roguing of those individuals which bolted under long days. 'Gold Spike', 'Spartan Fancy', and 'Spartan Sweet' are said to tolerate muck soils.

Dioscorea (yam). With perhaps more than 600 species, the genus *Dioscorea* may be grouped as Old World species, with $x = 10$, and New World species, with $x = 9$. High polyploids are reported among both cultivars and wild species, aneuploids are found among cultivars. Chromosome numbers vary, even within individuals. Separate domestication took place in Asia, leading to the main Asian yam *D. alata* L. ($2n = 30$ to 80), also *D. bulbifera* L., *D. dumetorum* (Kunth.) Pax, *D. esculenta* (Lour.) Burk., *D. hispida* Dennst., *D. japonica* Thunb., *D. nummularia* Lam., *D. opposita* L., and *D. pentaphylla* L. (mostly tetraploids or higher); in Africa, leading to *D. rotundata* Poir. ($2n = 40$) and *D. cayenensis* Lam. ($2n = 40, 60, 14$); and in America, leading to *D. trifida* L. f. ($2n = 54, 72, 81$). Shade tolerance might be sought among the forest species for intercropping stratagems; drought, aluminum, and poor-soil tolerance among the savanna species, salt tolerance where evapotranspiration exceeds precipitation, and cold tolerance at the upper limits of upland populations.

Diospyros (persimmon). Of the 500 species of the warm temperate genus *Diospyros* offering fruits of promise, the Japanese persimmon *Diospyros kaki* L. f. is most significant ($2n = 6x = 90$). It is cross-compatible with the American persimmon, *Diospyros virginiana* L. ($2n = 4x, 6x = 60, 90$) which is more tolerant of cold. Relatively hardy cultivars include: 'Edris', 'Garretson', 'Hicks', 'John Ricks', 'Juhl', 'Meader' ($-25°F$), and 'Peiping'. Among the hardier Japanese persimmons are 'Fuju' and 'Hachiya'. Several cultivars tolerate the low chill expected in central Florida: 'Fuyugaki', 'Hachiya',

'Hanafuyu', 'Hayakume', 'O'Gosho', 'Saijo', 'Tamopan', and 'Tanenashi'. McDaniel (personal communication, 1978) notes how many cultivars of *D. virginiana*, for example, 'Penland,' and all cultivars of *D. kaki* were killed by the hard Illinois winter of 1976–1977. Intolerant of shade *D. virginiana* appears to tolerate extreme pH, heat, and water stress.

Elaeis (oil palm). With $2n = 2x = 32$, Elaeis contains three species, the West African oil palm *Elaeis guineensis* Jacq., the American oil palm *E. oleifera* (H.B.K.) Cortes, and the Amazonian oil palm *E. odora* Tail. The African and American vicariads hybridize freely, producing fertile offspring. Within the genus, there is tolerance for both savanna and swamp, two agriculturally difficult environments. Development of salt tolerance might permit the establishment of this palm in periodically inundated tropical salt marshes. Oil palm is well adapted morphologically to drought. Drought resistance of seedlings is being investigated by submitting them to high osmotic pressures or high temperatures. Salt tolerance might be similarly tested (26). A salt-tolerant oil palm to replace the Nypa palms in Asian mangroves might further flood the world with palm oil, and further depress the market for soy oil.

Fagopyrum (buckwheat). Generally three diploid species ($2n = 2x = 16$) are recognized: *F. cymosum* Meissn., *F. esculentum* Moench. (including *F. emarginatum* Moench.), and *F. tataricum* (L.) Gaertn.). *Cymosum* may have been the progenitor of the other cultivated species. Cultivation originated in temperate Eurasia. Perennial *F. cymosum* is native to China and India, self-compatible *F. esculentum* to China and Russia, and self-compatible *F. tataricum* to India and China, under harsher climatic conditions. Tolerance of poor soils is one of buckwheat's attributes. In Nepal, common buckwheat ranges widely in altitude from 100 to 2000 m while tatary buckwheat grows only above 1300 m with highest yields at 3800 m. *Fagopyrum cymosum* ranged from 1500 to 3000 m, being a potential source of cold tolerance.

Ficus (fig). Most species of the genus *Ficus* (about 2000 spp.) are diploid with $2n = 2x = 26$, but triploids (*F. elastica* Roxb. var. *deora*) and tetraploids (five African species) occur. The common fig, *Ficus carica* L., has been crossed with *F. palmata* Forsk., *F. pseudo-carica* Miquel, and *F. pumila* L. Arabia is suspected to be the center of origin for the fig, a center having considerable drought stress on plants growing there. 'Hunt' is a cultivar said to be cold tolerant.

Fragaria (strawberries). *Fragaria* × *ananassa* Duch. is the most important cultivated species of *Fragaria*, a genus of some 15–50 species. *Fragaria vesca* L., the European strawberry, is the commonest wild diploid ($2n = 14$). Several diploids, like *F. vesca,* are monoecious, self-compatible, and mostly inbreeding. *Fragaria* × *ananassa* is octoploid ($2n = 8x = 56$) derived from crossing two American octoploids. *F. chiloensis* (L.) Duch. and *F. virginiana* Duch. "The cultivated strawberry displays a wide variation in adaptation to environmental conditions, and, therefore, is an excellent subject for genetic engineering" (56). *Fragaria virginiana* has been used for American cultivars resistant to drought and low temperatures. Transgressive segregation for many characters may be relatively frequent in strawberry breeding. Both 'Cheam' ('Sileta' × 'Puget Beauty') and WSU 1239 (['Sierra' × 'Northwest'] × 'Columbia') are hardier than their parents. Others suggesting transgression for winter-hardiness are WSU 1165 ('Northwest' × 'Columbia'), WSU 1172 & 1173 ('Cascade' × 'Puget Beauty'), 'BC 2' ('Siletz' × 'Puget Beauty'), 'BC 5' ('Northwest' × 'Puget Beauty'), 'BC 15' ('Northwest' × 'BC-59-22-35' ['Siletz' × 'Puget Beauty']), and 'BC 25' ('BC-59-22-35' × 'Northwest') (16). Extreme winter-hardiness is partially dominant over nonhardiness. One source of hardiness is *Fragaria ovalis* Duch. from Cheyenne. Selections having frost resistance when crossed with 'Midway' transmit a high degree of frost resistance to seedlings. One selection of *Fragaria virginiana* showed no flower injury at −5°C; 'Sheldon' showed only 10%. Late-ripening cultivars like 'Sparkle' are not necessarily more hardy; the small buds in the protecting crowns are less susceptible to injury compared to cultivars with flowers or young fruits. Although there are differences between field- and greenhouse-grown plants, the relative hardiness between cultivars remains (48). 'Earlidawn' and 'Howard 17' are early cultivars with blossom hardiness.

Among tolerances reported in strawberry:[1] 'Alaska Pioneer' is said to tolerate acid soils; 'Alaska Pioneer', 'Institute D4', 'Kasuga' to tolerate alkaline soils; 'Agassiz', 'Alaska Pioneer', 'Arapahoe', 'BC4', 'Dry Weather', 'Evermore', 'Great Bay', 'Guardsman', 'Macherauchs', 'Fruhernte', 'Nokomis', 'Prairie Belle', 'Ralph', 'Surecrop' to tolerate drought; 'Ft. Laramie' to tolerate hail; 'Arapahoe', 'Arrowhead', 'BC-15', 'BC-25', 'Burgundy', 'Carver Bell', 'Catskill', 'Cheam' (v), 'Chief Bemidji' (e), 'Cyclone', 'Darrow', 'Earlidawn', 'Earlimore', 'Early Cheyenne 1', 'Evermore', 'Fletcher', 'Fort Laramie' (e) (−30°F), 'Glenalice', 'Glenda' (v), 'Glenheart' (v), 'Glenman', 'Glenrich', 'Isabella', 'Jubilee' (e), 'Kanner King', 'Kristina', 'Locke Lake Ruby' (9), 'Lowridge', 'Northerner' (e), 'Ogallala' (e), 'Pam Ann' (v), 'Parker', 'Parkland' (v), 'Porter's Pride', 'Potagold', 'Protem', 'Radiance', 'Red Chief', 'Red Coat', 'Red Giant', 'Red Glow', 'Rossella', 'Shuksan' (v),' 'Sioux' (e), 'Stoplight' (v), 'Streamliner', 'Tardiva di Romagna', 'Vale' to tolerate cold; 'Armore' to tolerate heavy silt loam; 'Glamour', 'Institute D4', 'Puget Beauty', 'Saanich Belle' to tolerate heavy soil; 'Chief-Bemnidji', 'Fruhernte', 'Institute Z4' to tolerate mildew; 'Fresno', 'Heidi', 'Institute D4', 'Kasuga', 'Lassen', 'Solana', 'Tioga', 'Torrey', 'Vale' to tolerate salinity; 'Alaska Pioneer', 'Beaver' to tolerate sandy loam; 'Alaska Pioneer', 'Armore' to tolerate silt loam; 'Frost Proof', 'Earlidawn', 'Howard 17', and 'Macherauchs Fruhernte' to tolerate spring frost.

---

[1]In quoting from the literature, I use "v" for very tolerant, and "e" for extremely tolerant, fully aware that the comparitiveness is speculative.

Certain cultivars are said to tolerate a short rest period, perhaps a heritable tolerance: 'Benizuri', 'Dabreak', 'Festiva', 'Florida 90', 'Fresno', 'Fukuba', 'Missionary', 'Parfait', 'Sequoia, 'Tiogal', 'Yachivo'.

Glycine (soybean). With close to 300 described species, *Glycine* now is reduced to fewer than 10, mostly with $2n = 2x = 40$, except for the perennial soybean with $2n = 2x = 22$. Tetraploids occur in *Glycine wightii* (Grah. ex Wight & Arn.) Verdc. alone in the subgenus *Bracteata* and in *G. tabacina* (Labill.) Benth and *G. tomentella* Hayata ($2n = 2x = 40, 80$) of the subgenus *Glycine*. According to the taxonomic point of view, the subgenus *Soja* has one or two species, the wild soybean *G. soja* Sieb. & Zucc. and the cultivated *G. max* (L.) Merr. There are few, if any, genetic barriers to gene flow between these species. Crosses within the subgenera may be successful, but attempts to cross the soybean with members of the other subgenera have been largely unsuccessful (30).

Cultivars better adapted to the tropics include: 'Kahala', 'Kaikoo', 'Kailua', and 'Mokapu Summer'. 'Manchu', 'Mukden', and 'Richland' are much more tolerant of Al than 'A-100', 'Flambeau', 'JA53-7-6', and 'Lindarin'. Seedlings of 'Ada' can tolerate cold weather, 'Ada' and 'Wilins' heavy soil.

High P levels decreased growth of P-sensitive 'Lincoln' soybeans more than P-tolerant 'Chief'. Added Zn overcame the depressing effects of P on 'Chief' but not on 'Lincoln' (44). Soybean cultivars show differential response to Fe stress by developing severe ('Forrest'), moderate ('Hodgson'), or no Fe chlorosis ('Bragg'). A recessive factor controls iron uptake in Fe-inefficient 'T203' (P.I. 54169-51) (7). When Fe-efficient 'Hawkeye' was replaced by new cultivars in central and north-central Iowa, Fe chlorosis developed, indicating that the new cultivars, not tested for Fe efficiency, were Fe inefficient. Leaves drop from 'Forrest' and 'Bragg' soybeans as they develop severe Mn toxicity symptoms, while 'Lee' showed only slight symptoms. Most cultivars seem to tolerate Zn and Cu stresses and Al-toxic soils. Andrew (1) noted that *Glycine wightii* was more tolerant than *Macroptilium atropurpureum* (DC.) Urb. to Mn in solution. Seven other species were even more tolerant, *Centrosema pubescens* Benth. being most tolerant of the nine, followed by *Stylosanthes humilis* H.B.K., *Lotonotis bainesii* Baker, *Macroptilium lathyroides* (L.) Urb., *Leucaena leucocephala* (Lam.) de Wit, *Desmodium uncinatum* (Jacq.) DC., and *Medicago sativa* L. He reported another study ranking *Centrosema pubescens*, *Glycine javanica* 'Sp-1', *Stylosanthes gracilis*, *Glycine javanica* 'Tinaroo', and *Macroptilum atropurpureum* 'Siratro'.

Gossypium (cotton). *Gossypium hirsutum* L. ($2n = 4x = 52$), *G. barbadense* L. ($2n = 4x = 52$), *G. arboreum* L. ($2n = 2x = 26$), and *G. herbaceum* L. ($2n = 2x = 26$), are the species contributing the world's cotton fiber.

Drought tolerance is fairly common in the genus, but frost tolerance is far from achieved. In some cotton regions, late-season insect buildup dictated early planting to minimize the destructive effects of the insects. Hence, Buxton and Sprenger (9) are looking for cotton with tolerance to lower temperature for germination and seedling emergence. Lines developed for low elevations germinated better at both 15° and 25°C than did lines developed for higher elevations. Germination characteristics showed a marked correlation with the geographical area where the cotton cultivar was developed. The search for cold-tolerant cotton should be directed toward colder altitudes and latitudes.

Although cotton may generally be considered salt sensitive, the fiber yield may not be reduced as much as biomass. Cotton germplasm used in the United States is Cu and Zn inefficient and Mn tolerant.

Helianthus (sunflowers). One of the few crops native to North America, the annual sunflower (*Helianthus annuus* L. $2n = 2x = 34$) can hybridize with most of the annual species. In common with so many other domesticates, it introgresses with wild and/or weedy elements of its species.

Hybrids between *H. annuus* L. and perennial *H. tuberosus* L., known as sunchoke, can be propagated vegetatively like the artichokes.

Drought tolerance and cold tolerance already prevail in some races of sunflower, which has become Russia's major oil seed. Heiser (personal communication, 1978) notes that some of our southwestern Indian cultivars of the sunflower may be of some value for drought resistance; India's 'Armavirskij' and 'NP15' are said to tolerate Zn deficiency. The Jerusalem artichoke has some tolerance for shade, flooding, and heavy alluvial soils.

Hordeum (barley). Barley is one of the most tolerant cereals, reaching upper cultivated montane limits, even desert oases, being more salt tolerant than other cereals. As irrigated land salinized in Mesopotamia (B.C.), a monoculture of barley developed to tolerate the salt. Barley irrigated with undiluted sea water can yield 20% of conventional yields. Salt tolerance is a heritable trait in barley (20). In general the salt tolerance decreases after germination. Though salt may diminish vegetative growth, seed yields may not necessarily be diminished significantly (45). Some authors gauge salt tolerance of native grasses by the rapidity with which they root in saline media. In the Andes, the introduction of barley was important to fill the short season with a cold-tolerant cereal. A cool-season crop, barley tolerates high temperatures if humidity is low.

Recent work has shown wide variations in tolerance to Al. Six-rowed barleys are generally more Al-tolerant than two-rowed barleys, but the tolerance can be transferred to high yielding two-rowed cvs. In some winter barley cultivars, Al tolerance is due to a single dominant gene. Within some 2000 strains, about 10% show tolerance to Al and/or low pH (52). In transferring resistance to mildew and leafrust from Al-sensitive 'Franger' to Al-tolerant barley, plants resistant to both diseases were selected from crossing with the recurrent 'Dayton' parent, with no selection for Al response. In the resulting homozygous mildew- and leafrust-resistant lines, some were as tolerant of Al as 'Dayton'. A composite cross of world winter barley grown on Al-toxic soils increased its tolerance considerably in just one generation. Forage yields of 'Dayton' and 'Kearney' did not differ when pH was increased to 5.9 by liming, although 'Dayton' outyielded 'Kearney' more than twice at pH 5.2, probably due to Al tolerance. In India, 'Jyot' and 'Vijay' are said to tolerate both deficiencies and excesses of Mn. 'K19', 'Kesari', and 'Vijay' are said to tolerate B toxicity. Barley, most other cereals and grasses, large-seeded legumes, and potatoes are relatively tolerant to low Mo or Mo deficiencies; alfalfa and clovers are intermediate; while beets, brassicas, lettuce, and tomato are highly susceptible (13). In India, Zn deficiency is a major stress to which 'BG7', 'BH2', 'DL40', 'DL70', 'PL26', 'PL27', 'PL74', 'PL76', 'RD103', and 'Vijay' are relatively tolerant. Apparently, Al and Mn tolerances in 'Dayton' and 'Kearney' coincide, unlike certain cottons and wheats in which Al tolerance or sensitivity does not necessarily correlate with Mn tolerance or sensitivity (24).

'Barsoy' and 'Dayton' have shown tolerance to Al, 'Sutter' and 'Winter Tennessee' to cold soil, 'Primus' and 'Excelsior' to drought, 'Primus' to heat, 'Barsoy' and 'Hudson' to Mn, 'Sutter' and 'Winter Tennessee' to wet soil, and 'Kamiaki', 'Rapidan', and 'Wong' to winter cold.

Ipomoea (sweet potato). Mexico is the supposed center of origin for the sweet potato, *Ipomoea batatas* (L.) Lam., $(2n = 6x = 90)$ and center of diversity for the Batatas Section. Comparative study of cultivars in America, Asia, and Oceania revealed the widest array of diversity in America. Geographically separated populations exhibit great variation in such physiological characters as cold and disease reaction in addition to morphological characters. Hill cultures of New Guinea and South America have been selected for cold tolerance and the long-vine soil-holding tendency on their steep farms. Cold tolerance is suggested for 'Jewel' and 'M3-309', heavy soil tolerance in 'UPR #7', tropical tolerance in Don Juan 'LO-360' and 'LO-323'. Yen (personal communication, 1978) suggests that there are many abiotic tolerances in sweet potato, which he notes grows in every conceivable edaphic situation—post-strand beach sand, riverine alluvial, forest, and mountain soil—suggesting wide adaptive tolerance.

Juglans (walnut, butternut). With some 15 species, one of them approaching the endangered status, the genus *Juglans* contains useful frost-hardy germplasm for nut and timber production. *Juglans regia* L. $(2n = 2x = 32)$ is the most widely cultivated species. Many species of Juglans can be hybridized, and exhibit heterosis. Hybridization programs are directed toward cold tolerance, as well as improved nut quality. Hardy "heartnuts" include 'Fodermaier', 'Iona', 'Marvel', 'Rival', 'Rosedale', 'Walters', and 'Wright'. Hardy walnut cultivars include 'Aco', 'Adams', 'Arthur Benninger', 'Ashworth', 'Boone' (v), 'Burtner', 'Colby', 'Countryman 1, 2, 3, 4, 6x', 'Crath 1' (e), 'Dunoka', 'Fateley' (v), 'Fioka' (e), 'Freel', 'Gent' (v), 'Hansen', 'Helmle', 'Himalaya', 'Illinois', 'Kentucky Giant', 'Kentucky Papershell', 'Krouse' (v), 'Lake', 'Littlepage' (v), 'Little York', 'Metcalfe', 'Moyer', 'NuSchafer', 'Orth', 'Pinecrest' (v), Seeando', 'Sparrow', 'Stark Kwik Krop', 'Tasterite', 'Shelf Fungus', 'Valnur' (v), 'Wallick 16' (v), 'Weng', and 'Weschake'.

Seedlings of *Juglans hindsii* Jeps. ex. R. E. Sm. and *J. regia* L. were much more sensitive to waterlogging (at 33°C) than *Pterocarya stenoptera.* At 23°, *J. regia* showed symptoms quicker than *J. hindsii,* while hybrids between the two 'Paradox' plants, though still sensitive, were more tolerant than *J. hindsii.*

Hardy butternut cultivars (*Juglans cinerea* L.) include 'Barney', 'Chamberlain', and 'Werschcke'. Butternut is the most cold tolerant of the American *Juglans* species. Carpathian walnuts will tolerate −37 to −40°C, but due to early spring leafing out, they are often damaged by spring frosts (22).

Lactuca (lettuce).   Lettuce belongs to the temperate genus *Lactuca,* a genus of some 100 species. A self-fertilizing polymorphic species, lettuce ($2n = 2x = 18$) is the major salad crop in North America.

Lettuce has medium to high tolerance of saline soils, and most cultivars high frost tolerance. Many populations of *L. serriola* L. are highly tolerant of drought while cvs of *L. sativa* L are very susceptible to droughty conditions (Whitaker, personal communication, 1978).

Tolerance to adverse weather with short photoperiod has been developed in 'Wesgreen'; to cold in 'Climax', 'Imperial 404', 'Jade', 'Red Coach', and 'Winterhaven'; to heat in 'Burpeeana', 'Calmar', 'Cosberg', 'Empire', 'Great Lakes G59', 'Lakeland', 'Merit', 'Oakleaf', 'Pennlake', 'Primaverde Great Lakes', 'Salad Bowl', and 'Slobolt'; and to muck in 'Bibb', 'Chesibb', 'Empire 1957', 'Greenland', 'Minett', 'No 456', and 'Tendercrisp'.

Lens (lentil).   With 5 to 10 species, Lens is native to southwestern Asia and the Mediterranean. Diploid ($2n = 2x = 14$) *Lens culinaris* Medik. may have been derived from the very similar wild species, *Lens orientialis* Popow, a species of the Near East, where one finds the earliest archeological evidence of lentil cultivation (7000–6000 BC). As with other self-pollinated seed crops, lentil has been selected for numerous true-breeding lines in locally endemic land races (69).

Litchi (lychee).   An asian genus of 10–12 species, *Litchi* has only one important cultivated species, the lychee, *Litchi chinensis* Sonn. ($2n = 2x = 30$, 28). Attempts to move this species into temperate or humid tropical situations have failed. It fruits well only in cool dry seasons. 'Bengal' is said to tolerate alkaline soil.

Lupinus (lupines).   With perhaps 500 species or more, the genus *Lupinus* is largely American, secondarily Mediterranean. Compared with the wild lupines, the cultivated species have larger, lighter colored seeds (sometimes black in *Lupinus mutabilis* Sweet). *Lupinus albus* L. ($2n = 50, 30, 40$), *L.*

mutabilis (2*n* = 48), and *L. pilosus* Murray (2*n* = 42, 40) have been cultivated for thousands of years, the former in Eurasia, the latter in Andean
America. *L. luteus* L. (2*n* = 46, 48, 52) and *L. angustifolius* L. (2*n* = 40, 48)
are more recent domesticates. The telescoping of habitats in the Andes
might have resulted in clines of climatic land races.

Cold tolerance might best be sought among the Andean species and Russian cultivars, sand tolerance in *L. angustifolius,* waterlogging tolerance in
L. angustifolius, and Al tolerance in *L. luteus,* drought, acid and sandy soil
tolerance in *L. mutabilis.* Frost resistance is reported in 'Uniharvest' and
'Uniwhite' (*L. angustifolius*). Egyptian cultivars of *L. albus* seem to have
more tolerance to saline and tropical conditions, while some Yugoslavian
germplasm of *L. angustifolius* & subsp. *reticulatus* tolerates maritime sands.

Lycopersicon (tomato). Tomato is one of about ten closely related species
in the genus *Lycopersicon,* native to western South America. All species are
diploid (2*n* = 2*x* = 24).

Tolerant to a wide range of soils and climates, tomatoes demand a warm
season and well-drained soil, although some accessions of *L. esculentum*
var. *cerasiforme* (Dun.) A. Gray suggest tolerance to waterlogging. In
"tropical rainforests" of the eastern Andean slopes of Ecuador with rainfall
40–50 dm, much of the land surface may be covered by water. Tolerating
such circumstances, *cerasiforme's* very existence seems to imply resistance
to wilt and root-rotting fungi which often prevail in such environments.
Large-fruited cultivars of *L. esculentum* seldom survive the rainy season
there.

All cultivars differ in ability to germinate at low temperatures. 'P.I.
341988' (from 'Rhode Island Early' × a cold-germinating selection from 'P.I.
174263') has a recessive gene, *Ltg,* which enables it to germinate quickly at
10°C. A recessive factor controls the uptake of iron in Fe-inefficient
'T3238FER' (Brown, 7). 'T3238FER' is a good indicator plant for B availability in soil, growing well in soils containing 0.29 μg/g hot-water extractable
B. Use of B by plants is genetically controlled. 'T3238FER' shows B-
deficiency symptoms on Shano soil while 'Rutgers' grows normally (7).

Solanum pennellii Correll, *Lycopersicon chilense* Dunal, and *L. peruvianum* (L.) Mill. are suggested as sources of drought resistance; *L. cheesmanii* Riley f. *minor* accessions are possible sources of salt tolerance (Rick,
personal communication, 1978).

'Tuckcross K' and 'Westernred' are said to tolerate adverse conditions;
'Earlinorth', 'Pinklady', 'Red Cushion', 'Veebrite', 'Vision', and 'Wiscon-
sin Chief', cold at fruiting; 'Rhode Island Early', 'Veebrite', and 'Vision',
cold at germination; 'Farthest North' and 'Kenearly', cold; 'Solid Red Strain
A', 'Texto #1', and 'Viceroy-1960', cold; 'Golden Marglobe', 'Summer
Prolific', and 'Trent', drought; 'Caribe', 'Golden Marglobe', 'Kashmire',
'Louisiana All-Season', 'Ohio WR Brookston', 'Red Cloud', 'Red Global',
'Sioux', 'State Fair', 'Summerset', 'Texto #1', VF 145FS', and 'Vividi',

heat; 'Mustang Hybrid', prairie; 'Manalee', sand; 'Burgess Hybrid #1', 'Early Boy', 'Morden 1945', and 'Victor 1941', short season; 'Saturne', 'Supermarket', 'Turrialba', and 'Venus', subtropical conditions. Cultivars differ in their tolerance to Al.

Macadamia (macadamia). The genus *Macadamia*, largely from Oceania, has three nut-bearing species, all $2n = 2x = 28$. They are the Maroochy or Gympie nut, *Macadamia ternifolia* F. V. Muell, the smooth shell macadamia, *M. integrifolia* Maiden and Betche, and the rough shell macadamia, *M. tetraphylla* L. Johnson. These three produce fertile hybrids where their ranges overlap. The smooth-shelled Macadamia is more tolerant of tropical insular environs; the rough shell macadamia is more tolerant of a Mediterranean climate. Smooth-shelled macadamia is more tolerant of moisture and nutrient stress and high temperatures than is rough-shelled. 'Ikaika' and 'Kakea' hold some promise for cold tolerance. 'Ikaika' is said to be wind tolerant.

Malus (apple). Leading temperate fruit crops in world production (about 20,000,000 MT), apples belong to the genus *Malus,* a genus of some 25 species, mostly diploids. All species contributing to cultivars have been diploid ($2n = 34$), but many of the scions are triploid. Hybridization between wild species occurs readily. Intergeneric hybrids with *Sorbus* are easier attained than with *Pyrus*. Some workers recognize three apple species: *M. domestica* Borkh. (the cultivated apple), *M. pumila* Mill (including var. *paradisiaca,* the paradise apple), and *M. sylvestris* Mill (the wild crab apple); others lump them into *Malus baccata* (L.) Borkh. or *M. pumila* Mill, containing germplasm from most of the Eurasian species (e.g., *M. baccata, M. astracanica* Hort. ex Dum.-Cours, *M. prattii* (Hemsl.) Schneid., *M. prunifolia* (Willd.) Borkh., *M. pumila* and *M. spectabalis* (Ait.) Borkh. (62).

Apples are usually attributed to the EuroSiberian center of diversity and origin, with secondary centers of diversity and/or origin within the primary center [e.g., *M. pumila, M. trilobata* (Labill) R. Schneid.] or to the east [e.g., *M. baccata* (L.) Borkh., *M. ioensis*, (Wood) Britton.], or to the west [e.g., *M. florentina* (Zucc.) Schneid.].

Wild species have been bred in attempts to improve winterhardiness. Cummings and Aldwinckel (15) recommend 'M-26' and 'K-24' for low temperature in late fall. For midwinter cold, they recommend *M. baccata, M. fusca* (Raf.) O. Schneid., *M. ioensis* and *M. prunifolia*. 'Robusta 5', 'K-14', 'Budagovsky' rootstocks and numerous crabs, for example, 'Chestnut', 'Dolgo', and 'Kerr'. 'M-7', 'MM-111', and 'K-14' are recommended for drought; 'M-13' and *Malus fusca,* for "wet feet."

For hardy apple rootstocks, seedlings of related species are suggested: *Malus baccata, M. prunifolia, M. sargentii* Rehder, *M. virginiana,* and *M. × zumi* (Wats.) Rehder as the most hardy (64). All important apples are now grafted on special rootstocks, for example, *M. robusta* Rehder '5' *(M. baccata × M. prunifolia)* is widely used for cold tolerance in Canada.

Reportedly cold-tolerant cultivars include: 'Abbot', 'Acheson', 'Adanac'

(−50°F), 'Advance', 'Almata' (v), 'Almey', 'Alnarp' (v), 'Alred' (v), 'Altagold', 'Amsib' (v), 'Anaros', 'Andersen' (e), 'Anoka' (v), 'Antonovka', 'Atlas' (v), 'Barrie', 'Battleford', 'Beauty Blush', 'Blushed Caville', 'Bottle Greening', 'Bowyer', 'Breakey', 'Brightly Rosybloom', 'Brightness' (e), 'Britemac,' 'Brooks 27', 'Buchanan' (v), 'Calros', 'Caravel', 'Cardinal', 'Carma' (v), 'Carmine Queen Rosybloom', 'Centennial', 'Chesapeake', 'Chieftan', 'Chipman' (e), 'Christmas Red', 'Choka', 'Classic', 'Collet' (v), 'Connell Red' (v), 'Criteron', 'Cortland' (v), 'Custer' (v), 'Dakota', 'Dakota Beauty' (e), 'Daniels Redstreak', 'Davies', 'Dawn' (e), 'Delite', 'Discovery' (v), 'Dodd', 'Dolgo', 'Dr. Bill', 'Early', 'Eastman Sweet', 'Edith Smith' (v), 'Empire Red', 'Erickson' (v), 'Exeter', 'Fantazja', 'Florence', 'Folwell', 'Francis', 'Garnet', 'Garrison', 'Geneva', 'George', 'Gertie', 'Glenett', 'Glenmary' (v), 'Glenwal', 'Godfrey', 'Golden Anniversary' (v), 'Goodland' (v), 'Goosbey', 'Greene Spy', 'Greensweet', 'Hadlock Reinett', 'Haralson', 'Harcourt', 'Hardi-Spur Delicious' (e), 'Hardy', 'Harvester' (e), 'Harvest Special' (e), 'Hawkeye Greening' (v), 'Heart River' (v), 'Heaver', 'Heyer 12' (e), 'Heyer 14', 'Heyer 20', 'Hibernal', 'Holly', 'Hume', 'Jacques', 'Jenner Sweet', (e), 'Jesim', 'Joel's Red Delicious', 'Joey', 'Jonadel', 'Jonalicous', 'June Wealthy', 'Katja', 'Kendall', 'Kent', 'Kerr' (v), 'Killand', 'King Luscious', 'Kingscourt', 'Kress McIntosh' (e), 'Landland', 'Lambton', 'Law Rome Beauty', 'Leafland' (v), 'Lethonia', 'Linda Sweet', 'Linoloe', 'Lobo', 'Lubsk Queen', 'Luke' (e), 'Macfree', 'Macspur' (e), 'Manalta' (v), 'Manan', 'Manbee', 'Manchu' (e), 'Manden', 'Manitoba Spy', 'Manred', 'Mantet', 'Marvel', 'McClean', 'McCloy' (e), 'McLemore', 'Miller Sturdy Spur Delicious', 'Minjon' (v), 'Minnehaha', 'Minnetonka Beauty', 'Morden 347', 'Morden 361', 'Moris', 'Mount' (v), 'Mystery' (v), 'Northland', 'Nugget', 'Ogden', 'Ohlson' (v), 'Osimoe' (e), 'Osman', 'Pacific Pride', 'Park', 'Patterson' (e), 'Pattie', 'Paulared', 'Piotet', 'Piotosh', 'Pocomoke', 'Porcupine', 'Prairie Gold' (v), 'Prairie Spy', 'Prince', 'Prolific', 'Quality', 'Ranger', 'Redgold', 'Redheart', 'Red Heaver', 'Red of Eger', 'Redolgo' (−50°F), 'Red Prince', 'Red Sharon', 'Red Soviet' (e), 'Redwell', 'Redwine', 'Reid', 'Rescue' (v), 'Reta', 'Richland', 'Rodney', 'Rose Red Delicious', 'Rosthern 18', 'Rosthern 19', 'Royal-Red Delicious', 'Russian White', 'Rutherford', 'Sapinia', 'Saska', 'Scottie', 'Secor', 'Sharon', 'Shelley', 'Shortwell Delicious', 'Silvia', 'Snowy', 'South Dakota Macata', 'South Dakota Wendel', 'South Dakota Winter', 'Spangelo', 'Spencer' (v), 'Spuree Rome', 'Spur Mac', 'Stark Bounty', 'Stark Earliblaze', 'Stark Fullred Delicious', 'Starkspur Supreme', 'Starkspur Winesap', 'Stephens', 'Summerglo', 'Summerland', 'Super Starling Delicious', 'Tangowine', 'Thew Gold', 'Thorberg', 'Tolmo', 'YTony', 'Topred Delicious', 'Trail', 'Truax Greening', 'Ultrared Delicious', 'Unity' (−40°F), 'Viking' (e), 'Wakaga', 'Wakonda', 'Waldorf', 'Waubay', 'Wedge', 'West Virginia Red York', 'Winter Queen', 'Wisconsin Viking', 'Yakhon Towoye', 'Yellow Beauty', and 'Yellow Sweet'.

Attempts to move apples into the tropics have been complicated by chill requirements. Most cultivars require some chilling during the dormant season for proper development. Cultivars with low chill requirements are often of poor quality (6), but can be crossed with better quality cultivars. Early leafing is the criterion for selection, a character which seems to be polygenically controlled (about 30% of F_1 progeny and 40% of the back-crosses to 'Delicious' and 'Jonathan' were sufficiently early to be retained for further test). It may be possible to eliminate the chill requirement. Some of the earliest to leaf were even earlier after warm winters than after cooler ones (6). Apples reportedly tolerant to low chilling include: 'Anna', 'Beverly Hills', 'Dorsett Golden', 'Ein Shemer', 'Pettingill', 'Princess Noble', 'Red Dougherty', 'Rome Beauty', 'Vered', 'Vista Bella'. 'Muz' and 'Pocomoke' are reportedly tolerant to drought, 'Annalee', 'Muz' and 'Yellow Hackworth' to heat, 'Rose Red Delicious' to lead-arsenic, 'Muz' to neglect, 'Yellow Hackworth' to poor soil, 'Malling XV' to waterlogging, and 'Muz' and 'Pocomoke' to wind.

Mangifera (mango).   With 50 species or so, the genus *Mangifera* ($2n = 40$) belongs to the Indochinese-Indonesian-Malaysian center of diversity, cradle of many tropical fruits. The mango, *Mangifera indica* L., consists of outbred trees, producing great variability in seedlings, most of them useless. Mango seems to require a tropical dry season, though not so markedly as cashew. Although breeding programs aim at dwarfness, uniform maturation, and disease resistance, cold and drought tolerance would be extremely valuable in moving this excellent tropical fruit into higher latitudes. 'Bullock's Heart', 'Cogra', 'Langra', and 'Nimrod' are promising for cooler climate, 'Pope' for humid situations.

Manihot (cassava).   The genus *Manihot* is divided conveniently into two sections Arborae with primitive mesophytic tree species, and Fruticosae with more xerophytic shrub species, and germplasm for deserts, savannas, and steppe. Native to the Americas, the genus has Mexican and Brazilian centers of diversity. Most or all of the species can be intercrossed.

Cassava, *Manihot esculenta* Crantz ($2n = 36$), is closely related to *M. aesculifolia* Pohl, *M. pringlei* S. Wats. (low HCN), and *M. rubricaulis* I.M. Johnston. Introgression with wild species, coupled with man selecting in one direction, nature in the other, has produced a great diversity. Generally tolerant to and recommended for poor, acid soils, cassava is a good "low-input" crop. Low-input cultivars of cassava are tolerant to poor, acid, allic, infertile soils. In Colombia, trials were conducted with 138 cassava cultivars subjecting them to 0–6 MT lime/ha. Although subject to diseases, including superelongation, cercospora, and bacteriosis, plants developed normally in the first 3 months. Some were not responsive to lime at all, some were damaged by high lime (perhaps a Zn deficiency) and others were intermediate (60).

Two rubber-producing species, *M. dichotoma* Ule and *M. glaziovii* Muell.-Arg., have contributed improved vigor, drought resistance, and disease resistance to cassava, while *M. melanobasis* Muell.-Arg. modified the leaf display and root protein content. Jennings (31) summarizes: "There seems no reason why the ecological range of cassava could not be greatly extended by crossing with species adapted to drought (from areas of low rainfall or sandy soil), lower temperatures (high altitudes) or acid soils." *M. aesculifolia* Pohl is probably tolerant of high pH. *M. rubricaulis* spp. *rubricaulis* and ssp. *isoloba* are rather tolerant of cool weather. Drought tolerance is found in *M. augustiloba* (Torr.) Muell.-Arg. and *M. davisiae* Croizat.

Musa (bananas). Intolerant of frost, banana and its relatives are tropical species confined to latitudes below 40° until cold tolerance can be developed. Diploid wild bananas are from southeast Asia and Oceania. Most cultivars originated from section Eumusa with $x = 11$, but there are some relatively unimportant species with $x = 10$.

Edibility first evolved in *Musa acuminata* Colla in Malaya or nearby. Most of the cultivars are vigorous triploids. Genomes of the edible *Musa acuminata,* sometimes commingled with the genomes of nonedible *M. balbisiana* Colla, yielded the edible cultivars; *M. balbisiana* perhaps donated a little cold and drought tolerance.

Nicotiana (tobacco). Of some 60 species, *N. bigelovii* (Torr.) S. Wats., *N. rustica* L., and *N. tabacum* L. are polymorphic, particularly the latter. Soil type and climate no doubt contribute greatly to the diversity. Heritable low rates of photorespiration which characterize certain tobacco lines may contribute to heat resistance. Cultivars differ in susceptibility to Ca deficiency.

Olea (olive). In the Mediterranean, still relict from tropical mid-Tertiary, the olive persists. It frequently mirrors hatitat diversity with clonal diversity (59).

Oryza (rice). Asian rice, *Oryza sativa* L., and African rice, *Oryza glaberrima* Steud. are morphologically rather similar but hybrids between them are highly sterile. In addition to these, the "sativa complex" includes *Oryza barthii* A. Chev., *O. longistaminata* A. Chev. & Roehr., *O. nivara* Sharma & Shastry, and *O. rufipogon* Griff. with partial sterility and minor pairing aberrations in hybrids between them. *Oryza nivara* and *O. glaberrima* share the aquatic (waterlog-tolerant) tendency with the "spontanea" forms of rice ("*Oryza fatua*"). Considerable introgression has taken place, with gene flow mainly from the cultivars to the wild forms. Selection in rice may have decreased the aquatic habit (in dryland selection), photoperiod response, and cold sensitivity. The greatest diversity of plant characters and the more primitive cultivars are found among lowland (wetland) rices (11).

Rice cultivars vary in Al tolerance. Crosses between Al-susceptible and Al-tolerant cultivars showed that the tolerance can be incorporated into sensitive high-yielding cultivars. Introduction of high-yielding Philippine cultivars generated Zn-deficiency problems. Because Zn levels were apparently lower, Zn supplementation was required for the Zn-inefficient cultivars (44). India's 'Annapurna', 'Balmagna', 'BR24', 'Caloro', 'Cauvery', 'CEB24', 'CR10-113', 'Madhukar', 'Ratna', 'S 10-18' and 'Sabarmati' are said to tolerate Zn deficiency (51).

At IRRI, cultivars are being sought for tolerance of deep water (for aquatic rice), drought tolerance (for upland rice), problem soils, and tolerance of low temperatures. Wide adaptation of 'IR8', 'IR20', and 'IR22' enable the wide spread of the semidwarf habit (controlled by a single recessive gene). Photoperiod insensitivity of semidwarfs enabled South Korea to grow N-responsive 'Tongil' (IR667-98) in a short temperate growing season. Temperate cultivars are often called 'japonica', tropical 'indica', with a third tropical insular group called 'bulu'. Seedlings of cold-tolerant 'japonicas' may grow and develop faster than 'indica' seedling when temperatures or water are low. Some tolerant cultivars include: 'Caloro' and 'Japonica' for cold; 'Aznil', 'Azucena', 'Bala', 'IR5', 'IR442-58', 'MI 48', 'Norin 21', 'Rikuto' for drought; 'Arkrose', 'Bonnet 73', 'Caloro', 'Dawn', 'Nato', 'Novo 66', and 'PVR-1' for high pH; 'Bonnet 73' and 'Satum' for low N situations, and 'Vista' for rainstorms.

Salt tolerance in rice would be worth millions. Certain cultivars germinate well at 30–40 mmhos/cm but die in the seedling stage, tolerance diminishing after germination. Some authors maintain that sensitivity increases during flowering and seed set; others report that tolerance continues to increase after the seedling stage with no decrease during flowering and seed set.

Passiflora (passionfruit).   Most passionfruit juice of commercial importance comes from tropical *Passiflora edulis* Sims and *Passiflora edulis* f. *flavicarpa*. The North American maypop, *Passiflora incarnata* L., survives winter well into the north temperate zone, budding back from the roots, which grow deeper than the soil freezes.

Persea (avocado).   The diploid outbreeding avocado ($2n = 2x = 24$), *Persea americana* Mill., has been crossed only with *P. floccosa* Mex.

It is generally believed that Guatemalan (var. *guatemalensis*) and Mexican (var. *drymifolia*) races are more tolerant to low temperature than West Indian (var. *americana*) races. The three races cross freely, and separation is sometimes difficult. 'Lula' is one of the more cold-tolerant Guatemalan-West Indian hybrids, taking $-2.8°C$ without leaf or wood damage (68). West Indian material is more salt tolerant (4) and is the only material adapted to fruit in a lowland tropical environment. As might be expected, chlorosis resistance is also greatest in the West Indian variety. The Mexican race is generally most susceptible to high-lime chlorosis, the Guatemalan race to poor-drainage chlorosis. Although the West Indian races are consequently favored for both chloroses, their apparently poorer performance in water-logged soils during and after unusually wet winters makes it probable that their usefulness for less tropical regions will be limited to breeding.

For problem soils, sudden climatic variations, and perhaps other stress factors, such sturdy cultivars as 'Bacon' and 'Zutano' make good commercial selections and therefore are preferred parents. 'Hass' seems prone to injury from various environmental stresses, but paradoxically is widely adaptable. In California, Guatemala cultivars generally perform better near the coast and are less fruitful inland, while Mexican cultivars are just the reverse. 'Hass' does very well in nearly every microclimate tried, except too cold or hot. It is the indicated parent where performance in a broad spectrum of ecosystems is important. By contrast 'Fuerte' bears well only in limited parts of California (4). Salt tolerance seems better in Guatemalan and West Indian than in Mexican races; 'DuPuis 3' is reportedly tolerant to adverse weather; 'Alma', 'Arturo', 'Bacon', 'Bannon', 'Brogdon', 'Carolyn', 'Chapot's No. 43', 'Chapultepec Park', 'Clifton', 'Day', 'Dewey', 'Duke', 'Gainesville', 'Gottfried', 'Harms', 'Lula' ($-2.8°C$), 'Mexicola', 'Murray Red', 'Nabal', 'Olivia', 'Romain', 'Sharwil', 'Teague', 'Topa Topa', 'Winter Mexican', 'Yama', 'Yon', 'Young', and 'Zutano' to cold; MacArthur to high N; 'Benedict', 'Indio', 'Mayo', and 'Mexicola' to heat; 'Scott' to lime-induced chlorosis; 'Scott' to salt; 'Reed' to salt burn; and 'Benedict' and 'Duke' to wind. 'Irving', a Mexican-Guatemalan hybrid, has shown exceptional tolerance to desert heat and low humidity in California.

Phaseolus (bean).   With some 200–250 species, *Phaseolus* is largely a New World genus; most of the so-called Old World species, at least the cultivated ones, transferred to the genus Vigna. The four cultivated species are diploids with $2n = 2x = 22$: *P. acutifolius* A. Gray, *P. coccineus* L., *P. lunatus* L., and *P. vulgaris* L.

Among lima bean cultivars, the following tolerances are noted: 'Cowey Red', 'Winfield' for cold weather; 'Bixby', 'Butterbean', 'Fordhook 242', 'Peerless', 'Plump Champion', 'Triumph' for heat; 'Piloy' for short day; 'Early Thorogreen' for weather shock at pod set. Lima beans can be selected for cool-season germination (17).

'Criolla' is a rather heat-tolerant dry bean cultivar whereas 'Bonita', 'Borinquen', and 'Criolla' tolerate tropical conditions. Navy bean cultivars vary in their tolerance to low levels of available Zn, 'Sanilac' and 'Saginaw' both responding to added Zn, though 'Saginaw' did not appear to be severely Zn deficient (44). According to Spain (60), Colombian bean cultivars divide sharply into two groups, with the black beans generally much more tolerant to soil acidity than the non-black beans. In general, kidney beans are quite sensitive to salinity, 11 mmhos/cm decreasing growth in bright light (1.1 lux) but not dim (0.4). High humidity increases the tolerance to salinity of kidney bean.

Among snap bean cultivars, the following tolerances have been noted: 'Mild White Giant' for adverse conditions; 'Royalty' for cold soil; 'OSU 949', 'OSU 2065', 'SRS 1884' for cool weather, 'Longval' for drought; 'Alabama #1', 'Ashley Wax', 'Choctaw', 'Cooper Wax', 'Logan', 'Longval' for heat; 'Pacer' for short season; and 'Royalty' for wet soils. 'Dade' is more tolerant than 'Romano' to Al. Russian studies of five French bean cultivars showed 'Khar'Kov 65-90' to be more resistant to heat and drought (CAB PB abstr. 11034 [048]).

Phoenix (date).   Drought, salt, and frost tolerance are desirable breeding objectives in the date (*Phoenix dactylifera* L.) ($2n = 2x = 36$). 'Dairee' is reported to be tolerant to heat; 'Dairee', 'Halawy', 'Khadrawy', 'Medjhool', 'Tadala', and 'Thoory' to humidity; and 'Medjhool' to rain.

Piper (pepper).   The main cultivar in the complex genus (about 2000 spp) is the black pepper (*Piper nigrum* L.), one of the oldest of spices. Already shade tolerant, black pepper has some cultivars that tolerate heavy soils and laterite, and makes a good candidate for intercropping with tree crops.

Pistachio (pistachio).   Native to western Asia and the Near East, pistachio ($2n = 2x = 30$) has been cultivated there for thousands of years. Cold tolerance should be sought in higher elevations of the centers of diversity. Staminate Clone 'Israel 502', 'Kerman', and 'Lassen' showed remarkable drought resistance under runoff farming in the Negev Desert Highlands.

Pisum (pea).   *Pisum* probably has no more than two species. So called *P. elatius* Bieb. tolerates more humid climates; *P. sativus* L., more arid climates. Among pea cultivars, 'Bridger', 'Emerald', 'Louisiana Bayou', and 'Wando' are relatively cold tolerant; 'Kormovoi-24', 'Selkirk', and 'Thriftigreen' are relatively drought tolerant; 'Emerald', 'Lolo', 'Manoa Sugar', 'Thriftigreen', and 'Wando', heat tolerant; 'Sprite' is said to tolerate high rainfall; 'Asgrow 40', 'Burpeeana', 'Canner 75', 'Freezer 37', 'Greenfeast', 'Melting Sugar', 'Ronda', and 'World Record' are said to tolerate subtropical conditions; 'Tiny Tim', wind; and 'Hy-pak' is said to tolerate stress in general. Peas are cool-weather crops but 'Manoa Sugar', being somewhat heat tolerant, is a preferred cultivar for Hawaii.

Prunus.   Ranked in order of decreasing world production, peaches (5,500,000 MT), plums (4,000,000 MT), cherries (2,500,000 MT), apricots (1,200,000 MT), and almonds (750,000 MT) are important members of this genus. The subgenus *Prunophora* contains plums and apricots; subgenus *Amygdalus,* almonds and peaches; and the subgenus *Cerasus,* the cherries. *Amygdalus* and *Prunophora* transfer genes readily, but almonds and damson plums do not.

Almonds, in Mediterranean areas and Turkey, have, until recently, been grown almost exclusively from seed, whereas they are mostly propagated vegetatively in the United States. Hardy or cold-tolerant cultivars include 'Carrion', 'Monterey', 'Northland', 'Price Cluster', and 'Utah'. Salt tolerance occurs in 'Davey', sand tolerance in 'Titan'. Within cultivars of the genus *Prunus,* almond is said to be most tolerant of drought, relatively tolerant of high pH, least tolerant of waterlogging.

Prunus fenzliana Fritch is very hardy, *P. spartioides* (Spach) Schneid. is adapted to arid montane situations, like other species in Sect. Spartioides, for example, *P. scoparia* (Spach.) Schneid, *P. agrestis* (Boiss.) Grasselly, and *P. arabica* (Oliv.) Grasselly. More xerophytic species occur in sect. Lycioides (33), and almondlike American *P. fasciculata* Gray, *P. minutiflora* Engelm., *P. macrophylla* Hens., and *P. havardii* Wright. In Russia, emphasis has been on developing late-blooming, frost-tolerant almonds by crossing and back-crossing with *Prunus mira, P. bucharica,* and *P. nana.*

Apricots are often considered to be drought resistant and will thrive in areas with low atmospheric humidity; still, they are sensitive to lack of soil moisture (2). In a sense, Bailey and Hough (2) imply that tolerance is grounds for specific designation. Speaking of *Prunus ansu* as distinct from *P. armeniaca,* they say: "However since the cultivars are adapted to more humid growing conditions, grouping them in a separate species has meaning for pomologists as well as for botanists." *Prunus sibirica* and *P. mandshurica* are very cold-tolerant species, withstanding −40 to 50°C.

Prunus mume Sieb. & Zucc. and cultivars of *P. ansu* are widely grown in humid oriental areas where winters are mild, with *P. ansu* growing farther north. *P. mume* should be the best source of disease resistance and adaptation to humid growing areas, also for milder winters. After their introduction into Europe, late-flowering types (avoiding spring frost) and cold-hardiness became desirable objectives. Quite a few cultivars are reported to be hardy: 'Abutalibi', 'Alfred', 'Anban', 'Anda', 'Baradzhabi', 'Big Davis', 'Chow' (v), 'Coffing', 'Deatrick', 'Doty', 'Earliril' (−19°F), 'Early Gold', 'Erevani', 'Farmingdale', 'Fialkovyi', 'Franciscan', 'Geneva', 'Gold Dollar', 'Golden Amber', 'Golden Delicious', 'Golden Gem', 'Goldrich', 'Haggith' (v), 'Harbin' (v), 'Hardy Iowa' (v), 'Harriet', 'Hulan' (v), 'Hungarian', 'Khosroveni', 'Lalin' (v), 'Manchu', 'Mandarin' (v), 'Mantoy', 'Max-Gold', 'Morden 601', 'Morden 604', 'Ninguta', 'Perfection' (blossoms tender), 'Purpurovyi', 'Robust', 'Rosy Cheeked', 'Sansin' (v), 'Scout', 'Shamrock', 'Sing', 'Sino' (v), 'Starglow' (v), 'Sungold' (e), 'Sunshine', 'Tola', 'Ul'Yanishehev 28', 'Worley', 'Zard', 'Zolotoe Letoe', and 'Zun' (v). Transgressive inheritance for cold tolerance is reported with crosses involving 'Zard'.

Tolerance to adverse weather is reported in 'Haffer'; to cold in 'Franciscan', 'Haggith', and 'Zard'; to heat in 'Franciscan' and 'Nukuly'; to humidity in 'Goldcot'; to brief chill exposure in 'Early Royal'; and to wind in 'Goldcot' and 'Robust'.

Cherries are frequently classified as the diploid sweet cherry, *Prunus avium* (L.) L. and the tetraploid sour cherry, *P. cerasus* L., here in the United States. Hybrids between these two are called "Duke types". *Prunus mahaleb* L. rootstocks are often hardier and are more tolerant of irrigated slightly saline soils than *P. avium* (L.) L. *Prunus fruticosa* Pall.; *P. pensylvanica* L. f. have been rated hardier than *P. mahaleb* (64). Nonetheless, breeding and selection in cherries have been confined largely to commercial sweet and sour cherries, especially the former. Consequently, cherries have been more isolated from the gene pool than other *Prunus* fruit species.

Cultivars of cherry reported to be hardy or frost tolerant include 'Angela', 'Baton Rouge', 'Belaya 4', 'Bing', 'Black Russian', 'Chinook', 'Coronation', 'Drilea', 'Dropmore', 'Griot Ostgeimskii', 'Hudson', 'Kansas Sweet', 'Krasa Severa Beauty', 'Lambert', 'Lotovka', 'Mesabi' (v), 'Meteor' (v), 'Negrityanka negress', 'Northstar', 'Orient', 'Podarok 50', 'Salmo', 'Sam', 'Shubert', 'Skris Shavi', 'Starking Hardy Giant', 'Truzhenita Toilec', 'Unark', 'Victor', 'White 4', 'Zhukov's Griotte'.

Those singled out for their ability to tolerate spring frosts include: 'Angela' (evades), 'Bell Montmorency' (evades), 'Geftera Krasnaya', 'Hefter's Red', 'Krasna Severa', 'Olivier', 'Ol'skaya', 'Skris Shavi', 'Sopernitsa Rival', and 'Turzhenitsa'.

The Chinese Bush Cherry cultivar 'Drilea' is reported to tolerate dry conditions, 'Mistawis' to tolerate prairie conditions. The 'Nanking' or 'Hansen' Bush Cherry *P. tomentosa* Thunb. ($2n = 16$) tolerates areas too cold, arid, and windy for sweet or sour cherries (21).

Nectarine tolerance to frost in flower is found in 'Lexington', 'New Yorker', 'Pocahontas', and 'Redbud'; to winter in 'Cavalier', 'Flaming Gold', 'Francisco', 'Gold King', 'Harcol', 'Hardired', 'Lategold', 'Mericrest', 'Morton', and 'Stark'; to low chilling in 'Banquet', 'Fantasia', 'Flamekist', 'Granderli', 'Independence', 'May Queen', 'Panamint', 'Rosel', 'Silver Lode', and 'Sunred'; to low rainfall in 'Flamekist'; and to spring frost in 'Cherokee'.

Peaches probably originated in western China and spread to Europe during the past 2000 years. Commercially the peach is a temperate crop, the equatorial boundaries being limited by minimum accumulated chilling temperatures and the polar boundaries by minimum winter temperatures. 'Honey' and 'Peen-to' were principal kinds of peaches used in breeding for mild winters. In the United States, "native" seedlings with low-chilling requirements were crossed with commercial temperate-region cultivars to combine low-chilling with desirable horticultural characteristics. Chilling requirements appear to be genetically controlled by multiple genes and physiologically by a hormone balance. Selection for adaptation to new environments has been especially important in the United States, where short tree life is of concern.

Peach × almond hybrid rootstocks offer one approach to drought tolerance in peach. Plums are generally better adapted than peaches to heavy soils. Almond-peach hybrids give better adaptation than peaches to high calcium and high pH soils, an advantage not operative in most of the southern United States where soils are acid (57).

Hesse (28) summarizes "We have reported on the observed improvement of lime-induced chlorosis through the use of peach × almond hybrid stocks." Italians use myrobalan plums as stocks to overcome this high pH chlorosis (Fe inefficiency). If we look hard enough, we should find an Fe-efficient peach cultivar on its own root.

Temperature and photoperiod are primary factors regulating cold acclimation in woody perennials. In some cases, the rootstock may determine the cold-hardiness of such deciduous fruit trees as peach. For example, 'Siberian C' rootstocks inflence cold-hardiness of scion flowers. 'Harrow Blood' and 'Siberian C' rootstocks transmitted more hardiness to the scion cultivar than did 'Rutgers Red Leaf' (46). There is great potential for improving cold-hardiness of peach rootstocks by breeding and selecting within *P. persica* (L.) Batsch, or by hybridizing with other hardy species, for example, *P. americana* Marsh., *P. besseyi* L. H. Bailey, *P. davidiana* (Carr.) Franch., *P. nana* Stokes, *P. nigra* Ait., and *P. spinosa* L. Russian scientists have been investigating many of these hybrids for winter-hardiness.

'Redhaven' peach on five seedling rootstocks exhibited varying degrees of injury due to waterlogged soils. 'Rutger's Red Leaf' was most tolerant, 'Lovell' was least tolerant. It seems feasible to select and breed for "wet feet" resistance in peach rootstocks (12).

Peach cultivars tolerant of hard winters include 'Almena Hart', 'Bailey', 'Belle' ($-25°C$), 'Brilliant', 'Burbank's Orchid', 'Burbank's Santa Rosa', 'Calora' ($-25°C$), 'Carol' (v), 'Cherryred', 'Chili', 'Colora' ($-17°C$), 'Comanche' ($-25°C$), 'Cumberland', 'Dawne', 'Dinkle', 'Early Champion' (e), 'Early Fair Beauty', 'Early Profit', 'Early Shinn', 'Early Triogem', 'Eclipse', 'Elliott Special', 'Fowler', 'Freeland', 'Frostking (v)', 'Frostqueen', 'Golden Jubilee', 'Golden Masterpiece', 'Goldray', 'Grebing', 'Harbelle', 'Harbrite', 'Hardee', 'Hardy-Berta', 'Harkon' (v), 'Harriet' (v), 'Harrow Blood' (LT50 = -22), 'Herholdt's Summer Pride', 'Hope Farm', 'Johnson Early', 'Elberta', 'July Gold', 'June Bride', 'Kenlate Elberta', 'Kimbo', 'LaGold', 'Legacy', 'Lizzie', 'Manitou', 'Mary Jane', 'Massasoit', 'Melvin', 'Meredith', 'Meteor', 'Missouri', 'My-o-My', 'Osage', 'Palmer White' ($-24°F$), 'Pioneer', 'Polly' (v), 'Raritan Rose', 'Redkist' (v), 'Redqueene' (e), 'Reliance' (v), 'Richhaven', 'Rosebud', 'Siberian' (LT50 = 21.8), 'Somerset', 'Spoto Gem', 'Stark Lateglo', 'Stark Late Gold' (v), 'Stark Sure Crop', 'Sullivan I', 'Sunapee' (v), 'Sunhaven' ($-25°C$), 'Tremmel' (v), 'Walter', and 'Welcome Hale'.

Flower and foliage buds of peaches and nectarines may have different chilling requirements. When they differ, the foliage buds nearly always have the greater chill requirement. Multiple or polygenic control seems to predominate, but one or a few genes may have major effects. Those tolerant of mild winters with low chill factors include 'Albatross', 'Armgold', 'Aurora', 'Babcock', 'Bells November', 'Boland', 'Curlew', 'De Berard', 'Desertgold', 'Dixiland', 'Dorothy N', 'Duke of Georgia', 'Earligold', 'Earlihale', 'Early Amber', 'Fairtime', 'Flordabelle' (v), 'Flordagold', 'Flordared', 'Flordasun', 'Flordawon', 'Four Star', 'Franciscan', 'Frankie', 'Glenalton', 'Golden State', 'Hawaiian', 'Hermosa', 'Honey', 'Ingwe', 'Jewel', 'John S. Shaw', 'June Gold', 'Keimoes', 'Maluti', 'Maygold', 'McRed', 'Meadow Lark', 'Okinawa' (v), 'One Star', 'Orion', 'Peen-to', 'Prenda', 'Ramona', 'Red Ceylon', 'Redtop', 'Redwin', 'Rio Grande', 'Rochon', 'Rubidou', 'Safari', 'Saharanpur', 'Saturn', 'Sunred', 'Swatow',

'Three Star', 'Tokane', 'Two Star', 'Ventura', 'White Knight 1', 'White Knight 2', and 'Woltemade'. 'Palmer White' is tolerant of heat; 'Rubired' and 'Shalil' of heavy soils; 'Shalil' of waterlogging; 'Fairhaven' and 'Giant Elberta' of wind.

Plums have the greatest genetic diversity of any subgenus, and tend to be a genetic link to the other subgenera. North American breeding programs tend to emphasize environmental adaptiveness by adding useful characters to *Prunus salicina* Lindl. and *P. domestica* L. from several germplasm donors. At Pont-de-la-Maye, France, rootstocks are being bred for resistance to winter asphyxiation of the roots and for less sensitivity to chlorine excess, a condition induced by calcareous soils.

'Ackermann', 'Brussel', and 'Damascena' plums, regarded as tender, might be better tried for low-chill situations; 'Brompton', 'Kroosjes', and 'Myrobalan' are moderately tender; 'Marianna', 'Huttner IV', and 'St. Julien' are moderately hardy; 'Pershore' is hardy; and *Prunus americana* Marsh., *Prunus besseyi* L. H. Bailey, and *Prunus nigra* Ait. are considered hardy (63, 64). *Prunus nigra* Ait. is reported to transmit greater hardiness than *P. americana* Marsh.

Cold-tolerant cultivars of plum include 'Acme', 'Algoma' (e), 'Anderson', 'Anoka', 'Black Kelsey', 'Bounty' (v), 'Brilliant', 'Chemal Jibilee', 'Chemalskii Suvenir', 'Chilcott', 'Chinook', 'C.K.C.', 'Crystal Red', 'Dandy', 'Dropmore Blue', 'Ducie', 'Eclipse', 'Elite', 'Elephant Heart', 'Elliott' (v), 'Ember', 'Excell', 'Fiebing' (v), 'Flaming Delicious', 'Geddes' (e), 'Gilbert', 'Gilman', 'Gloucester' (v), 'Gracious' (v), 'Great Yellow', 'Grenville' (v), 'Hennepin', 'Hinchliff', 'Hoffman', 'Ivanovka', 'Kota', 'La Crescent', 'Mandarin', 'Mason Early Italian', 'McRobert', 'Mendota', 'Meredith', 'Mina', 'Monitor' (v), 'Mordel', 'Morden 124' (v), 'Morden 125', 'Mound', 'Newport', 'Nobel French', 'Norseman' (−55°F), 'Norther', 'Novinka', 'Olson', 'Opata', 'Oranzhevaya', 'Parkside', 'Patterson', 'Pembina', 'Perfection', 'Pershore', 'Pipestone', 'Prairie', 'Ptitsin 9', 'Purple Heart' (v), 'Radisson', 'Red Ace', 'Redcoat', 'Redglow', 'Redtalian', 'Red Wing', 'Robinson', 'Russian Green Gage', 'Saskatchewan', 'South Dakota' (v), 'Stevens' (v), 'Superb', 'Superior', 'Supreme', 'Surprise', 'Tawena', 'Tecumseh', 'Tonka', 'Underwood' (v), 'Verity', 'Victor Large French', 'Wilbur', 'Wilson River' (v), 'Winered', 'Winona'. Hardy cherry plums include 'Alace', 'Beta', 'Convoy', 'Cooper', 'Deep Purple' (v), 'Delta', 'Dura' (v), 'Eileen', 'Epsilon', 'Gamma', 'Genevieve', 'Giant', 'Heaver', 'Hiawatha', 'Honey Dew', 'Kappa', 'Little Eagles', 'Manor', 'Manorette', 'Mansan', 'Mistawasis' (−50°F), 'Mordena', 'Nicollet', 'Oka', 'Omega', 'Prolific' (v), 'Ruby' (v), 'Sacagawea', 'Sanoba', 'Sapalta', 'Sigma', 'Skinner Favorite', 'St. Anthony', 'Wessex' (e), 'Zeta', and 'Zumbra' (v). 'Myrobalan' types have shown tolerance to heavy soils and waterlogging. 'Victoria Myrobalan' is highly resistant to high salt concentrations. Myrobalan and Marianna have had selections vegetatively propagated for hardiness, tolerance to wet soil conditions, drought resistance, and so forth (63).

Sandcherries belong to *Prunus besseyi* and *P. pumila*. Hardy cultivars in this complex include 'Black Beauty', 'Brooks', 'Honeywood', 'Leafland', 'Mando', 'Manmoor', 'Manorett' (−60°C), 'Oahe', and 'Ruby'.

Psidium (guava).   Guava, *Psidium guajava* L. ($2n = 2x = 22$), is one of many vitamin-rich edible-fruited species in an American genus of 100–150 species. Limited to the tropics by frost sensitivity, guava cold tolerance might be sought among some of its tropical montane relatives. 'Hart' is said to be relatively cold hardy.

Punica (pomegranate).   Alone in an Old World family of one genus, *Punica* has two species, the pomegranate, *Punica granatum* L., and the Socotro pomegranate, *Punica protopunica* Ralf. With $2n = 2x = 16, 18, 19$, pomegranate is an ancient Old World fruit, tolerant of the Mediterranean climate.

Pyrus (pear).   Pear belongs to the genus *Pyrus*, with about 21 species. All species examined have $2n = 34$ but there are a few polyploid pear cultivars. Hybridization between most wild species occurs readily. *Pyrus communis* L., *P. nivalis* Jacq., and *P. serotina* Rehder are considered to be the major contributors to the gene pool of the cultivated pear, with some contributions from *P. longipes* Coss. & Dur., *P. syriaca* Boiss., and *P. ussuriensis* Maxim., as well as other wild species. The cultivated pear, *Pyrus communis* L., is self-incompatible and crosses readily with other species of its genus, but hybridization with the apple genus *Malus* rarely produces surviving offsprings. *Pyrus* more readily hybridizes with *Cydonia* and *Sorbus*.

Several species of *Pyrus* can be tapped for cold-hardy rootstocks (*Pyrus betulaefolia* Bunge, *P. ovoidea* Rehder, and *P. ussuriensis* Maxim. as well as quince cultivar 'Melitopolskaya'). *P. ussuriensis* Maxim. in its native Siberian habitat of constantly cold winters, easily survives −50°F without injury. All *Pyrus* species are graft compatible, ranging from subartic hardiness (*P. ussuriensis* Maxim., −50°F) to tropical (*P. koehnei* Schneider, killed at 10°F) (64). *Cydonia* is more frequently used as a rootstock in Europe, especially in efforts to improve dwarfing. However it may be better to develop dwarfing *Pyrus* rootstocks than to attempt further improvement of the quince rootstocks because more incompatibility is encountered between intergeneric than intrageneric grafts. Furthermore, some of the *Pyrus* species are also more winter hardy, disease resistant, drought tolerant, waterlog tolerant, and have better anchorage than the quince rootstocks (38). Inheritance of cold-hardiness is polygenic and additive with little evidence for epistasis or dominance being important components of the genetic variance for winter-hardiness. Greatest efforts toward cold tolerance have been in North America and the Soviet Union, especially in the steppe and prairie regions where winters are severe. The most promising cultivars that combine cold-hardiness with quality fruit are hybrids of *P. communis* × *P. ussuriensis*. The selections have better size and quality than the *P. ussuriensis* parents and greater hardiness than the *P. communis* L. parents. Promising for the prairie are 'David', 'Golden Spice', 'John', 'Olia', 'Peter', 'Philip', 'Pioneer 3', 'Tait Dropmore', and 'Tioma'.

Pears grafted on *Pyrus salicifolia* do better on arid soils than those on *P. communis* rootstocks. *P. salicina* is also tolerant of extreme temperature changes and saline soil. One can select for drought resistance in both scions and rootstocks (38).

Cold tolerance is reported for several cultivars: 'Andrew', 'Bantam', 'Canal Red', 'Cashman', 'David', 'Dropmore', 'Dymond', 'Enie', 'Finland' (v), 'Finsib' (v), 'Finsib Sister', 'Funks Colorado', 'Golden Spice', 'Harbin', 'James', 'John', 'Lacock', 'Lady Marion', 'Luscious', 'Melitopolskaya' (rootstock), 'Menie', 'Mericourt' (−23°F), 'Miney', 'Ming' (v), 'Moe' (v), 'Olia', 'Patten', 'Peter', 'Phileson', 'Philip', 'Pioneer 3', 'Rosired Bartlett', 'Simon', 'Southworth' (e), 'Spartlet', 'Stark Jumbo', 'Tait 1' (−42°F), 'Tait 2', 'Tait-Dropmore', 'Tanya', 'Thomas', 'Tioma', 'Willard'; drought tolerance for 'General Lee'; low chill tolerance for 'Hood' and 'Pineapple'; and wind tolerance for 'Harbin'.

Raphanus (radish).   With about eight species, *Raphanus* has only one economically important crop, the radish, *Raphanus sativus* L. According to Banga (3), there are four types (all $2n = 2x = 18$), var. *radicula* the small cool-season root, var. *niger,* the heat-tolerant large root, var. *mougri,* grown as a leaf vegetable, and var. *oleifera,* a fodder-radish. Taxonomically, one of these should perhaps be called var. *sativa,* perhaps the first. Cold tolerance prevails in var. *radicula,* heat tolerance in var. *niger.* Selection for early cultivars has lead to a smaller root, rapid growth and an annual life cycle. Annual types may flower under any photoperiod.

'Red Boy' and 'White Spike' tolerate muck; 'Newari', which tolerates salt, is said to grow successfully only with brackish irrigation.

Ribes (currants, gooseberries).   The formerly recognized genetic distinction between currants and gooseberries, roughly *Ribes* and *Grossularia* (all diploid, $2n = 16$), will not stand the test of crossing sterility. Considerable interfertility exists among the various species.

Drought resistance occurs in the 'Wapago Golden' currant; cold tolerance in numerous cultivars, among them 'Abundance', 'Altaj', 'Brodtorp', 'Clark', 'Downig', 'Hibinskaja Rannjaja', 'Howton', 'Lenskaja', 'Ojebyn', 'Pembina Pride', 'Perkins', 'Perry', 'Pixwell', 'Poorman', and 'Silvia' gooseberries, and the 'Red Dutch', 'Stephens', and 'Varzuga' currants. 'Pankiw' is reportedly tolerant to heavy soil. 'Pankiw' is better adapted to poor soils. Cultivars 'Amos Black', 'Consort', and 'Roodknop' are relatively resistant to spring frost. Most cultivars of the 'Versailles' group are said to be easily broken off by wind (Wind prone cultivars include 'Fay's Prolific', 'Heinemanns' Rote Spatlese', 'Laxton's No. 1', 'Laxton's Perfection', and 'Versailles'). Wind tolerant cultivars include 'Ayrshire Queen', 'Erstling aus

Vierlanden', 'Jokheer van Tets', 'New Red Dutch', 'Prince Albert', 'Red Lake', and 'Victoria'.

Rubus (blackberries, loganberries, raspberries). Temperate blackberries and raspberries grow weakly and fruit sparsely in Florida. Few introductions have both a low-chilling requirement and the cold-hardiness to survive the Florida winter.

Rubus argutus Link appears to be adapted to hilly sandy soils of central Florida. Many blackberry species tolerate very poor soils. The subtropical upland species *Rubus glaucus* Benth. and *R. shankii* Standl. and L. Wms. do not thrive in Florida highs or lows, winter killing at 1.7°C. They do not flower well at higher latitudes. The Queensland raspberry, *R. probus* Bailey, does well in tropical America. Four important raspberry species for heat tolerance are *R. biflorus* Smith, *R. coreanus* L., *R. kuntzeanus* Hemsl., and *R. parviflorus* Nutt. Many progenies from backcrosses of these with *R. idaeus* L. tolerate high summer temperature and drought. Important blackberry species for heat tolerance are *R. glaucus* Benth. and *R. trivialis* Michx (Jennings, personal communication, 1978). Cold tolerance is available in the Cylactis section; for example, wild forms of *R. arcticus* L. and *R. stellatus* Sm. (the arctic raspberries) and *R. chaemaemorus* L. (cloudberry) are popular in Scandinavia, especially for making liqueurs.

Hardy blackberries include 'Alfred', 'Black Satin', 'Brainerd' (v), 'Darrow' (v), 'Delicious' (v), 'Eldorado', 'Gem', 'Lowden', 'May Hardiest' (v), and 'Snyder'; 'Nessberry' is said to tolerate drought; 'Brazos', 'Cherokee', 'Cheyenne', 'Comanche', 'Flordagrand', and 'Oklawaha', heat; 'Flordagrand', humidity; 'Brazos', 'Flordagrand' and 'Oklawaha', low chilling; and 'Monroe', sandy soil.

Among raspberries, the following have been called hardy: 'Alaska', 'Amethys', 'Anelma', 'Black Hawk' (v), 'Boyne', 'Bristol', 'Canby' (v), 'Carnival', 'Chief', 'Clyde', 'Comet', 'Creston' (v), 'Dike', 'Durham', 'Early Red', 'Festival' (v), 'Fraser' (e), 'Gatineau', 'Glen Clova', 'Hilton', 'Honey King', 'Honeyqueen' (v), 'Iowa', 'Itasca', 'Jewel', 'John Robertson', 'Killarney' (v), 'Lake Geneva', 'Latham' (v), 'Liberty', 'Madawaska', 'Milton', 'Muskoka' (v), 'Napler', 'Newburgh', 'New Hampshire', 'Novost Kuzmina', 'Ottawa', 'Phoenix', 'Polaris', 'Puyallup', 'Rideau', 'Ruddy' (v), 'Seneca', 'September', 'Sodus', 'Success' (v), 'Sumner' (v), 'Sunrise' (v), 'Tohoma', 'Taxpayer', 'Trent', 'Turner', 'Tweed', and 'Van Fleet'. 'Black Pearl', 'Latham', 'Marcy', 'Sodus', 'Starking Black Giant', and 'Taxpayer' are said to tolerate drought; 'Cherokee', 'Pocahontas', 'Reveille', 'Scepter', and 'Sentinel' to tolerate fluctuating winter temperatures; 'Dormanred', 'Mandarin', 'Mysore', 'Sonoma', 'Southland', and 'Starking Black Giant' to tolerate heat; 'Enterprise', 'Fairview', 'Malling Enterprise', 'Malling Exploit', 'Newburgh', and 'Sumner', heavy soil; 'Itasca' and 'MySore', humidity; 'Cuthbert' and

'Puyallup', light soils; 'Bonanza', 'Mysore', and 'Queensland', low chilling; 'Malling Exploit', poorly drained soil; 'Muskoka', prairie; and 'Canby', sandy soil.

Saccharum (sugarcane). More important germplasm resides with *S. officinarum* L., *S. robustum* Brandes and Jeswiet ex Grassl, *S. sinense* Robs., and *S. spontaneum* L. Following tolerances are reported: 'C.P. 50-28' and 'C.P. 52-48' for cold, 'C.P. 56-63' and 'C.P. 62-374' for cold organic soils, 'H 50-7209' for drought, 'C.P. 67-500' for early freezes, 'C.P. 56-63' and 'C.P. 62-374' for hot organic soils, and 'Pindar' for poor soil.

Sambucus (elderberry). A rather cosmopolitan genus of some 40 species, *Sambucus* is now showing up among some of the cultivated fruit listings. For cold climates, 'Adams', 'Johns', 'Kent', 'Nova', 'Scotia', and 'Victoria' are recommended.

Secale (rye). Rye, *Secale cereale* L., ($2n = 2x = 14$) is superior to many other cereals in its capacity to yield in areas with hot dry summers and cold winters. It can also tolerate poor light soils. As a weed among large-grained cereals, it may have been selected out as the only one to surive a particularly harsh or dry growing season. Several rye cultivars are tolerant of acid soils and aluminum. Unlike wheat and barley breeding, rye breeding has not been dominated by disease resistance, but by yield, quality, and cold tolerance. Cold tolerance is recorded in such cultivars as 'Athens Abruzzi', 'Cougar', 'Frontier', 'Kodiak', 'Okema', 'Petkus', 'Puma', and 'Rymin'.

Sesamum (sesame). With about 30 species, the genus *Sesamum* contains at least three cultivated species (basically $2n = 2x = 26$). Yield has been the main objective of breeding programs, disease a secondary goal, and adaptation to mechanized harvesting a recent goal. Land races seem to tolerate semidesert conditions, clay soils, and black cotton soils of India.

Sinapis (white mustard). *Sinapis alba* L., the white mustard ($2n = 24$), with an eastern Mediterranean center of origin, is the only crop species in a genus of about 10 species. Like *Brassica, Sinapis* is well endowed with cold-tolerant germplasm.

Solanum (potato, eggplant). More than 50 tuber-bearing species can be crossed with the potato *S. tuberosum* L. ($2n = 4x = 48$).

In Peru clones are being tested to derive from their wide genetic diversity the ability to adapt to extreme temperature conditions. Crossing genetically diverse parents, previously selected for adaptation, might yield heterozygosity to maximize the performance of the hybrids. Mendoza (42) reports a screening of 6000 clones for germplasm adapted to high temperatures and humidity. Hybrids of tuberosum × neotuberosum and tuberosum × phureja showed promise given the short growing season and the stress imposed not only by the weather but also by weeds, insects, and diseases.

Many Andean peasants plant small insurance crops of wild-type potatoes, of poor quality, but good frost resistance. Potato breeders are charged to combine frost resistance with quality and yield potential of *Solanum tuberosum*. *Solanum demissum* Lindl. and *S. commersonii* Dunal ex Poir. are among the frost-resistant species. Richardson and Weiser (53) list *S. acaule* Bitter, *S. chromatophilum*, *S. commersonnii* Dunal ex Poir., *S.*, × *juzepczukii* Bukas., and *S. multidissectum* Hawkes as highly resistant; *S. ajanhuiri* Fozepczuk & Bukasov, *S.* × *curtilobum* Juz. and Buk., *S. demissum* Lindl., *S. megistacrolobum* Bitter, *S. microdontum* Bitter, and *S. vernei* Bitter & Wittm. as very resistant. *S. acaule* has been reported to tolerate −3 to −9°C. *S. brevicaule* Bitt. × *S. phureja* Juz. & Buk. hybrids have survived −4 to −9°. Frost resistance, like heat resistance, of potatoes can be improved (53). *Solanum megistacrolobum* Bitter is an effective donor of frost tolerance.

Among tolerances reported are 'Dakota Chief', 'Fillmore', 'Rila', and 'Viking' tolerant to drought; 'Belchip', to fluctuating temperatures; 'Sebago', to frost; 'Alpha', 'Arka', and 'Up-to-Date', to hail; 'Superior' and 'Virgil', to heat; 'Kasota' and 'Yampa', to heavy soil; 'Cherokee' and 'Osage', to muck; 'York', to organic soils; 'Mesaba' and 'Waseca', to peat; 'Atlantic', 'Red Lasoda', and 'Red Pontiac', to photoperiod; 'Yampa', to sand; and 'Osseo', to short season. Cultivars differ in Al tolerance.

The eggplant, *S. melongena* L., presumed native to India, belongs to a group of closely related spiny diploids in India. 'Apple Green' is said to be tolerant to adverse northern conditions, 'New Hampshire' to cool weather, 'Burpee Hybrid F_1' to drought, and 'Puerto Rican Beauty' and 'Rosita' to tropical conditions.

Sorghum (sorghum). Like certain millets, sorghum is well adapted for drought tolerance and arid climates, sorghum for heavier soils, bulrush millets for lighter soils. As a C-4 plant, sorghum has both heat- and drought-tolerant potential and holds promise as a major cereal for the semiarid tropics.

When reviewing the table published by the American Society of Agronom (18), members of the Plant Stress Laboratory were surprised by reports of sorghum at pH greater than 8. Usually iron chlorosis due to iron inefficiency precludes successful culture of sorghum in such alkaline soils. Consequently I queried our computer for a mailing list of all correspondents (about 500 at that time) who had reported sorghum at pH 7.5 or greater (about 15 correspondents). Letters were sent to 10 correspondents. Four provided seed, and one lot from India was iron efficient. An Australian correspondent reminded us that they were using an American strain, seeds of which also proved to be iron efficient. This demonstrates the value of the computerized data bank for seeking out stress-tolerant germplasm.

Studying 15 sorghum lines, representing diverse germplasm currently used in the United States, Brown and Jones (8) found differential cultivar response to Cu, Fe, and Zn stresses, and Al and Mn toxicities. 'Redland B' is regarded as Al tolerant. Randhawa and Takkar (51) list India's '2077 × 1151' and 'Swarna' as tolerating zinc deficiency.

Spinacia (spinach).   Spinach ($2n = 2x = 12$) is a relatively recent cultivar, native to southwestern Asia, perhaps India or Nepal. A short-day, cool-season, leafy vegetable, spinach has given rise to few long-day cultivars like 'Viking'.

Theobroma (cacao).   In the diploid genus ($2n = 2x = 20$) of 22 species, *Theobroma cacao* L. is the only economically important species, although others may contribute important germplasm. Selection has been directed toward resistance or tolerance to diseases, and increased productivity, not toward abiotic tolerance.

Triticosecale (triticale).   Triticale was developed to combine the cold tolerance of rye with the commercial potential of wheat. Hexaploid triticales derive from tetraploid wheat × diploid rye; octaploids from hexaploid diploid. CIMMYT eliminated day-length problems and developed day-length-insensitive triticales in Mexico. Russian studies indicate that most of the winter-hardy forms are 6x. Frost-tolerant strains include: 'AD71-176', 'AD71-307', 'AD71-403', and '8A310' (Russian numbers). In general, most triticale cultivars are highly tolerant to soil acidity and probably to aluminum toxicity, perhaps due to the addition of the acid-tolerant rye genome (52).

Triticum (wheat).   Most wheat cultivars belong to the hexaploid *T. aestivum* L. var. *aestivum*, ($2n = 6x = 42$) but drought tolerance is more prevalent in the tetraploid *T. turgidum* L. var. *durum*. The cultivated polyploids exhibit a wide range of genetic flexibility and can adapt themselves to a wide variety of environments. Wild "einkorn," occurring naturally in steppes, has innate drought resistance as do the tetraploid "emmers." Furthermore, they are mostly tolerant of poor, thin, rocky soils. *Monococcum* wheats show much cold and rust tolerance. The *tauschii* genome permitted adaptation to more rigorous continental climates than the mediterranean climates of tetraploids. *Aegilops speltoides* Tausch (K1000), *A. tauschii* Coss., and *A. cylindrica* Host are among the most cold-tolerant *Aegilops*.

Cold limits production of winter annual cereals. Studying several strains, Fowler et al. (23) reported LT50's (lethal temperatures) of −5 to −9°C for *Triticum boeoticum*; −5 to −9°C for *T. urartu*; 9°C for one cultivar of *T. timopheevii* var. *araraticum*; −6°C for *T. turgidum* var. *dicoccoides*; −5 to −9°C for *T. turgidum* var. *turgidum*; −5 to −11°C for *T. turgidum* var. *durum*; −8 to −20°C for *T. aestivum* L.; −13 to −18°C for *T. tauschii*, but −19 to −29°C for *Secale cereale* L.

Hexaploid wheats have a higher tolerance to soil acidity; diploids and tetraploids have little or no tolerance, and are highly sensitive to Al. Hexaploids vary in their Al tolerance. Many cultivars developed in the eastern United States seem to have been selected unconsciously for Al tolerance. Similarly, Brazilian wheat cultivars developed under acid soil conditions frequently did not give a yield response to liming, presumably due to their tolerance to high levels of exchangeable Al and possibly Mn (40). Tolerance to Al does not necessarily imply tolerance to Mn or vice versa. 'Atlas 66' is more tolerant to Al than is 'Monon' which is more tolerant to Mn (36) like India's 'K816'. Similar relationships have been observed among some cotton cultivars (24). Summarizing Brazilian studies, da Silva (58) notes that tolerance to Al allows production of wheat with less or no lime. This is important in opening new territories where the cost of fertilizers and lime would be prohibitive. Tolerance to high Al has allowed previously unproductive areas in Brazil to produce millions of tons of wheat. Konzak et al. (35) tabulate wheat cultivars with their Al tolerance. Among Al-tolerant cultivars are 'Atlas 66', 'BH1146', 'Blueboy', 'Butler', 'Cajeme 7', 'Carazinho', 'Cinquentenario', 'Coker 47-27', 'Colonias', 'Cotipora', 'Erechim', 'Fulcaster', 'Fulton', 'Fultz', 'Georgia 1123', 'Hadden', 'IAS20', 'Lagoa', 'Pennoll', 'P.I. 344161', '344162', '193916', '353388-Pel', '353389-Pel', '344192-Pel', 'Potomac', 'Preludio', 'Ruler', 'Seneca', 'Thorne', 'Toropi', 'Trintani', 'Trumbul', 'Vermelha', and 'Wakeland'; India's 'Vijay' and 'WG357' are said to tolerate B toxicity; among the cold tolerant are 'Alabasskaya', 'Froid', 'Hume', 'IAS54', 'IAS55', 'Kharkov NC22', 'Logan', 'Londrina', 'Martin', 'Michurinka', 'Minter', 'Mironovka 808', 'Nebraska 60', 'Nova Michurinka', 'Reliable', 'Ruler', 'Seneca', 'Sundance', 'Thorne', 'Ulyanovka', 'Winalta', 'Windsor', 'Winoka', and 'Yogo'. Among heat tolerant are 'Erythrospermum 841', 'Kzyl Sark Red East', and 'Saratov 38'. India's 'UP30', 'WG377' and 'WL212' are said to tolerate Zn deficiency (51).

Some drought-resistant cultivars include 'Flora', 'Florence', 'Klein Lucero', 'Pusa 4', 'Red River 68', and 'Waratah'. Cold-tolerant cultivars include: 'Albidium 114', 'Bezenchuk 5', 'Erythrospermum 917', 'Froid', 'Kharkov 4', 'Kharkov MC22', 'Lutescens 226, 230, 121', 'Martin', 'Michigan Amber', 'Milturum 527', 'Mironovka 808', 'Nebraska', 'Odessa 16', 'Omsk 6', 'Oro et Fulhard', 'Priekuli 48', 'Reliable', 'Sundance', 'Turkey Red', 'Ul'Yanovka', 'Winalta', 'Windsor', 'Yogo', 'Voronezh 34/7 & 42/4'. Russian drought-tolerant strains include: 'Bezenchuk 123', 'Candeal 17', 'Falcon', 'Florelle', 'Erythrospermum 841', 'K45185', 'K43283', 'K38531', 'K44275', 'K42620', 'K45747', 'Kzyl Sark Red East', 'Milturum 45', 'Pusa 52', 'Pyrothrix 28', 'Saratov 29', and 'Vesna Spring'.

Russians claim to have combined heat and drought tolerance in such cvs as 'Candeal 17', 'Erythrospermum 841', 'Falcon', 'Florell', 'Kzyl Sark Red East', 'Pusa 52', 'Rescue × Regent', and 'Saratov 29'.

Vaccinium (blueberry). The temperate genus *Vaccinium*, with some 400 species, embraces the blueberries, cranberries, and whortleberries. Many are noted for their tolerances, indeed requirements, for low pH soils.

Hardier cultivars are required to extend blueberry production into colder regions. *Vaccinium angustifolium* Ait., growing wild in Canada and the northern United States, is a potential source of germplasm. Although *V. angustifolium* and its hybrids may withstand temperatures lower than the highbush species, its ability to survive the northern winter may partially result from its low stature and the protection provided by persistent snow. *Vaccinium membranaceum* Dougl. and *V. constablaei* A. Gray were superior in hardiness to other highbush species (49). Galletta (25) also lists *V. brittonii, V. myrtilloides* Michx., *V. myrtillus* L., and *V. uliginosum* as possible donors of cold tolerance.

Breeding potential for low-chill blueberries started with *Vaccinium ashei* Reade (6x) and *V. darrowi* Camp. (2x) from around Winter Haven, Florida, where chill is about 200 hrs. *Vaccinium myrsinites* is also a short-chilling parental species. These studies also involved *Vaccinium tenellum* Ait. (2x) and *V. angustifolium* Ait. (4x). *V. atrococcum* (A. Gray) A. Heller and *V. elliottii* are potential donors of tolerance to short chilling. Some of the Florida germplasm lines include selections that are probably more drought and heat tolerant than other tetraploid blueberries of good quality. The related genus *Polycodium* fruits well in dry sandy soils. Galletta (25) lists *V. alto-montanum, V. ashei* Reade, *V. darrowi, V. membranaceum, V. myrsinites, V. pallidum, V. tenellum,* and *V. vacillans* Torr. as potential donors of drought tolerance.

Blueberry improvement should utilize germplasm from throughout the genus to incorporate broader edaphic and climatic tolerances. Adaptation to mineral soils and continental climates will be the key to widespread blueberry culture. 'Bluecrop', 'Herbert', 'Jersey', 'Tifblue', and 'US 11-93' have broad climatic tolerances; and *V. ashei* Reade, *V. atrococcum, V. corymbosum, V. pallidum,* and *V. vacillans* are potential donors of broad climatic tolerance. *Vaccinium ashei* is the main source of germplasm for upland mineral soil adaptation, but *V. angustifolium, V. atrococcum, V. brittonii, V. darrowi, V. elliottii, V. myrsinites, V. myrtilloides, V. myrtillus, V. pallidum, V. tenellum, V. vacillans,* and *V. virgatum* are listed as potential donors.

Blueberry tolerances include: 'Ashworth' (−50°F), 'Bluecrop', 'Blue Haven', 'Blueray', 'Burlington', 'Collins', 'Earliblue', 'Elliott', 'Gem' (9), 'Rancocas', and 'Wareham' tolerate cold; 'Bluecrop' tolerates drought; 'Ashworth' and 'Bluetta' tolerate frost (in flower); 'Avonblue', 'Bluegem', 'Flordablue' (v), 'Sharpblue' (v) tolerate low chill.

Cranberry improvement has been accomplished at the diploid level, mostly within the one species. Hybridization, and intercrossing natural and artificial polyploid forms, and polyploid with diploid species might combine the superior fruits of the diploid with the shade and moisture tolerance of the hexaploid.

Vicia (vetches, tares). With some 150 north temperate and South American species, Vicia contains several forage species and one grain species. The 2n number is 10, 12, or 14, and none of them, not even the $2n = 12$ species, can be hybridized with the grain species *V. faba* L. Berthelem (5) discusses the French program for breeding fodder broadbeans (*Vicia faba*). Synthetic cultivars and hybrid cultivars, using male sterility, are being developed to improve resistance to cold, lodging, and disease.

In general, legumes are more sensitive than nonlegumes to excess soil Mn. *Vicia faba* (like *Trifolium pratense* L. and *T. repens* L.) is relatively tolerant of Mn excess but *V. sativa* L. (like *Medicago sativa* L. and *Phaseolus vulgaris* L.) is very sensitive (1).

Vigna (cowpea). *Vigna aconitifolia* (Jacq.) Marechal, like some cultivars of *unguiculata,* is tolerant of drought, heat, and sandy soils. *Vigna angularis* (Willd.) Ohwi & Ohashi may be nearly as tolerant of heat and drought, and intolerant of waterlogging. The more cold-tolerant northern ecotypes in Japan tend to be erect bushes whereas the southern type tends to be clambering or much branched. It is said to be a good candidate for acidic subtropical soils.

Within mung bean and blackgram, *V. radiata* (L.) Wilczek. and *V. mungo* (L.) Hepper, there are photoinsensitive cultivars that can flower 30 days after planting. *V. radiata* is said to be drought tolerant, yet sensitive to waterlogging and salt. V. *mungo* tolerates rain during flowering and sets seed. Root growth of mung bean seems even more sensitive to salt than cowpeas, which do not, like mung bean, exhibit a fall in N-fixing efficiency with increasing salinity. 'Khargaon B', 'T55', and 'T65' are Zn-efficient cultivars of *V. mungo*.

Some or all cultivars of cowpeas, *V. unguiculata* (L.) Walp., are hybrids (61).

The sesquipedalis cultivars tolerate high rainfall and humidity better than the others. Nearly all cowpeas tested in Colombia (60) show excellent tolerance to soil acidity.

Vitis (grapes). *V. riparia* Michx. is much more resistant to phylloxera and tolerant of cold than *V. vinifera* L. *Muscadinia* is much more tolerant of heat than many northern *Vitis,* which reciprocally can not tolerate the low chilling period of the *Muscadinia* territory. *Vitis monticola* Buckl, is noted for its tolerance of high soil lime like *V. berlandieri* Planch. as well as tolerance to phylloxera and certain diseases; *Vitis candicans* Engelm. is noted for its tolerance of hot dry conditions, but it is sensitive to high lime. Hybrids of *berlandieri* and *rupestris* show best promise for drought tolerance. Cultivars differ markedly in their tolerance to ozone and 2, 4-D (19).

Hungarians, after 20 years breeding work with *Vitis amurensis* Rupr., concluded that it confers frost resistance as well as Plasmopara hypersensitivity to the progeny. *Vitis labrusca* L. and *V. riparia* Michx. are good American donors of cold tolerance. *Vitis berlandieri* shows tolerance of calcareous soils.

Some tolerances reported in *Vitis* are: 'Chilcott', 'Cimarron', 'Keating', and 'Osborn' and rootstocks of '110 Richter', '1103 Paulsen', '196-17 Millardet', '333 Ecolede', 'Montpellier', and '44-53 Malegue', to drought; 'Seibel 7053 & 8745' to spring frosts; 'Adams', 'Aligote', 'Aurora', 'Bath', 'Bluebell', 'Bluejay', 'Buffalo', 'Carlos', 'Catawba', 'Cimarron', 'Concord', 'Cook', 'Early', 'Fredonia', 'Grey Friar', 'Little Girl', 'Pride', 'Red Amber', 'Red Traminer', 'Roubidoux', 'Seibel 1000' ($-35°F$), 'Seibel 5279', 'Seibel 10878', 'Seibel 13047', 'Seibel 13053', 'Seneca', and 'Seyve-Villard 20-473' for cold; 'Angelina', 'Everglades', 'Fairchild', 'Indaiatuba', 'Largo', 'Ligia', 'Marta', and 'Ruby Cabernet' to heat; 'Everglades', 'Fairchild', 'Florida Concord', 'Largo', 'Masters', 'Myakka', 'Seibel 5279' to humidity; 'Millardet 41B' and 'Richter 99' to lime chlorosis; 'Blue Lake', 'Everglades', 'Fairchild', 'Florida Concord', 'Lake Emerald', 'Largo', 'Liberty', 'Masters', 'Myakka', 'Seminole', 'Stover', and 'Tropico' to low chilling (Muscadine cultivars adapted to Florida listed more or less in decreasing chill requirement, 'Creek', 'Dearing', 'Fry', 'Higgins', 'Magnolia', 'Topsail', 'Welder', 'Chief', 'Cowart', 'Dixie', 'Hunt', 'Magoon', 'Southland', and 'Thomas'); 'Canada Muscat', 'Delaware', 'Dutchess', 'Hanover', 'Himrod', 'Iona', 'Kendalia', and 'Seneca' to air pollution, especially oxidant.

Zea (corn). Mock and Bakri (43) report that recurrent selection for cold tolerance in Iowa maize has resulted in improved seedling vigor, earlier flowering data, and reduced harvest moisture. More vigorous seedling growth may reflect a reduction in temperature sensitivity, that is, a tolerance to cold.

Single crosses of acid-tolerant with acid-sensitive cultivars showed heterosis in high pH trials, but in a low pH, tolerance to acidity seemed to be dominant, even allowing for heterosis. In general, corn is poorly adapted to extremely acid soils, and in Colombia is usually grown on alluvial soils or after cutting and burning the forest (60). A white brachytic selection from Carimagua in 1972 showed promise for acid subsistence farms. In general sweet corn is more tolerant to high Al than sorghum (52). A recessive factor controls iron uptake in Fe-inefficient ysl/ysl corn (7).

Aluminum tolerance is reported for '64A X M15', 'B57', 'Caribi', 'CAR73A-CARS Sel AM', 'CAR73A-BR2AM', 'PA36', 'W182BN', and 'W703'. Cold tolerance is reported in 'Arctic Frost', 'Earligem 64', 'Earlitreat', 'Seneca Raider', 'Sugar Frost', and 'Tendersure'; drought tolerance in 'Early Triumph', 'Golden Fancy', 'Heart of Dixie', 'Silver King', 'Tendersure', 'Tendertreat', and 'Ultratender; heat tolerance in 'Eversweet', 'Heart of Dixie', 'Tenderchoice', 'Tendersure', 'Tendertreat'; wind tolerance in 'Goldenchief', 'Tenderchoice', 'Ultrasweet', and 'Ultratender'; "stress" tolerance in 'Mainliner'; 'Gatorgold' is recommended for muck soils; 'Gatorgold' for sandy soils, and 'Earlytreat' for wet soils at planting. 'Ganga 2', 'Ganga 3', and 'Ganga 4', 'JML22', 'T1W', and 'T41' are said to tolerate Zn deficiency in India.

REFERENCES

1   Andrew, C. S. 1977. Screening tropical legumes for manganese tolerance. Pp. 329–340 in *Plant Adaptation to Mineral Stress in Problem Soils*, M. J. Wright, Ed. Proc. Workshop, NAL, Beltsville, MD. November 22–23, 1976. AID, Washington, D.C.

2   Bailey, C. H., and L. F. Hough. 1975. Apricots. Pp. 367–383 in *Advances in Fruit Breeding*, J. Janick and J. N. Moore, Eds. Purdue Univ. Press, West Lafayette, Indiana.

3   Banga, O. 1976. Carrot. Pp. 291–293 in *Evolution of Crop Plants*, N. W. Simmonds, Ed. Longman's, London.

4   Bergh, B. O. 1975. Avocados. Pp. 541–567 in *Advances in Fruit Breeding*, J. Janick and J. N. Moore, Eds. Purdue Univ. Press, West Lafayette, Indiana.

5   Berthelem, P. 1977. *Rev. de l'Agric.* 30(2): No. 89, pp. 48–53.

6   Brown, A. G. 1975. Apples, Pp. 3–37 in *Advances in Fruit Breeding*, J. Janick and J. N. Moore, Eds. Purdue Univ. Press, West Lafayette, Indiana.

7   Brown, J. C. 1977. Genetic potentials for solving problems of soil mineral stress: iron efficiency and boron toxicity in alkaline soils. Pp. 83–94 in *Plant Adaptation to Mineral Stress in Problem Soils*, M. J. Wright, Ed. Proc. Workshop, NAL, Beltsville, MD. November 22–23, 1976. AID, Washington, D.C.

8   Brown, J. C., and W. E. Jones. 1977. *Agron. J.* 69(3):410–414.

9   Buxton, D. R., and P. J. Sprenger. 1976. *Crop Sci.* 16(2):243–246.

10 Cameron, J. W., and R. K. Soost. 1976. Citrus. Pp. 261–265 in *Evolution of Crop Plants*, N. W. Simmonds, Ed. Longman's, London.

11 Chang, T. T. 1976. Rice. Pp. 98–104 in *Evolution of Crop Plants*, N. W. Simmonds, Ed. Longman's, London.

12 Chaplin, C. E., G. W. Schneider, and D. C. Martin. 1974. *HortScience* 9(1):28–29.

13 Clark, R. B. 1977. Plant efficiencies in the use of calcium, magnesium, and molybdenum. Pp. 175–191 in *Plant Adaptation to Mineral Stress in Problem Soils*, M. J. Wright, Ed. Proc. Workshop, NAL, Beltsville, MD. November 22–23, 1976. AID, Washington, D.C.

14 Clift, C. 1977. *Fruit Vars. J.* 31(2):45–47.

15 Cummings, J. N., and H. S. Aldwinckle. 1974. *HortScience* 9(4):367–372.

16 Daubeny, H. A., R. A. Norton, C. D. Schwartze, and B. H. Barritt. 1970. *HortScience* 5(3):152–153.

17 Dickson, M. H. 1973. *HortScience* 8(5):410.

18 Duke, J. A. 1978. The quest for tolerant germplasm. Pp. 1–62 in *Crop Tolerance to Suboptimal Land Conditions*. ASA Spec. Publ. 32, ASA, Madison, Wisconsin.

19 Einset, J., and C. Pratt. 1975. Grapes. Pp. 130–153 in *Advances in Fruit Breeding*, J. Janick and J. N. Moore, Eds. Purdue Univ. Press, West Lafayette, Indiana.

20 Epstein, E. 1977. Genetic potentials for solving problems of soil mineral stress: adaptation of crops to salinity. Pp. 73–82 in *Plant Adaptation to Mineral Stress in Problem Soils*, M. J. Wright, Ed. Proc. Workshop, NAL, Beltsville, MD. November 22–23, 1976. AID, Washington, D.C.

21 Fogle, H. W. 1975. Cherries. Pp. 348–366 in *Advances in Fruit Breeding*, J. Janick and J. N. Moore, Eds. Purdue Univ. Press, West Lafayette, Indiana.

22 Forde, H. I. 1975. Walnuts. Pp. 439–455 in *Advances in Fruit Breeding*, J. Janick and J. N. Moore, Eds. Purdue Univ. Press, West Lafayette, Indiana.

23 Fowler, D. B., J. Dvorak, and L. V. Gusta. 1977. *Crop Sci.* 17(6):941–943.

24 Foy, C. D. 1977. General principles involved in screening plants for aluminum and manganese tolerance. Pp. 255–267 in *Plant Adaptation to Mineral Stress in Problem Soils*, M. J. Wright, Ed. Proc. Workshop, NAL, Beltsville, MD. November 22–23, 1976. AID, Washington, D.C.

25 Galletta, G. J. 1975. Blueberries and cranberries. Pp. 154–196 in *Advances in Fruit Breed-ing*, J. Janick and J. N. Moore, Eds. Purdue Univ. Press, West Lafayette, Indiana.

26 Hartley, C. W. S. 1977. *The Oil Palm*. Longman's, London, 806 pp.

27 Hearn, C. J., D. J. Hutchinson, and H. C. Barrett. 1974. *HortScience* 9(4):357–358.

28 Hesse, C. O. 1975. Peaches. Pp. 285–335 in *Advances in Fruit Breeding*, J. Janick and J. N. Moore, Eds. Purdue Univ. Press, West Lafayette, Indiana.

29 Holden, J. W. H. 1976. Oats. In *Evolution of Crop Plants*, N. W. Simmonds, Ed. Longman's, London. 339 pp.

30 Hymowitz, T. 1976. Soybeans. Pp. 159–162 in *Evolution of Crop Plants*, N. W. Simmonds, Ed. Longman's, London.

31 Jennings, D. L. 1976. Cassava. Pp. 81–84 in *Evolution of Crop Plants*, N. W. Simmonds, Ed. Longman's, London.

32 Keser, M., B. F. Neubauer, F. E. Hutchinson, and D. B. Verrill. 1977. *Agron. J.* 69(3):347–350.

33 Kester, D. E., and R. Asay. 1975. Almonds. Pp. 387–419 in *Advances in Fruit Breeding*, J. Janick and J. N. Moore, Eds. Purdue Univ. Press, West Lafayette, Indiana.

34 Knowles, P. F. 1977. *Calif. Agric.* 31(9):12–13.

35 Konzak, C. F., E. Polle, and J. A. Kittrick. 1977. Screening several crops for aluminum tolerance. Pp. 311–327 in *Plant Adaptation to Mineral Stress in Problem Soils*, M. J. Wright, Ed. Proc. Workshop, NAL, Beltsville, MD. November 22–23, 1976. AID, Washington, D.C.

36 Lafever, H. N., L. G. Campbell, and C. D. Foy. 1977. *Agron. J.* 69(4):563–568.

37 Lagerstedt, H. B. 1975. Filberts. Pp. 456–489 in *Advances in Fruit Breeding*, J. Janick and J. N. Moore, Eds. Purdue Univ. Press, West Lafayette, Indiana.

38 Layne, R. E. C., and H. A. Quamme. 1975. Pears. Pp. 38–70 in *Advances in Fruit Breed-ing*, J. Janick and J. N. Moore, Eds. Purdue Univ. Press, West Lafayette, Indiana.

39 Marshall, H. G. 1976. *Crop Sci.* 16(1):9–15.

40 Martini, J. A., R. A. Kochhann, E. P. Gomes, and F. Langer. 1977. *Agron. J.* 69(4):612–616.

41 McCollum, G. D. 1976. Onion and allies. Pp. 186–190 in *Evolution of Crop Plants*, N. W. Simmonds, Ed. Longman's, London.

42 Mendoza, H. A. 1977. Adaptation of cultivated potatoes to the lowland tropics. *Proc. 4th Symp. Internat. Soc. Trop. Root Crops Sect. 1*. pp. 50–53.

43 Mock, J. J., and A. A. Bakri. 1976. *Crop Sci.* 16(2):230–233.

44 Mortvedt, J. J. 1977. Soil chemical constraints in tailoring plant to fit problem soils 2 Alkaline soils. Pp. 141–149 in *Plant Adaptation to Mineral Stress in Problem Soils*, M. J. Wright, Proc. Workshop, NAL, Beltsville, MD. November 22–23, 1976. AID, Washington D.C.

45 Nieman, R. H., and M. C. Shannon. 1977. Screening plants for salinity tolerance. Pp. 359–367 in *Plant Adaptation to Mineral Stress in Problem Soils*, M. J. Wright, Ed. Proc Workshop, NAL, Beltsville, MD. November 22–23, 1976. AID, Washington, D.C.

46 Ormrod, D. P., and R. E. C. Layne. 1974. *HortScience* 9(5):451–453.

47 Ourecky, D. K. 1975. Brambles. Pp. 98–129 in *Advances in Fruit Breeding*, J. Janick and J. N. Moore, Eds. Purdue Univ. Press, West Lafayette, Indiana.

48 Ourecky, D. K., and J. E. Reich. 1976. *HortScience* 11(4):413–414.

49 Quamme, H. A., C. Stushnoff, and C. J. Weiser. 1972. *HortScience* 7(5):500–502.

50 Ramanujam, S. 1976. Chickpea. Pp. 157–159 in N. W. Simmonds, Ed. *Evolution of Crop Plants*, Longman's, London.

51 Randhawa, N. S., and P. N. Takkar. 1977. Screening of crop varieties with respect to micronutrient stresses in India. Pp. 393–400 in *Advances in Fruit Breeding*, M. J. Wright, Ed. Purdue Univ. Press, West Lafayette, Indiana.

52 Reid, D. A. 1977. Genetic potentials for solving problems of soil mineral proc. workshop stress: aluminum and manganese toxicities in the cereal grains. Pp. 55–64 in *Plant Adaptation to Mineral Stress in Problem Soils*, M. J. Wright, Ed. Proc. Workshop. NAL, Beltsville, MD. November 22–23, 1976. AID, Washington, D.C.

53 Richardson, D. G., and C. J. Weiser. 1972. *HortScience* 7(1):19–22.

54 Rogers, W. V. 1976. Pigeon pea. Pp. 154–156 in *Evolution of Crop Plants*, N. W. Simmonds, Ed. Longman's, London.

55 Sauer, J. D. 1976. Grain amaranthus. Pp. 4–7 in *Evolution of Crop Plants*, N. W. Simmonds, Ed. Longman's, London.

56 Scott, D. H., and F. J. Lawrence. 1975. Strawberries. Pp. 71–95 in *Advances in Fruit Breeding*, J. Janick and J. N. Moore, Eds. Purdue Univ. Press, West Lafayette, Indiana.

57 Sharpe, R. H. 1974. *HortScience* 9(4):362–363.

58 Silva, A. R. da. 1977. Application of the genetic approach to wheat culture in Brazil. Pp. 223–236 in *Plant Adaptation to Mineral Stress in Problem Soils*, M. J. Wright, Ed. Proc. Workshop, NAL, Beltsville, MD. November 22–23, 1976. AID, Washington, D.C.

59 Simmonds, N. W. 1976. *Evolution of Crop Plants*. Longman's, London, 339 pp.

60 Spain, J. M. 1977. Field studies on tolerance of plant species and cultivars to acid soil conditions in Colombia. Pp. 213–222 in *Plant Adaptation to Mineral Stress in Problem Soils*, M. J. Wright, Ed. Proc. Workshop, NAL, Beltsville, MD. November 22–23, 1976. AID, Washington, D.C.

61 Summerfield, R. J., P. A. Huxley, and W. Steele. 1974. *Field Crop Abstr.* 27(7):301–312.

62 Watkins, R. 1976. Cherry, plum, peach, apricot and almond. Pp. 243–247 in *Evolution of Crop Plants*, N. W. Simmonds, Ed. Longman's, London.

63 Weinberger, J. H. 1975. Plums. Pp. 336–347 in *Advances in Fruit Breeding*, J. Janick and J. N. Moore, Ed. Purdue Univ. Press, West Lafayette, Indiana.

64 Westwood, M. N. 1970. *HortScience* 5:418–421.

65 Whitaker, T. W., and W. P. Bemis. 1976. Cucurbits. Pp. 64–69 in *Evolution of Crop Plants*, N. W. Simmonds, Ed. Longman's, London.

66 Whitehead, R. A. 1976. Coconut. Pp. 221–225 in *Evolution of Crop Plants*, N. W. Simmonds, Ed. Longman's, London.

67 Young, R. 1970. *HortScience* 5(5):411–413.

68 Zauberman, G., M. Schiffmann-Nadel, and V. Yanko. 1973. *HortScience* 8(6):511–512.

69 Zohary, D. 1976. Lentils. Pp. 163–164 in *Evolution of Crop Plants*, N. W. Simmonds, Ed. Longman's, London.

Chapter 13

GENETIC ENGINEERING FOR IMPROVING ENVIRONMENTAL RESILIENCY IN CROP SPECIES

C. F. LEWIS

U.S. Department of Agriculture, Science and Education Administration, Agricultural Research, National Program Staff, Plant Genetics and Breeding, (Retired) Beltsville, Maryland

It is an axiom in biology that crop productivity can be improved in only two fundamental ways: (*a*) by improving the genotypes of the plants to better fit them to the environments in which they are grown, and (*b*) by modifying the environment to minimize stresses that impair crop productivity. The two approaches are not mutually exclusive. Therefore, to optimize the interaction between genotype and environment, soil scientists, agronomists, and physiologists should cooperate to modify both the genotypes of plants and the environments in which they are grown.

The need to enhance crop productivity has been emphasized at various conferences on world food and nutrition, including those sponsored by the National Research Council (12, 13), Agricultural Research Policy Advisory Committee (1), American Society of Agronomy (2), the United Nations (19), and others. Some say food production must be doubled in the next 25–30 years if the demands of an increasing world population are to be met. This production cannot be achieved simply by bringing new land into cultivation. Part of the increase must come from increasing the efficiency of production on land already in cultivation and part from bringing into cultivation marginal land not in cultivation now. The need to improve plants genetically to cope better with environmental stresses is clear.

Classical plant breeding uses the sexual cycle to recombine DNA through independent assortment of chromosomes and through crossing-over. Breeding has made and continues to make important contributions to crop improvement by providing plants that can tolerate environmental stresses (9). Several reviews of breeding for stress environments are available:

Wright (22), Stone (17), Treshow (18), and Jung (8). Environmental stresses are considered to be nonliving stresses such as hot and cold temperatures, too much water and drought, mineral deficiencies and toxicities, salinity, acidity, alkalinity, and air pollution. Environmental stresses are not those caused by living organisms such as weeds, insects, nematodes, pathogens, or the toxins produced by them.

Reigar et al. (15) define genetic engineering as ". . . genetic manipulations (by-passing the sexual cycle) by which an individual having a new combination of inherited properties is established." They specify two major approaches to genetic engineering: (a) the cellular approach involves the *in vitro* culturing of haploid cells and the hybridization of somatic cells, and (b) the molecular approach involves the direct manipulation of DNA. Direct manipulation involves recombinant DNA molecules, which the Department of Health, Education, and Welfare has defined as either "(i) molecules which are constructed outside living cells by joining natural or synthetic DNA segments to DNA molecules, or (ii) DNA molecules that result from the replication of those described in (i) above."

Genetic engineering, including the cellular approach and the direct molecular manipulation of DNA, has been reviewed by many authors, including Carlson (3), Smith (16), Vasil (20), Nabors (11), Day (6), Widholm (21), Nickell (14), Dale (5), and others.

This chapter is not a general review of the improvement of environmental resiliency in crop species or a general review of genetic engineering. I do not speculate unduly about the future potential of genetic engineering. I do consider those special cases in which genetic engineering has been used to improve the environmental resiliency of crop plants. This double restriction of approach and objectives sets a rather narrow scope to the presentation.

Widholm (21) observed that reviews "generally contain more ideas than accomplishments"; Vasil (20) thought that "many of us . . . have generally failed to adequately realize the variety and severity of problems that must be resolved before today's dreams can be translated into tomorrow's realities."

CELLULAR APPROACH

The cellular approach includes protoplast fusion and culturing protoplasts, tissue, callus, and anthers, with or without treatment with mutagenic agents.

Dix and Street (7) chilled callus cultures of *Nicotiana sylvestris* Speg. and Comes, and *Capsicum annuum* L. with and without treatment with ethyl methyl-suphonate (EMS). Callus lines were selected from the mutagen-treated material that retained their resistance to chilling in subcultures. Plants were not regenerated from the callus, however, and the relationship between callus culture and field response of plants has not been established.

Croughan et al. (4) selected salt-tolerant cell lines of alfalfa (*Medicago sativa* L.) by exposing suspension cultures of alfalfa cells to agar-solidified media containing 1% (w/v) NaCl. They concluded that the salt-tolerant cell lines had shifted toward a halophyte mode of salt tolerance. Proof that salt tolerance of cell lines is genetic and that it can be expressed under field conditions awaits tests with regenerated plants from these cells. Alfalfa plantlets have been regenerated, but they have not been subjected to field testing for salt tolerance nor has a genetic analysis been made of salt tolerance.

Widholm (21) has reported on the published work of Nabors et al. (10) on salt-tolerant lines of *Nicotiana tabacum* L. selected from cell suspension cultures. Plants have been regenerated from cultures growing in 0.64% NaCl, and self-pollinated progeny of regenerated plants retain their tolerance to salt (Nabors, personal communications). Genetic analysis of salt tolerance is underway.

Nickell (14) found that cell cultures of three surgarcane (*Saccharum officinarum* L.) clones had the same responses to cold temperature as did the clones themselves growing under field conditions. He suggested that "if this correlation is widespread, cell culture techniques could be readily utilized, both to determine the temperature response of sugarcane varieties, and to determine the effect of mutation, genetic manipulation, and other changes on progeny of varieties in which such changes are desired."

MOLECULAR APPROACH

Smith (16) and Day (6) review attempts to modify genetic expression of plants through recombinant DNA techniques. It can be said that specific genes can be inserted into plants and can then be expressed.

The recombinant DNA approach is still in the early stages, and investigators are searching primarily for a model system that works. Our understanding of host-vector systems has been dominated by research on that of *E. coli* K12, which is dissimilar to other systems that can also dependably transfer DNA from donor to host cells in higher plants. Until more reliable techniques are developed for the manipulation of DNA in higher plants, it does not seem likely that specific attempts to modify plants for greater resiliency to environmental stress have good probability of success. More basic research is needed before practical applications can be realized. As Lewis and Christiansen (9) point out, "The genetic manipulation by recombinant DNA techniques has not resulted in any stress-resistant varieties."

DISCUSSION

Reviews of breeding for resistance to environmental stresses in plants are extensive, as are reviews of genetic engineering in general. Specific cases of real accomplishments for improving environmental resiliency in crop species by genetic engineering techniques are scarce. It is not uncommon in science for new technology to be over-enthusiastically touted at the outset as the approach to solving problems. Generally, the initial excitement is followed by a long quiet period, during which scientists patiently and painstakingly overcome the many obstacles to application that were not fully appreciated in the beginning. My estimate is that the first overly enthusiastic phase of genetic engineering is over and that the long, quiet phases have begun. In due time the accomplishments will come, and this technology can then be set into scientific and historic perspective alongside the discoveries of Mendel in 1865 and Darwin's theories of evolution.

Zirkle (23) has said that "practical knowledge of heredity precedes the dawn of history." Historically, man has domesticated and improved plants and animals using not much more than the observation that like begets like. Zirkle (23) has also pointed out that the good breeding techniques "were generally imbedded in much irrelevant nonsense." Mendel's laws and the relationship of genes to chromosomes comprise classical genetics. Its application in plant and animal breeding has resulted in spectacular advances in the efficiency of crop and livestock production. In the meantime molecular geneticists have discovered the double helix nature of the DNA molecule and have broken the genetic code. Techniques for cleaving and annealing DNA molecules with precision and the development of host-vector systems for carrying DNA into host cells have had great advances in bacteria. So far these techniques have not been applicable in higher plants.

The culture of protoplasts, cells, and tissue is likely to be the bridging technology that allows molecular genetics to merge with classical genetics.

REFERENCES

1   Agricultural Policy Advisory Committee (ARPAC). 1975. Research to meet U.S. and world food needs. Report of a Working Conference, sponsored by ARPAC, Kansas City, Missouri, July 9–11, Vol. 1, 326 p.

2   American Society of Agronomy. 1965. *World population and food supplies*. Madison, Wisconsin. ASA special publication 6, 50 p.

3   Carlson, P. S. 1973. *Proc. Natl. Acad. Sci., U.S.* 70:598–602.

4   Crougham, T. P., S. J. Stavarer, and D. W. Rains. 1978. *Crop Sci.* 18:959–963.

5   Dale, P. J. 1976. Tissue culture in plant breeding. Pp. 101–113 in *Report for 1975*. Welsh Plant Breed. Sta., Univ. Coll. of Wales, Aberystwyth, England.

6   Day, P. R. 1977. *Science* 197:1334–1339.

7   Dix, P. J., and H. E. Street. *Ann. Bot.* 40:903–910.

8   Jung. Gerald A., Ed. 1978. Crop tolerance to suboptimal land conditions. ASA Spec. Publ. 32. Am. Soc. of Agron., Madison, Wisconsin.

9   Lewis, C. F., and M. N. Christiansen. 1981. Breeding plants for stress environments. *Plant Breeding II*. Iowa State Univ., Ames, Iowa.

10   Nabors, M. W., A. Daniels, L. Nadolny, and C. Brown. 1975. *Plant Sci. Let.* 4:155–159.

11   Nabors, M. W. 1976. *BioScience* 26:761–768.

12 National Research Council. 1975. World food and nutrition study: enhancement of food production for the United States. *Report of the Board of Agriculture and Renewable Resources, Commission on Natural Resources*, NRC, 174 p. (Available from: National Academy of Sciences, Washington, D.C.)

13 National Research Council. 1977. World food and nutrition study: the potential contributions of research. Prepared by the steering committee, *NRC Study on World Food and Nutrition of the Commission on International Relations*, NRC, 192 p. (Available from: National Academy of Sciences, Washington, D.C.)

14 Nickell, L. G. 1977. *Crop Sci.* 17:717–719.

15 Reiger, R., A. Michaelis, and M. M. Green. 1976. *Glossary of Genetics and Cytogenetics—Classical and Molecular*. Fourth ed. Springer-Verlag, New York.

16 Smith, H. 1974. *Span* 17:61–63.

17 Stone, John F., Ed. 1975. *Plant Modification for More Efficient Water Use*. Elsevier, New York.

18 Treshow, Michael. 1970. *Environments and Plant Response*. McGraw-Hill, New York.

19 United Nations, Food and Agriculture Organization. 1974. *World Food Conference: Assessment of the World Food Situation, Present and Future*. Rome.

20 Vasil, I. K. 1978. *Plant Tissue Culture and Crop Improvement—Fact and Fancy*. Intnal. Assoc. for Plant Tissue Culture Newsl. 26.

21 Widholm, Jack M. 1978. The selection of agriculturally desirable traits with cultured plant cells. In *Propagation of Higher Plants Through Tissue Culture*, K. W. Hughes, R. Henke, and M. Constantin, Eds. Proc. of an Intnal. Symp. at Univ. Tennessee, Knoxville, April 16–18. (Available from NTIS, Springfield, Virginia; Conf-7804111).

22 Wright, M. J., Ed. 1976. Proceedings of Workshop on Plant Adaptation to Mineral Stress in Problem Soils, Beltsville, Maryland, Nov. 22–23.

23 Zirkle, Conway. 1951. The knowledge of heredity before 1900. Pp. 35–37 in *Genetics in the 20th Century*, L. C. Dunn, Ed. Macmillan Co., New York.

INDEX

Abiotic stress, 10
Abscissic acid, 23, 203, 307, 320
Absorbed solar energy, 214, 236
Acclimation, 19, 23, 37
Acetylcholinesterase, 22
Acidity tolerance, 155, 164
Acid mine spoils sites, 163
Acid precipitation, 260
Acid soils, 76-80
  breeding progress in, 166-167
  deficiencies of, 76-78
  genetic adaptation to, 127, 128
  toxicity, 76, 78
  world distribution of, 80
Acid sulfate soil (Asia), 82, 84, 85
Acrisols, 5, 74
Activation energy, 20, 27-28
Adaptation, genetic, 127, 128, 435
  to flooding, 319-320
  to marginal lands, 391-430
  to mineral stresses, 71, 121-134
  to soil stress, 165
  to specific environments, 206-207
Adenine, 34-35
  nucleotides, mitochrondrial, 30
Adenosine and increased heat tolerance,
  34-35
Adenylate energy charge, cellular, 30
Adventitious root formation, 319-320
Aeration, soil, 296-297
Aerenchyma development, 319
Aerisols, 72
Age, plant, and air pollution, 270-271
Agricultural chemicals and resistance,
  353-354
Agricultural productivity, 1, 2, 3
Agricultural Research Service, U.S.
  Department of Agriculture, 6
Agronomic Crops and air pollutants,
  264
Agrostis tenuis and heavy metal toxicity,
  149

Airflow, criterion for drought, 197
Air movement, 9
Air pollutant, 260-261
  defined, 261-262
  exposures, 267-270
  gene control of, 275-284
  genetic plant variability, 262-267
  as limit to crop productivity, 3, 9
  mixtures, 274-275
  sensitivity screens, 267-272
  tolerance for, 259-286
Air pollutant resistance, 262-264, 352-353
  genetic research in, 276-286
Air temperature and drought, 195
"Albidum" (Russian) wheat as winter
  hardy, 49
Alfalfa, and air pollutant resistance, 263
  breeding for cold tolerance in, 50, 52,
  58
  liming and boron deficiencies, 83
  manganese tolerance, 151
  and molybdenum deficiencies, 86, 107
  mosaic, 353
  salt tolerance, 153
  stem nematode, 369
  as waterlog sensitive, 315
Alfisols, 76
Alkaline soil, 5, 76
  adaptation to, 127
  allic soils, 147
  deficiencies in, 76, 81
  tolerance, 167-168
  toxicities in, 81
Almond germplasm, 416-420
  waterlogging in, 317, 318
"Alpha," sorghum and stomatal
  conductance, 184
Alternaria alternata, association of
  nematodes with, 380
Aluminum deficiencies, U.S., 87, 108
  oxides, 77
  visual symptoms of excess levels of, 116

441